ANALYSIS OF ELECTRIC MACHINERY AND DRIVE SYSTEMS

IEEE Press
445 Hoes Lane, P.O. Box 1331
Piscataway, NJ 08855-1331

BOOKS IN THE IEEE PRESS SERIES ON POWER ENGINEERING

Analysis of Faulted Power Systems
P. M. Anderson
1995   Hardcover   536pp   0-7803-1145-0

Power System Protection
P. M. Anderson
1999   Hardcover   1344pp   0-7803-3427-2

Power and Communication Cables: Theory and Applications
Edited by R. Bartnikas and K. D. Srivastava
2000   Hardcover   896pp   0-7803-1196-5

Understanding Power Quality Problems: Voltage Sags and Interruptions
Math H. J. Bollen
2000   Hardcover   576pp   0-7803-4713-7

Electric Power Applications of Fuzzy Systems
Edited by M. E. El-Hawary
1998   Hardcover   384pp   0-7803-1197-3

ANALYSIS OF ELECTRIC MACHINERY AND DRIVE SYSTEMS

Second Edition

PAUL C. KRAUSE
OLEG WASYNCZUK
SCOTT D. SUDHOFF
Purdue University

IEEE Power Engineering Society, *Sponsor*

IEEE Press Power Engineering Series
Mohamed E. El-Hawary, *Series Editor*

IEEE
PRESS

A JOHN WILEY & SONS, INC. PUBLICATION

For ordering and customer service, call 1-800-CALL WILEY.

Library of Congress Cataloging-in-Publication is available.

ISBN 0-471-14326-X

Printed in the United States of America.

10 9 8 7 6 5 4 3 2

To our families

CONTENTS

PREFACE   xi

**Chapter 1   BASIC PRINCIPLES FOR ELECTRIC
            MACHINE ANALYSIS**   1

1.1   Introduction / 1
1.2   Magnetically Coupled Circuits / 1
1.3   Electromechanical Energy Conversion / 11
1.4   Machine Windings and Air-Gap MMF / 35
1.5   Winding Inductances and Voltage Equations / 47
References / 58
Problems / 58

Chapter 2   DIRECT-CURRENT MACHINES   67

2.1   Introduction / 67
2.2   Elementary Direct-Current Machine / 68
2.3   Voltage and Torque Equations / 76
2.4   Basic Types of Direct-Current Machines / 78
2.5   Dynamic Characteristics of Permanent-Magnet and Shunt dc Motors / 88
2.6   Time-Domain Block Diagrams and State Equations / 92
2.7   Solution of Dynamic Characteristics by Laplace Transformation / 98
References / 104
Problems / 105

Chapter 3   REFERENCE-FRAME THEORY   **109**

3.1   Introduction / 109

3.2   Background / 109

3.3   Equations of Transformation: Changes of Variables / 111

3.4   Stationary Circuit Variables Transformed to the Arbitrary
Reference Frame / 115

3.5   Commonly Used Reference Frames / 123

3.6   Transformation Between Reference Frames / 124

3.7   Transformation of a Balanced Set / 126

3.8   Balanced Steady-State Phasor Relationships / 127

3.9   Balanced Steady-State Voltage Equations / 130

3.10 Variables Observed from Several Frames of Reference / 133

References / 137

Problems / 138

Chapter 4   SYMMETRICAL INDUCTION MACHINES   **141**

4.1   Introduction / 141

4.2   Voltage Equations in Machine Variables / 142

4.3   Torque Equation in Machine Variables / 146

4.4   Equations of Transformation for Rotor Circuits / 147

4.5   Voltage Equations in Arbitrary Reference-Frame Variables / 149

4.6   Torque Equation in Arbitrary Reference-Frame Variables / 153

4.7   Commonly Used Reference Frames / 154

4.8   Per Unit System / 155

4.9   Analysis of Steady-State Operation / 157

4.10 Free Acceleration Characteristics / 165

4.11 Free Acceleration Characteristics Viewed from Various
Reference Frames / 172

4.12 Dynamic Performance During Sudden Changes in Load Torque / 174

4.13 Dynamic Performance During a 3-Phase Fault at the
Machine Terminals / 181

4.14 Computer Simulation in the Arbitrary Reference Frame / 184

References / 187

Problems / 188

Chapter 5   SYNCHRONOUS MACHINES   **191**

5.1   Introduction / 191

5.2   Voltage Equations in Machine Variables / 192

5.3   Torque Equation in Machine Variables / 197

5.4 Stator Voltage Equations in Arbitrary Reference-Frame Variables / 198

5.5 Voltage Equations in Rotor Reference-Frame Variables: Park's Equations / 200

5.6 Torque Equations in Substitute Variables / 206

5.7 Rotor Angle and Angle Between Rotors / 207

5.8 Per Unit System / 209

5.9 Analysis of Steady-State Operation / 210

5.10 Dynamic Performance During a Sudden Change in Input Torque / 219

5.11 Dynamic Performance During a 3-Phase Fault at the Machine Terminals / 225

5.12 Approximate Transient Torque Versus Rotor Angle Characteristics / 229

5.13 Comparison of Actual and Approximate Transient Torque–Angle Characteristics During a Sudden Change in Input Torque: First Swing Transient Stability Limit / 232

5.14 Comparison of Actual and Approximate Transient Torque–Angle Characteristics During a 3-Phase Fault at the Terminals: Critical Clearing Time / 239

5.15 Equal-Area Criterion / 242

5.16 Computer Simulation / 246

References / 255

Problems / 256

Chapter 6 THEORY OF BRUSHLESS dc MACHINES   261

6.1 Introduction / 261

6.2 Voltage and Torque Equations in Machine Variables / 261

6.3 Voltage and Torque Equations in Rotor Reference-Frame Variables / 264

6.4 Analysis of Steady-State Operation / 266

6.5 Dynamic Performance / 274

References / 281

Problems / 281

Chapter 7 MACHINE EQUATIONS IN OPERATIONAL IMPEDANCES AND TIME CONSTANTS   283

7.1 Introduction / 283

7.2 Park's Equations in Operational Form / 284

7.3 Operational Impedances and $G(p)$ for a Synchronous Machine with Four Rotor Windings / 284

7.4 Standard Synchronous Machine Reactances / 288

7.5 Standard Synchronous Machine Time Constants / 290

7.6 Derived Synchronous Machine Time Constants / 291

7.7   Parameters from Short-Circuit Characteristics / 294

7.8   Parameters from Frequency-Response Characteristics / 301

References / 307

Problems / 308

Chapter 8   LINEARIZED MACHINE EQUATIONS   311

8.1   Introduction / 311

8.2   Machine Equations to Be Linearized / 312

8.3   Linearization of Machine Equations / 313

8.4   Small-Displacement Stability: Eigenvalues / 323

8.5   Eigenvalues of Typical Induction Machines / 324

8.6   Eigenvalues of Typical Synchronous Machines / 327

8.7   Transfer Function Formulation / 330

References / 335

Problems / 335

Chapter 9   REDUCED-ORDER MACHINE EQUATIONS   337

9.1   Introduction / 337

9.2   Reduced-Order Equations / 338

9.3   Induction Machine Large-Excursion Behavior Predicted by
       Reduced-Order Equations / 343

9.4   Synchronous Machine Large-Excursion Behavior Predicted
       by Reduced-Order Equations / 350

9.5   Linearized Reduced-Order Equations / 354

9.6   Eigenvalues Predicted by Linearized Reduced-Order Equations / 354

9.7   Simulation of Reduced-Order Models / 355

9.8   Closing Comments and Guidelines / 358

References / 358

Problems / 359

Chapter 10   SYMMETRICAL AND UNSYMMETRICAL
             2-PHASE INDUCTION MACHINES   361

10.1   Introduction / 361

10.2   Analysis of Symmetrical 2-Phase Induction Machines / 362

10.3   Voltage and Torque Equations in Machine Variables for
        Unsymmetrical 2-Phase Induction Machines / 371

10.4   Voltage and Torque Equations in Stationary Reference-Frame
        Variables for Unsymmetrical 2-Phase Induction Machines / 373

10.5   Analysis of Steady-State Operation of Unsymmetrical
        2-Phase Induction Machines / 377

10.6   Single-Phase Induction Machines / 383

References / 393

Problems / 393

Chapter 11   SEMICONTROLLED BRIDGE CONVERTERS   **395**

11.1   Introduction / 395

11.2   Single-Phase Load Commutated Converter / 395

11.3   3-Phase Load Commutated Converter / 406

References / 425

Problems / 425

Chapter 12   dc MACHINE DRIVES   **427**

12.1   Introduction / 427

12.2   Solid-State Converters for dc Drive Systems / 427

12.3   Steady-State and Dynamic Characteristics of ac/dc Converter Drives / 431

12.4   One-Quadrant dc/dc Converter Drive / 443

12.5   Two-Quadrant dc/dc Converter Drive / 460

12.6   Four-Quadrant dc/dc Converter Drive / 463

12.7   Machine Control with Voltage-Controlled dc/dc Converter / 466

12.8   Machine Control with Current-Controlled dc/dc Converter / 468

References / 476

Problems / 476

**Chapter 13   FULLY CONTROLLED 3-PHASE BRIDGE
              CONVERTERS**   **481**

13.1   Introduction / 481

13.2   The 3-Phase Bridge Converter / 481

13.3   180° Voltage Source Operation / 487

13.4   Pulse-Width Modulation / 494

13.5   Sine-Triangle Modulation / 499

13.6   Third-Harmonic Injection / 503

13.7   Space-Vector Modulation / 506

13.8   Hysteresis Modulation / 510

13.9   Delta Modulation / 512

13.10   Open-Loop Voltage and Current Control / 513

13.11   Closed-Loop Voltage and Current Controls / 516

References / 520

Problems / 521

Chapter 14   INDUCTION MOTOR DRIVES   525

14.1   Introduction / 525

14.2   Volts-Per-Hertz Control / 525

14.3   Constant Slip Current Control / 532

14.4   Field-Oriented Control / 540

14.5   Direct Rotor-Oriented Field-Oriented Control / 544

14.6   Robust Direct Field-Oriented Control / 546

14.7   Indirect Rotor Field-Oriented Control / 550

14.8   Conclusions / 554

References / 554

Problems / 555

Chapter 15   BRUSHLESS dc MOTOR DRIVES   557

15.1   Introduction / 557

15.2   Voltage-Source Inverter Drives / 558

15.3   Equivalence of VSI Schemes to Idealized Source / 560

15.4   Average-Value Analysis of VSI Drives / 568

15.5   Steady-State Performance of VSI Drives / 571

15.6   Transient and Dynamic Performance of VSI Drives / 574

15.7   Consideration of Steady-State Harmonics / 578

15.8   Case Study: Voltage-Source Inverter-Based Speed Control / 582

15.9   Current-Regulated Inverter Drives / 586

15.10   Voltage Limitations of Current-Source Inverter Drives / 590

15.11   Current Command Synthesis / 591

15.12   Average-Value Modeling of Current-Regulated Inverter Drives / 595

15.13   Case Study: Current-Regulated Inverter-Based Speed Controller / 597

References / 600

Problems / 600

**Appendix A   Trigonometric Relations, Constants and
             Conversion Factors, and Abbreviations**   603

INDEX   605

PREFACE

The first edition of this book was written by Paul C. Krause and published in 1986 by McGraw-Hill. Eight years later the same book was republished by IEEE Press with Oleg Wasynczuk and Scott D. Sudhoff added as co-authors. The focus of the first edition was the analysis of electric machines using reference frame theory, wherein the concept of the arbitrary reference frame was emphasized. Not only has this approach been embraced by the vast majority of electric machine analysts, it has also become the approach used in the analysis of electric drive systems. The use of reference-frame theory to analyze the complete drive system (machine, converter, and control) was not emphasized in the first edition. The goal of this edition is to fill this void and thereby meet the need of engineers whose job it is to analyze and design the complete drive system. For this reason the words "and Drive Systems" have been added to the title.

Although some of the material has been rearranged or revised, and in some cases eliminated, such as 3-phase symmetrical components, most of the material presented in the first ten chapters were taken from the original edition. For the most part, the material in Chapters 11–15 on electric drive systems is new. In particular, the analysis of converters used in electric drive systems is presented in Chapters 11 and 13 while dc, induction, and brushless dc motor drives are analyzed in Chapters 12, 14, and 15, respectively.

Central to the analysis used in this text is the transformation to the arbitrary reference frame. All real and complex transformations used in machine and drive analyses can be shown to be special cases of this general transformation. The modern electric machine and drive analyst must understand reference frame theory. For this reason, the complete performance of all electric machines and drives considered are illustrated by computer traces wherein variables are often portrayed in different frames of reference so that the student is able to appreciate the advantages and significance of the transformation used.

The material presented in this text can be used most beneficially if the student has had an introductory course in electric machines. However, a senior would be comfortable using this textbook as a first course. For this purpose, considerable time should be devoted to the basic principles discussed in Chapter 1, perhaps some of Chapter 2 covering basic dc machines, most of Chapter 3 covering reference frame theory, and the beginning sections of Chapters 4, 5, and 6 covering induction, synchronous, and brushless dc machines.

Some of the material that would be of interest only to the electric power engineer has been reduced or eliminated from that given in the first edition. However, the material found in the final sections in Chapters 4 and 5 on induction and synchronous machines as well as operational impedances (Chapter 7), and reduced-order modeling (Chapter 9) provide an excellent background for the power utility engineer.

We would like to acknowledge the efforts and assistance of the reviewers, in particular Mohamed E. El-Hawary, and the staff of IEEE Press and John Wiley & Sons.

PAUL C. KRAUSE
OLEG WASYNCZUK
SCOTT D. SUDHOFF

West Lafayette, Indiana
November 2001

ANALYSIS OF ELECTRIC MACHINERY AND DRIVE SYSTEMS

Chapter 1

BASIC PRINCIPLES FOR ELECTRIC MACHINE ANALYSIS

1.1 INTRODUCTION

There are several basic concepts that must be established before the analysis of electric machines can begin. The principle of electromechanical energy conversion is perhaps the cornerstone of machine analysis. This theory allows us to establish an expression of electromagnetic torque in terms of machine variables, generally the currents and the displacement of the mechanical system. Other principles that must be established are (1) the derivation of equivalent circuit representations of magnetically coupled circuits, (2) the concept of a sinusoidally distributed winding, (3) the concept of a rotating air-gap magnetomotive force (MMF), and (4) the derivation of winding inductances. The above-mentioned basic principles are presented in this chapter, concluding with the voltage equations of a 3-phase synchronous machine and a 3-phase induction machine. It is shown that the equations, which describe the behavior of alternating-current (ac) machines, contain time-varying coefficients due to the fact that some of the machine inductances are functions of the rotor displacement. This establishes an awareness of the complexity of these voltage equations and sets the stage for the change of variables (Chapter 3), which reduces the complexity of the voltage equations by eliminating the time-dependent inductances.

1.2 MAGNETICALLY COUPLED CIRCUITS

Magnetically coupled electric circuits are central to the operation of transformers and electric machines. In the case of transformers, stationary circuits are

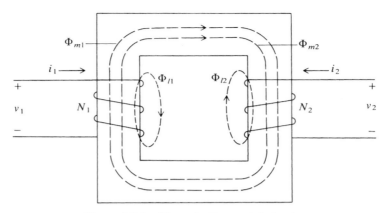

Figure 1.2-1 Magnetically coupled circuits.

magnetically coupled for the purpose of changing the voltage and current levels. In the case of electric machines, circuits in relative motion are magnetically coupled for the purpose of transferring energy between mechanical and electrical systems. Because magnetically coupled circuits play such an important role in power transmission and conversion, it is important to establish the equations that describe their behavior and to express these equations in a form convenient for analysis. These goals may be achieved by starting with two stationary electric circuits that are magnetically coupled as shown in Fig. 1.2-1. The two coils consist of turns N_1 and N_2, respectively, and they are wound on a common core that is generally a ferromagnetic material with permeability large relative to that of air. The permeability of free space, μ_0, is $4\pi \times 10^{-7}$ H/m. The permeability of other materials is expressed as $\mu = \mu_r \mu_0$ where μ_r is the relative permeability. In the case of transformer steel the relative permeability may be as high 2000 to 4000.

In general, the flux produced by each coil can be separated into two components: a leakage component denoted with an l subscript and a magnetizing component denoted by an m subscript. Each of these components is depicted by a single streamline with the positive direction determined by applying the right-hand rule to the direction of current flow in the coil. Often, in transformer analysis, i_2 is selected positive out of the top of coil 2, and a dot is placed at that terminal.

The flux linking each coil may be expressed as

$$\Phi_1 = \Phi_{l1} + \Phi_{m1} + \Phi_{m2} \tag{1.2-1}$$
$$\Phi_2 = \Phi_{l2} + \Phi_{m2} + \Phi_{m1} \tag{1.2-2}$$

The leakage flux Φ_{l1} is produced by current flowing in coil 1, and it links only the turns of coil 1. Likewise, the leakage flux Φ_{l2} is produced by current flowing in coil 2, and it links only the turns of coil 2. The magnetizing flux Φ_{m1} is produced by current flowing in coil 1, and it links all turns of coils 1 and 2. Similarly, the magnetizing flux Φ_{m2} is produced by current flowing in coil 2, and it also links all turns of coils 1 and 2. With the selected positive direction of current flow and the manner in which the coils are wound (Fig. 1.2-1), magnetizing flux produced by positive current in one coil adds to the magnetizing flux produced by positive current in the other coil. In other words, if both currents are actually flowing in the same direction, the magnetizing fluxes produced by each coil are in the same direction, making the total magnetizing flux or the total core flux the sum of the instantaneous magnitudes of the individual magnetizing fluxes. If the actual currents are in opposite directions, the magnetizing fluxes are in opposite directions. In this case, one coil is said to be magnetizing the core, the other demagnetizing.

Before proceeding, it is appropriate to point out that this is an idealization of the actual magnetic system. Clearly, all of the leakage flux may not link all the turns of the coil producing it. Likewise, all of the magnetizing flux of one coil may not link all of the turns of the other coil. To acknowledge this practical aspect of the magnetic system, the number of turns is considered to be an equivalent number rather than the actual number. This fact should cause us little concern because the inductances of the electric circuit resulting from the magnetic coupling are generally determined from tests.

The voltage equations may be expressed in matrix form as

$$\mathbf{v} = \mathbf{ri} + \frac{d\boldsymbol{\lambda}}{dt} \tag{1.2-3}$$

where $\mathbf{r} = \text{diag}\,[r_1\ r_2]$, a diagonal matrix, and

$$(\mathbf{f})^T = [f_1\ f_2] \tag{1.2-4}$$

where f represents voltage, current, or flux linkage. The resistances r_1 and r_2 and the flux linkages λ_1 and λ_2 are related to coils 1 and 2, respectively. Because it is assumed that Φ_1 links the equivalent turns of coil 1 and Φ_2 links the equivalent turns of coil 2, the flux linkages may be written as

$$\lambda_1 = N_1 \Phi_1 \tag{1.2-5}$$
$$\lambda_2 = N_2 \Phi_2 \tag{1.2-6}$$

where Φ_1 and Φ_2 are given by (1.2-1) and (1.2-2), respectively.

Linear Magnetic System

If saturation is neglected, the system is linear and the fluxes may be expressed as

$$\Phi_{l1} = \frac{N_1 i_1}{\mathscr{R}_{l1}} \tag{1.2-7}$$

$$\Phi_{m1} = \frac{N_1 i_1}{\mathscr{R}_m} \tag{1.2-8}$$

$$\Phi_{l2} = \frac{N_2 i_2}{\mathscr{R}_{l2}} \tag{1.2-9}$$

$$\Phi_{m2} = \frac{N_2 i_2}{\mathscr{R}_m} \tag{1.2-10}$$

where $\mathscr{R}_{l1}$ and $\mathscr{R}_{l2}$ are the reluctances of the leakage paths and $\mathscr{R}_m$ is the reluctance of the path of the magnetizing fluxes. The product of N times i (ampere-turns) is the MMF, which is determined by application of Ampere's law. The reluctance of the leakage paths is difficult to express and impossible to measure. In fact, a unique determination of the inductances associated with the leakage flux cannot be made by tests; instead, it is either calculated or approximated from design considerations. The reluctance of the magnetizing path of the core shown in Fig. 1.2-1 may be computed with sufficient accuracy from the well-known relationship

$$\mathscr{R} = \frac{l}{\mu A} \tag{1.2-11}$$

where l is the mean or equivalent length of the magnetic path, A is the cross-sectional area, and μ is the permeability.

Substituting (1.2-7)–(1.2-10) into (1.2-1) and (1.2-2) yields

$$\Phi_1 = \frac{N_1 i_1}{\mathscr{R}_{l1}} + \frac{N_1 i_1}{\mathscr{R}_m} + \frac{N_2 i_2}{\mathscr{R}_m} \tag{1.2-12}$$

$$\Phi_2 = \frac{N_2 i_2}{\mathscr{R}_{l2}} + \frac{N_2 i_2}{\mathscr{R}_m} + \frac{N_1 i_1}{\mathscr{R}_m} \tag{1.2-13}$$

Substituting (1.2-12) and (1.2-13) into (1.2-5) and (1.2-6) yields

$$\lambda_1 = \frac{N_1^2}{\mathscr{R}_{l1}} i_1 + \frac{N_1^2}{\mathscr{R}_m} i_1 + \frac{N_1 N_2}{\mathscr{R}_m} i_2 \tag{1.2-14}$$

$$\lambda_2 = \frac{N_2^2}{\mathscr{R}_{l2}} i_2 + \frac{N_2^2}{\mathscr{R}_m} i_2 + \frac{N_2 N_1}{\mathscr{R}_m} i_1 \tag{1.2-15}$$

When the magnetic system is linear, the flux linkages are generally expressed in terms of inductances and currents. We see that the coefficients of the first two terms on the right-hand side of (1.2-14) depend upon the turns of coil 1 and the reluctance of the magnetic system, independent of the existence of coil 2. An analogous statement may be made regarding (1.2-15). Hence the self-inductances are defined as

$$L_{11} = \frac{N_1^2}{\mathscr{R}_{l1}} + \frac{N_1^2}{\mathscr{R}_m} = L_{l1} + L_{m1} \tag{1.2-16}$$

$$L_{22} = \frac{N_2^2}{\mathscr{R}_{l2}} + \frac{N_2^2}{\mathscr{R}_m} = L_{l2} + L_{m2} \tag{1.2-17}$$

where L_{l1} and L_{l2} are the leakage inductances and L_{m1} and L_{m2} are the magnetizing inductances of coils 1 and 2, respectively. From (1.2-16) and (1.2-17) it follows that the magnetizing inductances may be related as

$$\frac{L_{m2}}{N_2^2} = \frac{L_{m1}}{N_1^2} \tag{1.2-18}$$

The mutual inductances are defined as the coefficient of the third term of (1.2-14) and (1.2-15).

$$L_{12} = \frac{N_1 N_2}{\mathscr{R}_m} \tag{1.2-19}$$

$$L_{21} = \frac{N_2 N_1}{\mathscr{R}_m} \tag{1.2-20}$$

Obviously, $L_{12} = L_{21}$. The mutual inductances may be related to the magnetizing inductances. In particular,

$$L_{12} = \frac{N_2}{N_1} L_{m1} = \frac{N_1}{N_2} L_{m2} \tag{1.2-21}$$

The flux linkages may now be written as

$$\boldsymbol{\lambda} = \mathbf{L}\mathbf{i} \tag{1.2-22}$$

where

$$\mathbf{L} = \begin{bmatrix} L_{11} & L_{12} \\ L_{21} & L_{22} \end{bmatrix} = \begin{bmatrix} L_{l1} + L_{m1} & \frac{N_2}{N_1} L_{m1} \\ \frac{N_1}{N_2} L_{m2} & L_{l2} + L_{m2} \end{bmatrix} \tag{1.2-23}$$

Although the voltage equations with the inductance matrix $\mathbf{L}$ incorporated may be used for purposes of analysis, it is customary to perform a change of variables that yields the well-known equivalent T circuit of two magnetically coupled coils. To set the stage for this derivation, let us express the flux linkages from (1.2-22) as

$$\lambda_1 = L_{l1} i_1 + L_{m1}\left(i_1 + \frac{N_2}{N_1} i_2\right) \tag{1.2-24}$$

$$\lambda_2 = L_{l2} i_2 + L_{m2}\left(\frac{N_1}{N_2} i_1 + i_2\right) \tag{1.2-25}$$

Now we have two choices. We can use a substitute variable for $(N_2/N_1)i_2$ or for $(N_1/N_2)i_1$. Let us consider the first of these choices

$$N_1 i_2' = N_2 i_2 \tag{1.2-26}$$

whereupon we are using the substitute variable i_2' that, when flowing through coil 1, produces the same MMF as the actual i_2 flowing through coil 2. This is said to be referring the current in coil 2 to coil 1 whereupon coil 1 becomes the reference coil. On the other hand, if we use the second choice, then

$$N_2 i_1' = N_1 i_1 \tag{1.2-27}$$

Here, i_1' is the substitute variable that produces the same MMF when flowing through coil 2 as i_1 does when flowing in coil 1. This change of variables is said to refer the current of coil 1 to coil 2.

We will demonstrate the derivation of the equivalent T circuit by referring the current of coil 2 to coil 1; thus from (1.2-26) we obtain

$$i_2' = \frac{N_2}{N_1} i_2 \tag{1.2-28}$$

Power is to be unchanged by this substitution of variables. Therefore,

$$v_2' = \frac{N_1}{N_2} v_2 \tag{1.2-29}$$

whereupon $v_2 i_2 = v_2' i_2'$. Flux linkages, which have the units of volt-second, are related to the substitute flux linkages in the same way as voltages. In particular,

$$\lambda_2' = \frac{N_1}{N_2} \lambda_2 \tag{1.2-30}$$

If we substitute (1.2-28) into (1.2-24) and (1.2-25) and then multiply (1.2-25) by N_1/N_2 to obtain λ_2' and we further substitute $(N_2^2/N_1^2)L_{m1}$ for L_{m2} into (1.2-24), then

$$\lambda_1 = L_{l1} i_1 + L_{m1}(i_1 + i_2') \tag{1.2-31}$$

$$\lambda_2' = L_{l2}' i_2' + L_{m1}(i_1 + i_2') \tag{1.2-32}$$

where

$$L_{l2}' = \left(\frac{N_1}{N_2}\right)^2 L_{l2} \tag{1.2-33}$$

The voltage equations become

$$v_1 = r_1 i_1 + \frac{d\lambda_1}{dt} \tag{1.2-34}$$

$$v_2' = r_2' i_2' + \frac{d\lambda_2'}{dt} \tag{1.2-35}$$

where

$$r_2' = \left(\frac{N_1}{N_2}\right)^2 r_2 \tag{1.2-36}$$

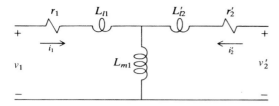

Figure 1.2-2   Equivalent circuit with coil 1 selected as the reference coil.

The above voltage equations suggest the T equivalent circuit shown in Fig. 1.2-2. It is apparent that this method may be extended to include any number of coils wound on the same core.

Example 1A   It is instructive to illustrate the method of deriving an equivalent T circuit from open- and short-circuit measurements. For this purpose let us assume that when coil 2 of the two-winding transformer shown in Fig. 1.2-1 is open-circuited, the power input to coil 2 is 12 W with an applied voltage is 100 V (rms) at 60 Hz and the current is 1 A (rms). When coil 2 is short-circuited, the current flowing in coil 1 is 1 A when the applied voltage is 30 V at 60 Hz. The power during this test is 22 W. If we assume $L_{l1} = L'_{l2}$, an approximate equivalent T circuit can be determined from these measurements with coil 1 selected as the reference coil.

The power may be expressed as

$$P_1 = \left|\tilde{V}_1\right|\left|\tilde{I}_1\right|\cos\phi \tag{1A-1}$$

where $\tilde{V}$ and $\tilde{I}$ are phasors and ϕ is the phase angle between $\tilde{V}_1$ and $\tilde{I}_1$ (power-factor angle). Solving for ϕ during the open-circuit test, we have

$$\phi = \cos^{-1}\frac{P_1}{\left|\tilde{V}_1\right|\left|\tilde{I}_1\right|} = \cos^{-1}\frac{12}{110 \times 1} = 83.7° \tag{1A-2}$$

With $\tilde{V}_1$ as the reference phasor and assuming an inductive circuit where $\tilde{I}_1$ lags $\tilde{V}_1$, we obtain

$$Z = \frac{\tilde{V}_1}{\tilde{I}_1} = \frac{110/0°}{1/-83.7°} = 12 + j109.3\,\Omega \tag{1A-3}$$

If we neglect hysteresis (core) losses, which in effect assumes a linear magnetic system, then $r_1 = 12\,\Omega$. We also know from the above calculation that $X_{l1} + X_{m1} = 109.3\,\Omega$.

For the short-circuit test we will assume that $i_1 = -i'_2$ because transformers are designed so that $X_{m1} \gg \left|r'_2 + jX'_{l2}\right|$. Hence using (1A-1) again we obtain

$$\phi = \cos^{-1}\frac{22}{30 \times 1} = 42.8° \tag{1A-4}$$

In this case the input impedance is $(r_1 + r_2') + j(X_{l1} + X_{l2}')$. This may be determined as follows:

$$Z = \frac{30 / \underline{0^\circ}}{1 / \underline{-42.8^\circ}} = 22 + j\,20.4\,\Omega \qquad (1\text{A-5})$$

Hence, $r_2' = 10\,\Omega$ and, because it is assumed that $X_{l1} = X_{l2}'$, both are $10.2\,\Omega$. Therefore $X_{m1} = 109.3 - 10.2 = 99.1\,\Omega$. In summary,

$$r_1 = 12\,\Omega \qquad L_{m1} = 262.9\,\text{mH} \qquad r_2' = 10\,\Omega$$
$$L_{l1} = 27.1\,\text{mH} \qquad\qquad\qquad L_{l2}' = 27.1\,\text{mH}$$

Nonlinear Magnetic System

Although the analysis of transformers and electric machines is generally performed assuming a linear magnetic system, economics dictate that in the practical design of these devices some saturation occurs and that heating of the magnetic material exists due to hysteresis losses. The magnetization characteristics of transformer or machine materials are given in the form of the magnitude of flux density versus magnitude of field strength (B–H curve) as shown in Fig. 1.2-3. If it is assumed that the magnetic flux is uniform through most of the core, then B is proportional to Φ and H is proportional to MMF. Hence a plot of flux versus current is of the same shape as

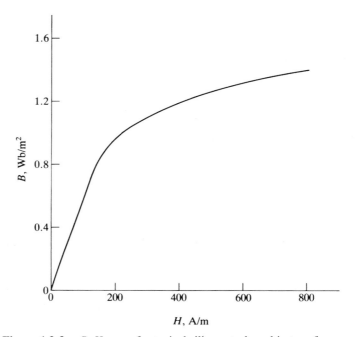

Figure 1.2-3 B–H curve for typical silicon steel used in transformers.

the *B–H* curve. A transformer is generally designed so that some saturation occurs during normal operation. Electric machines are also designed similarly in that a machine generally operates slightly in the saturated region during normal, rated operating conditions. Because saturation causes coefficients of the differential equations describing the behavior of an electromagnetic device to be functions of the coil currents, a transient analysis is difficult without the aid of a computer. Our purpose here is not to set forth methods of analyzing nonlinear magnetic systems. This procedure is quite straightforward for steady-state operation, but it cannot be used when analyzing the dynamics of electromechanical devices [1]. A method of incorporating the effects of saturation into a computer representation is of interest.

Computer Simulation of Coupled Circuits

Formulating the voltage equations of stationary coupled windings appropriate for computer simulation is straightforward and yet this technique is fundamental to the computer simulation of ac machines. Therefore it is to our advantage to consider this method here. For this purpose let us first write (1.2-31) and (1.2-32) as

$$\lambda_1 = L_{l1} i_1 + \lambda_m \tag{1.2-37}$$
$$\lambda_2' = L_{l2}' i_2' + \lambda_m \tag{1.2-38}$$

where

$$\lambda_m = L_{m1}(i_1 + i_2') \tag{1.2-39}$$

Solving (1.2-37) and (1.2-38) for the currents yields

$$i_1 = \frac{1}{L_{l1}}(\lambda_1 - \lambda_m) \tag{1.2-40}$$

$$i_2' = \frac{1}{L_{l2}'}(\lambda_2' - \lambda_m) \tag{1.2-41}$$

If (1.2-40) and (1.2-41) are substituted into the voltage equations (1.2-34) and (1.2-35) and we solve the resulting equations for flux linkages, the following equations are obtained:

$$\lambda_1 = \int \left[v_1 + \frac{r_1}{L_{l1}}(\lambda_m - \lambda_1) \right] dt \tag{1.2-42}$$

$$\lambda_2' = \int \left[v_2' + \frac{r_2'}{L_{l2}'}(\lambda_m - \lambda_2') \right] dt \tag{1.2-43}$$

Substituting (1.2-40) and (1.2-41) into (1.2-39) yields

$$\lambda_m = L_a \left(\frac{\lambda_1}{L_{l1}} + \frac{\lambda_2'}{L_{l2}'} \right) \tag{1.2-44}$$

where

$$L_a = \left(\frac{1}{L_{m1}} + \frac{1}{L_{l1}} + \frac{1}{L_{l2}'} \right)^{-1} \tag{1.2-45}$$

We now have the equations expressed with λ_1 and λ_2' as state variables. In the computer simulation, (1.2-42) and (1.2-43) are used to solve for λ_1 and λ_2' and (1.2-44) is used to solve for λ_m. The currents can then be obtained from (1.2-40) and (1.2-41). It is clear that (1.2-44) could be substituted into (1.2-40)–(1.2-43) and that λ_m could be eliminated from the equations, whereupon it would not appear in the computer simulation. However, we will find λ_m (the mutual flux linkages) an important variable when we include the effects of saturation.

If the magnetization characteristics (magnetization curve) of the coupled winding is known, the effects of saturation of the mutual flux path may be readily incorporated into the computer simulation. Generally, the magnetization curve can be adequately determined from a test wherein one of the windings is open-circuited (winding 2, for example), and the input impedance of the other winding (winding 1) is determined from measurements as the applied voltage is increased in magnitude from zero to say 150% of the rated value. With information obtained from this type of test, we can plot λ_m versus $(i_1 + i_2')$ as shown in Fig. 1.2-4 wherein the slope of the linear portion of the curve is L_{m1}. From Fig. 1.2-4, it is clear that in the region of saturation we have

$$\lambda_m = L_{m1}(i_1 + i_2') - f(\lambda_m) \tag{1.2-46}$$

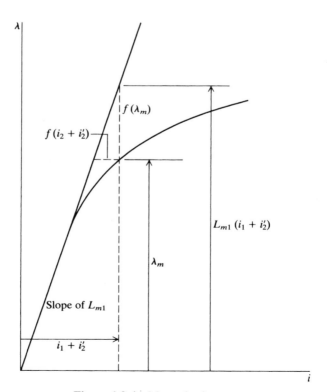

Figure 1.2-4 Magnetization curve.

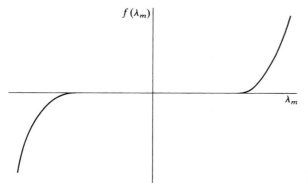

$f(\lambda_m)$

λ_m

Figure 1.2-5 $f(\lambda_m)$ versus λ_m from Fig. 1.2-4.

where $f(\lambda_m)$ may be determined from the magnetization curve for each value of λ_m. In particular, $f(\lambda_m)$ is a function of λ_m as shown in Fig. 1.2-5. Therefore, the effects of saturation of the mutual flux path may be taken into account by replacing (1.2-39) with (1.2-46) for λ_m. Substituting (1.2-40) and (1.2-41) for i_1 and i_2', respectively, into (1.2-46) yields the following computer equation for λ_m:

$$\lambda_m = L_a \left(\frac{\lambda_1}{L_{l1}} + \frac{\lambda_2'}{L_{l2}'} \right) - \frac{L_a}{L_{m1}} f(\lambda_m) \qquad (1.2\text{-}47)$$

Hence the computer simulation for including saturation involves replacing λ_m given by (1.2-44) with (1.2-47) where $f(\lambda_m)$ is a generated function of λ_m determined from the plot shown in Fig. 1.2-5.

1.3   ELECTROMECHANICAL ENERGY CONVERSION

Although electromechanical devices are used in some manner in a wide variety of systems, electric machines are by far the most common. It is desirable, however, to establish methods of analysis that may be applied to all electromechanical devices.

Energy Relationships

Electromechanical systems are comprised of an electrical system, a mechanical system, and a means whereby the electrical and mechanical systems can interact. Interaction can take place through any and all electromagnetic and electrostatic fields that are common to both systems, and energy is transferred from one system to the other as a result of this interaction. Both electrostatic and electromagnetic coupling fields may exist simultaneously, and the electromechanical system may have any number of electrical and mechanical systems. However, before considering an involved system it is helpful to analyze the electromechanical system in a simplified form. An electromechanical system with one electrical system, one mechanical system, and

Figure 1.3-1   Block diagram of an elementary electromechanical system.

one coupling field is depicted in Fig. 1.3-1. Electromagnetic radiation is neglected, and it is assumed that the electrical system operates at a frequency sufficiently low so that the electrical system may be considered as a lumped parameter system.

Losses occur in all components of the electromechanical system. Heat loss will occur in the mechanical system due to friction, and the electrical system will dissipate heat due to the resistance of the current-carrying conductors. Eddy current and hysteresis losses occur in the ferromagnetic material of all magnetic fields, whereas dielectric losses occur in all electric fields. If W_E is the total energy supplied by the electrical source and W_M is the total energy supplied by the mechanical source, then the energy distribution could be expressed as

$$W_E = W_e + W_{eL} + W_{eS} \tag{1.3-1}$$

$$W_M = W_m + W_{mL} + W_{mS} \tag{1.3-2}$$

In (1.3-1), W_{eS} is the energy stored in the electric or magnetic fields that are not coupled with the mechanical system. The energy W_{eL} is the heat losses associated with the electrical system. These losses occur due to the resistance of the current-carrying conductors as well as the energy dissipated from these fields in the form of heat due to hysteresis, eddy currents, and dielectric losses. The energy W_e is the energy transferred to the coupling field by the electrical system. The energies common to the mechanical system may be defined in a similar manner. In (1.3-2), W_{mS} is the energy stored in the moving member and compliances of the mechanical system, W_{mL} is the energy losses of the mechanical system in the form of heat, and W_m is the energy transferred to the coupling field. It is important to note that with the convention adopted, the energy supplied by either source is considered positive. Therefore, $W_E(W_M)$ is negative when energy is supplied to the electrical source (mechanical source).

If W_F is defined as the total energy transferred to the coupling field, then

$$W_F = W_f + W_{fL} \tag{1.3-3}$$

where W_f is energy stored in the coupling field and W_{fL} is the energy dissipated in the form of heat due to losses within the coupling field (eddy current, hysteresis, or dielectric losses). The electromechanical system must obey the law of conservation of energy; thus

$$W_f + W_{fL} = (W_E - W_{eL} - W_{eS}) + (W_M - W_{mL} - W_{mS}) \tag{1.3-4}$$

which may be written

$$W_f + W_{fL} = W_e + W_m \tag{1.3-5}$$

This energy relationship is shown schematically in Fig. 1.3-2.

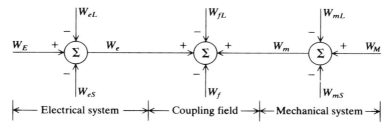

Figure 1.3-2   Energy balance.

The actual process of converting electrical energy to mechanical energy (or vice versa) is independent of (1) the loss of energy in either the electrical or the mechanical systems (W_{eL} and W_{mL}), (2) the energies stored in the electric or magnetic fields that are not common to both systems (W_{eS}), or (3) the energies stored in the mechanical system (W_{mS}). If the losses of the coupling field are neglected, then the field is conservative and (1.3-5) becomes

$$W_f = W_e + W_m \qquad (1.3\text{-}6)$$

Examples of elementary electromechanical systems are shown in Figs. 1.3-3 and 1.3-4. The system shown in Fig. 1.3-3 has a magnetic coupling field while the electromechanical system shown in Fig. 1.3-4 employs an electric field as a means of transferring energy between the electrical and mechanical systems. In these systems, v is the voltage of the electric source and f is the external mechanical force applied to the mechanical system. The electromagnetic or electrostatic force is denoted by f_e. The resistance of the current-carrying conductors is denoted by r, and l is the inductance of a linear (conservative) electromagnetic system that does not couple the mechanical system. In the mechanical system, M is the mass of the movable member while the linear compliance and damper are represented by a spring constant K and a

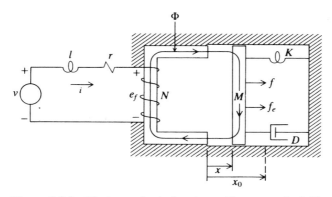

Figure 1.3-3   Electromechanical system with a magnetic field.

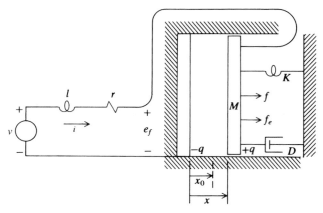

Figure 1.3-4   Electromechanical system with an electric field.

damping coefficient D, respectively. The displacement x_0 is the zero force or equilibrium position of the mechanical system which is the steady-state position of the mass with f_e and f equal to zero. A series or shunt capacitance may be included in the electrical system wherein energy would also be stored in an electric field external to the electromechanical process.

The voltage equation that describes both electrical systems may be written as

$$v = ri + l\frac{di}{dt} + e_f \tag{1.3-7}$$

where e_f is the voltage drop across the coupling field. The dynamic behavior of the translational mechanical system may be expressed by employing Newton's law of motion. Thus,

$$f = M\frac{d^2x}{dt^2} + D\frac{dx}{dt} + K(x - x_0) - f_e \tag{1.3-8}$$

The total energy supplied by the electric source is

$$W_E = \int vi\, dt \tag{1.3-9}$$

The total energy supplied by the mechanical source is

$$W_M = \int f\, dx \tag{1.3-10}$$

which may also be expressed as

$$W_M = \int f\frac{dx}{dt}\, dt \tag{1.3-11}$$

Substituting (1.3-7) into (1.3-9) yields

$$W_E = r \int i^2 dt + l \int i \, di + \int e_f i \, dt \qquad (1.3\text{-}12)$$

The first term on the right-hand side of (1.3-12) represents the energy loss due to the resistance of the conductors (W_{eL}). The second term represents the energy stored in the linear electromagnetic field external to the coupling field (W_{eS}). Therefore, the total energy transferred to the coupling field from the electrical system is

$$W_e = \int e_f i \, dt \qquad (1.3\text{-}13)$$

Similarly, for the mechanical system we have

$$W_M = M \int \frac{d^2 x}{dt^2} dx + D \int \left(\frac{dx}{dt} \right)^2 dt + K \int (x - x_0) \, dx - \int f_e \, dx \qquad (1.3\text{-}14)$$

Here, the first and third terms on the right-hand side of (1.3-14) represent the energy stored in the mass and spring, respectively (W_{mS}). The second term is the heat loss due to friction (W_{mL}). Thus, the total energy transferred to the coupling field from the mechanical system is

$$W_m = - \int f_e \, dx \qquad (1.3\text{-}15)$$

It is important to note that a positive force, f_e, is assumed to be in the same direction as a positive displacement, dx. Substituting (1.3-13) and (1.3-15) into the energy balance relation, (1.3-6), yields

$$W_f = \int e_f i \, dt - \int f_e \, dx \qquad (1.3\text{-}16)$$

The equations set forth may be readily extended to include an electromechanical system with any number of electrical and mechanical inputs to any number of coupling fields. Considering the system shown in Fig. 1.3-5, the energy supplied to the coupling fields may be expressed as

$$W_f = \sum_{j=1}^{J} W_{ej} + \sum_{k=1}^{K} W_{mk} \qquad (1.3\text{-}17)$$

wherein J electrical and K mechanical inputs exist. The total energy supplied to the coupling field at the electrical inputs is

$$\sum_{j=1}^{J} W_{ej} = \int \sum_{j=1}^{J} e_{fj} i_j \, dt \qquad (1.3\text{-}18)$$

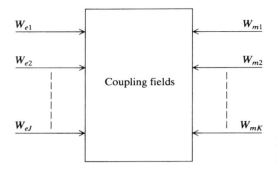

Figure 1.3-5 Multiple electrical and mechanical inputs.

The total energy supplied to the coupling field from the mechanical inputs is

$$\sum_{k=1}^{K} W_{mk} = -\int \sum_{k=1}^{K} f_{ek}\, dx_k \tag{1.3-19}$$

The energy balance equation becomes

$$W_f = \int \sum_{j=1}^{J} e_{fj} i_j\, dt - \int \sum_{k=1}^{K} f_{ek}\, dx_k \tag{1.3-20}$$

In differential form we obtain

$$dW_f = \sum_{j=1}^{J} e_{fj} i_j\, dt - \sum_{k=1}^{K} f_{ek}\, dx_k \tag{1.3-21}$$

Energy in Coupling Fields

Before using (1.3-21) to obtain an expression for the electromagnetic force f_e, it is necessary to derive an expression for the energy stored in the coupling fields. Once we have an expression for W_f, we can take the total derivative to obtain dW_f, which can then be substituted into (1.3-21). When expressing the energy in the coupling fields it is convenient to neglect all losses associated with the electric and magnetic fields whereupon the fields are assumed to be conservative and the energy stored therein is a function of the state of the electrical and mechanical variables. Although the effects of the field losses may be functionally accounted for by appropriately introducing a resistance in the electric circuit, this refinement is generally not necessary because the ferromagnetic material is selected and arranged in laminations so as to minimize the hysteresis and eddy current losses. Moreover, nearly all of the energy stored in the coupling fields is stored in the air gaps of the electromechanical device. Because air is a conservative medium, all of the energy stored therein can be returned to the electrical or mechanical systems. Therefore, the assumption of loss-less coupling fields is not as restrictive as it might first appear.

The energy stored in a conservative field is a function of the state of the system variables and not the manner in which the variables reached that state. It is convenient to take advantage of this feature when developing a mathematical expression for the field energy. In particular, it is convenient to fix mathematically the position of the mechanical systems associated with the coupling fields and then excite the electrical systems with the displacements of the mechanical systems held fixed. During the excitation of the electrical systems, W_{mk} is zero even though electromagnetic or electrostatic forces occur. Therefore, with the displacements held fixed the energy stored in the coupling fields during the excitation of the electrical systems is equal to the energy supplied to the coupling fields by the electrical systems. Thus, with $W_{mk} = 0$, the energy supplied from the electrical system may be expressed from (1.3-20) as

$$W_f = \int \sum_{j=1}^{J} e_{fj} i_j \, dt \qquad (1.3\text{-}22)$$

It is instructive to consider a singly excited electromagnetic system similar to that shown in Fig. 1.3-3. In this case, $e_f = d\lambda/dt$ and (1.3-22) becomes

$$W_f = \int i \, d\lambda \qquad (1.3\text{-}23)$$

Here $J = 1$; however, the subscript is omitted for the sake of brevity. The area to the left of the λ–i relationship (shown in Fig. 1.3-6) for a singly excited electromagnetic device is the area described by (1.3-23). In Fig. 1.3-6, this area represents the energy stored in the field at the instant when $\lambda = \lambda_a$ and $i = i_a$. The λ–i relationship need not be linear; it need only be single-valued, a property that is characteristic to a conservative or lossless field. Moreover, because the coupling field is conservative, the energy stored in the field with $\lambda = \lambda_a$ and $i = i_a$ is independent of the excursion of the electrical and mechanical variables before reaching this state.

The area to the right of the λ–i curve is called the *coenergy* and is expressed as

$$W_c = \int \lambda \, di \qquad (1.3\text{-}24)$$

which may also be written as

$$W_c = \lambda i - W_f \qquad (1.3\text{-}25)$$

Although the coenergy has little or no physical significance, we will find it a convenient quantity for expressing the electromagnetic force. It should be clear that for a linear magnetic system where the λ–i plots are straight-line relationships, $W_f = W_c$.

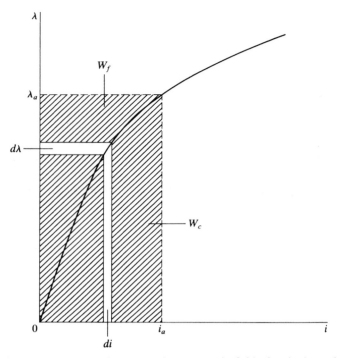

Figure 1.3-6   Stored energy and coenergy in a magnetic field of a singly excited electro-magnetic device.

The displacement x defines completely the influence of the mechanical system upon the coupling field; however, because λ and i are related, only one is needed in addition to x in order to describe the state of the electromechanical system. Therefore, either λ and x or i and x may be selected as independent variables. If i and x are selected as independent variables, it is convenient to express the field energy and the flux linkages as

$$W_f = W_f(i, x) \tag{1.3-26}$$

$$\lambda = \lambda(i, x) \tag{1.3-27}$$

With i and x as independent variables we must express $d\lambda$ in terms of di before substituting into (1.3-23). Thus from (1.3-27)

$$d\lambda(i, x) = \frac{\partial \lambda(i, x)}{\partial i}\, di + \frac{\partial \lambda(i, x)}{\partial x}\, dx \tag{1.3-28}$$

In the derivation of an expression for the energy stored in the field, dx is set equal to zero. Hence, in the evaluation of field energy, $d\lambda$ is equal to the first term on the right-hand side of (1.3-28). Substituting into (1.3-23) yields

$$W_f(i,x) = \int i \frac{\partial \lambda(i,x)}{\partial i}\, di = \int_0^i \xi \frac{\partial \lambda(\xi,x)}{\partial \xi}\, d\xi \qquad (1.3\text{-}29)$$

where ξ is the dummy variable of integration. Evaluation of (1.3-29) gives the energy stored in the field of the singly excited system. The coenergy in terms of i and x may be evaluated from (1.3-24) as

$$W_c(i,x) = \int \lambda(i,x)\, di = \int_0^i \lambda(\xi,x)\, d\xi \qquad (1.3\text{-}30)$$

With λ and x as independent variables we have

$$W_f = W_f(\lambda,x) \qquad (1.3\text{-}31)$$
$$i = i(\lambda,x) \qquad (1.3\text{-}32)$$

The field energy may be evaluated from (1.3-23) as

$$W_f(\lambda,x) = \int i(\lambda,x)\, d\lambda = \int_0^\lambda i(\xi,x)\, d\xi \qquad (1.3\text{-}33)$$

In order to evaluate the coenergy with λ and x as independent variables, we need to express di in terms of $d\lambda$; thus from (1.3-32) we obtain

$$di(\lambda,x) = \frac{\partial i(\lambda,x)}{\partial \lambda}\, d\lambda + \frac{\partial i(\lambda,x)}{\partial x}\, dx \qquad (1.3\text{-}34)$$

Because $dx = 0$ in this evaluation, (1.3-24) becomes

$$W_c(\lambda,x) = \int \lambda \frac{\partial i(\lambda,x)}{\partial \lambda}\, d\lambda = \int_0^\lambda \xi \frac{\partial i(\xi,x)}{\partial \xi}\, d\xi \qquad (1.3\text{-}35)$$

For a linear electromagnetic system the λ–i plots are straight-line relationships; thus for the singly excited system we have

$$\lambda(i,x) = L(x)i \qquad (1.3\text{-}36)$$

or

$$i(\lambda,x) = \frac{\lambda}{L(x)} \qquad (1.3\text{-}37)$$

Let us evaluate $W_f(i, x)$. From (1.3-28), with $dx = 0$

$$d\lambda(i, x) = L(x)\, di \tag{1.3-38}$$

Hence, from (1.3-29)

$$W_f(i, x) = \int_0^i \xi L(x)\, d\xi = \frac{1}{2} L(x)\, i^2 \tag{1.3-39}$$

It is left to the reader to show that $W_f(\lambda, x)$, $W_c(i, x)$, and $W_c(\lambda, x)$ are equal to (1.3-39) for this magnetically linear system.

The field energy is a state function, and the expression describing the field energy in terms of system variables is valid regardless of the variations in the system variables. For example, (1.3-39) expresses the field energy regardless of the variations in $L(x)$ and i. The fixing of the mechanical system so as to obtain an expression for the field energy is a mathematical convenience and not a restriction upon the result.

In the case of a multiexcited, electromagnetic system, an expression for the field energy may be obtained by evaluating the following relation with $dx_k = 0$:

$$W_f = \int \sum_{j=1}^{J} i_j\, d\lambda_j \tag{1.3-40}$$

Because the coupling fields are considered conservative, (1.3-40) may be evaluated independently of the order in which the flux linkages or currents are brought to their final values. To illustrate the evaluation of (1.3-40) for a multiexcited system, we will allow the currents to establish their final states one at a time while all other currents are mathematically fixed in either their final or unexcited state. This procedure may be illustrated by considering a doubly excited electric system with one mechanical input. An electromechanical system of this type could be constructed by placing a second coil, supplied from a second electrical system, on either the stationary or movable member of the system shown in Fig. 1.3-3. In this evaluation it is convenient to use currents and displacement as the independent variables. Hence, for a doubly excited electric system we have

$$W_f(i_1, i_2, x) = \int \left[i_1\, d\lambda_1(i_1, i_2, x) + i_2\, d\lambda_2(i_1, i_2, x) \right] \tag{1.3-41}$$

In this determination of an expression for W_f, the mechanical displacement is held constant $(dx = 0)$; thus (1.3-41) becomes

$$\begin{aligned}
W_f(i_1, i_2, x) = \int &\left\{ i_1 \left[\frac{\partial \lambda_i(i_1, i_2, x)}{\partial i_1}\, di_1 + \frac{\partial \lambda_1(i_1, i_2, x)}{\partial i_2}\, di_2 \right] \right. \\
&\left. + i_2 \left[\frac{\partial \lambda_2(i_1, i_2, x)}{\partial i_1}\, di_1 + \frac{\partial \lambda_2(i_1, i_2, x)}{\partial i_2}\, di_2 \right] \right\}
\end{aligned} \tag{1.3-42}$$

We will evaluate the energy stored in the field by employing (1.3-42) twice. First we will mathematically bring the current i_1 to the desired value while holding i_2 at zero. Thus, i_1 is the variable of integration and $di_2 = 0$. Energy is supplied to the coupling field from the source connected to coil 1. As the second evaluation of (1.3-42), i_2 is brought to its desired current while holding i_1 at its desired value. Hence, i_2 is the variable of integration and $di_1 = 0$. During this time, energy is supplied from both sources to the coupling field because $i_1 d\lambda_1$ is nonzero. The total energy stored in the coupling field is the sum of the two evaluations. Following this two-step procedure the evaluation of (1.3-42) for the total field energy becomes

$$W_f(i_1, i_2, x) = \int i_1 \frac{\partial \lambda_i(i_1, i_2, x)}{\partial i_1} \, di_1$$
$$+ \int \left[i_1 \frac{\partial \lambda_1(i_1, i_2, x)}{\partial i_2} \, di_2 + i_2 \frac{\partial \lambda_2(i_1, i_2, x)}{\partial i_2} \, di_2 \right] \quad (1.3\text{-}43)$$

which should be written as

$$W_f(i_1, i_2, x) = \int_0^{i_1} \xi \frac{\partial \lambda_1(\xi, i_2, x)}{\partial \xi} \, d\xi$$
$$+ \int_0^{i_2} \left[i_1 \frac{\partial \lambda_1(i_1, \xi, x)}{\partial \xi} \, d\xi + \xi \frac{\partial \lambda_2(i_1, \xi, x)}{\partial \xi} \, d\xi \right] \quad (1.3\text{-}44)$$

The first integral on the right-hand side of (1.3-43) or (1.3-44) results from the first step of the evaluation with i_1 as the variable of integration and with $i_2 = 0$ and $di_2 = 0$. The second integral comes from the second step of the evaluation with $i_1 = i_1$, $di_1 = 0$ and i_2 as the variable of integration. It is clear that the order of allowing the currents to reach their final state is irrelevant; that is, as our first step, we could have made i_2 the variable of integration while holding i_1 at zero $(di_1 = 0)$ and then let i_1 become the variable of integration while holding i_2 at its final variable. The results would be the same. It is also clear that for three electrical inputs the evaluation procedure would require three steps, one for each current to be brought mathematically to its final state.

Let us now evaluate the energy stored in a magnetically linear electromechanical system with two electrical inputs and one mechanical input. For this let

$$\lambda_1(i_1, i_2, x) = L_{11}(x)i_1 + L_{12}(x)i_2 \quad (1.3\text{-}45)$$
$$\lambda_2(i_1, i_2, x) = L_{21}(x)i_1 + L_{22}(x)i_2 \quad (1.3\text{-}46)$$

With that mechanical displacement held constant $(dx = 0)$, we obtain

$$d\lambda_1(i_1, i_2, x) = L_{11}(x) \, di_1 + L_{12}(x) \, di_2 \quad (1.3\text{-}47)$$
$$d\lambda_2(i_1, i_2, x) = L_{12}(x) \, di_1 + L_{22}(x) \, di_2 \quad (1.3\text{-}48)$$

It is clear that the coefficients on the right-hand side of (1.3-47) and (1.3-48) are the partial derivatives. For example, $L_{11}(x)$ is the partial derivative of $\lambda_1(i_1, i_2, x)$ with respect to i_1. Appropriate substitution into (1.3-44) gives

$$W_f(i_1, i_2, x) = \int_0^{i_1} \xi L_{11}(x)\, d\xi + \int_0^{i_2} [i_1 L_{12}(x) + \xi L_{22}(x)]\, d\xi \qquad (1.3\text{-}49)$$

which yields

$$W_f(i_1, i_2, x) = \frac{1}{2}L_{11}(x)i_1^2 + L_{12}(x)i_1 i_2 + \frac{1}{2}L_{22}(x)i_2^2 \qquad (1.3\text{-}50)$$

The extension to a linear electromagnetic system with J electrical inputs is straightforward whereupon the following expression for the total field energy is obtained:

$$W_f(i_1, \ldots, i_J, x) = \frac{1}{2}\sum_{p=1}^{J}\sum_{q=1}^{J} L_{pq} i_p i_q \qquad (1.3\text{-}51)$$

It is left to the reader to show that the equivalent of (1.3-22) for a multiexcited electrostatic system is

$$W_f = \int \sum_{j=1}^{J} e_{fj}\, dq_j \qquad (1.3\text{-}52)$$

Graphical Interpretation of Energy Conversion

Before proceeding to the derivation of expressions for the electromagnetic force, it is instructive to consider briefly a graphical interpretation of the energy conversion process. For this purpose let us again refer to the elementary system shown in Fig. 1.3-3 and let us assume that as the movable member moves from $x = x_a$ to $x = x_b$, where $x_b < x_a$, the λ–i characteristics are given by Fig. 1.3-7. Let us further assume that as the member moves from x_a to x_b the λ–i trajectory moves from point A to point B. It is clear that the exact trajectory from A to B is determined by the combined dynamics of the electrical and mechanical systems. Now, the area $OACO$ represents the original energy stored in field; area $OBDO$ represents the final energy stored in the field. Therefore, the change in field energy is

$$\Delta W_f = \text{area } OBDO - \text{area } OACO \qquad (1.3\text{-}53)$$

The change in W_e, denoted as ΔW_e, is

$$\Delta W_e = \int_{\lambda_A}^{\lambda_B} i\, d\lambda = \text{area } CABDC \qquad (1.3\text{-}54)$$

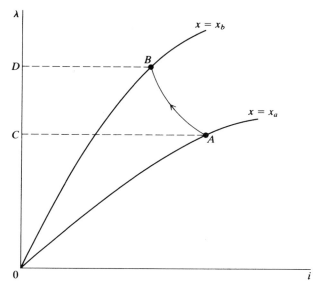

Figure 1.3-7 Graphical representation of electromechanical energy conversion for λ–i path A to B.

We know that

$$\Delta W_m = \Delta W_f - \Delta W_e \qquad (1.3\text{-}55)$$

Hence,

$$\Delta W_m = \text{area } OBDO - \text{area } OACO - \text{area } CABDC = -\text{area } OABO \qquad (1.3\text{-}56)$$

The change in W_m, denoted as ΔW_m, is negative; energy has been supplied to the mechanical system from the coupling field part of which came from the energy stored in the field and part from the electrical system. If the member is now moved back to x_a, the λ–i trajectory may be as shown in Fig. 1.3-8. Hence the ΔW_m is still area $OABO$ but it is positive, which means that energy was supplied from the mechanical system to the coupling field, part of which is stored in the field and part of which is transferred to the electrical system. The net ΔW_m for the cycle from A to B back to A is the shaded area shown in Fig. 1.3-9. Because ΔW_f is zero for this cycle

$$\Delta W_m = -\Delta W_e \qquad (1.3\text{-}57)$$

For the cycle shown the net ΔW_e is negative, thus ΔW_m is positive; we have generator action. If the trajectory had been in the counterclockwise direction, the net ΔW_e would have been positive and the net ΔW_m would have been negative, which would represent motor action.

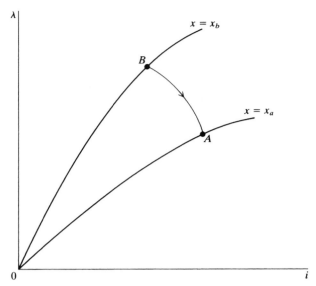

Figure 1.3-8 Graphical representation of electromechanical energy conversion for λ–i path B to A.

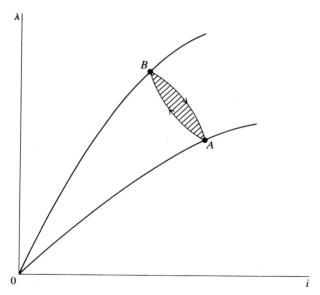

Figure 1.3-9 Graphical representation of electromechanical energy conversion for λ–i path A to B to A.

Electromagnetic and Electrostatic Forces

The energy balance relationships given by (1.3-21) may be arranged as

$$\sum_{k=1}^{K} f_{ek}\, dx_k = \sum_{j=1}^{J} e_{fj}\, i_j\, dt - dW_f \tag{1.3-58}$$

In order to obtain an expression for f_{ek}, it is necessary to first express W_f and then take its total derivative. One is tempted to substitute the integrand of (1.3-22) into (1.3-58) for the infinitesimal change of field energy. This procedure is, of course, incorrect because the integrand of (1.3-22) was obtained with all mechanical displacements held fixed ($dx_k = 0$), where the total differential of the field energy is required in (1.3-58).

The force or torque in any electromechanical system may be evaluated by employing (1.3-58). In many respects, one gains a much better understanding of the energy conversion process of a particular system by starting the derivation of the force or torque expressions with (1.3-58) rather than selecting a relationship from a table. However, for the sake of completeness, derivation of the force equations will be set forth and tabulated for electromechanical systems with K mechanical inputs and J electrical inputs [2].

For an electromagnetic system, (1.3-58) may be written as

$$\sum_{k=1}^{K} f_{ek}\, dx_k = \sum_{j=1}^{J} i_j\, d\lambda_j - dW_f \tag{1.3-59}$$

With i_j and x_k selected as independent variables we have

$$W_f = W_f(i_1, ..., i_J; x_1, ..., x_K) \tag{1.3-60}$$

$$\lambda_j = \lambda_j(i_1, ..., i_J; x_1, ..., x_K) \tag{1.3-61}$$

From (1.3-60) and (1.3-61) we obtain

$$dW_f = \sum_{j=1}^{J} \frac{\partial W_f(i_j, x_k)}{\partial i_j}\, di_j + \sum_{k=1}^{K} \frac{\partial W_f(i_j, x_k)}{\partial x_k}\, dx_k \tag{1.3-62}$$

$$d\lambda_j = \sum_{n=1}^{J} \frac{\partial \lambda_j(i_j, x_k)}{\partial i_n}\, di_n + \sum_{k=1}^{K} \frac{\partial \lambda_j(i_j, x_k)}{\partial x_k}\, dx_k \tag{1.3-63}$$

In (1.3-62) and (1.3-63) and hereafter in this development the functional notation of $(i_1, ..., i_J; x_1, ..., x_K)$ is abbreviated as (i_j, x_k). The index n is used so as to avoid confusion with the index j because each $d\lambda_j$ must be evaluated for changes in all currents in order to account for mutual coupling between electrical systems.

[Recall that we did this in (1.3-42) for $J = 2$.] Substituting (1.3-62) and (1.3-63) into (1.3-59) yields

$$\sum_{k=1}^{K} f_{ek}(i_j, x_k)\, dx_k = \sum_{j=1}^{J} i_j \left[\sum_{n=1}^{J} \frac{\partial \lambda_j(i_j, x_k)}{\partial i_n}\, di_n + \sum_{k=1}^{K} \frac{\partial \lambda_j(i_j, x_k)}{\partial x_k}\, dx_k \right]$$

$$- \sum_{j=1}^{J} \frac{\partial W_f(i_j, x_k)}{\partial i_j}\, di_j - \sum_{k=1}^{K} \frac{\partial W_f(i_j, x_k)}{\partial x_k}\, dx \qquad (1.3\text{-}64)$$

By gathering terms, we obtain

$$\sum_{k=1}^{K} f_{ek}(i_j, x_k)\, dx_k = \sum_{k=1}^{K} \left[\sum_{j=1}^{J} i_j \frac{\partial \lambda_j(i_j, x_k)}{\partial x_k} - \frac{\partial W_f(i_j, x_k)}{\partial x_k} \right] dx_k$$

$$+ \sum_{j=1}^{J} \left[\sum_{n=1}^{J} i_j \frac{\partial \lambda_j(i_j, x_k)}{\partial i_n}\, di_n - \frac{\partial W_f(i_j, x_k)}{\partial i_j}\, di_j \right] \qquad (1.3\text{-}65)$$

When we equate coefficients, we obtain

$$f_{ek}(i_j, x_k) = \sum_{j=1}^{J} i_j \frac{\partial \lambda_j(i_j, x_k)}{\partial x_k} - \frac{\partial W_f(i_j, x_k)}{\partial x_k} \qquad (1.3\text{-}66)$$

$$0 = \sum_{j=1}^{J} \left[\sum_{n=1}^{J} i_j \frac{\partial \lambda_j(i_j, x_k)}{\partial i_n}\, di_n - \frac{\partial W_f(i_j, x_k)}{\partial i_j}\, di_j \right] \qquad (1.3\text{-}67)$$

Although (1.3-67) is of little practical importance, (1.3-66) can be used to evaluate the force at the kth mechanical terminal of an electromechanical system with only magnetic coupling fields and with i_j and x_k selected as independent variables. A second force equation with i_j and x_k as independent variables may be obtained from (1.3-66) by incorporating the expression for coenergy. For a multiexcited system the coenergy may be expressed as

$$W_c = \sum_{j=1}^{J} i_j \lambda_j - W_f \qquad (1.3\text{-}68)$$

Because i_j and x_k are independent variables, the partial derivative with respect to x is

$$\frac{\partial W_c(i_j, x_k)}{\partial x_k} = \sum_{j=1}^{J} i_j \frac{\partial \lambda_j(i_j, x_k)}{\partial x_k} - \frac{\partial W_f(i_j, x_k)}{\partial x_k} \qquad (1.3\text{-}69)$$

Hence, substituting (1.3-69) into (1.3-66) yields

$$f_{ek}(i_j, x_k) = \frac{\partial W_c(i_j, x_k)}{\partial x_k} \qquad (1.3\text{-}70)$$

Table 1.3-1   Electromagnetic Force at kth Mechanical Input^a

$$f_{ek}(i_j, x_k) = \sum_{j=1}^{J} \left[i_j \frac{\partial \lambda_j(i_j, x_k)}{\partial x_k} \right] - \frac{\partial W_f(i_j, x_k)}{\partial x_k}$$

$$f_{ek}(i_j, x_k) = \frac{\partial W_c(i_j, x_k)}{\partial x_k}$$

$$f_{ek}(\lambda_j, x_k) = -\frac{\partial W_f(\lambda_j, x_k)}{\partial x_k}$$

$$f_{ek}(\lambda_j, x_k) = -\sum_{j=1}^{J} \left[\lambda_j \frac{\partial i_j(\lambda_j, x_k)}{\partial x_k} \right] + \frac{\partial W_c(\lambda_j, x_k)}{\partial x_k}$$

^{a}For rotational systems replace f_{ek} with T_{ek} and x_k with θ_k.

It should be recalled that positive f_{ek} and positive dx_k are in the same direction. Also, if the magnetic system is linear, then $W_c = W_f$.

By a procedure similar to that used above, force equations may be developed for magnetic coupling with λ_j and x_k as independent variables. These relations are given in Table 1.3-1 without proof. In Table 1.3-1 the independent variables to be used are designated in each equation by the abbreviated functional notation. Although only translational mechanical systems have been considered, all force relationships developed herein may be modified for the purpose of evaluating the torque in rotational systems. In particular, when considering a rotational system, f_{ek} is replaced with the electromagnetic torque T_{ek}, and x_k is replaced with the angular displacement θ_k. These substitutions are justified because the change of mechanical energy in a rotational system is expressed as

$$dW_{mk} = -T_{ek}\, d\theta_k \tag{1.3-71}$$

The force equation for an electromechanical system with electric coupling fields may be derived by following a procedure similar to that used in the case of magnetic coupling fields. These relationships are given in Table 1.3-2 without explanation.

It is instructive to derive the expression for the electromagnetic force of a singly excited electrical system as shown in Fig. 1.3-3. It is clear that the expressions given in Table 1.3-1 are valid for magnetically linear or nonlinear systems. If we assume that the magnetic system is linear, then $\lambda(i, x)$ is expressed by (1.3-36) and $W_f(i, x)$ is expressed by (1.3-39), which is also equal to the coenergy. Hence, either the first or second entry of Table 1.3-1 can be used to express f_e. In particular,

$$f_e(i, x) = \frac{\partial W_c(i, x)}{\partial x} = \frac{1}{2} i^2 \frac{dL(x)}{dx} \tag{1.3-72}$$

Table 1.3-2   Electrostatic Force at kth Mechanical Input^a

$$f_{ek}(e_{fj}, x_k) = \sum_{j=1}^{J}\left[e_{fj}\frac{\partial q_j(e_{fj}, x_k)}{\partial x_k}\right] - \frac{\partial W_f(e_{fj}, x_k)}{\partial x_k}$$

$$f_{ek}(e_{fj}, x_k) = \frac{\partial W_c(e_{fj}, x_k)}{\partial x_k}$$

$$f_{ek}(q_j, x_k) = -\frac{\partial W_f(q_j, x_k)}{\partial x_k}$$

$$f_{ek}(q_j, x_k) = -\sum_{j=1}^{J}\left[e_j\frac{\partial e_{fj}(q_j, x_k)}{\partial x_k}\right] + \frac{\partial W_c(q_j, x_k)}{\partial x_k}$$

^{a}For rotational systems replace f_{ek} with T_{ek} and x_k with θ_k.

With the convention established, a positive electromagnetic force is assumed to act in the direction of increasing x. Thus with (1.3-15) expressed in differential form as

$$dW_m = -f_e\,dx \tag{1.3-73}$$

we see that energy is supplied to the coupling field from the mechanical system when f_e and dx are opposite in sign, and energy is supplied to the mechanical system from the coupling field when f_e and dx are the same in sign.

From (1.3-72) it is apparent that when the change of $L(x)$ with respect to x is negative, f_e is negative. In the electromechanical system shown in Fig. 1.3-3 the change $L(x)$ with respect to x is always negative; therefore, the electromagnetic force is in the direction so as to pull the movable member to the stationary member. In other words, an electromagnetic force is set up so as to maximize the inductance of the coupling system, or, since reluctance is inversely proportional to the inductance, the force tends to minimize the reluctance. Because f_e is always negative in the system shown in Fig. 1.3-3, energy is supplied to the coupling field from the mechanical system (generator action) when dx is positive and is supplied from the coupling field to the mechanical system (motor action) when dx is negative.

Steady-State and Dynamic Performance of an Electromechanical System

It is instructive to consider the steady-state and dynamic performance of the elementary electromagnetic system shown in Fig. 1.3-3. The differential equations that describe this system are given by (1.3-7) for the electrical system and by (1.3-8) for the mechanical system. The electromagnetic force f_e is expressed by (1.3-72). If the applied voltage, v, and the applied mechanical force, f, are constant, all derivatives with respect to time are zero during steady-state operation, and the behavior can be predicted by

$$v = ri \tag{1.3-74}$$
$$f = K(x - x_0) - f_e \tag{1.3-75}$$

Equation (1.3-75) may be written as

$$-f_e = f - K(x - x_0) \tag{1.3-76}$$

The magnetic core of the system in Fig. 1.3-3 is generally constructed of ferromagnetic material with a relative permeability on the order of 2000 to 4000. In this case the inductance $L(x)$ can be adequately approximated by

$$L(x) = \frac{k}{x} \tag{1.3-77}$$

In the actual system the inductance will be a large finite value rather than infinity, as predicted by (1.3-77), when $x = 0$. Nevertheless, (1.3-77) is quite sufficient to illustrate the action of the system for $x > 0$. Substituting (1.3-77) into (1.3-72) yields

$$f_e(i, x) = -\frac{ki^2}{2x^2} \tag{1.3-78}$$

A plot of (1.3-76), with f_e replaced by (1.3-78), is shown in Fig. 1.3-10 for the following system parameters:

$$r = 10\,\Omega \qquad x_0 = 3\,\text{mm}$$
$$K = 2667\,\text{N/m} \qquad k = 6.293 \times 10^{-5}\,\text{H} \cdot \text{m}$$

In Fig. 1.3-10, the plot of the negative of the electromagnetic force is for an applied voltage of 5 V and a steady-state current of 0.5 A. The straight lines represent the right-hand side of (1.3-76) with $f = 0$ (lower straight line) and $f = 4\,\text{N}$ (upper straight line). Both lines intersect the $-f_e$ curve at two points. In particular, the upper line intersects the $-f_e$ curve at 1 and 1'; the lower line intersects at 2 and 2'. Stable operation occurs at only points 1 and 2. The system will not operate stably at points 1' and 2'. This can be explained by assuming that the system is operating at one of these points (1' and 2') and then showing that any system disturbance whatsoever will cause the system to move away from these points. If, for example, x increases slightly from its value corresponding to point 1', the restraining force $f - K(x - x_0)$ is larger in magnitude than $-f_e$, and x will continue to increase until the system reaches operating point 1. If x increases beyond its value corresponding to operating point 1, the restraining force is less than the electromagnetic force. Therefore, the system will establish steady-state operation at 1. If, on the other hand, x decreases

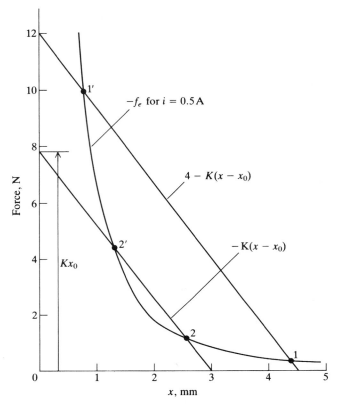

Figure 1.3-10   Steady-state operation of electromechanical system shown in Fig. 1.3-3.

from point $1'$, the electromagnetic force is larger than the restraining force. Therefore, the movable member will move until it comes in contact with the stationary member $(x = 0)$. It is clear that the restraining force that yields a straight line below the $-f_e$ curve will not permit stable operation with $x > 0$.

The dynamic behavior of the system during step changes in the source voltage v is shown in Fig. 1.3-11, and Figs. 1.3-12 and 1.3-13 show dynamic behavior during step changes in the applied force f. The following system parameters were used in addition to those given previously:

$$l = 0 \qquad M = 0.055\,\text{kg} \qquad D = 4\,\text{N} \cdot \text{s/m}$$

The computer traces shown in Fig. 1.3-11 depict the dynamic performance of the example system when the applied voltage is stepped from zero to 5 V and then back to zero with the applied mechanical force held equal to zero. The system variables are e_f, λ, i, f_e, x, W_e, W_f, and W_m. The energies are plotted in millijoules (mJ). Initially the mechanical system is at rest with $x = x_0$ (3 mm). When the source

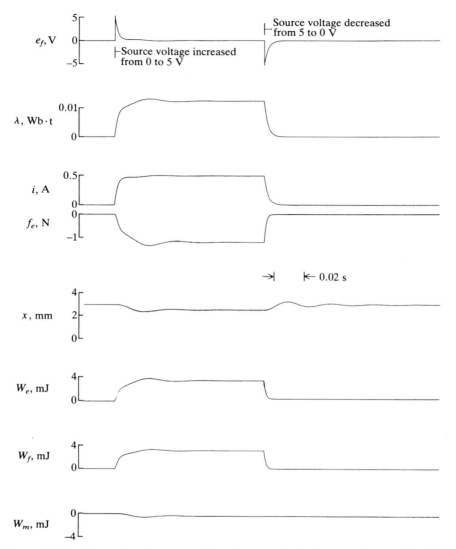

Figure 1.3-11   Dynamic performance of the electromechanical system shown in Fig. 1.3-3 during step changes in the source voltage.

voltage is applied, x decreases; and when steady-state operation is reestablished, x is approximately 2.5 mm. Energy enters the coupling field via W_e. The bulk of this energy is stored in the field (W_f) with a smaller amount transferred to the mechanical system, some of which is dissipated in the damper during the transient period while the remainder is stored in the spring. When the applied voltage is removed, the electrical and mechanical systems return to their original states. The change in W_m

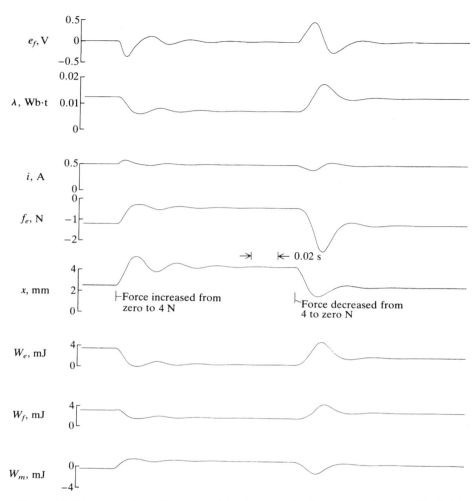

Figure 1.3-12 Dynamic performance of the electromechanical system shown in Fig. 1.3-3 during step changes in the applied force.

is small, increasing only slightly. Hence, during the transient period there is an interchange of energy between the spring and mass which is finally dissipated in the damper. The net change in W_f during the application and removal of the applied voltage is zero; hence the net change in W_e is positive and equal to the negative of the net change in W_m. The energy transferred to the mechanical system during this cycle is dissipated in the damper, because f is fixed at zero, and the mechanical system returns to its initial rest position with zero energy stored in the spring.

   In Fig. 1.3-12, the initial state is that shown in Fig. 1.3-11 with 5 V applied to the electrical system. The mechanical force f is increased from zero to 4 N, whereupon energy enters the coupling field from the mechanical system. Energy is transferred

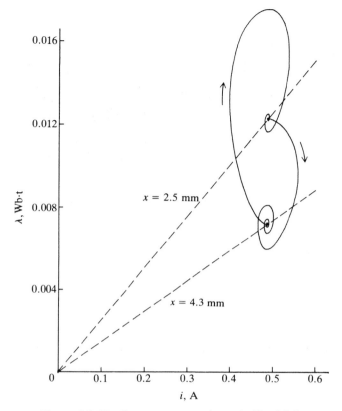

Figure 1.3-13   System response shown in Fig. 1.3-3.

from the coupling field to the electrical system, some coming from the mechanical system and some from the energy originally stored in the magnetic field. Next the force is stepped back to zero from 4 N. The electrical and mechanical systems return to their original states. During the cycle a net energy has been transferred from the mechanical system to the electrical system which is dissipated in the resistance. This cycle is depicted on the λ–i plot shown in Fig. 1.3-13.

Example 1B   It is helpful to formulate an expression for the electromagnetic torque of the elementary rotational device shown in Fig. 1B-1. This device consists of two conductors. Conductor 1 is placed on the stationary member (stator); conductor 2 is fixed on the rotating member (rotor). The crossed lines inside a circle indicate that the assumed direction of positive current flow is into the paper (we are seeing the tail of the arrow), whereas a dot inside a circle indicates positive current flow is out of the paper (the point of the arrow). The length of the air gap between the stator and rotor is shown exaggerated relative to the inside diameter of the stator.

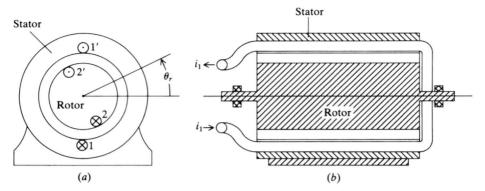

Figure 1B-1 Elementary rotational electromechanical device. (*a*) End view; (*b*) cross-sectional view.

The voltage equations may be written as

$$v_1 = i_1 r_1 + \frac{d\lambda_i}{dt} \tag{1B-1}$$

$$v_2 = i_2 r_2 + \frac{d\lambda_2}{dt} \tag{1B-2}$$

where r_1 and r_2 are the resistances of conductor 1 and 2, respectively. The magnetic system is assumed linear; therefore the flux linkages may be expressed as

$$\lambda_1 = L_{11} i_1 + L_{12} i_2 \tag{1B-3}$$

$$\lambda_2 = L_{21} i_1 + L_{22} i_2 \tag{1B-4}$$

The self-inductances L_{11} and L_{22} are constant. Let us assume that the mutual inductance may be approximated by

$$L_{12} = L_{21} = M \cos \theta_r \tag{1B-5}$$

where θ_r is defined in Fig. 1B-1. The reader should be able to justify the form of (1B-5) by considering the mutual coupling between the two conductors as θ_r varies from 0 to 2π radians.

$$T_e(i_1, i_2, \theta_r) = \frac{\partial W_c(i_1, i_2, \theta_r)}{\partial \theta_r} \tag{1B-6}$$

Because the magnetic system is assumed to be linear, we have

$$W_c(i_1, i_2, \theta_r) = \frac{1}{2} L_{11} i_1^2 + L_{12} i_1 i_2 + \frac{1}{2} L_{22} i_2^2 \tag{1B-7}$$

Substituting into (1B-6) yields

$$T_e = -i_1 i_2 M \sin \theta_r \tag{1B-8}$$

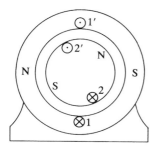

Figure 1B-2   Stator and rotor poles for constant currents.

Consider for a moment the form of the torque if i_1 and i_2 are both constant. For the positive direction of current shown, the torque is of the form

$$T_e = -K \sin \theta_r \qquad (1\text{B-}9)$$

where K is a positive constant. We can visualize the production of torque by considering the interaction of the magnetic poles produced by the current flowing in the conductors. If both i_1 and i_2 are positive, the poles produced are as shown in Fig. 1B-2. One should recall that flux issues from a north pole. Also, the stator and rotor each must be considered as separate electromagnetic systems. Thus, flux produced by the 1–1′ winding issues from the north pole of the stator into the air gap. Similarly, the flux produced by the 2–2′ winding enters the air gap from the north pole of the rotor. It is left to the reader to justify the fact that the range of θ_r over which stable operation can occur for the expression of electromagnetic torque given by (1B-9) is $-\pi/2 \leq \theta_r \leq \pi/2$.

1.4   MACHINE WINDINGS AND AIR-GAP MMF

For the purpose of discussing winding configurations in rotating machines and the resulting air-gap MMF as well as the calculations of machine inductances, it is convenient to begin with the elementary 2-pole, 3-phase, wye-connected salient-pole synchronous machine shown in Fig. 1.4-1. Once these concepts are established for this type of a machine, they may be readily modified to account for all types of induction machines and easily extended to include the synchronous machine with short-circuited windings on the rotor (damper windings).

The stator windings of the synchronous machine are embedded in slots around the inside circumference of the stationary member. In the 2-pole machine, each phase winding of the 3-phase stator winding is displaced 120° with respect to each other as illustrated in Fig. 1.4-1. The field or *fd* winding is wound on the rotating member. The *as*, *bs*, *cs*, and *fd* axes denote the positive direction of the flux produced by each of the windings. The *as*, *bs*, and *cs* windings are identical in that each winding has the same resistance and the same number of turns. When a machine has three identical stator windings arranged as shown in Fig. 1.4-1, it is often referred to

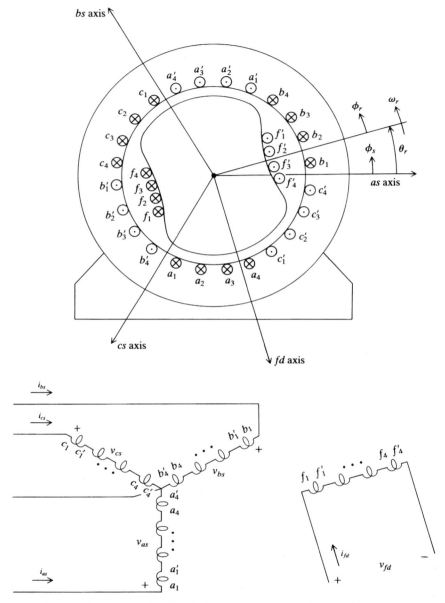

Figure 1.4-1 Elementary, 2-pole, 3-phase, wye-connected salient-pole synchronous machine.

as a machine with symmetrical stator windings. We will find that the symmetrical induction machine has identical multiphase stator windings and identical multiphase rotor windings. An unsymmetrical induction machine has nonidentical multiphase stator windings (generally 2-phase) and symmetrical multiphase rotor windings.

In Fig. 1.4-1, it is assumed that each coil spans π radians of the stator for a 2-pole machine. One side of the coil (coil side) is represented by a $\otimes$ indicating that the assumed positive direction of current is down the length of the stator (into the paper). The $\odot$ indicates that the assumed positive direction of current is out of the paper. Each coil contains n_c conductors. Therefore, in the case of the *as* winding, positive current flows in a conductor of coil a_1, then axially down the length of the stator and back the length of the stator and out at coil side a_1'. This is repeated for n_c conductors. The last conductor of the coil a_1-a_1' is then placed in the appropriate slot so as to start coil a_2-a_2' wherein the current flows down the stator via coil side a_2 and then back through a_2' and so on until a_4'. The *bs* and *cs* windings are arranged similarly, and the last conductors of coil sides a_4', b_4', and c_4' are connected together to form the wye-connected stator. The end turns (looping of the coils) at both ends of the stator so as to achieve the span of π radians are not shown in Fig. 1.4-1. As mentioned, each coil consists of n_c conductors, each of which makes up an individual single conductor coil within the main coil. Thus the number of turns of each winding is determined by the product of n_c and the number of coils or the product of n_c and the number of coil sides carrying current in the same direction. In the case of the *fd* winding, each coil (f_1-f_1', for example) consists of n_f conductors. It should be mentioned that in Example 1B, the stator and rotor coils each consisted of only one coil side with one conductor ($n_c = 1$) in each coil side.

One must realize that the winding configuration shown in Fig. 1.4-1 is an oversimplification of a practical machine. The coil sides of each phase winding are considered to be distributed uniformly over 60° of the stator circumference. Generally, the coil sides of each phase are distributed over a larger area, perhaps as much as 120°, in which case it is necessary for some of the coil sides of two of the phase windings to occupy the same slot. In some cases, the coil sides may not be distributed uniformly over the part of the circumference that it occupies. For example, it would not be uncommon, in the case of the machine shown in Fig. 1.4-1, to have more turns in coil sides a_2 and a_3 than in a_1 and a_4. (Similarly for the *bs* and *cs* windings.) We will find that this winding arrangement produces an air-gap MMF which more closely approximates a sinusoidal air-gap MMF with respect to the angular displacement about the air gap. Another practical consideration is the so-called fractional-pitch winding. The windings shown in Fig. 1.4-1 span π radians for the 2-pole machine. This is referred to as full-pitch winding. In order to reduce voltage and current harmonics, the windings are often wound so that they span slightly less than π radians for a 2-pole machine. This is referred to as a fractional-pitch winding. All of the above-mentioned practical variations from the winding arrangement shown in Fig. 1.4-1 are very important to the machine designer; however, these features are of less importance in machine analysis, where in most cases a simplified approximation of the winding arrangement is sufficient.

A salient-pole synchronous machine is selected for consideration because the analysis of this type of machine may be easily modified to account for other machine types. However, a salient-pole synchronous machine would seldom be a 2-pole machine except in the case of small reluctance machines which are of the synchronous class but which do not have a field winding. Generally, 2- and 4-pole machines are round-rotor machines with the field winding embedded in a solid steel (nonlaminated) rotor. Salient-pole machines generally have a large number of poles composed of laminated steel whereupon the field winding is wound around the poles similar to that shown in Fig. 1.4-1.

For the purposes of deriving an expression for the air-gap MMF, it is convenient to employ the so-called developed diagram of the cross-sectional view of the machine shown in Fig. 1.4-1. The developed diagram is shown in Fig. 1.4-2. The length of the air gap between the stator and rotor is exaggerated in Figs. 1.4-1 and 1.4-2 for clarity. The fact that the air-gap length is small relative to the inside diameter of the stator permits us to employ the developed diagram for analysis purposes. In order to relate the developed diagram to the cross-sectional view of the machine, it is helpful to define a displacement to the left of the origin as positive. The angular displacement along the stator circumference is denoted ϕ_s and ϕ_r along the rotor circumference. The angular velocity of the rotor is ω_r, and θ_r is the angular displacement of the rotor. For a given angular displacement relative to the as axis we can relate ϕ_s, ϕ_r, and θ_r as

$$\phi_s = \phi_r + \theta_r \qquad (1.4\text{-}1)$$

Our analysis may be simplified by considering only one of the stator windings at a time. Figure 1.4-3 is a repeat of Figs. 1.4-1 and 1.4-2 with only the as winding shown. Due to the high permeability of the stator and rotor steel the magnetic fields

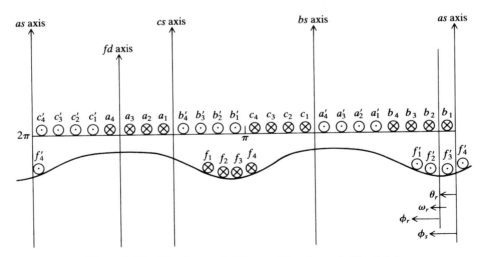

Figure 1.4-2   Development of the machine shown in Fig. 1.4-1.

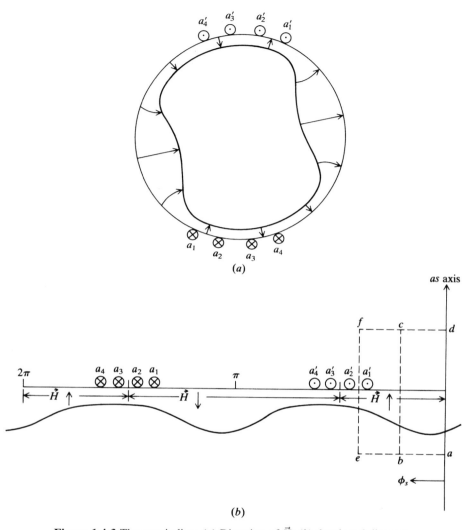

Figure 1.4-3 The *as* winding. (*a*) Direction of $\vec{H}$; (*b*) developed diagram.

essentially exist only in the air gap and tend to be radial in direction due to the short length of the air gap relative to the inside stator diameter. Therefore, in the air gap the magnetic field intensity $\vec{H}$ and the flux density $\vec{B}$ have a component only in the $\vec{a}_r$ direction, and the magnitude is a function of the angle ϕ_s whereupon

$$\vec{H}(r, \phi_s, z) = H_r(\phi_s)\vec{a}_r \tag{1.4-2}$$

and

$$B_r(\phi_s) = \mu_0 H_r(\phi_s) \tag{1.4-3}$$

With the assumed direction of the current i_{as}, the magnetic field intensity $\vec{H}$ in the air gap due to the *as* winding is directed from the rotor to the stator for $-\pi/2 < \phi_s < \pi/2$ and from the stator to the rotor for $\pi/2 < \phi_s < \frac{3}{2}\pi$ (Fig. 1.4-3).

Ampere's law may now be used to determine the form of the air-gap MMF due to the *as* winding. In particular, Ampere's law states that

$$\oint \vec{H} \cdot d\vec{L} = i \tag{1.4-4}$$

where i is the net current enclosed within the closed path of integration. Let us consider the closed path of integration depicted in Fig. 1.4-3*b*. Applying Ampere's law around this closed path denoted as *abcda*, where the path *bc* is at $\phi_s = \pi/4$, and neglecting the field intensity within the stator and rotor steel, we can write and evaluate (1.4-4) as

$$\int_{r(\pi/4)}^{r(\pi/4)+g(\pi/4)} H\left(\frac{\pi}{4}\right) dL + \int_{r(0)+g(0)}^{r(0)} H(0)\,dL = 0$$
$$H_r\left(\frac{\pi}{4}\right) g\left(\frac{\pi}{4}\right) - H_r(0)g(0) = 0 \tag{1.4-5}$$

where $r(\pi/4)$ and $r(0)$ are the radius of the rotor at the respective paths of integration, and $g(\pi/4)$ and $g(0)$ are the corresponding air-gap lengths.

The magnetomotive force is defined as the line integral of $\vec{H}$. Therefore, the terms on the left-hand side of (1.4-5) may be written as MMFs. In particular, (1.4-5) may be written as

$$\text{MMF}\left(\frac{\pi}{4}\right) + \text{MMF}(0) = 0 \tag{1.4-6}$$

In (1.4-6) the MMF includes sign and magnitude; that is, $\text{MMF}(0) = -H_r(0)g(0)$.

Let us now consider the closed path *aefda* where path *ef* occurs at $\phi_s = \frac{7}{16}\pi$; here Ampere's law gives

$$\text{MMF}\left(\frac{7}{16}\pi\right) + \text{MMF}(0) = -n_c i_{as} \tag{1.4-7}$$

The right-hand side of (1.4-7) is negative in accordance with the right-hand or *corkscrew* rule.

We can now start to plot the air-gap MMF due to the *as* winding. For the lack of a better guess let us assume that $\text{MMF}(0) = 0$. With this assumption, (1.4-6) and (1.4-7) tell us that the MMF is zero from $\phi_s = 0$ to where our path of integration encircles the first coil side (a_1'). If we continue to perform line integrations starting and ending at a and each time progressing further in the ϕ_s direction, we will obtain the plot shown in Fig. 1.4-4. Therein, a step change in MMF is depicted at the center

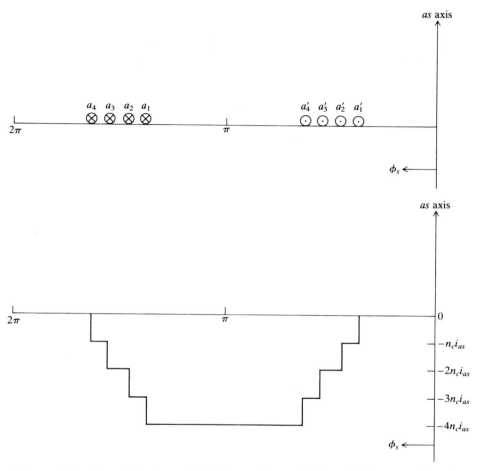

Figure 1.4-4 Plot of the air-gap MMF due to the *as* winding with the assumption that MMF(0) is zero.

of the conductors; actually, there would be a finite slope as the path of integration passes through the conductors.

There are two items left to be considered. First, Gauss's law states that

$$\int_s \vec{B} \cdot d\vec{S} = 0 \qquad (1.4\text{-}8)$$

Hence, no net flux may travel across the air gap because the flux density is assumed to exist only in the radial direction:

$$\int_0^{2\pi} B_r(\phi_s) rl \, d\phi_s = 0 \qquad (1.4\text{-}9)$$

where l is the axial length of the air gap of the machine and r is the mean radius of the air gap. It follows that $rl \, d\phi_s$ is the incremental area of an imaginary cylindrical surface within the air gap of the machine. This fact in itself makes us suspicious of our assumption that $MMF(0) = 0$ because the air-gap MMF shown in Fig. 1.4-4 is unidirectional, which would give rise to a flux in only one direction across the air gap. The second fact is one that is characteristic of electrical machines. In any practical electric machine, the air-gap length is either (a) constant as in the case of a round-rotor machine or (b) a periodic function of displacement about the air gap as in the case of salient-pole machine. In particular, for a 2-pole machine

$$g(\phi_r) = g(\phi_r + \pi) \qquad (1.4\text{-}10)$$

or

$$g(\phi_s - \theta_r) = g(\phi_s - \theta_r + \pi) \qquad (1.4\text{-}11)$$

Equations (1.4-8) and (1.4-10) are satisfied if the air-gap MMF has zero average value and, for a 2-pole machine,

$$MMF(\phi_s) = -MMF(\phi_s + \pi) \qquad (1.4\text{-}12)$$

Hence, the air-gap MMF wave for the *as* winding, which is denoted as MMF_{as}, is shown in Fig. 1.4-5. It is clear that the MMF due to the *bs* winding, MMF_{bs}, is identical to MMF_{as} but displaced to the left by 120°. MMF_{cs} is also identical but displaced 240° to the left. The significance of the *as*, *bs*, and *cs* axes is now apparent. These axes are positioned at the center of maximum positive MMF corresponding to positive current into the windings.

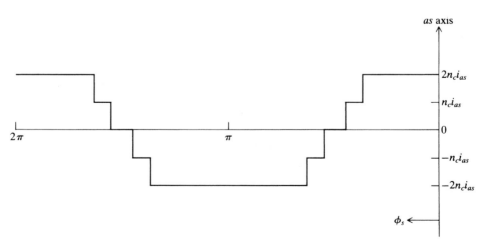

Figure 1.4-5   Air-gap MMF due to *as* winding.

The waveform of the MMF's produced by the stator phase windings shown in Fig. 1.4-1 may be considered as coarse approximations of sinusoidal functions of ϕ_s. Actually most electric machines, especially large machines, are designed so that the stator windings produce a relatively good approximation of a sinusoidal air-gap MMF with respect to ϕ_s so as to minimize the voltage and current harmonics. In order to establish a truly sinusoidal MMF waveform (often referred to as a space sinusoid) the winding must also be distributed sinusoidally. Except in cases where the harmonics due to the winding configuration are of importance, it is typically assumed that all windings may be approximated as sinusoidally distributed windings. We will make this same assumption in our analysis.

A sinusoidally distributed as winding and a sinusoidal air-gap MMF_{as} are depicted in Fig. 1.4-6. The distribution of the as winding may be written as

$$N_{as} = N_p \sin \phi_s, \qquad 0 \leq \phi_s \leq \pi \qquad (1.4\text{-}13)$$

$$N_{as} = -N_p \sin \phi_s, \qquad \pi \leq \phi_s \leq 2\pi \qquad (1.4\text{-}14)$$

where N_p is the maximum turn or conductor density expressed in turns per radian. If N_s represents the number of turns of the equivalent sinusoidally distributed

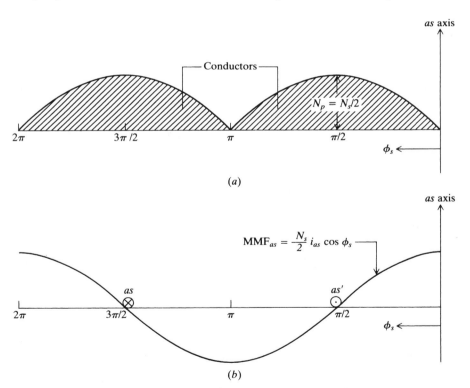

Figure 1.4-6 Sinusoidal distribution. (a) Equivalent distribution of as winding; (b) MMF due to equivalent as winding.

winding, then

$$N_s = \int_0^\pi N_p \sin\phi_s \, d\phi_s = 2N_p \tag{1.4-15}$$

It is important to note that the number of turns is obtained by integrating N_{as} over π radians rather than 2π radians. Also, N_s is not the total number of turns of the winding; instead, it is an equivalent number of turns of a sinusoidally distributed winding which would give rise to the same fundamental component as the actual winding distribution. For example, if we represent the air-gap MMF wave given in Fig. 1.4-5 as a sinusoidal air-gap MMF as given in Fig. 1.4-6, it can be readily shown by Fourier analysis that $N_s/2$ is equal to $2.43n_c$. It is clear that the influence of design features that have not been considered in the analysis such as fractional-pitch windings are accounted for in N_s. (See problem at end of chapter.) Information regarding machine windings may be found in reference 3.

Current flowing out of the paper is assumed negative so as to comply with the convention established earlier when applying Ampere's law. The circles, which are labeled *as* and *as'* and positioned at the point of maximum turn density in Fig. 1.4-6b, will be used hereafter to signify schematically a sinusoidally distributed winding. It is interesting to note in passing that some authors choose to incorporate the direction of current flow in the expression for the turns whereupon the turns for the *as* winding are expressed as

$$N_{as} = -\frac{N_s}{2} \sin\phi_s \tag{1.4-16}$$

The air-gap MMF waveform is also shown in Fig. 1.4-6b. This waveform is readily established by applying Ampere's law. The MMF waveform of the equivalent *as* winding is

$$\mathrm{MMF}_{as} = \frac{N_s}{2} i_{as} \cos\phi_s \tag{1.4-17}$$

It follows that

$$\mathrm{MMF}_{bs} = \frac{N_s}{2} i_{bs} \cos\left(\phi_s - \frac{2\pi}{3}\right) \tag{1.4-18}$$

$$\mathrm{MMF}_{cs} = \frac{N_s}{2} i_{cs} \cos\left(\phi_s + \frac{2\pi}{3}\right) \tag{1.4-19}$$

Let us now express the total air-gap MMF produced by the stator currents. This can be obtained by adding the individual MMFs, given by (1.4-17)–(1.4-19):

$$\mathrm{MMF}_s = \frac{N_s}{2}\left[i_{as}\cos\phi_s + i_{bs}\cos\left(\phi_s - \frac{2\pi}{3}\right) + i_{cs}\cos\left(\phi_s + \frac{2\pi}{3}\right)\right] \tag{1.4-20}$$

For balanced, steady-state conditions the stator currents may be expressed as

$$I_{as} = \sqrt{2}I_s \cos[\omega_e t + \theta_{ei}(0)] \tag{1.4-21}$$

$$I_{bs} = \sqrt{2}I_s \cos\left[\omega_e t - \frac{2\pi}{3} + \theta_{ei}(0)\right] \tag{1.4-22}$$

$$I_{cs} = \sqrt{2}I_s \cos\left[\omega_e t + \frac{2\pi}{3} + \theta_{ei}(0)\right] \tag{1.4-23}$$

where $\theta_{ei}(0)$ is the phase angle at time zero. Substituting the currents into (1.4-20) and using the trigonometric relations given in Appendix A yields

$$\text{MMF}_s = \left(\frac{N_s}{2}\right)\sqrt{2}I_s\left(\frac{3}{2}\right)\cos[\omega_e t + \theta_{ei}(0) - \phi_s] \tag{1.4-24}$$

If the argument is set equal to a constant while the derivative is taken with respect to time, we see that the above expression describes a sinusoidal air-gap MMF wave with respect to ϕ_s, which rotates about the stator at an angular velocity of ω_e in the counterclockwise direction and which may be thought of as a rotating magnetic pole pair. If, for example, the phase angle $\theta_{ei}(0)$ is zero, then at the instant $t = 0$ the rotating air-gap MMF is positioned in the as axis with the north pole at $\phi_s = 180°$ and the south pole at $\phi_s = 0$. (The north pole is by definition the stator pole from which the flux issues into the air gap.)

The rotating air-gap MMF of a P-pole machine can be determined by considering a 4-pole machine. The arrangement of the windings is shown in Fig. 1.4-7. Each phase winding consists of two series connected windings, which will be considered as sinusoidally distributed windings. The air-gap MMF established by each phase is now a sinusoidal function of $2\phi_s$ for a 4-pole machine or, in general, $(P/2)\phi_s$ where P is the number of poles. In particular,

$$\text{MMF}_{as} = \frac{N_s}{P} i_{as} \cos\frac{P}{2}\phi_s \tag{1.4-25}$$

$$\text{MMF}_{bs} = \frac{N_s}{P} i_{bs} \cos\left(\frac{P}{2}\phi_s - \frac{2\pi}{3}\right) \tag{1.4-26}$$

$$\text{MMF}_{cs} = \frac{N_s}{P} i_{cs} \cos\left(\frac{P}{2}\phi_s + \frac{2\pi}{3}\right) \tag{1.4-27}$$

where N_s is the total equivalent turns per phase. With balanced steady-state stator currents as given previously, the air-gap MMF becomes

$$\text{MMF}_s = \left(\frac{2N_s}{P}\right)\sqrt{2}I_s\left(\frac{3}{2}\right)\cos\left[\omega_e t + \theta_{ei}(0) - \frac{P}{2}\phi_s\right] \tag{1.4-28}$$

Here we see that the MMF produced by balanced steady-state stator currents rotates about the air gap in the counterclockwise direction at a velocity of $(2/P)\omega_e$. It may

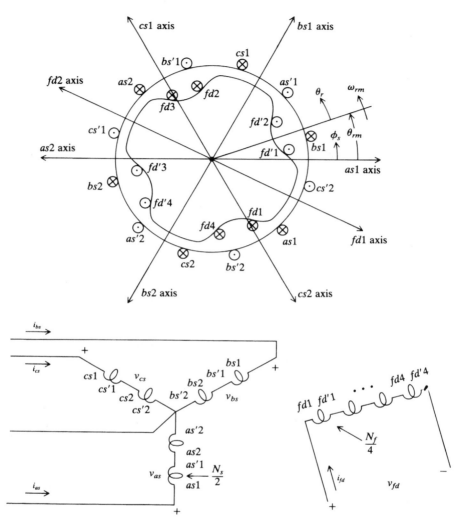

Figure 1.4-7 Winding arrangement of an elementary 4-pole, 3-phase, wye-connected, salient-pole synchronous machine.

at first appear necessary to modify extensively the analysis of a 2-pole machine in order to analyze a *P*-pole machine. Fortunately, we will find that the modification amounts to a simple change of variables.

We can now start to see the mechanism by which torque is produced in a synchronous machine. The stator windings are arranged so that with balanced steady-state currents flowing in these windings, an air-gap MMF is produced which rotates about the air gap as a set of magnetic poles at an angular velocity corresponding to the frequency of the stator currents and the number of poles. During steady-state operation the voltage applied to the field winding is constant. The resulting constant field current produces a set of magnetic poles, stationary with respect to the rotor. If the rotor rotates at the same speed, or in synchronism with the rotating MMF established by the stator currents, torque is produced due to the interaction of these poles. It follows that a steady-state average torque is produced only when the rotor and the stator air-gap MMFs are rotating in synchronism, thus the name synchronous machine.

> **Example 1C**   When the stator currents of a multiphase electric machine, which is equipped with symmetrical stator windings, are unbalanced in amplitude and/or in phase, the total air-gap MMF consists of two oppositely rotating MMFs. This can be demonstrated by assuming that during steady-state operation, the 3-phase stator currents are unbalanced in amplitude. In particular,
>
> $$I_{as} = \sqrt{2}I_a \cos \omega_e t \qquad (1C\text{-}1)$$
>
> $$I_{bs} = \sqrt{2}I_b \cos \left(\omega_e t - \frac{2\pi}{3} \right) \qquad (1C\text{-}2)$$
>
> $$I_{cs} = \sqrt{2}I_c \cos \left(\omega_e t + \frac{2\pi}{3} \right) \qquad (1C\text{-}3)$$
>
> where I_a, I_b, and I_c are unequal constants. Substituting into (1.4-20) yields
>
> $$\begin{aligned} \text{MMF}_s = \frac{N_s}{2} \sqrt{2} \Bigg[& \left(\frac{I_a + I_b + I_c}{2} \right) \cos \left(\omega_e t - \phi_s \right) \\ & + \left(\frac{2I_a - I_b - I_c}{4} \right) \cos \left(\omega_e t + \phi_s \right) \\ & + \frac{\sqrt{3}}{4} (I_b - I_c) \sin \left(\omega_e t + \phi_s \right) \Bigg] \end{aligned} \qquad (1C\text{-}4)$$
>
> The first term is an air-gap MMF rotating in the counterclockwise direction at ω_e. The last two terms are air-gap MMFs (which may be combined into one) that rotate in the clockwise direction at ω_e. Note that when the steady-state currents are balanced, I_a, I_b, and I_c are equal whereupon the last two terms of (1C-4) disappear.

1.5   WINDING INDUCTANCES AND VOLTAGE EQUATIONS

Once the concepts of the sinusoidally distributed winding and the sinusoidal air-gap MMF have been established, the next step is to determine the self- and mutual inductances of the machine windings. As in the previous section, it is advantageous to use the elementary 2-pole, 3-phase synchronous machine to develop these inductance relationships. This development may be readily modified to account for additional windings (damper windings) placed on the rotor of the synchronous machine or for a synchronous machine with a uniform air gap (round rotor). Also, we will show that these inductance relationships may be easily altered to describe the winding inductances of an induction machine.

Synchronous Machine

Figure 1.5-1 is Fig. 1.4-1 redrawn with the windings portrayed as sinusoidally distributed windings. In a magnetically linear system the self-inductance of a winding is the ratio of the flux linked by a winding to the current flowing in the winding with all other winding currents zero. Mutual inductance is the ratio of flux linked by one winding due to current flowing in a second winding with all other winding currents zero including the winding for which the flux linkages are being determined. For this analysis it is assumed that the air-gap length may be approximated

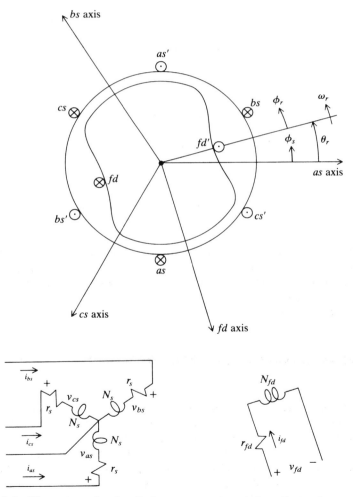

Figure 1.5-1 Elementary 2-pole, 3-phase, wye-connected, salient-pole synchronous machine.

as (Fig. 1.4-2)

$$g(\phi_r) = \frac{1}{\alpha_1 - \alpha_2 \cos 2\phi_r} \tag{1.5-1}$$

or

$$g(\phi_s - \theta_r) = \frac{1}{\alpha_1 - \alpha_2 \cos(2\phi_s - \theta_r)} \tag{1.5-2}$$

where the minimum air-gap length is $(\alpha_1 + \alpha_2)^{-1}$ and the maximum is $(\alpha_1 - \alpha_2)^{-1}$.
Recall from (1.4-5) and (1.4-6) that MMF is defined as the line integral of $\vec{H}$. Thus, from (1.4-3)

$$B_r = \mu_0 \frac{\text{MMF}}{g} \tag{1.5-3}$$

or, because it is convenient to express the stator MMF in terms of MMF(ϕ_s), we can write

$$B_r(\phi_s, \theta_r) = \mu_0 \frac{\text{MMF}(\phi_s)}{g(\phi_s - \theta_r)} \tag{1.5-4}$$

The air-gap flux density due to current in the *as* winding (i_{as}) with all other currents zero may be obtained by substituting (1.4-17) and (1.5-2) into (1.5-4). In particular, the flux density with all currents zero except i_{as} may be expressed as

$$B_r(\phi_s, \theta_r) = \mu_0 \frac{\text{MMF}_{as}(\phi_s)}{g(\phi_s - \theta_r)} = \mu_0 \frac{N_s}{2} i_{as} \cos \phi_s [\alpha_1 - \alpha_2 \cos 2(\phi_s - \theta_r)] \tag{1.5-5}$$

Similarly, the flux density with all currents zero except i_{bs} is

$$B_r(\phi_s, \theta_r) = \mu_0 \frac{N_s}{2} i_{bs} \cos\left(\phi_s - \frac{2\pi}{3}\right)[\alpha_1 - \alpha_2 \cos 2(\phi_s - \theta_r)] \tag{1.5-6}$$

With all currents zero except i_{cs} we have

$$B_r(\phi_s, \theta_r) = \mu_0 \frac{N_s}{2} i_{cs} \cos\left(\phi_s + \frac{2\pi}{3}\right)[\alpha_1 - \alpha_2 \cos 2(\phi_s - \theta_r)] \tag{1.5-7}$$

In the case of salient-pole synchronous machines the field winding is generally uniformly distributed and the poles are shaped to approximate a sinusoidal distribution of air-gap flux due to current flowing in the field winding. In the case of round-rotor synchronous machines the field winding is arranged to approximate more closely a sinusoidal distribution [4–6]. In any event it is sufficient for our purposes to assume as a first approximation that the field winding is sinusoidally distributed with N_f equivalent turns. Later we will assume that the damper windings (additional rotor windings) may also be approximated by sinusoidally distributed windings. Thus, the air-gap MMF due to current i_{fd} flowing in the fd winding may be expressed from Fig. 1.4-2 as

$$\text{MMF}_{fd} = -\frac{N_f}{2} i_{fd} \sin \phi_r \tag{1.5-8}$$

Hence the air-gap flux density due to i_{fd} with all other currents zero may be expressed from (1.5-3) as

$$B_r(\phi_r) = -\mu_0 \frac{N_f}{2} i_{fd} \sin \phi_r (\alpha_1 - \alpha_2 \cos 2\phi_r) \tag{1.5-9}$$

In the determination of self-inductance it is necessary to compute the flux linking a winding due to its own current. To determine mutual inductance it is necessary to compute the flux linking one winding due to current flowing in another winding. Let us consider the flux linkages of a single turn of a stator winding which spans π radians and which is located at an angle ϕ_s. In this case the flux is determined by performing a surface integral over the open surface of the single turn. In particular,

$$\Phi(\phi_s, \theta_r) = \int_{\phi_s}^{\phi_s + \pi} B_r(\xi, \theta_r) rl \, d\xi \tag{1.5-10}$$

where Φ is the flux linking a single turn oriented ϕ_s from the as axis, l is the axial length of the air gap of the machine, r is the radius to the mean of the air gap (essentially to the inside circumference of the stator), and ξ is a dummy variable of integration. In order to obtain the flux linkages of an entire winding, the flux linked by each turn must be summed. Because the windings are considered to be sinusoidally distributed and the magnetic system is assumed to be linear, this summation may be accomplished by integrating over all coil sides carrying current in the same direction. Hence, computation of the flux linkages of an entire winding involves a double integral. As an example, let us determine the total flux linkages of the as winding due to current flowing only in the as winding. Here

$$\lambda_{as} = L_{ls} i_{as} + \int N_{as}(\phi_s) \Phi(\phi_s, \theta_r) \, d\phi = L_{ls} i_{as} + \int N_{as}(\phi_s) \int_{\phi_s}^{\phi_s + \pi} B_r(\xi, \theta_r) rl \, d\xi \, d\phi_s \tag{1.5-11}$$

In (1.5-11), L_{ls} is the stator leakage inductance due primarily to leakage flux at the end turns. Generally, this inductance accounts for 5% to 10% of the maximum self-inductance. Substituting (1.4-14) (with N_p replaced by $N_s/2$) and (1.5-5) for $B_r(\xi, \theta_r)$ into (1.5-11) yields

$$\lambda_{as} = L_{ls} i_{as} - \int_{\pi}^{2\pi} \frac{N_s}{2} \sin \phi_s \int_{\phi_s}^{\phi_s + \pi} \mu_0 \frac{N_s}{2} i_{as} \cos \xi [\alpha_1 - \alpha_2 \cos 2(\xi - \theta_r)] rl \, d\xi \, d\phi_s$$

$$= L_{ls} i_{as} + \left(\frac{N_s}{2}\right)^2 \pi \mu_0 rl \left(\alpha_1 - \frac{\alpha_2}{2} \cos 2\theta_r\right) i_{as} \tag{1.5-12}$$

The interval of integration is taken from π to 2π so as to comply with the convention that positive flux linkages are obtained in the direction of the positive *as* axis by circulation of the assumed positive current in the clockwise direction about the coil (right-hand rule). The self-inductance of the *as* winding is obtained by dividing (1.5-12) by i_{as}. Thus

$$L_{asas} = L_{ls} + \left(\frac{N_s}{2}\right)^2 \pi\mu_0 rl \left(\alpha_1 - \frac{\alpha_2}{2}\cos 2\theta_r\right) \qquad (1.5\text{-}13)$$

The mutual inductance between the *as* and *bs* windings may be determined by first computing the flux linking the *as* winding due to current flowing only in the *bs* winding. In this case it is assumed that the magnetic coupling that might occur at the end turns of the windings may be neglected. Thus

$$\lambda_{as} = \int N_{as}(\phi_s) \int_{\phi_s}^{\phi_s + \pi} B_r(\xi, \theta_r) rl \, d\xi \, d\phi_s \qquad (1.5\text{-}14)$$

Substituting (1.4-14) and (1.5-6) into (1.5-14) yields

$$\lambda_{as} = -\int_{\pi}^{2\pi} \frac{N_s}{2}\sin\phi_s \int_{\phi_s}^{\phi_s+\pi} \mu_0 \frac{N_s}{2} i_{bs}\cos\left(\xi - \frac{2\pi}{3}\right)[\alpha_1 - \alpha_2 \cos 2(\xi - \theta_r)] rl \, d\xi \, d\phi$$

$$(1.5\text{-}15)$$

Therefore, the mutual inductance between the *as* and *bs* windings is obtained by evaluating (1.5-15) and dividing the result by i_{bs}. This gives

$$L_{asbs} = -\left(\frac{N_s}{2}\right)^2 \frac{\pi}{2}\mu_0 rl \left[\alpha_1 + \alpha_2 \cos 2\left(\theta_r - \frac{\pi}{3}\right)\right] \qquad (1.5\text{-}16)$$

The mutual inductance between the *as* and *fd* windings is determined by substituting (1.5-9), expressed in terms of $\phi_s - \theta_r$, into (1.5-14). Thus

$$\lambda_{as} = \int_{\pi}^{2\pi} \frac{N_s}{2}\sin\phi_s \int_{\phi_s}^{\phi_s+\pi} \mu_0 \frac{N_f}{2} i_{fd}\sin(\xi - \theta_r)[\alpha_1 - \alpha_2 \cos 2(\xi - \theta_r)] rl \, d\xi \, d\phi_s$$

$$(1.5\text{-}17)$$

Evaluating and dividing by i_{fd} yields

$$L_{asfd} = \left(\frac{N_s}{2}\right)\left(\frac{N_f}{2}\right)\pi\mu_0 rl \left(\alpha_1 + \frac{\alpha_2}{2}\right)\sin\theta_r \qquad (1.5\text{-}18)$$

The self-inductance of the field winding may be obtained by first evaluating the flux linking the *fd* winding with all currents equal to zero except i_{fd}. Thus, with the *fd* winding considered as sinusoidally distributed and the air-gap flux density expressed by (1.5-9) we can write

$$\lambda_{fd} = L_{lfd}i_{fd} + \int_{\pi/2}^{3\pi/2} \frac{N_f}{2}\cos\phi_r \int_{\phi_r}^{\phi_r+\pi} \mu_0 \frac{N_f}{2} i_{fd}\sin\xi(\alpha_1 - \alpha_2\cos2\xi)rl\,d\xi\,d\phi_r$$

(1.5-19)

from which

$$L_{fdfd} = L_{lfd} + \left(\frac{N_f}{2}\right)^2 \pi\mu_0 rl\left(\alpha_1 + \frac{\alpha_2}{2}\right)$$

(1.5-20)

where L_{lfd} is the leakage inductance of the field winding.

The remaining self- and mutual inductances may be calculated using the same procedure. We can express these inductances compactly by defining

$$L_A = \left(\frac{N_s}{2}\right)^2 \pi\mu_0 rl\alpha_1$$

(1.5-21)

$$L_B = \frac{1}{2}\left(\frac{N_s}{2}\right)^2 \pi\mu_0 rl\alpha_2$$

(1.5-22)

$$L_{sfd} = \left(\frac{N_s}{2}\right)\left(\frac{N_f}{2}\right)\pi\mu_0 rl\left(\alpha_1 + \frac{\alpha_2}{2}\right)$$

(1.5-23)

$$L_{mfd} = \left(\frac{N_f}{2}\right)^2 \pi\mu_0 rl\left(\alpha_1 + \frac{\alpha_2}{2}\right)$$

(1.5-24)

The machine inductances may now be expressed as

$$L_{asas} = L_{ls} + L_A - L_B\cos2\theta_r$$ (1.5-25)

$$L_{bsbs} = L_{ls} + L_A - L_B\cos2\left(\theta_r - \frac{2\pi}{3}\right)$$ (1.5-26)

$$L_{cscs} = L_{ls} + L_A - L_B\cos2\left(\theta_r + \frac{2\pi}{3}\right)$$ (1.5-27)

$$L_{fdfd} = L_{lfd} + L_{mfd}$$ (1.5-28)

$$L_{asbs} = -\frac{1}{2}L_A - L_B\cos2\left(\theta_r - \frac{\pi}{3}\right)$$ (1.5-29)

$$L_{ascs} = -\frac{1}{2}L_A - L_B\cos2\left(\theta_r + \frac{\pi}{3}\right)$$ (1.5-30)

$$L_{bscs} = -\frac{1}{2}L_A - L_B\cos2(\theta_r + \pi)$$ (1.5-31)

$$L_{asfd} = L_{sfd}\sin\theta_r$$ (1.5-32)

$$L_{bsfd} = L_{sfd}\sin\left(\theta_r - \frac{2\pi}{3}\right)$$ (1.5-33)

$$L_{csfd} = L_{sfd}\sin\left(\theta_r + \frac{2\pi}{3}\right)$$ (1.5-34)

In later chapters we will consider a practical synchronous machine equipped with short-circuited rotor windings (damper windings). The expressions for the inductances of these additional windings can be readily ascertained from the work presented here. Also, high-speed synchronous machines used with steam turbines are round rotor devices. The $2\theta_r$ variation is not present in the inductances of a uniform air-gap machine. Therefore, the winding inductances may be determined from the above relationships by simply setting $\alpha_2 = 0$ in (1.5-22)–(1.5-24). It is clear that with $\alpha_2 = 0$, $L_B = 0$.

The voltage equations for the elementary synchronous machine shown in Fig. 1.5-1 are

$$v_{as} = r_s i_{as} + \frac{d\lambda_{as}}{dt} \tag{1.5-35}$$

$$v_{bs} = r_s i_{bs} + \frac{d\lambda_{bs}}{dt} \tag{1.5-36}$$

$$v_{cs} = r_s i_{cs} + \frac{d\lambda_{cs}}{dt} \tag{1.5-37}$$

$$v_{fd} = r_{fd} i_{fd} + \frac{d\lambda_{fd}}{dt} \tag{1.5-38}$$

where r_s is the resistance of the stator winding and r_{fd} the resistance of the field winding. The flux linkages are expressed as

$$\lambda_{as} = L_{asas} i_{as} + L_{asbs} i_{bs} + L_{ascs} i_{cs} + L_{asfd} i_{fd} \tag{1.5-39}$$

$$\lambda_{bs} = L_{bsas} i_{as} + L_{bsbs} i_{bs} + L_{bscs} i_{cs} + L_{bsfd} i_{fd} \tag{1.5-40}$$

$$\lambda_{cs} = L_{csas} i_{as} + L_{csbs} i_{bs} + L_{cscs} i_{cs} + L_{csfd} i_{fd} \tag{1.5-41}$$

$$\lambda_{fd} = L_{fdas} i_{as} + L_{fdbs} i_{bs} + L_{fdcs} i_{cs} + L_{fdfd} i_{fd} \tag{1.5-42}$$

One starts to see the complexity of the voltage equations due to the fact that some of the machine inductances are functions of θ_r and therefore a function of the rotor speed ω_r. Hence, the coefficients of the voltage equations are time-varying except when the rotor is stalled. Moreover, rotor speed is a function of the electromagnetic torque, which we will find to be a product of machine currents. Clearly, the solution of the voltage equations is very involved. In fact, a computer is required in order to solve for the transient response of the combined electrical and mechanical systems. In Chapter 3 we will see that a change of variables allows us to eliminate the time-varying inductances and thereby markedly reduce the complexity of the voltage equations. However, the resulting differential equations are still nonlinear, requiring a computer to solve for the electromechanical transient behavior.

Induction Machine

The winding arrangement of a 2-pole, 3-phase, wye-connected symmetrical induction machine is shown in Fig. 1.5-2. The stator windings are identical with

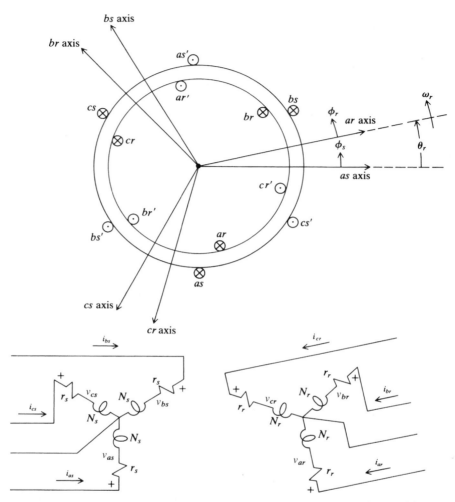

Figure 1.5-2 Two-pole, 3-phase, wye-connected symmetrical induction machine.

equivalent turns N_s and resistance r_s. The rotor windings, which may be wound or forged as a squirrel cage winding, can be approximated as identical windings with equivalent turns N_r and resistance r_r. The air gap of an induction machine is uniform and it is assumed that the stator and rotor windings may be approximated as sinusoidally distributed windings.

In nearly all applications the induction machine is operated as a motor with the stator windings connected to a balanced 3-phase source and the rotor windings short-circuited. The principle of operation in this mode is quite easily deduced [4–6]. With balanced 3-phase currents flowing in the stator windings, a rotating air-gap MMF is established, as in the case of the synchronous machine, which rotates about the air gap at a speed determined by the frequency of the stator currents and the number of poles. If the rotor speed is different from the speed of this rotating MMF, balanced 3-phase currents will be induced (thus the name induction) in the short-circuited rotor windings. The frequency of the rotor currents corresponds to the difference in the speed of the rotating MMF due to the stator currents and the speed of the rotor. The induced rotor currents will, in turn, produce an air-gap MMF that rotates relative to the rotor at a speed corresponding to the frequency of the rotor currents. The speed of the rotor air-gap MMF superimposed upon the rotor speed is the same speed as that of the air-gap MMF established by the currents flowing in the stator windings. These two air-gap MMFs rotating in unison may be thought of as two synchronously rotating sets of magnetic poles. Torque is produced due to an interaction of the two magnetic systems. It is clear, however, that torque is not produced when the rotor is running in synchronism with the air-gap MMF due to the stator currents because in this case currents are not induced in the short-circuited rotor windings.

The winding inductances of the induction machine may be expressed from the inductance relationships given for the salient-pole synchronous machine. In the case of the induction machine the air gap is uniform. Thus, $2\theta_r$ variations in the self- and mutual inductances do not occur. This variation may be eliminated by setting $\alpha_2 = 0$ in the inductance relationship given for the salient-pole synchronous machine. All stator self-inductances are equal; that is, $L_{asas} = L_{bsbs} = L_{cscs}$ with

$$L_{asas} = L_{ls} + L_{ms} \tag{1.5-43}$$

where L_{ms} is the stator magnetizing inductance that corresponds to L_A in (1.5-25). It may be expressed as

$$L_{ms} = \left(\frac{N_s}{2}\right)^2 \frac{\pi\mu_0 rl}{g} \tag{1.5-44}$$

where g is the length of the uniform air gap. Likewise all stator-to-stator mutual inductances are the same. For example,

$$L_{asbs} = -\frac{1}{2} L_{ms} \tag{1.5-45}$$

which corresponds to (1.5-29) with $L_B = 0$.

It follows that the rotor self-inductances are equal with

$$L_{arar} = L_r + L_{mr} \tag{1.5-46}$$

where the rotor magnetizing inductance may be expressed as

$$L_{mr} = \left(\frac{N_r}{2}\right)^2 \frac{\pi\mu_0 rl}{g} \tag{1.5-47}$$

The rotor-to-rotor mutual inductances are

$$L_{arbr} = -\frac{1}{2}L_{mr} \tag{1.5-48}$$

Expressions for the mutual inductances between stator and rotor windings may be written by noting the form of the mutual inductances between the field and stator windings of the synchronous machine given by (1.5-32)–(1.5-34). Here we see that L_{asar}, L_{bsbr}, and L_{cscr} are equal with

$$L_{asar} = L_{sr}\cos\theta_r \tag{1.5-49}$$

Also, L_{asbr}, L_{bscr}, and L_{csar} are equal with

$$L_{asbr} = L_{sr}\cos\left(\theta_r + \frac{2\pi}{3}\right) \tag{1.5-50}$$

Finally, L_{ascr}, L_{bsar}, and L_{csbr} are equal with

$$L_{ascr} = L_{sr}\cos\left(\theta_r - \frac{2\pi}{3}\right) \tag{1.5-51}$$

where

$$L_{sr} = \left(\frac{N_s}{2}\right)\left(\frac{N_r}{2}\right)\frac{\pi\mu_0 rl}{g} \tag{1.5-52}$$

The voltage equations for the induction machine shown in Fig. 1.5-2 are

$$v_{as} = r_s i_{as} + \frac{d\lambda_{as}}{dt} \tag{1.5-53}$$

$$v_{bs} = r_s i_{bs} + \frac{d\lambda_{bs}}{dt} \tag{1.5-54}$$

$$v_{cs} = r_s i_{cs} + \frac{d\lambda_{cs}}{dt} \tag{1.5-55}$$

$$v_{ar} = r_r i_{ar} + \frac{d\lambda_{ar}}{dt} \tag{1.5-56}$$

$$v_{br} = r_r i_{br} + \frac{d\lambda_{br}}{dt} \tag{1.5-57}$$

$$v_{cr} = r_r i_{cr} + \frac{d\lambda_{cr}}{dt} \tag{1.5-58}$$

where r_s is the resistance of each stator phase winding and r_r is the resistance of each rotor phase winding. The flux linkages may be written following the form of λ_{as}:

$$\lambda_{as} = L_{asas}i_{as} + L_{asbs}i_{bs} + L_{ascs}i_{cs} + L_{asar}i_{ar} + L_{asbr}i_{br} + L_{ascr}i_{cr} \tag{1.5-59}$$

Here again we see the complexity of the voltage equations due to the time-varying mutual inductances between stator and rotor circuits (circuits in relative motion). We will see in Chapter 3 that a change of variables eliminates the time-varying inductances resulting in voltage equations which are still nonlinear but much more manageable.

Example 1D   The winding inductances of a P-pole machine may be determined by considering the elementary 4-pole, 3-phase synchronous machine shown in Fig. 1.4-7. Let us determine the self-inductance of the as winding. The air-gap flux density due to current only in the as winding may be expressed from (1.5-4); in particular,

$$B_r(\phi_s, \theta_{rm}) = \mu_0 \frac{\text{MMF}_{as}}{g(\phi_s - \theta_{rm})} \tag{1D-1}$$

where for a P-pole machine, MMF_{as} is given by (1.4-25) and θ_{rm} is defined in Fig. 1.4-7. From (1.5-2) and Fig. 1.4-7 we obtain

$$g(\phi_s - \theta_{rm}) = \frac{1}{\alpha_1 - \alpha_2 \cos(P/2)2(\phi_s - \theta_{rm})} \tag{1D-2}$$

Substituting into (1D-1) yields

$$B_r(\phi_s, \theta_{rm}) = \mu_0 \frac{N_s}{P} i_{as} \cos \frac{P}{2}\phi_s \left[\alpha_1 - \alpha_2 \cos\left(\frac{P}{2}\right)2(\phi_s - \theta_{rm})\right] \tag{1D-3}$$

Following (1.5-11) for a P-pole machine we have

$$\lambda_{as} = L_{ls}i_{as} - \frac{P}{2}\int_{2\pi/P}^{4\pi/P} N_{as}(\phi_s) \int_{\phi_s}^{\phi_s + 2\pi/P} B_r(\xi, \theta_{rm}) rl \, d\xi \, d\phi_s \tag{1D-4}$$

where

$$N_{as}(\phi_s) = -\frac{N_s}{P}\sin\frac{P}{2}\phi_s, \qquad \frac{2\pi}{P} \le \phi_s \le \frac{4\pi}{P} \tag{1D-5}$$

The double integral is multiplied by $P/2$ to account for the flux linkages of the complete as winding. Evaluating (1D-4) and dividing by i_{as} yields

$$L_{asas} = L_{ls} + \frac{N_s^2}{2P}\pi\mu_0 rl \left[\alpha_1 - \frac{\alpha_2}{2}\cos 2\left(\frac{P}{2}\right)\theta_{rm}\right] \tag{1D-6}$$

If, in the above relation, we substitute

$$\theta_r = \frac{P}{2}\theta_{rm} \tag{1D-7}$$

then (1D-3) is of the same form as for the 2-pole machine, (1.5-13). Moreover, if we evaluate all of the machine inductances for the P-pole machine and make this same substitution, we will find that the winding inductances of a P-pole machine are of the same form as those of a 2-pole machine. Also, ω_r is defined as

$$\omega_r = \frac{P}{2}\omega_{rm} \tag{1D-8}$$

and the substitute variables θ_r and ω_r are referred to as the electrical angular displacement and the electrical angular velocity of the rotor, respectively. This connotation stems from the fact that $\theta_r(\omega_r)$ refers a rotor displacement (velocity) to an electrical displacement (velocity). It is clear that, at synchronous speed, ω_r is equal to the angular velocity of the electrical system connected to the stator windings regardless of the number of poles.

Because the winding inductances of a P-pole machine are identical in form to those of a 2-pole machine if θ_r and ω_r are defined by (1D-7) and (1D-8), respectively, all machines can be treated as if they were 2-pole devices as far as the voltage equations are concerned. We will find that it is necessary to multiply the torque equation of a 2-pole machine by $P/2$ in order to express the electromagnetic torque of a P-pole machine correctly.

REFERENCES

[1] C. M. Ong, *Dynamic Simulation of Electric Machinery*, Prentice-Hall PTR, Upper Saddle River, NJ, 1998.

[2] D. C. White and H. H. Woodson, *Electromechanical Energy Conversion*, John Wiley and Sons, New York, 1959.

[3] M. S. Say, *Alternating Current Machines*, A Halsted Press Book, John Wiley and Sons, New York, 1976.

[4] A. E. Fitzgerald, C. Kingsley, Jr., and S. D. Umans, *Electric Machinery*, 5th ed., McGraw-Hill, New York, 1990.

[5] S. J. Chapman, *Electric Machinery Fundamentals*, 3rd ed., McGraw-Hill, New York, 1999.

[6] G. McPherson and R. D. Laramore, *An Introduction to Electrical Machines and Transformers*, 2nd ed., John Wiley and Sons, New York, 1990.

PROBLEMS

1 A two-winding, iron-core transformer is shown in Fig. 1.P-1. $N_1 = 50$ turns, $N_2 = 100$ turns, and $\mu_R = 4000$. Calculate L_{m1}, L_{m2}, and L_{12}. The resistance of coils 1 and 2 is $2\,\Omega$ and $4\,\Omega$, respectively. If the voltage applied to coil 1 is $v_1 = 10\cos 400t$, calculate the voltage appearing at the terminals of coil 2 when coil 2 is open-circuited.

2 Repeat Problem 1 if the iron core has an air gap of 0.2 cm in length and is cut through the complete cross section. Assume that fringing does not occur in the air gap, that is, the effective cross-sectional area of the air gap is $25\,\mathrm{cm}^2$.

3 Two coupled coils have the following parameters

$$
\begin{aligned}
L_{11} &= 100\,\mathrm{mH} & r_1 &= 10\,\Omega \\
L_{22} &= 25\,\mathrm{mH} & r_2 &= 2.5\,\Omega \\
N_1 &= 1000\,\mathrm{turns} & N_2 &= 500\,\mathrm{turns} \\
L_{l1} &= 0.1 L_{11} & L_{l1} &= 0.1 L_{11}
\end{aligned}
$$

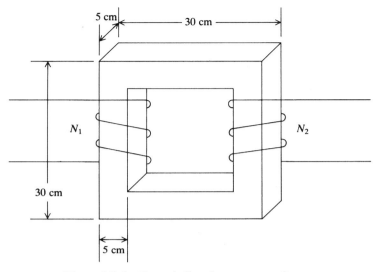

Figure 1.P-1 Two-winding, iron-core transformer.

(a) Develop a T equivalent circuit with coil 1 as the reference coil. (b) Repeat with coil 2 as the reference coil. Determine the input impedance of the coupled circuits if the applied frequency to coil 1 is 60 Hz with coil 2 (c) open-circuited; (d) short-circuited. Repeat (d) with the current flowing in the magnetizing reactance neglected.

4 A constant 10 V is suddenly applied to coil 1 of the coupled circuits given in Problem 3. Coil 2 is short-circuited. Calculate the transient and steady-state current flowing in each coil.

5 Consider the coils given in Problem 3. Assume that the coils are wound so that, with the assigned positive direction of currents i_1 and i_2, the mutual inductance is negative. (a) If the coils are connected in series so that $i_1 = i_2$, calculate the self-inductance of the series combination. (b) Repeat (a) with the coils connected in series so that $i_1 = -i_2$.

6 A third coil is wound on the ferromagnetic core shown in Fig. 1.2-1. The resistance is r_3 and the leakage and magnetizing inductances are L_{l3} and L_{m3}, respectively. The coil is wound so that positive current (i_3) produces Φ_{m3} in the opposite direction as Φ_{m1} and Φ_{m2}. Derive the T equivalent circuit for this three-winding transformer. Actually one should be able to develop the equivalent circuit without derivation.

7 Use Σ and $1/p$ to denote summation and integration, respectively. Draw a time-domain block diagram for two coupled windings with saturation.

8 A resistor and an inductor are connected as shown in Fig. 1.P-2 with $R = 15\,\Omega$ and $L = 250\,\text{mH}$. Determine the energy stored in the inductor W_{eS} and the energy dissipated by the resistor W_{eL} for $i > 0$ if $i(0) = 10$ A.

9 Consider the spring–mass–damper system shown in Fig. 1.P-3. At $t = 0$, $x(0) = x_0$ (rest position) and $dx/dt = 1.5\,\text{m/s}$. $M = 0.8\,\text{kg}$, $D = 10\text{N} \cdot \text{s/m}$, and $K = 120\text{N} \cdot \text{m}$. For $t > 0$, determine the energy stored in the spring W_{mS1}, the kinetic energy of the mass W_{mS2}, and the energy dissipated by the damper W_{mL}.

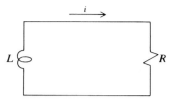

Figure 1.P-2 R–L circuit.

10   Express $W_f(i,x)$ and $W_c(i,x)$ for (a) $\lambda(i,x) = i^{3/2}x^3$; (b) $\lambda(i,x) = xi^2 + ki\cos x$.

11   The energy stored in the coupling field of a magnetically linear system with two electrical inputs may be expressed as

$$W_f(\lambda_1,\lambda_2,x) = \frac{1}{2}B_{11}\lambda_1^2 + B_{12}\lambda_1\lambda_2 + \frac{1}{2}B_{22}\lambda_2^2$$

Express B_{11}, B_{12}, and B_{22} in terms of inductances L_{11}, L_{12}, and L_{22}.

12   An electromechanical system has two electrical inputs. The flux linkages may be expressed as

$$\lambda_1(i_1,i_2,x) = x^2 i_1^2 + x i_2$$
$$\lambda_2(i_1,i_2,x) = x^2 i_2^2 + x i_1$$

Express $W_f(i_1,i_2,x)$ and $W_c(i_1,i_2,x)$.

13   Express $f_e(i,x)$ for the electromechanical systems described by the relations given in Problem 10.

14   Express $f_e(i_1,i_2,x)$ for the electromechanical system given in Problem 12.

15   Refer to Fig. 1.3-7. As the system moves from x_a to x_b the λ–i trajectory moves from A to B. Does the voltage v increase or decrease? Does the applied force f increase or decrease? Explain.

16   Refer to Fig. 1.3-11. Following the system transients due to the application of the source voltage ($v = 5\text{V}$), the system assumes steady-state operation. During this steady-state operation calculate W_{eS}, W_f, W_c, and W_{mS}.

17   Refer to Fig. 1.3-12. Repeat Problem 16 for steady-state operation following the application of $f = 4N$.

18   Refer to Fig. 1.3-13. Identify the area corresponding to ΔW_m when (a) x moves from 2.5 mm to 4.3 mm, and (b) x moves from 4.3 mm to 2.5 mm.

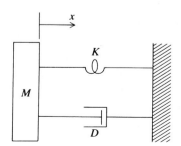

Figure 1.P-3   Spring–mass–damper system.

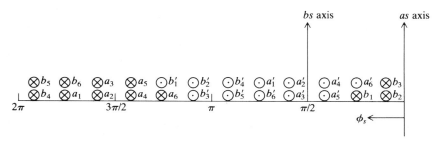

Figure 1.P-4 Stator winding arrangement of a 2-pole, 2-phase machine.

19 Assume the steady-state currents flowing in the conductors of the device shown in Fig. 1B-1 are

$$I_1 = I_s \cos[\omega_1 t + \theta_1(0)]$$
$$I_2 = I_r \cos[\omega_2 t + \theta_2(0)]$$

where $\theta_1(0)$ and $\theta_2(0)$ are the time-zero displacements of I_1 and I_2, respectively. Assume also that during steady-state operation the rotor speed is constant; thus

$$\theta_r = \omega_r t + \theta_r(0)$$

where $\theta_r(0)$ is the rotor displacement at time zero. Determine the rotor speeds at which the device produces a nonzero average torque during steady-state operation if (a) $\omega_1 = \omega_2 = 0$; (b) $\omega_1 = \omega_2 \neq 0$; (c) $\omega_2 = 0$. Express the instantaneous torque at each of these rotor speeds.

20 The developed diagram shown in Fig. 1.P-4 shows the arrangement of the stator windings of a 2-pole, 2-phase machine. Each coil side has n_c conductors, and i_{as} flows in the as winding and i_{bs} flows in the bs winding. Draw the air-gap MMF (a) due to i_{as}(MMF$_{as}$) and (b) due to i_{bs}(MMF$_{bs}$).

21 Assume that each coil side of the winding shown in Fig. 1.P-4 contains n_c conductors. The windings are to be described as sinusoidally distributed windings with N_p the maximum turns density and N_s the number of equivalent turns in each winding, which is a function of n_c. Express (a) N_{as} and N_{bs}, (b) MMF$_{as}$ and MMF$_{bs}$, and (c) the total air-gap MMF (MMF$_s$) produced by the stator windings.

22 A fractional-pitch stator winding (a phase) is shown in developed form in Fig. 1.P-5. Each coil side contains n_c conductors. The winding is to be described as a sinusoidally distributed winding. Express the number of equivalent turns N_s in terms of n_c.

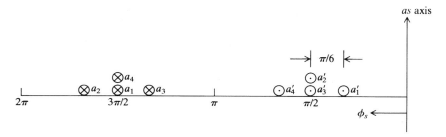

Figure 1.P-5 Stator winding arrangement for fractional-pitch winding.

23 Consider the 2-phase stator windings described in Problem 21. (*a*) With $I_{as} = \sqrt{2}I_s$ $\cos[\omega_e t + \theta_{es}(0)]$ determine I_{bs} in order to obtain a constant amplitude MMF (MMF$_s$) which rotates clockwise around the air gap of the machine. Repeat (*a*) with (*b*) $I_{as} = -\sqrt{2}I_s \cos\omega_e t$, (*c*) $I_{as} = \sqrt{2}I_s \sin\omega_e t$, and (*d*)$I_{as} = -\sqrt{2}I_s \sin[\omega_e t + \theta_{es}(0)]$.

24 A single-phase source is connected between terminals *as* and *bs* of the 4-pole, 3-phase machine shown in Fig. 1.4-7. The *cs* terminal is open-circuited. The current into the *as* winding is $I_{as} = \sqrt{2}I\cos\omega_e t$. Express the total air-gap MMF.

25 An elementary 2-pole, 2-phase, salient-pole synchronous machine is shown in Fig. 1.P-6. The winding inductances may be expressed as

$$L_{asas} = L_{ls} + L_A - L_B \cos 2\theta_r$$
$$L_{bsbs} = L_s + L_A + L_B \cos 2\theta_r$$
$$L_{asbs} = -L_B \sin 2\theta_r$$
$$L_{fdfd} = L_{lfd} + L_{mfd}$$
$$L_{asfd} = L_{sfd} \sin \theta_r$$
$$L_{bsfd} = -L_{sfd} \cos \theta_r$$

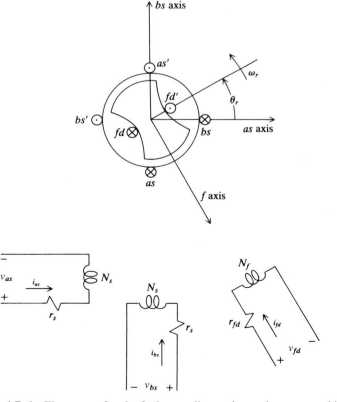

Figure 1.P-6   Elementary 2-pole, 2-phase, salient-pole synchronous machine.

Verify these relationships and give expressions for the coefficients L_A, L_B, L_{mfd}, and L_{sfd} similar in form to those given for a 3-phase machine. Modify these inductance relationships so that they will describe a 2-phase, uniform air-gap synchronous machine.

26   Write the voltage equations for the elementary 2-pole, 2-phase, salient-pole synchronous machine shown in Fig. 1.P-6 and derive the expression for $T_e(i_{as}, i_{bs}, i_{fd}, \theta_r)$.

27   An elementary 4-pole, 2-phase, salient-pole synchronous machine is shown in Fig. 1.P-7. Use this machine as a guide to derive expressions for the winding inductances of a P-pole synchronous machine. Show that these inductances are of the same form as those given in Problem 25 if $(P/2)\theta_{rm}$ is replaced by θ_r.

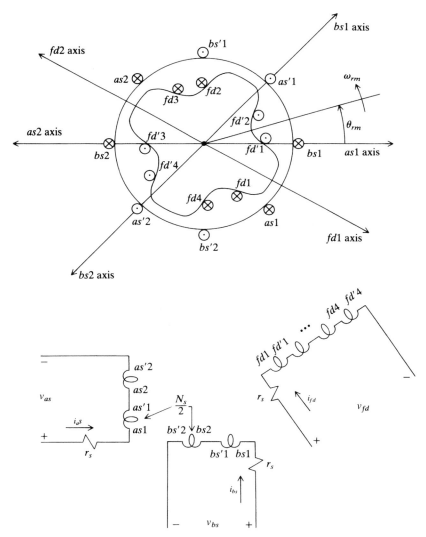

Figure 1.P-7   Elementary 4-pole, 2-phase, salient-pole synchronous machine.

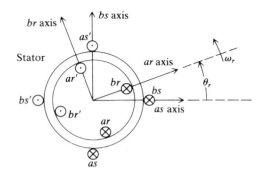

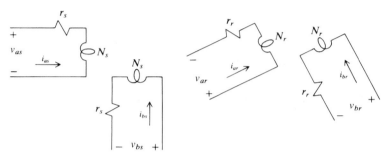

Figure 1.P-8   Elementary 2-pole, 2-phase symmetrical induction machine.

28 Derive an expression for the electromagnetic torque, $T_e(i_{as}, i_{bs}, i_{fd}, \theta_r)$, for a P-pole, 2-phase, salient-pole synchronous machine. This expression should be identical in form to that obtained in Problem 26 multiplied by $P/2$.

29 A reluctance machine has no field winding on the rotor. Modify the inductance relationships given in Problem 25 so as to describe the winding inductances of a 2-pole, 2-phase, reluctance machine. Write the voltage equations and derive an expression for $T_e(i_{as}, i_{bs}, \theta_r)$.

30 An elementary 2-pole, 2-phase, symmetrical induction machine is shown in Fig. 1.P-8. Derive expressions for the winding inductances. If you have worked Problem 25, you may modify those results accordingly.

31 Write the voltage equations for the induction machine shown in Fig. 1.P-8 and derive an expression for the electromagnetic torque $T_e(i_{as}, i_{bs}, i_{ar}, i_{br}, \theta_r)$.

32 An elementary 4-pole, 2-phase, symmetrical induction machine is shown in Fig. 1.P-9. Use this machine as a guide to derive expressions for the winding inductance of a P-pole induction machine. Show that these inductances are of the same form as those given in Problem 30 if $(P/2)\theta_{rm}$ is replaced by θ_r.

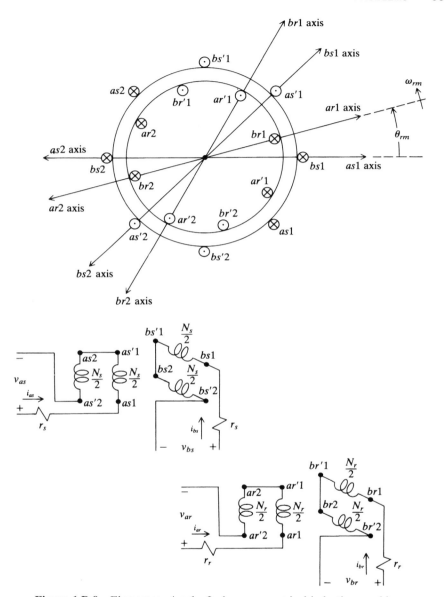

Figure 1.P-9 Elementary 4-pole, 2-phase symmetrical induction machine.

Chapter 2

DIRECT-CURRENT MACHINES

2.1 INTRODUCTION

The direct-current (dc) machine is not as widely used today as it was in the past. For the most part, the dc generator has been replaced by solid-state rectifiers that convert alternating current into direct current with provisions to control the magnitude of the dc voltage. Nevertheless, it is still desirable to devote some time to the dc machine in a course on electromechanical devices because it is still used as a drive motor, especially at the low-power level. This chapter is an attempt to treat dc machines, perhaps a bit more in detail than necessary, but with the flexibility so that one can focus upon the topics that are still of interest to the control or energy systems engineer. In this regard, the dc generator is not considered in detail and, for that matter, neither are the series and compound machines. Today's focus is on the shunt-connected dc motor, and the permanent-magnet dc motor, which have similar operating characteristics.

A simplified method of analysis is presented rather than an analysis wherein commutation is treated in detail. With this type of analytical approach, the dc machine is considered to be the most straightforward to analyze of all electromechanical devices. Numerous textbooks have been written over the last century on the design, theory, and operation of dc machines. One can add little to the analytical approach that has been used for years [1–3]. In this chapter, the well-established theory of dc machines is set forth and the dynamic characteristics of the shunt and permanent-magnet machines are illustrated. The time-domain block diagrams and state equations are then developed for these two types of motors and, in the final section of this chapter, the linear differential equations which describe the dynamic characteristics of these devices are solved by using Laplace transformation techniques.

2.2 ELEMENTARY DIRECT-CURRENT MACHINE

It is instructive to discuss the elementary machine shown in Fig. 2.2-1 prior to a formal analysis of the performance of a practical dc machine. The 2-pole elementary machine is equipped with a field winding wound on the stator poles, a rotor coil (a–a'), and a commutator. The commutator is made up of two semicircular copper segments mounted on the shaft at the end of the rotor and insulated from one another as well as from the iron of the rotor. Each terminal of the rotor coil is connected to a copper segment. Stationary carbon brushes ride upon the copper segments whereby the rotor coil is connected to a stationary circuit by a near frictionless contract.

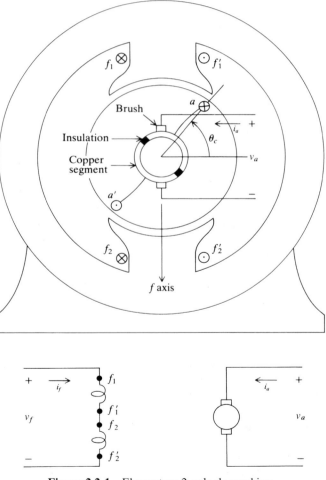

Figure 2.2-1 Elementary 2-pole dc machine.

The voltage equations for the field winding and rotor coil are

$$v_f = r_f i_f + \frac{d\lambda_f}{dt} \tag{2.2-1}$$

$$v_{a-a'} = r_a i_{a-a'} + \frac{d\lambda_{a-a'}}{dt} \tag{2.2-2}$$

where r_f and r_a are the resistance of the field winding and armature coil, respectively. The rotor of a dc machine is commonly referred to as the *armature*; rotor and armature will be used interchangeably. At this point in the analysis it is sufficient to express the flux linkages as

$$\lambda_f = L_{ff} i_f + L_{fa} i_{a-a'} \tag{2.2-3}$$

$$\lambda_{a-a'} = L_{af} i_f + L_{aa} i_{a-a'} \tag{2.2-4}$$

As a first approximation, the mutual inductance between the field winding and an armature coil may be expressed as a sinusoidal function of θ_r as

$$L_{af} = L_{fa} = -L\cos\theta_r \tag{2.2-5}$$

where L is a constant. As the rotor revolves, the action of the commutator is to switch the stationary terminals from one terminal of the rotor coil to the other. For the configuration shown in Fig. 2.2-1, this switching or commutation occurs at $\theta_r = 0$, π, 2π, ... At the instant of switching, each brush is in contact with both copper segments whereupon the rotor coil is short-circuited. It is desirable to commutate (short-circuit) the rotor coil at the instant the induced voltage is a minimum. The waveform of the voltage induced in the open-circuited armature coil, during constant-speed operation with a constant field winding current, may be determined by setting $i_{a-a'} = 0$ and i_f equal to a constant. Substituting (2.2-4) and (2.2-5) into (2.2-2) yields the following expression for the open-circuit voltage of coil a–a' with the field current i_f a constant:

$$v_{a-a'} = \omega_r L I_f \sin\theta_r \tag{2.2-6}$$

where $\omega_r = d\theta_r/dt$ is the rotor speed. The open-circuit coil voltage $v_{a-a'}$ is zero at $\theta_r = 0$, π, 2π, ..., which is the rotor position during commutation. Commutation is illustrated in Fig. 2.2-2. The open-circuit terminal voltage, v_a corresponding to the rotor positions denoted as θ_{ra}, $\theta_{rb}(\theta_{rb} = 0)$, and θ_{rc} are indicated. It is important to note that, during one revolution of the rotor, the assumed positive direction of armature current i_a is down coil side a and out coil side a' for $0 < \theta_r < \pi$. For $\pi < \theta_r < 2\pi$, positive current is down coil side a' and out of coil side a. In Chapter 1 we let positive current flow into the winding denoted without a prime and out the winding denoted with a prime. We will not be able to adhere to this relationship in the case of the armature windings of a dc machine since commutation is involved.

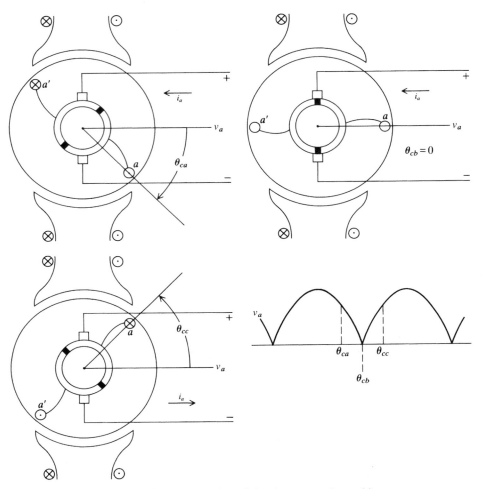

Figure 2.2-2 Commutation of the elementary dc machine.

The machine shown in Fig. 2.2-1 is not a practicable machine. Although it could be operated as a generator supplying a resistive load, it could not be operated effectively as a motor supplied from a voltage source owing to the short-circuiting of the armature coil at each commutation. A practicable dc machine, with the rotor equipped with an a winding and an A winding, is shown schematically in Fig. 2.2-3. At the rotor position depicted, coils a_4–a'_4 and A_4–A'_4 are being commutated. The bottom brush short-circuits the a_4–a'_4 coil while the top brush short-circuits the A_4–A'_4 coil. Figure 2.2-3 illustrates the instant when the assumed direction of positive current is into the paper in coil sides a_1, A_1; a_2, A_2; ..., and out in coil sides a'_1, A'_1; a'_2, A'_2; .... It is instructive to follow the path of current through one of the parallel paths from one brush to the other. For the angular position shown

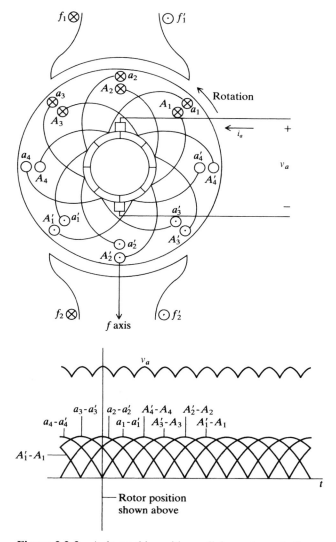

Figure 2.2-3   A dc machine with parallel armature windings.

in Fig. 2.2-3, positive current enters the top brush and flows down the rotor via a_1 and back through a'_1; down a_2 and back through a'_2; down a_3 and back through a'_3 to the bottom brush. A parallel current path exists through A_3–A'_3, A_2–A'_2, and A_1–A'_1. The open-circuit or induced armature voltage is also shown in Fig. 2.2-3; however, these idealized waveforms require additional explanation. As the rotor advances in the counterclockwise direction, the segment connected to a_1 and A_4 moves from under the top brush, as shown in Fig. 2.2-4. The top brush then rides only on the segment connecting A_3 and A'_4. At the same time the bottom brush is riding on

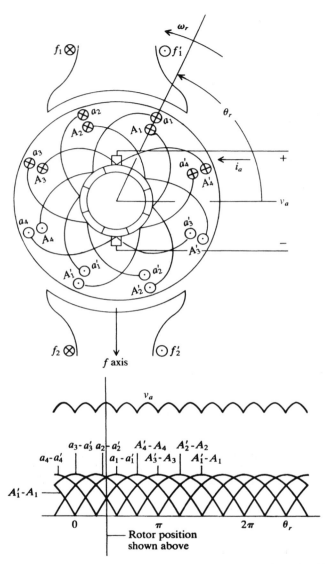

Figure 2.2-4   Same as Fig. 2.2-3 with rotor advanced approximately 22.5° counter-clockwise.

the segment connecting a_4 and a'_3. With the rotor so positioned, current flows in A_3 and A'_4 and out a_4 and a'_3. In other words, current flows down the coil sides in the upper one-half of the rotor and out of the coil sides in the bottom one-half. Let us follow the current flow through the parallel paths of the armature windings shown in Fig. 2.2-4. Current now flows through the top brush into A'_4 out A_4, into a_1 out a'_1, into a_2 out a'_2, into a_3 out a'_3 to the bottom brush. The parallel path beginning at the top brush is A_3–A_3', A_2–A_2', A_1–A_1', and a_4'–a_4 to the bottom brush. The voltage induced in the coils is shown in Figs. 2.2-3 and 2.2-4 for the first parallel path described. It is noted that the induced voltage is plotted only when the coil is in this parallel path.

In Figs. 2.2-3 and 2.2-4, the parallel windings consist of only four coils. Usually the number of rotor coils is substantially more than four, thereby reducing the harmonic content of the open-circuit armature voltage. In this case, the rotor coils may be approximated as a uniformly distributed winding, as illustrated in Fig. 2.2-5. Therein the rotor winding is considered as current sheets that are fixed in space due to the action of the commutator and that establish a magnetic axis positioned orthogonal to the magnetic axis of the field winding. The brushes are shown positioned on the current sheet for the purpose of depicting commutation. The small angular displacement, denoted by 2γ, designates the region of commutation wherein the coils are short-circuited. However, commutation cannot be visualized from Fig. 2.2-5; one must refer to Figs. 2.2-3 and 2.2-4.

In our discussion of commutation, it was assumed that the armature current was zero. With this constraint, the sinusoidal voltage induced in each armature coil

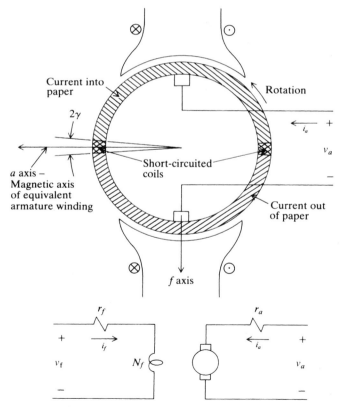

Figure 2.2-5   Idealized dc machine with uniformly distributed rotor winding.

crosses through zero when the coil is orthogonal to the field flux. Hence, the commutator was arranged so that the commutation would occur when an armature coil was orthogonal to the field flux. When current flows in the armature winding, the flux established therefrom is in an axis orthogonal to the field flux. Thus, a voltage will be induced in the armature coil that is being commutated as a result of "cutting" the flux established by the current flowing in the other armature coils. Arcing at the brushes will occur, and the brushes and copper segments may be damaged with even a relatively small armature current. Although the design of dc machines is not a subject of this text, it is important to mention that brush arcing may be substantially reduced by mechanically shifting the position of the brushes as a function of armature current or by means of interpoles. Interpoles or commutating poles are small stator poles placed over the coil sides of the winding being commutated, midway between the main poles of large-horsepower machines. The action of the interpole is to oppose the flux produced by the armature current in the region of the short-circuited coil. Because the flux produced in this region is a function of the armature current, it is desirable to make the flux produced by the interpole a function of the armature current. This is accomplished by winding the interpole with a few turns of the conductor carrying the armature current. Electrically, the interpole winding is between the brush and the terminal. It may be approximated, in the voltage equations, by increasing slightly the armature resistance and inductance (r_a and L_{AA}).

In dc motors, which are subjected to relatively large armature currents or to severe load cycles as in steel mill applications, arcing may occur between commutator segments. This undesirable situation may be minimized by embedding a compensating or pole face winding in the face of each stator pole. It consists of several turns connected in series with the armature circuit. It is wound so that its flux opposes the flux produced by the armature windings under the poles. Electrically the compensating winding may be accounted for, as before, by increasing r_a and L_{AA}.

Another aspect of dc machines, which is important but which cannot be considered in detail in this brief look at dc machines, is the method of winding the armature of multipole machines. Generally, multipole dc machines are used to supply large currents or generate high voltages. To carry large currents, the lap type of winding is used because this method of winding requires a pair of brushes for each pair of poles. The brushes are connected in parallel; consequently, large armature currents may be tolerated. The lap winding is similar to the type of winding described in the previous section. For high voltages, a so-called wave winding is used in multipole machines. This method of winding employs only two brushes, but is arranged so that a large number of coils are connected in series, resulting in a higher armature voltage. Regardless of the type of winding or the number of poles, the equations set forth in the following section for a 2-pole machine, wherein ω_r is the rotor angular velocity, may be applied directly to any multipole machine without modification.

Before proceeding to the development of the equations portraying the operating characteristics of these devices, it is instructive to take a brief look at the arrangement of the armature windings and the method of communication used in many low-power permanent-magnet dc motors. Small dc motors used in low-power control systems are often the permanent-magnet type, wherein a constant field flux is established by a permanent magnet rather than by a current flowing in a field winding.

Three rotor positions of a typical low-power permanent-magnet dc motor are shown in Fig. 2.2-6. The rotor is turning in the counterclockwise direction, and the rotor position advances from left to right. Physically, these devices may be an inch or less in diameter and in axial length with brushes sometimes as small as a pencil lead. They are mass-produced and relatively inexpensive. Although the brushes actually ride on the outside of the commutator, for convenience they are shown on the inside in Fig. 2.2-6. The armature windings consist of a large number of turns of fine wire; hence, each circle shown in Fig. 2.2-6 represents many conductors. Note that the position of the brushes is shifted approximately 40° relative to a line drawn between the center of the north and south poles. This shift in the brushes

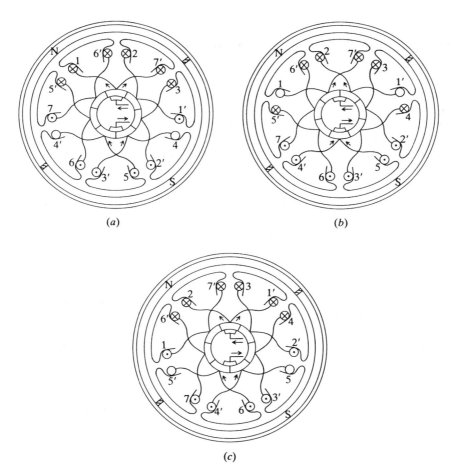

(a)

(b)

(c)

Figure 2.2-6   Commutation of a permanent-magnet dc motor.

was probably determined experimentally by minimizing brush arcing for normal load conditions. Note also that the windings do not span π radians but more like 140°, and there is an odd number of commutator segments.

In Fig. 2.2-6a, only winding 4 is being commutated. As the rotor advances from the position shown in Fig. 2.2-6a, both windings 4 and 1 are being commutated. In Fig. 2.2-6b, winding 1 is being commutated, then windings 1 and 5; and finally, in Fig. 2.2-6c, only winding 5 is being commutated. The windings are being commutated when the induced voltage is nonzero, and one would expect some arcing to occur at the brushes. We must realize, however, that the manufacture and sale of these devices is very competitive, and one often must compromise when striving to produce an acceptable motor at the least cost possible. Although we realize that in some cases it may be a rather crude approximation, we will consider the permanent-magnet dc motor as having current sheets on the armature with orthogonal armature and field magnetic axes as shown in Fig. 2.2-5.

2.3 VOLTAGE AND TORQUE EQUATIONS

Although rigorous derivation of the voltage and torque equations is possible, it is rather lengthy and little is gained because these relationships may be deduced. The armature coils revolve in a magnetic field established by a current flowing in the field winding. We have established that voltage is induced in these coils by virtue of this rotation. However, the action of the commutator causes the armature coils to appear as a stationary winding with its magnetic axis orthogonal to the magnetic axis of the field winding. Consequently, voltages are not induced in one winding due to the time rate of change of the current flowing in the other (transformer action). Mindful of these conditions, we can write the field and armature voltage equations in matrix form as

$$\begin{bmatrix} v_f \\ v_a \end{bmatrix} = \begin{bmatrix} r_f + pL_{FF} & 0 \\ \omega_r L_{AF} & r_a + pL_{AA} \end{bmatrix} \begin{bmatrix} i_f \\ i_a \end{bmatrix} \tag{2.3-1}$$

where L_{FF} and L_{AA} are the self-inductances of the field and armature windings, respectively, and p is the short-hand notation for the operator d/dt. The rotor speed is denoted as ω_r, and L_{AF} is the mutual inductance between the field and the rotating armature coils. The above equation suggests the equivalent circuit shown in Fig. 2.3-1. The voltage induced in the armature circuit, $\omega_r L_{AF} i_f$, is commonly referred to as the counter or back emf. It also represents the open-circuit armature voltage.

There are several other forms in which the field and armature voltage equations are often expressed. For example, L_{AF} may also be written as

$$L_{AF} = \frac{N_a N_f}{\mathscr{R}} \tag{2.3-2}$$

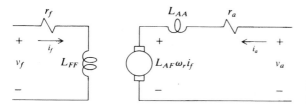

Figure 2.3-1 Equivalent circuit of a dc machine.

where N_a and N_f are the equivalent turns of the armature and field windings, respectively, and $\mathscr{R}$ is the reluctance. Thus,

$$L_{AF}i_f = N_a \frac{N_f i_f}{\mathscr{R}} \tag{2.3-3}$$

If we now replace $N_f i_f / \mathscr{R}$ with Φ_f, the field flux per pole, then $N_a \Phi_f$ may be substituted for $L_{AF}i_f$ in the armature voltage equation.

Another substitute variable often used is

$$k_v = L_{AF}i_f \tag{2.3-4}$$

We will find that this substitute variable is particularly convenient and frequently used. Even though a permanent-magnet dc machine has no field circuit, the constant field flux produced by the permanent magnet is analogous to a dc machine with a constant k_v.

We can take advantage of previous work to obtain an expression for the electromagnetic torque. In particular, the expression for torque given by (1B-8) may be used directly to express the torque for the dc machine. If we fix θ_r in Fig. 1B-1 at $-\frac{1}{2}\pi$, the same relationship exists between the magnetic axes of a dc machine (Fig. 2.2-5) and the magnetic axes of the two-coil machine. Hence, (1B-8) may be written for the dc machine as

$$T_e = L_{AF}i_f i_a \tag{2.3-5}$$

Here again the variable k_v is often substituted for $L_{AF}i_f$. In some instances, k_v is multiplied by a factor less than unity when substituted into (2.3-5) so as to approximate the effects of rotational losses. It is interesting that the field winding produces a stationary MMF and, owing to commutation, the armature winding also produces a stationary MMF that is displaced $\frac{1}{2}\pi$ electrical degrees from the MMF produced by the field winding. It follows then that the interaction of these two MMFs produces the electromagnetic torque.

The torque and rotor speed are related by

$$T_e = J \frac{d\omega_r}{dt} + B_m \omega_r + T_L \tag{2.3-6}$$

where J is the inertia of the rotor and, in some cases, the connected mechanical load. The units of the inertia are $\text{kg} \cdot \text{m}^2$ or $\text{J} \cdot \text{s}^2$. A positive electromagnetic torque T_e acts to turn the rotor in the direction of increasing θ_r. The load torque T_L is positive for a torque, on the shaft of the rotor, which opposes a positive electromagnetic torque T_e. The constant B_m is a damping coefficient associated with the mechanical rotational system of the machine and it has the units of $\text{N} \cdot \text{m} \cdot \text{s}$.

2.4   BASIC TYPES OF DIRECT-CURRENT MACHINES

The field and armature windings may be excited from separate sources or from the same source with the windings connected differently to form the basic types of dc machines, such as the shunt-connected, the series-connected, and the compound-connected dc machines. The equivalent circuits for each of these machines are given in this section along with an analysis and discussion of their steady-state operating characteristics.

Separate Winding Excitation

When the field and armature windings are supplied from separate voltage sources, the device may operate as either a motor or a generator; it is a motor if it is driving a torque load and a generator if it is being driven by some type of prime mover. The equivalent circuit for this type of machine is shown in Fig. 2.4-1. It differs from that shown in Fig. 2.3-1 in that external resistance r_{fx} is connected in series with the field winding. This resistance, which is often referred to as a *field rheostat*, is used to adjust the field current if the field voltage is supplied from a constant source.

The voltage equations that describe the steady-state performance of this device may be written directly from (2.3-1) by setting the operator p to zero ($p = d/dt$), whereupon

$$V_f = R_f I_f \tag{2.4-1}$$

$$V_a = r_a I_a + \omega_r L_{AF} I_f \tag{2.4-2}$$

where $R_f = r_{fx} + r_f$ and capital letters are used to denote steady-state voltages and currents. We know, from the torque relationship given by (2.3-6), that during

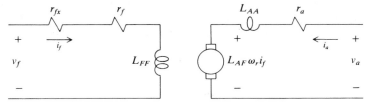

Figure 2.4-1   Equivalent circuit for separate field and armature excitation.

steady-state operation $T_e = T_L$ if B_m is assumed to be zero. Analysis of steady-state performance is straightforward.

A permanent-magnet dc machine fits into this class of dc machines. As we have mentioned, the field flux is established in these devices by a permanent magnet. The voltage equation for the field winding is eliminated and $L_{AF}i_f$ is replaced by a constant k_v, which can be measured if it is not given by the manufacturer. Most small, hand-held, fractional-horsepower dc motors are of this type, and speed control is achieved by controlling the amplitude of the applied armature voltage.

Shunt-Connected dc Machine

The field and armature windings may be connected as shown schematically in Fig. 2.4-2. With this connection, the machine may operate as either a motor or a generator. Because the field winding is connected between the armature terminals, $V_a = V_f$. This winding arrangement is commonly referred to as a *shunt-connected dc machine* or simply a shunt machine. During steady-state operation, the armature circuit voltage equation is (2.4-2) and, for the field circuit,

$$V_a = R_f I_f \qquad (2.4\text{-}3)$$

The total current I_t is

$$I_t = I_f + I_a \qquad (2.4\text{-}4)$$

Solving (2.4-2) for I_a and (2.4-3) for I_f and substituting the results in (2.3-5) yields the following expression for the steady-state electromagnetic torque, positive for motor action, for this type of dc machine:

$$T_e = \frac{L_{AF} V_a^2}{r_a R_f} \left(1 - \frac{L_{AF}}{R_f} \omega_r \right) \qquad (2.4\text{-}5)$$

The shunt-connected dc machine may operate as either a motor or a generator when connected to a dc source. It may also operate as an isolated self-excited

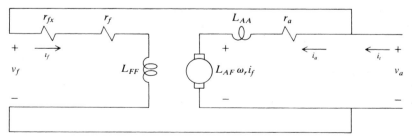

Figure 2.4-2 Equivalent circuit of a shunt-connected dc machine.

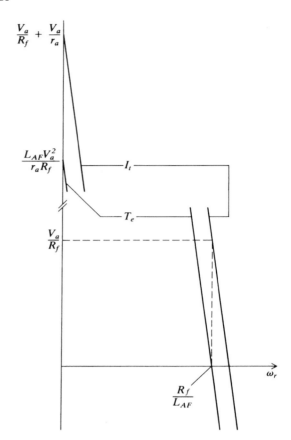

Figure 2.4-3 Steady-state operating characteristics of a shunt-connected dc machine with constant source voltage.

generator, supplying an electric load such as a dc motor or a static load. When the shunt machine is operated from a constant-voltage source, the steady-state operating characteristics are those shown in Fig. 2.4-3. Several features of these characteristics warrant discussion. At stall $(\omega_r = 0)$, the steady-state armature current I_a is limited only by the armature resistance. In the case of small, permanent-magnet motors, the armature resistance is quite large so that the starting armature current, which results when rated voltage is applied, is generally not damaging. However, larger-horse-power machines are designed with a small armature resistance. Therefore, an excessively high armature current will occur during the starting period if rated voltage is applied to the armature terminals. To prevent high starting current, resistance may be inserted into the armature circuit at stall and may be decreased either manually or automatically to zero as the machine accelerates to normal operating speed. When silicon-controlled rectifiers (SCRs) or thyristors are used to convert an ac source voltage to dc to supply the dc machine, they may be controlled to provide a reduced voltage during the starting period, thereby preventing a high starting current and eliminating the need to insert resistance into the armature circuit. Other features of the shunt machine with a small armature resistance are the steep torque versus speed characteristics. In other words, the speed of the shunt machine does not change appreciably as the load torque is varied from zero to rated.

When a shunt machine is operated as an isolated generator, it is referred to as a self-excited generator because the field is supplied by the voltage generated by the armature. Although this type of machine is rapidly being replaced by solid-state ac to dc converters, some attention will be given to the self-excited generator in an attempt to satisfy the curiosity of the interested reader. During generator operation, the armature current is negative relative to the positive direction indicated in Fig. 2.4-2. When operated as a generator (negative T_e and I_a), it is assumed that the prime mover maintains the speed of the generator constant regardless of the electric load supplied by the generator.

It would appear from Fig. 2.4-2 that, if the armature terminals of a self-excited generator are short-circuited while the rotor is being driven, the armature current I_a will be zero because the field current I_f and, consequently, the induced voltage $\omega_r L_{AF} I_f$ are all zero during steady-state short-circuit conditions. In practice this is not the case due to the permanent-magnet effect of the iron of the stator poles. In fact, this residual flux may produce a relatively large short-circuit current, depending upon the magnetic characteristics of the iron and the parameters of the machine. Because the existence of residual flux forms the basis of the self-excited mode of operation, it is appropriate to depart from the present analysis to discuss this phenomenon.

A terminal voltage could not be established in the case of a self-excited generator without the residual flux giving rise to an armature voltage once the rotor is driven. This feature may be explained by starting with the magnetization curve of a

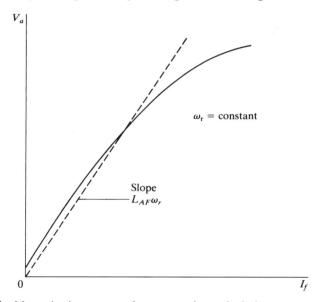

Figure 2.4-4 Magnetization curve of a separately excited dc generator: open-circuit armature voltage versus field current.

separately excited dc generator, shown in Fig. 2.4-4. The magnetization curve is a plot of the steady-state open-circuit armature voltage versus field current with the rotor driven at a constant speed. The dashed line in Fig. 2.4-4 illustrates the linear relationship between the steady-state open-circuit armature voltage and the field current as predicted by the analysis employed heretofore. The departure of the magnetization curve from the linear relationship at low values of field current is due to the permanent-magnet effect of the stator poles (residual flux). The departure at the higher values of field current is due to the saturation of the iron in the magnetic circuit. The effect of the residual flux upon the self-excited generator wherein the field current is determined by the armature voltage is our immediate interest.

To illustrate this feature, it is convenient to assume that the machine is being driven at a constant speed with the field winding open-circuited. In this mode of operation, the open-circuit terminal voltage is determined by the residual flux and the speed of the rotor. If now the field winding is connected to the armature terminals and if this connection is made so that the field current, resulting from the residual voltage initially appearing at the terminals of the machine, produces a flux which aids the residual flux, the armature voltage will increase. This "building-up" process will continue until the field voltage (terminal voltage) versus field current characteristic intersects the plot of the generated terminal voltage versus field current. In this case, the generated terminal voltage versus field characteristic differs only slightly from the magnetization curve owing to the voltage drop across the armature resistance. The slope of the straight-line relationship between the field voltage (terminal voltage) and field current is the total field resistance R_f. Hence, for a given rotor speed, the building up of the self-excited generator is predicted, first, upon the sense of the flux relative to the residual flux and, second, by the value of field resistance. The relationship between the steady-state terminal voltage with $I_t = 0$ ($I_f = -I_a$) and the field current of a self-excited generator is shown in Fig. 2.4-5, where $R_{f2} > R_{f1}$. A total field resistance of R_{f1} will yield a terminal voltage in the normal operating range; however, the terminal voltage will not build up to a normal value if the total field resistance is too large. For example, if the total field resistance is R_{f2}, the terminal voltage will fail to build up beyond a relatively small value.

After consideration of the nonlinear characteristics of the magnetization curve, one wonders if the performance of the dc machine may be accurately predicted with a linear approximation of the magnetization curve. The linear approximation is not valid if the field current is varied over a wide range, causing the machine to operate in the linear region as well as in the saturated region of the magnetization curve. The nonlinear features of the magnetic system may be taken into account by graphical methods, for analysis of steady-state performance, or a computer may be employed to study either a steady-state or transient performance with the effects of saturation included.

Series-Connected dc Machine

When the field is connected in series with the armature circuit as shown in Fig. 2.4-6, the machine is referred to as a *series-connected dc machine* or a series machine. It is convenient to add the subscript s to denote quantities associated with the series field.

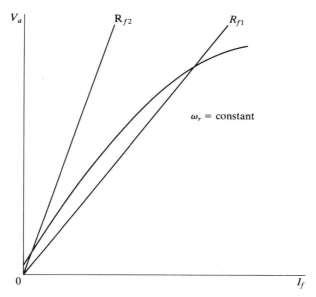

Figure 2.4-5 Influence of field resistance upon terminal voltage of self-excited dc generator ($I_t = 0$).

It is important to mention the physical difference between the field winding of a shunt machine and that of a series machine. If the field winding is to be a shunt-connected winding, it is wound with a large number of turns of small-diameter wire, making the resistance of the field winding quite large. However, because the series-connected field winding is in series with the armature, it is designed to minimize the voltage drop across it. Thus, the winding is wound with a few turns of low-resistance wire.

Although the series machine does not have wide application, a series field is often used in conjunction with a shunt field to form a compound-connected dc machine that is more common. In the case of a series machine (Fig. 2.4-6), we have

$$v_t = v_{fs} + v_a \tag{2.4-6}$$

$$i_a = i_{fs} \tag{2.4-7}$$

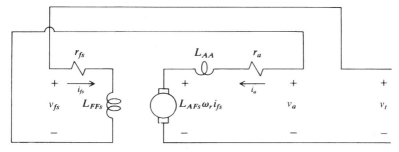

Figure 2.4-6 Equivalent circuit for a series-connected dc machine.

where v_{fs} and i_{fs} denote the voltage and current associated with the series field. The subscript s is added to avoid confusion with the shunt field when both fields are used in a compound machine.

If the constraints given by (2.4-6) and (2.4-7) are substituted into the armature voltage equation, the steady-state performance of the series-connected dc machine may be described by

$$V_t = (r_a + r_{fs} + L_{AFs}\omega_r)I_a \qquad (2.4\text{-}8)$$

From (2.3-5), we obtain

$$T_e = L_{AFs}I_a^2 = \frac{L_{AFs}V_t^2}{(r_a + r_{fs} + L_{AFs}\omega_r)^2} \qquad (2.4\text{-}9)$$

The steady-state torque–speed characteristic for a typical series machine is shown in Fig. 2.4-7. The stall torque is quite high because it is proportional to the square of the armature current for a linear magnetic system. However, saturation of the magnetic system due to large armature currents will cause the torque to be less than that calculated from (2.4-9). At high rotor speeds, the torque decreases less rapidly with increasing speed. In fact, if the load torque is small, the series motor may accelerate to speeds large enough to cause damage to the machine. Consequently, the series motor is used in applications such as traction motors for trains and buses or in hoists and cranes where high starting torque is required and an appreciable load torque exists under normal operation.

Compound-Connected dc Machine

A compound-connected or compound dc machine, which is equipped with both a shunt and a series field winding, is illustrated in Fig. 2.4-8. In most compound

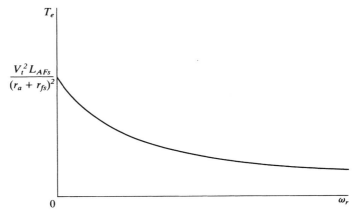

Figure 2.4-7   Steady-state torque–speed characteristics of a series-connected dc machine.

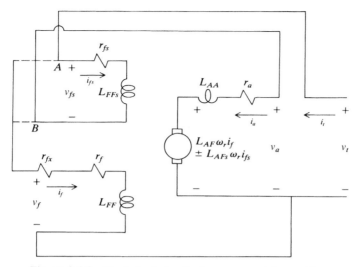

Figure 2.4-8 Equivalent circuit of a compound dc machine.

machines, the shunt field dominates the operating characteristics while the series field, which consists of a few turns of low-resistance wire, has a secondary influence. It may be connected so as to aid or oppose the flux produced by the shunt field. If the compound machine is to be used as a generator, the series field is connected so as to aid the shunt field (cumulative compounding). Depending upon the strength of the series field, this type of connection can produce a "flat" terminal voltage versus load current characteristic, whereupon a near-constant terminal voltage is achieved from no load to full load. In this case, the machine is said to be "flat-compounded." An "overcompounded" machine occurs when the strength of the series field causes the terminal voltage at full load to be larger than at no load. The meaning of the "under-compound" machine is obvious. In the case of compound dc motors, the series field is often connected to oppose the flux produced by the shunt field (differential compounding). If properly designed, this type of connection can provide a near-constant speed from no-load to full-load torque.

The voltage equations for a compound dc machine may be written as

$$\begin{bmatrix} v_f \\ v_t \end{bmatrix} = \begin{bmatrix} R_f + pL_{FF} & \pm pL_{FS} & 0 \\ \omega_r L_{AF} \pm pL_{FS} & \pm \omega_r L_{AFs} + r_{fs} + pL_{FFs} & r_a + pL_{AA} \end{bmatrix} \begin{bmatrix} i_f \\ i_{fs} \\ i_a \end{bmatrix} \quad (2.4\text{-}10)$$

where L_{FS} is the mutual inductance between the shunt and the series fields. The plus and minus signs are used so that either a cumulative or a differential connection may be described.

The shunt field may be connected ahead of the series field (long-shunt connection) or behind the series field (short-shunt connection) as shown by *A* and B, respectively, in Fig. 2.4-8. The long-shunt connection is commonly used. In this case

$$v_t = v_f = v_{fs} + v_a \qquad (2.4\text{-}11)$$
$$i_t = i_f + i_{fs} \qquad (2.4\text{-}12)$$

where

$$i_{fs} = i_a \qquad (2.4\text{-}13)$$

The steady-state performance of a long-shunt-connected compound machine may be described by the following equation:

$$V_t = \left[\frac{r_a + r_{fs} \pm L_{AFs}\omega_r}{1 - (L_{AF}/R_f)\omega_r}\right] I_a \qquad (2.4\text{-}14)$$

The torque for the long-shunt connection may be obtained by employing (2.3-5) for each field winding. In particular,

$$T_e = L_{AF}I_f I_a \pm L_{AF}I_{fs}I_a$$
$$= \frac{L_{AF}V_t^2[1 - (L_{AF}/R_f)\omega_r]}{R_f(r_a + r_{fs} \pm L_{AFs}\omega_r)} \pm \frac{L_{AFs}V_t^2[1 - (L_{AF}/R_f)\omega_r]^2}{(r_a + r_{fs} \pm L_{AFs}\omega_r)^2} \qquad (2.4\text{-}15)$$

Example 2A   A permanent-magnet dc motor similar to that shown in Fig. 2.2-6 is rated at 6 V with the following parameters: $r_a = 7\,\Omega$, $L_{AA} = 120$ mH, $k_T = 2$ oz $\cdot$ in./A, $J = 150\,\mu$oz $\cdot$ in. $\cdot$ s^2. According to the motor information sheet, the no-load speed is approximately 3350 r/min and the no-load armature current is approximately 0.15 A. Let us attempt to interpret this information.

First, let us convert k_T and J to units that we have been using in this text. In this regard, we will convert the inertia to kg $\cdot$ m^2, which is the same as N $\cdot$ m $\cdot$ s^2. To do this, we must convert ounces to newtons and inches to meters (Appendix A). Thus,

$$J = \frac{150 \times 10^{-6}}{(3.6)(39.37)} = 1.06 \times 10^{-6} \text{ kg} \cdot \text{m}^2 \qquad (2A\text{-}1)$$

We have not seen k_T before. It is the torque constant and, if expressed in the appropriate units, it is numerically equal to k_v. When k_v is used in the expression for $T_e(T_e = k_v i_a)$, it is often referred to as the *torque constant* and denoted k_T. When used in the voltage equation, it is always denoted k_v. Now we must convert oz $\cdot$ in. to N $\cdot$ m, whereupon k_T equals our k_v; hence,

$$k_v = \frac{2}{(16)(0.225)(39.37)} = 1.41 \times 10^{-2} \text{N} \cdot \text{m/A} = 1.41 \times 10^{-2} \text{V} \cdot \text{s/rad}$$
$$(2A\text{-}2)$$

What do we do about the no-load armature current? What does it represent? Well, probably it is a measure of the friction and windage losses. We could neglect it, but we will not. Instead, let us include it as B_m. First, however, we must calculate the no-load speed. We can solve for the no-load rotor speed from the steady-state armature voltage equation for the shunt machine, (2.4-2), with $L_{AF}i_f$ replaced by k_v:

$$\omega_r = \frac{V_a - r_a I_a}{k_v} = \frac{6 - (7)(0.15)}{1.41 \times 10^{-2}} = 351.1 \text{ rad/s}$$

$$= \frac{(351.1)(60)}{2\pi} = 3353 \text{ r/min} \qquad (2A\text{-}3)$$

Now at this no-load speed,

$$T_e = k_v i_a = (1.41 \times 10^{-2})(0.15) = 2.12 \times 10^{-3} \text{N} \cdot \text{m} \qquad (2A\text{-}4)$$

Because T_L and $J(d\omega_r/dt)$ are zero for this steady-state no-load condition, (2.3-6) tells us that (2A-4) is equal to $B_m\omega_r$; hence,

$$B_m = \frac{2.12 \times 10^{-3}}{\omega_r} = \frac{2.12 \times 10^{-3}}{351.1} = 6.04 \times 10^{-6} \text{N} \cdot \text{m} \cdot \text{s} \qquad (2A\text{-}5)$$

Example 2B The permanent-magnet dc machine described in Example 2A is operating with rated applied armature voltage and load torque T_L of 0.5 oz·in. Our task is to determine the efficiency where percent eff = (power output/power input) 100.

First let us convert oz·in. into N·m:

$$T_L = \frac{0.5}{(16)(0.225)(39.37)} = 3.53 \times 10^{-3} \text{N} \cdot \text{m} \qquad (2B\text{-}1)$$

In Example 2A we determined k_v to be 1.41×10^{-2} V·s/rad and determined B_m to be 6.04×10^{-6} N·m·s.

During steady-state operation, (2.3-6) becomes

$$T_e = B_m\omega_r + T_L \qquad (2B\text{-}2)$$

From (2.3-5), with $L_{AF}i_f$ replaced by k_v, the steady-state electromagnetic torque is

$$T_e = k_v I_a \qquad (2B\text{-}3)$$

Substituting (2B-3) into (2B-2) and solving for ω_r yields

$$\omega_r = \frac{k_v}{B_m} I_a - \frac{1}{B_m} T_L \qquad (2B\text{-}4)$$

From (2.4-2) with $L_{AF}i_f = k_v$, we obtain

$$V_a = r_a I_a + k_v\omega_r \qquad (2B\text{-}5)$$

Substituting (2B-4) into (2B-5) and solving for I_a yields

$$I_a = \frac{V_a + (k_v/B_m)T_L}{r_a + (k_v^2/B_m)}$$

$$= \frac{6 + [(1.41 \times 10^{-2})/(6.04 \times 10^{-6})](3.53 \times 10^{-3})}{7 + (1.41 \times 10^{-2})^2/(6.04 \times 10^{-6})} = 0.357 \text{ A} \qquad (2B\text{-}6)$$

From (2B-4), we obtain

$$\omega_r = \frac{1.41 \times 10^{-2}}{6.04 \times 10^{-6}} 0.357 - \frac{1}{6.04 \times 10^{-6}} (3.53 \times 10^{-3})$$

$$= 249 \text{ rad/s} \qquad (2B\text{-}7)$$

The power input is

$$P_{\text{in}} = V_a I_a = (6)(0.357) = 2.14 \text{ W} \qquad (2B\text{-}8)$$

The power output is

$$P_{\text{out}} = T_L \omega_r = (3.53 \times 10^{-3})(249) = 0.88 \text{ W} \qquad (2B\text{-}9)$$

The efficiency is

$$\text{Percent eff} = \frac{P_{\text{out}}}{P_{\text{in}}} 100 = \frac{0.88}{2.14} 100 = 41.1\% \qquad (2B\text{-}10)$$

The low efficiency is characteristic of low-power dc motors due to the relatively large armature resistance. In this regard, it is interesting to determine the losses due to $i^2 r$, friction, and windage.

$$P_{i^2 r} = r_a I_a^2 = (7)(0.357)^2 = 0.89 \text{ W} \qquad (2B\text{-}11)$$

$$P_{fw} = (B_m \omega_r)\omega_r = (6.04 \times 10^{-6})(249)^2 = 0.37 \text{ W} \qquad (2B\text{-}12)$$

Let us check our work:

$$P_{\text{in}} = P_{i^2 r} + P_{fw} + P_{\text{out}} = 0.89 + 0.37 + 0.88 = 2.14 \text{ W} \qquad (2B\text{-}13)$$

which is equal to (2B-8).

2.5 DYNAMIC CHARACTERISTICS OF PERMANENT-MAGNET AND SHUNT dc MOTORS

The permanent-magnet and shunt dc motors are widely used and thus are appropriate candidates to illustrate the dynamic performance of typical dc machines. Two modes of dynamic operation are of interest—starting from stall and changes in load torque with the machine supplied from a constant-voltage source.

Dynamic Performance During Starting

In the previous section, it was pointed out that if the armature resistance is small, damaging armature current could result if rated voltage is applied to the armature terminals when the machine is stalled ($\omega_r = 0$). With the machine at stall, the counter emf is zero and the armature current is opposed only by the voltage drop across the armature resistance and inductance. However, we have noted in Examples 2A and 2B that low-power permanent-magnet dc motors characteristically have a relatively large armature resistance making it possible to "direct-line" start these devices without damaging the brushes and armature windings. The no-load starting characteristics ($T_L = 0$) of the permanent-magnet dc motor described in Example 2A are shown in Fig. 2.5-1. The armature voltage v_a, the armature current i_a, and the rotor speed ω_r are plotted. Initially the motor is at stall and, at time zero, 6 V is applied to the armature terminals. The peak transient armature current is limited to approximately 0.55 A by the inductance and resistance of the armature and the fact that the rotor is accelerating from stall, thereby developing the voltage $k_v \omega_r$ which opposes the applied voltage. After about 0.25 s, steady-state operation is achieved with the no-load armature current of 0.15 A. (From Example 2A, $B_m = 6.04 \times 10^{-6}$ N·m·s.) It is noted that the rotor speed is slightly oscillatory (underdamped), as illustrated by the small overshoot of the final value.

When starting larger-horsepower dc machines, it is generally necessary to limit the starting current. As mentioned, this can be accomplished either by phase-controlling an ac-to-dc converter, if it is used to provide the armature voltage, or by inserting resistance in series with the armature circuit during starting, if the motor is supplied from a constant-voltage source. However, because resistance starting of dc machines is rapidly being replaced by ac-to-dc converters, we will not consider this method further. For those who wish more information regarding this method of starting, it is treated in detail in references 4 and 5.

Dynamic Performance During Sudden Changes in Load Torque

In Example 2B, we calculated the efficiency of the permanent-magnet dc motor given in Example 2A with a load torque of 0.5 oz·in. (3.53×10^{-3} N·m). Let us assume that this load torque was suddenly applied with the motor initially operating at the no-load condition ($I_a = 0.15$ A). The dynamic characteristics following a step change in load torque T_L from zero to 0.5 oz·in. are shown in Fig. 2.5-2. The armature current i_a and the rotor speed ω_r are plotted. Because $T_e = k_v i_a$ and because k_v is constant, T_e differs from i_a by a constant multiplier. It is noted that the system is slightly oscillatory. Also, it is noted that the change in the steady-state rotor speed is quite large. From Example 2A or Fig. 2.5-2, we see that no-load speed is 351.1 rad/s. With $T_L = 0.5$ oz·in., the rotor speed is 249 rad/s from Example 2B or Fig. 2.5-2. There has been an approximately 30% decrease in speed for this increase in load torque. This is characteristic of a low-power permanent-magnet motor owing to the high armature resistance.

The dynamic performance of a shunt motor during a step decrease in load torque is shown in Fig. 2.5-3. The parameters and rated conditions of this machine are as

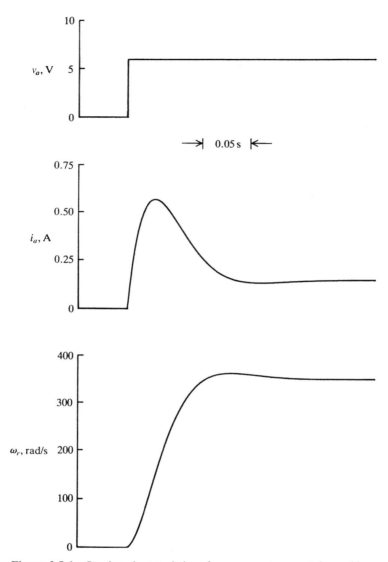

Figure 2.5-1   Starting characteristics of a permanent-magnet dc machine.

follows: $R_f = 240\ \Omega$, $L_{FF} = 120$ H, $L_{AF} = 1.8$ H, $r_a = 0.6\ \Omega$, $L_{AA} = 0.012$ H. The machine is a 240-V 5-hp dc shunt motor. The inertia of the machine and connected load is 1 kg·m². Also, it is clear that $R_f = r_{fx} + r_f$. Rated armature current is 16.2 A, and rated rotor speed is 127.7 rad/s. Here, rated armature current and rated speed are determined by setting v_a and v_f equal to 240 V and setting the power output equal to 5 hp, where the power output is the electromagnetic torque multiplied by the rotor speed.

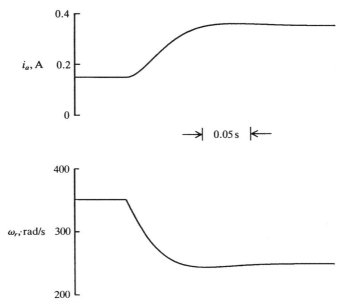

Figure 2.5-2   Dynamic performance of a permanent-magnet dc motor following a sudden increase in load torque from zero to 0.5 oz·in.

Initially, the motor is operating steadily with rated load torque (29.2 N·m) applied to the rotor shaft. The load torque is then stepped to 50% of rated value, whereupon the motor speeds up and reestablishes steady-state operation at the reduced load conditions. The armature current i_a and rotor speed ω_r are plotted. Because the field current is constant, the electromagnetic torque T_e is identical to the armature current, differing only by a constant multiplier. It is important to note the small change in rotor speed from full load to 50% of full-load torque. At full load, the rotor speed is 127.7 rad/s and 130.4 rad/s at 50% of full load. This is approximately a 2% change in rotor speed for a 50% change in load torque. We first noted this "steep" torque–speed characteristic of a shunt machine, which is due to

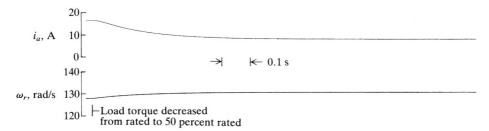

Figure 2.5-3   Dynamic performance of a 5-hp shunt motor following a sudden decrease in load torque from rated to 50% rated.

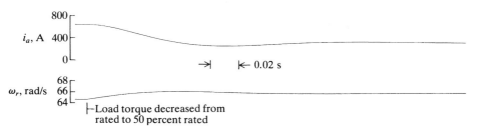

Figure 2.5-4   Dynamic performance of a 200-hp shunt motor following a sudden change in load torque from rated to 50% rated.

the low value of armature resistance in Fig. 2.4-3. With the field current constant, the voltage and torque equations that describe the shunt motor are linear (neglecting saturation); hence, the response of the machine variables when the load torque is stepped back to rated from 50% rated would be the reverse of that shown in Fig. 2.5-3.

The operating characteristics and dynamic response of a larger-horsepower shunt motor shall be illustrated using a 200-hp machine with the following parameters: 200 hp, 250 V, 600 r/min, $R_f = 12\ \Omega$, $L_{FF} = 9$ H, $L_{AF} = 0.18$ H, $r_a = 0.012\ \Omega$, $L_{AA} = 0.00035$ H. The total inertia of rotor and load is 30 kg$\cdot$m^2. When rated horsepower, voltage, and speed are all given, as in this case, the values are generally approximate and account for various machine losses. Not all conditions can be met, especially when we consider the machine as an idealized machine. If, for example, we select 600 r/min (62.8 rad/s) as rated speed, rated load torque is then 2375 N$\cdot$m [(200) (746)/62.8]. However, if this load torque is applied to the rotor and if v_a and v_f are both 250 V, then the calculated steady-state armature current is 633 A and the rotor speed is 617 r/min (64.6 rad/s). The response of this 200-hp machine to a decrease (step) in load torque from rated to 50% rated is shown in Fig. 2.5-4. Here, again, we see the small change in speed (approximately 1.6%) from full load to 50% of full load due to the small armature resistance.

2.6   TIME-DOMAIN BLOCK DIAGRAMS AND STATE EQUATIONS

Although the analysis of control systems is not our intent, it is worthwhile to set the stage for this type of analysis by means of a "first look" at time-domain block diagrams and state equations. In this section, we will consider only the shunt and permanent-magnet dc machines. The series and compound machines are treated in problems at the end of the chapter.

Shunt-Connected dc Machine

Block diagrams, which portray the interconnection of the system equations, are used extensively in control system analysis and design. Although block diagrams are generally depicted by using the Laplace operator, we shall work with the time-domain equations, for now, using the p operator to denote differentiation with respect to time and the operator $1/p$ to denote integration. Section 2.7 is devoted to Laplace transformation solution for the dynamics of the permanent-magnet and shunt machines.

Arranging the equations of a shunt machine into a block diagram representation is straightforward. The field and armature voltage equations, (2.3-1), and the relationship between torque and rotor speed, (2.3-6), may be written as

$$v_f = R_f(1 + \tau_f p)i_f \tag{2.6-1}$$

$$v_a = r_a(1 + \tau_a p)i_a + \omega_r L_{AF} i_f \tag{2.6-2}$$

$$T_e - T_L = (B_m + Jp)\omega_r \tag{2.6-3}$$

where the field time constant τ_f equals L_{FF}/R_f and the armature time constant τ_a equals L_{AA}/r_a. Here, again, p denotes d/dt and $1/p$ will denote integration. Solving (2.6-1) for i_f, (2.6-2) for i_a, and (2.6-3) for ω_r yields

$$i_f = \frac{1/R_f}{\tau_f p + 1} v_f \tag{2.6-4}$$

$$i_a = \frac{1/r_a}{\tau_a p + 1}(v_a - \omega_r L_{AF} i_f) \tag{2.6-5}$$

$$\omega_r = \frac{1}{Jp + B_m}(T_e - T_L) \tag{2.6-6}$$

A few comments are in order regarding these expressions. In (2.6-4), we see that the field voltage v_f is multiplied by the operator $(1/R_f)/(\tau_f p + 1)$ to obtain the field current i_f. The operator $(1/R_f)/(\tau_f p + 1)$ may also be interpreted as a transfer function relating the field voltage and current. The fact that we are multiplying the voltage by an operator to obtain current is in no way indicative of the procedure that we might actually use to calculate the current i_f given the voltage v_f. We are simply expressing the dynamic relationship between the field voltage and current in a form convenient for drawing block diagrams. However, to calculate i_f given v_f, we may prefer to express (2.6-4) in its equivalent form (2.6-1) and solve the resulting first-order differential equation.

The time-domain block diagram portraying (2.6-4) through (2.6-6) with $T_e = L_{AF}i_f i_a$ is shown in Fig. 2.6-1. This diagram consists of a set of linear blocks, wherein the relationship between the input and corresponding output variable is depicted in transfer function form and a pair of multipliers which represent nonlinear blocks. Because the system is nonlinear, it is not possible to apply previously described techniques for solving the differential equations implied by this block diagram. For this, we would use a computer. However, for certain dc machines—for example, permanent-magnet machines or separately excited shunt machines, where the field current i_f is maintained at a constant value—the multipliers in Fig. 2.6-1 are no longer needed, and conventional methods of analyzing linear systems may be applied with relative ease.

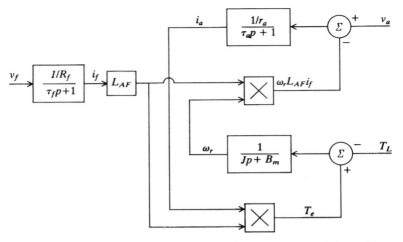

Figure 2.6-1   Time-domain block diagram of a shunt connected dc machine.

The so-called state equations of a system represent the formulation of the state variables into a matrix form convenient for computer implementation, particularly for linear systems. The state variables of a system are defined as a minimal set of variables such that knowledge of these variables at any initial time t_0 plus information on the input excitation subsequently applied is sufficient to determine the state of the system at any time $t > t_0$ [2]. In the case of dc machines, the field current i_f, the armature current i_a, the rotor speed ω_r, and the rotor position θ_r are the state variables. The rotor position θ_r can be established from ω_r by

$$\omega_r = \frac{d\theta_r}{dt} \tag{2.6-7}$$

Because θ_r is considered a state variable only when the shaft position is a controlled variable, we will omit θ_r from consideration in this development.

The formulation of the state equations for the shunt machine can be readily achieved by straightforward manipulation of the field and armature voltage equations given by (2.3-1) and the equation relating torque and rotor speed given by (2.3-6). In particular, solving the field voltage equation, (2.3-1), for di_f/dt yields

$$\frac{di_f}{dt} = -\frac{R_f}{L_{FF}} i_f + \frac{1}{L_{FF}} v_f \tag{2.6-8}$$

Solving the armature voltage equation, (2.3-1), for di_a/dt yields

$$\frac{di_a}{dt} = -\frac{r_a}{L_{AA}} i_a - \frac{L_{AF}}{L_{AA}} i_f \omega_r + \frac{1}{L_{AA}} v_a \tag{2.6-9}$$

If we wish, we could use k_v for $L_{AF}i_f$; however, we shall not make this substitution. Solving (2.3-6) for $d\omega_r/dt$ with $T_e = L_{AF}i_fi_a$ yields

$$\frac{d\omega_r}{dt} = -\frac{B_m}{J}\omega_r + \frac{L_{AF}}{J}i_fi_a - \frac{1}{J}T_L \qquad (2.6\text{-}10)$$

All we have done here is to solve the equations for the highest derivative of the state variables while substituting (2.3-5) for T_e into (2.3-6). Now let us write the state equations in matrix (or vector matrix) form as

$$p\begin{bmatrix} i_f \\ i_a \\ \omega_r \end{bmatrix} = \begin{bmatrix} -\frac{R_f}{L_{FF}} & 0 & 0 \\ 0 & -\frac{r_a}{L_{AA}} & 0 \\ 0 & 0 & -\frac{B_m}{J} \end{bmatrix} \begin{bmatrix} i_f \\ i_a \\ \omega_r \end{bmatrix} + \begin{bmatrix} 0 \\ -\frac{L_{AF}}{L_{AA}}i_f\omega_r \\ \frac{L_{AF}}{J}i_fi_a \end{bmatrix}$$

$$+ \begin{bmatrix} \frac{1}{L_{FF}} & 0 & 0 \\ 0 & \frac{1}{L_{AA}} & 0 \\ 0 & 0 & -\frac{1}{J} \end{bmatrix} \begin{bmatrix} v_f \\ v_a \\ T_L \end{bmatrix} \qquad (2.6\text{-}11)$$

where p is the operator d/dt. Equation (2.6-11) is the state equation(s); however, note that the second term (vector) on the right-hand side contains the products of state variables causing the system to be nonlinear.

Permanent-Magnet dc Machine

As we have mentioned previously, the equations that describe the operation of a permanent-magnet dc machine are identical to those of a shunt-connected dc machine with the field current constant. Thus, the work in this section applies to both. For the permanent-magnet machine, $L_{AF}i_f$ is replaced by k_v, which is a constant determined by the strength of the magnet, the reluctance of the iron, and the number of turns of the armature winding. The time-domain block diagram may be developed for the permanent-magnet machine by using (2.6-2) and (2.6-3) with k_v substituted for $L_{AF}i_f$. The time-domain block diagram for a permanent-magnet dc machine is shown in Fig. 2.6-2.

Because k_v is constant, the state variables are now i_a and ω_r. From (2.6-9), for a permanent-magnet machine we have

$$\frac{di_a}{dt} = -\frac{r_a}{L_{AA}}i_a - \frac{k_v}{L_{AA}}\omega_r + \frac{1}{L_{AA}}v_a \qquad (2.6\text{-}12)$$

From (2.6-10), we obtain

$$\frac{d\omega_r}{dt} = -\frac{B_m}{J}\omega_r + \frac{k_v}{J}i_a - \frac{1}{J}T_L \qquad (2.6\text{-}13)$$

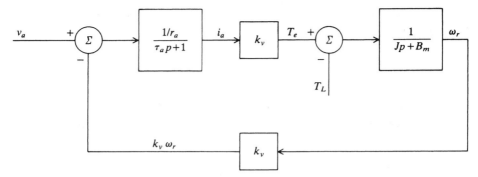

Figure 2.6-2   Time-domain block diagram of a permanent-magnet dc machine.

The system is described by a set of linear differential equations. In matrix form, the state equations become

$$p\begin{bmatrix} i_a \\ \omega_r \end{bmatrix} = \begin{bmatrix} -\frac{r_a}{L_{AA}} & -\frac{k_v}{L_{AA}} \\ \frac{k_v}{J} & -\frac{B_m}{J} \end{bmatrix}\begin{bmatrix} i_a \\ \omega_r \end{bmatrix} + \begin{bmatrix} \frac{1}{L_{AA}} & 0 \\ 0 & -\frac{1}{J} \end{bmatrix}\begin{bmatrix} v_a \\ T_L \end{bmatrix} \tag{2.6-14}$$

The form in which the state equations are expressed in (2.6-14) is called the fundamental form. In particular, the previous matrix equation may be expressed symbolically as

$$p\mathbf{x} = \mathbf{Ax} + \mathbf{Bu} \tag{2.6-15}$$

which is called the fundamental form, where p is the operator d/dt, $\mathbf{x}$ is the state vector (column matrix of state variables), and $\mathbf{u}$ is the input vector (column matrix of inputs to the system). We see that (2.6-14) and (2.6-15) are identical in form. Methods of solving equations of the fundamental form given by (2.6-15) are well known. Consequently, it is used extensively in control system analysis [6].

> **Example 2C**   Once the permanent-magnet dc motor is portrayed in block diagram form (Fig. 2.6-2), it is often advantageous, for control design purposes, to express transfer functions between state and input variables. Our task is to derive transfer functions between the state variables (i_a and ω_r) and the input variables (v_a and T_L) for the permanent-magnet dc machine. From (2.6-10), with $L_{AF}i_f$ replaced by k_v, we have
>
> $$i_a = \frac{1/r_a}{\tau_a p + 1}(v_a - k_v\omega_r) \tag{2C-1}$$
>
> From (2.6-11), with $T_e = k_v i_a$, we obtain
>
> $$\omega_r = \frac{1}{Jp + B_m}(k_v i_a - T_L) \tag{2C-2}$$

It is apparent that we could have obtained these same equations from the block diagram given in Fig. 2.6-2. If (2C-1) is substituted into (2C-2), we obtain, after considerable work,

$$\omega_r = \frac{(1/k_v\tau_a\tau_m)v_a - (1/J)(p + 1/\tau_a)T_L}{p^2 + (1/\tau_a + B_m/J)p + (1/\tau_a)(1/\tau_m + B_m/J)} \qquad (2C\text{-}3)$$

where a new time constant has been introduced. The inertia time constant, which is what τ_m is called, is defined as

$$\tau_m = \frac{Jr_a}{k_v^2} \qquad (2C\text{-}4)$$

The transfer function between ω_r and v_a may be obtained from (2C-3) by setting T_L equal to zero in (2C-3) and dividing both sides by v_a. Similarly, the transfer function between ω_r and T_L is obtained by setting v_a to zero and dividing by T_L. To calculate ω_r given v_a and T_L, we note that p is d/dt and p^2 is d^2/dt^2 and, if we multiply each side of (2C-3) by the denominator of the right side of the equation, we would have a second-order differential equation in terms of the state variable ω_r.

The characteristic or force-free equation for this linear system is obtained by setting the denominator equal to zero. It is of the general form

$$p^2 + 2\alpha p + \omega_n^2 = 0 \qquad (2C\text{-}5)$$

We are aware that α is the exponential damping coefficient and ω_n is the undamped natural frequency. The damping factor is defined as

$$\zeta = \frac{\alpha}{\omega_n} \qquad (2C\text{-}6)$$

Let us denote b_1 and b_2 as the negative values of the roots of this second-order equation,

$$b_1, b_2 = \zeta\omega_n \pm \omega_n\sqrt{\zeta^2 - 1} \qquad (2C\text{-}7)$$

If $\zeta > 1$, the roots are real and the natural response consists of two exponential terms with negative real exponents. When $\zeta < 1$, the roots are a conjugate complex pair and the natural response consists of an exponentially decaying sinusoid.

Now, the transfer function relationship between i_a and the input variables v_a and T_L may be obtained by substituting (2C-2) into (2C-1). After some work, we obtain

$$i_a = \frac{(1/\tau_a r_a)(p + B_m/J)v_a + (1/k_v\tau_a\tau_m)T_L}{p^2 + (1/\tau_a + B_m/J)p + (1/\tau_a)(1/\tau_m + B_m/J)} \qquad (2C\text{-}8)$$

2.7 SOLUTION OF DYNAMIC CHARACTERISTICS BY LAPLACE TRANSFORMATION

This section is devoted to examples of Laplace transformation solution of several of the dynamic characteristics of the permanent-magnet and shunt machines illustrated in the previous section.

Example 2D The purpose of this example is to illustrate the Laplace transformation solution of the starting characteristics of the permanent-magnet dc motor depicted in Fig. 2.5-1. The rotor is initially at stall with $T_L = 0$ and $B_m = 6.04 \times 10^{-6}\,\text{N} \cdot \text{m} \cdot \text{s}$. The armature voltage is stepped from zero to 6 V while the change in T_L is zero. Let us construct the block diagram in terms of the Laplace operator s for this situation. This block diagram is shown in Fig. 2D-1, and the Δ is used to denote changes from the predisturbance, steady-state values. We see that, because the change in T_L is zero, $\Delta T_L(s) = 0$ and, consequently, it does not appear in the block diagram or in the transfer functions.

We can express $\Delta i_a(s)$ and $\Delta \omega_r(s)$ from (2C-8) and (2C-3), respectively, by replacing the operator p with the Laplace operator s and by replacing all time-domain state variables (i_a and ω_r) and input variables (v_a and T_L) with the corresponding Δ variables as functions of the Laplace operator. In particular, i_a becomes $\Delta i_a(s)$, ω_r becomes $\Delta \omega_r(s)$, and v_a becomes $\Delta v_a(s)$, and as we mentioned previously, the change in T_L is zero; thus $\Delta T_L(s) = 0$. Hence, from (2C-8), we obtain

$$\Delta i_a(s) = \frac{(1/\tau_a r_a)(s + B_m/J)\Delta v_a(s)}{s^2 + (1/\tau_a + B_m/J)s + (1/\tau_a)(1/\tau_m + B_m/J)} \qquad (2D\text{-}1)$$

From (2C-3), we have

$$\Delta \omega_r(s) = \frac{(1/k_v \tau_a \tau_m)\Delta v_a(s)}{s^2 + (1/\tau_a + B_m/J)s + (1/\tau_a)(1/\tau_m + B_m/J)} \qquad (2D\text{-}2)$$

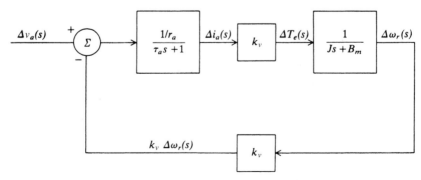

Figure 2D-1   Block diagram of a permanent-magnet dc machine for a step change in applied voltage.

Also,

$$\Delta T_e(s) = k_v \Delta i_a(s) \tag{2D-3}$$

We are now ready to solve for the roots of the characteristic equation. For this purpose, it is convenient to make a few preliminary calculations. Recall that the parameters for this motor are given in Example 2A.

$$\tau_a = \frac{L_{AA}}{r_a} = \frac{120 \times 10^{-3}}{7} = 0.017 \text{ sec} \tag{2D-4}$$

$$\tau_m = \frac{Jr_a}{k_v^2} = \frac{(1.06 \times 10^{-6})(7)}{(1.41 \times 10^{-2})^2} = 0.037 \text{ sec} \tag{2D-5}$$

Here, we used sec rather than s to denote seconds to avoid confusion with the Laplace operator. The characteristic equation is obtained by setting the denominator of (2D-1) or (2D-2) equal to zero. It is of the form of (2C-5), from which

$$2\alpha = \frac{1}{\tau_a} + \frac{B_m}{J} = \frac{1}{0.017} + \frac{6.04 \times 10^{-6}}{1.06 \times 10^{-6}}$$
$$= 58.82 + 5.7 = 64.52 \text{ sec}^{-1} \tag{2D-6}$$

$$\omega_n^2 = \frac{1}{\tau_a}\left(\frac{1}{\tau_m} + \frac{B_m}{J}\right) = \frac{1}{0.017}\left(\frac{1}{0.037} + \frac{6.04 \times 10^{-6}}{1.06 \times 10^{-6}}\right)$$
$$= 58.82(27.03 + 5.7) = 1925 \text{ sec}^{-2} \tag{2D-7}$$

From (2D-6), we obtain

$$\alpha = \frac{1}{2}(64.52) = 32.26 \text{ sec}^{-1} \tag{2D-8}$$

From (2D-7), we have

$$\omega_n = (1925)^{1/2} = 43.9 \text{ rad/sec} \tag{2D-9}$$

The damping factor is

$$\zeta = \frac{\alpha}{\omega_n} = \frac{32.26}{43.9} = 0.735 \tag{2D-10}$$

Because $\zeta < 1$, we know that the roots of the characteristic equation will be a conjugate complex pair and that the transient response will contain damped sinusoids. From (2C-7), we obtain

$$b_1, b_2 = \zeta\omega_n \pm \omega_n\sqrt{\zeta^2 - 1}$$
$$= 0.735 \times 43.9 \pm 43.9\sqrt{(0.735)^2 - 1}$$
$$= 32.3 \pm j29.8 \text{ sec}^{-1} \tag{2D-11}$$

which is of the form

$$b_1, b_2 = \gamma \pm j\beta \tag{2D-12}$$

Now for a step in applied voltage,

$$\Delta v_a(s) = \frac{K}{s} \tag{2D-13}$$

where $K = 6\,\text{V}$. Hence (2D-1) becomes

$$\Delta i_a(s) = \frac{(K/\tau_a r_a)(s + a_1)}{s(s + b_1)(s + b_2)} \tag{2D-14}$$

where, depending upon the damping factor, b_1 and b_2 may be real or a conjugate complex pair and

$$a_1 = \frac{B_m}{J} = 5.7\,\text{sec}^{-1} \tag{2D-15}$$

Therefore, (2D-2) becomes

$$\Delta\omega_r(s) = \frac{K/k_v\tau_a\tau_m}{s(s + b_1)(s + b_2)} \tag{2D-16}$$

We must now obtain the inverse Laplace transform of (2D-14) and (2D-16). For this, we will use the partial-fraction expansion. In particular, (2D-14) and (2D-16) may be written as

$$\Delta i_a(s) = \frac{K}{\tau_a r_a}\left(\frac{A}{s} + \frac{B}{s + b_1} + \frac{C}{s + b_2}\right) \tag{2D-17}$$

$$\Delta\omega_r(s) = \frac{K}{k_v\tau_a\tau_m}\left(\frac{D}{s} + \frac{E}{s + b_1} + \frac{F}{s + b_2}\right) \tag{2D-18}$$

Equating (2D-14) and (2D-17) yields

$$A(s + b_1)(s + b_2) + Bs(s + b_2) + Cs(s + b_1) = s + a_1 \tag{2D-19}$$

Similarly, from (2D-16) and (2D-18) we have

$$D(s + b_1)(s + b_2) + Es(s + b_2) + Fs(s + b_1) = 1 \tag{2D-20}$$

By equating coefficients of (2D-19), we obtain

$$A = \frac{a_1}{b_1 b_2} \tag{2D-21}$$

$$B = \frac{a_1 - b_1}{b_1(b_1 - b_2)} \tag{2D-22}$$

$$C = \frac{a_1 - b_2}{b_2(b_2 - b_1)} \tag{2D-23}$$

Similarly, from (2D-20) we have

$$D = \frac{1}{b_1 b_2} \tag{2D-24}$$

$$E = \frac{1}{b_1(b_1 - b_2)} \tag{2D-25}$$

$$F = \frac{1}{b_2(b_2 - b_1)} \tag{2D-26}$$

Taking the inverse Laplace transform of (2D-17) and (2D-18) yields

$$i_a = i_a(0) + \frac{K}{\tau_a r_a}\left[\frac{a_1}{b_1 b_2} + \frac{a_1 - b_1}{b_1(b_1 - b_2)}e^{-b_1 t} + \frac{a_1 - b_2}{b_2(b_2 - b_1)}e^{-b_2 t}\right] \tag{2D-27}$$

$$\omega_r = \omega_r(0) + \frac{K}{k_v \tau_a \tau_m}\left[\frac{1}{b_1 b_2} + \frac{1}{b_1(b_1 - b_2)}e^{-b_1 t} + \frac{1}{b_2(b_2 - b_1)}e^{-b_2 t}\right] \tag{2D-28}$$

where $i_a(0)$ and $\omega_r(0)$ are the predisturbance steady-state values that, in this case, are both zero.

If b_1 and b_2 are real, we would be essentially done; however, $\zeta < 1$ for this machine, hence b_1 and b_2 form a conjugate complex pair as expressed by (2D-12). Substituting (2D-12) for b_1 and b_2 into (2D-27) and (2D-28), and after a little exercise in complex algebra and making use of Euler's identity a couple of times, we obtain

$$i_a = i_a(0) + \frac{K}{\tau_a r_a}\frac{a_1}{\gamma^2 + \beta^2}\left[-e^{-\gamma t}\left(\cos\beta t - \frac{\gamma^2 + \beta^2 - a_1\gamma}{a_1\beta}\sin\beta t\right)\right] \tag{2D-29}$$

$$\omega_r = \omega_r(0) + \frac{K}{k_v \tau_a \tau_m}\frac{1}{\gamma^2 + \beta^2}\left[1 - e^{-\gamma t}\left(\cos\beta t + \frac{\gamma}{\beta}\sin\beta t\right)\right] \tag{2D-30}$$

Substituting in the appropriate values yields

$$i_a = 0.149[1 - e^{-32.3t}(\cos 29.8t - 10.3\sin 29.8t)] \tag{2D-31}$$

$$\omega_r = 350.3[1 - e^{-32.3t}(\cos 29.8t + 1.08\sin 29.8t)] \tag{2D-32}$$

Example 2E   Figures 2.5-3 and 2.5-4 illustrate the dynamic response to a load torque change for a 5- and 200-hp dc shunt motor, respectively. Let us use the material presented in Examples 2C and 2D to obtain these responses analytically. Because the terminal voltage is constant, equal to its predisturbance value, $\Delta v_a(s) = 0$; and because $B_m = 0$, (2C-3) may be written as

$$\Delta\omega_r(s) = \frac{-(1/J)(s + 1/\tau_a)\Delta T_L(s)}{s^2 + (1/\tau_a)s + 1/\tau_a\tau_m} \tag{2E-1}$$

where Δ is used to denote change from the predisturbance steady-state value and the operator p in (2C-3) has been replaced by the Laplace operator s. With

$\Delta v_a(s) = 0$, (2C-8) yields

$$\Delta i_a(s) = \frac{(1/k_v \tau_a \tau_m)\Delta T_L(s)}{s^2 + (1/\tau_a)s + 1/\tau_a \tau_m} \tag{2E-2}$$

A step change in load torque is expressed as

$$\Delta T_L(s) = \frac{K}{s} \tag{2E-3}$$

where, for the moment, we will let K be a constant with magnitude and sign; however, it is not the same K used in Example 2D. In Fig. 2.5-3, for the 5-hp motor, $K = -\frac{1}{2}(29.2)\,\text{N} \cdot \text{m}$; and in Fig. 2.5-4, for the 200-hp motor, $K = -\frac{1}{2}(2375)\,\text{N} \cdot \text{m}$. Hence, (2E-1) becomes

$$\Delta \omega_r(s) = \frac{-(K/J)(s + 1/\tau_a)}{s[s^2 + (1/\tau_a)s + 1/\tau_a \tau_m]} \tag{2E-4}$$

and (2E-2) becomes

$$\Delta i_a(s) = \frac{K/k_v \tau_a \tau_m}{s[s^2 + (1/\tau_a)s + 1/\tau_a \tau_m]} \tag{2E-5}$$

The characteristic equation is obtained by setting the denominator of (2E-1) or (2E-2) equal to zero. Comparing the result with (2C-5) we can write

$$\omega_n = \sqrt{\frac{1}{\tau_a \tau_m}} \tag{2E-6}$$

$$\alpha = \frac{1}{2\tau_a} \tag{2E-7}$$

$$\zeta = \frac{\sqrt{\tau_a \tau_m}}{2\tau_a} \tag{2E-8}$$

The negative values of the roots of the characteristic equation may be written in the form given by (2C-7) or as

$$b_1, b_2 = \frac{1}{2\tau_a} \pm \sqrt{\frac{1}{4\tau_a^2} - \frac{1}{\tau_a \tau_m}} \tag{2E-9}$$

Equations (2E-4) and (2E-5) may now be written as

$$\Delta \omega_r(s) = \frac{-(K/J)(s + a_1)}{s(s + b_1)(s + b_2)} \tag{2E-10}$$

$$\Delta i_a(s) = \frac{K/k_v \tau_a \tau_m}{s(s + b_1)(s + b_2)} \tag{2E-11}$$

where

$$a_1 = \frac{1}{\tau_a} \tag{2E-12}$$

It is interesting to note that (2E-10) and (2E-11) are similar in form to (2D-14) and (2D-16), respectively. Because we have taken the inverse Laplace transform of these expressions in Example 2D, we shall make use of that work. First, however, let us do some preliminary calculations. For the 5-hp machine,

$$\tau_a = \frac{L_{AA}}{r_a} = \frac{0.012}{0.6} = 0.02 \text{ sec} \tag{2E-13}$$

$$\tau_m = \frac{Jr_a}{k_v^2} = \frac{Jr_a}{(L_{AF}i_f)^2}$$

$$= \frac{(1)(0.6)}{[(1.8)(240/240)]^2} = 0.185 \text{ sec} \tag{2E-14}$$

Note that in the above calculation, i_f is obtained by dividing the rated terminal voltage (240 V) by the field resistance (240 Ω). As in Example 2D, sec is used to denote seconds to avoid confusion with the Laplace operator. From (2E-6) through (2E-9),

$$\omega_n = \sqrt{\frac{1}{(0.02)(0.185)}} = 16.44 \text{ rad/sec} \tag{2E-15}$$

$$\alpha = \frac{1}{(2)(0.02)} = 25 \text{ sec}^{-1} \tag{2E-16}$$

$$\zeta = \frac{\sqrt{(0.02)(0.185)}}{(2)(0.02)} = 1.52 \tag{2E-17}$$

$$a_1 = \frac{1}{\tau_a} = \frac{1}{0.02} = 50 \text{ sec}^{-1} \tag{2E-18}$$

$$b_1, b_2 = \frac{1}{(2)(0.02)} \pm \sqrt{\frac{1}{4(0.02)^2} - \frac{1}{(0.02)(0.185)}}$$

$$= 6.17, \ 43.83 \text{ sec}^{-1} \tag{2E-19}$$

We can use (2D-27) and (2D-28) as guides to express i_a and ω_r, respectively. In particular,

$$\omega_r = \omega_r(0) - \frac{K}{J}\left[\frac{a_1}{b_1b_2} + \frac{a_1 - b_1}{b_1(b_1 - b_2)}e^{-b_1t} + \frac{a_1 - b_2}{b_2(b_2 - b_1)}e^{-b_2t}\right] \tag{2E-20}$$

$$i_a = i_a(0) + \frac{K}{k_v\tau_a\tau_m}\left[\frac{1}{b_1b_2} + \frac{1}{b_1(b_1 - b_2)}e^{-b_1t} + \frac{1}{b_2(b_2 - b_1)}e^{-b_2t}\right] \tag{2E-21}$$

We have previously determined that $\omega_r(0) = 127.7$ rad/sec and $i_a(0) = 16.2$ A. Thus, because b_1 and b_2 are real($\zeta > 1$), (2E-20) becomes

$$\omega_r = 127.7 + \frac{29.2}{(2)(1)}\left[\frac{50}{(6.17)(43.83)} + \frac{50 - 6.17}{6.17(6.17 - 43.83)}e^{-6.17t}\right.$$

$$\left. + \frac{50 - 43.83}{43.83(43.83 - 6.17)}e^{-43.83t}\right]$$

$$= 130.4 - 2.75e^{-t/0.162} + 0.05e^{-t/0.0228} \tag{2E-22}$$

The current, (2E-21), can be written as

$$i_a = 8.1 + 9.43e^{-t/0.162} - 1.33e^{-t/0.0228} \qquad (2E\text{-}23)$$

Note that the dynamic responses consist of two exponential terms; one time constant is approximately equal to τ_m and the other is approximately equal to τ_a. The response is dominated by the exponential term containing τ_m.

In the case of the 200-hp motor, we have

$$\tau_a = 0.029 \text{ sec} \qquad (2E\text{-}24)$$

$$\tau_m = 0.0256 \text{ sec} \qquad (2E\text{-}25)$$

$$\omega_n = 36.7 \text{ rad/sec} \qquad (2E\text{-}26)$$

$$\alpha = 17.2 \text{ sec}^{-1} \qquad (2E\text{-}27)$$

$$\zeta = 0.47 \qquad (2E\text{-}28)$$

$$a_1 = 34.5 \text{ sec}^{-1} \qquad (2E\text{-}29)$$

$$b_1, b_2 = 17.2 \pm j32.4 \text{ sec}^{-1} = \gamma \pm j\beta \qquad (2E\text{-}30)$$

Here, we see that b_1 and b_2 are a conjugate complex pair, hence (2D-29) and (2D-30) may be used as guides for i_a and ω_r, respectively. In particular,

$$i_a = i_a(0) + \frac{K}{k_v \tau_a \tau_m} \frac{1}{\gamma^2 + \beta^2} \left[1 - e^{-\gamma t} \left(\cos\beta t + \frac{\gamma}{\beta} \sin\beta t \right) \right] \qquad (2E\text{-}31)$$

$$\omega_r = \omega_r(0) - \frac{K}{J} \frac{a_1}{\gamma^2 + \beta^2} \left[1 - e^{-\gamma t} \left(\cos\beta t - \frac{\gamma^2 + \beta^2 - a_1\gamma}{a_1\beta} \sin\beta t \right) \right] \qquad (2E\text{-}32)$$

Substituting in the appropriate values yields

$$i_a = 633 - 317[1 - e^{-17.2t}(\cos 32.4t + 0.531\sin 32.4t)] \qquad (2E\text{-}33)$$

$$\omega_r = 64.64 + 1.02[1 - e^{-17.2t}(\cos 32.4t - 0.673\sin 32.4t)] \qquad (2E\text{-}34)$$

which can be expressed more compactly as

$$i_a = 316 + 359e^{-t/0.058}\cos(32.4t - 28°) \qquad (2E\text{-}35)$$

$$\omega_r = 65.65 - 1.22e^{-t/0.058}\cos(32.4t + 33.9°) \qquad (2E\text{-}36)$$

REFERENCES

[1] A. E. Fitzgerald, C. Kingsley, Jr., and S. D. Umans, *Electric Machinery*, 5th ed., McGraw-Hill New York, 1990.

[2] S. J. Chapman, *Electric Machinery Fundamentals*, 3rd ed., McGraw-Hill New York, 1999.

[3] G. McPherson and R. D. Laramore, *An Introduction to Electrical Machines and Transformers*, 2nd ed., John Wiley and Sons, New York, 1990.

[4] P. C. Krause, O. Wasynczuk, and S. D. Sudhoff, *Analysis of Electric Machinery*, IEEE Press, Piscataway, NJ, 1995.

[5] P. C. Sen, *Thyristor DC Drives*, John Wiley and Sons, New York, 1981.

[6] B. C. Kuo, *Automatic Control Systems*, Prentice-Hall Englewood Cliffs, NJ, 1987.

PROBLEMS

1 The parameters of a dc shunt machine are $R_f = 240\,\Omega$, $L_{FF} = 120\,\text{H}$, $L_{AF} = 1.8\,\text{H}$, $r_a = 0.6\,\Omega$, $L_{AA} = 0$. The load torque is 5 N·m and $V_a = V_f = 240\,\text{V}$. Calculate the steady-state rotor speed.

2 The power input to a dc shunt motor during rated-load conditions is 100 W. The rotor speed is 2000 r/min and the armature voltage is 100 V. The armature resistance is 2 Ω and $R_f = 200\,\Omega$. Calculate the no-load rotor speed.

3 A permanent-magnet dc motor has the following parameters: $r_a = 8\,\Omega$ and $k_v = 0.01\,\text{V}\cdot\text{s/rad}$. The shaft load torque is approximated as $T_L = K\omega_r$, where $K = 5 \times 10^{-6}\,\text{N}\cdot\text{m}\cdot\text{s}$. The applied voltage is 6 V and $B_m = 0$. Calculate the steady-state rotor speed ω_r in rad/s.

4 A 250-V 600-r/min 200-hp dc shunt motor is delivering rated horsepower at rated speed. $R_f = 12\,\Omega$, $L_{AF} = 0.18\,\text{H}$, and $r_a = 0.012\,\Omega$.
(*a*) Calculate the terminal voltage that must be applied to this machine to satisfy this load condition.
(*b*) Calculate the full-load ohmic losses and determine the efficiency.

5 Losses in a dc machine include ohmic, no-load rotational, and stray-load losses. The no-load rotational losses range from approximately 2% to 14% of the rated output. These losses include windage, friction, and core losses. Stray-load losses include (i) the increase in core losses due to load and (ii) eddy current losses induced by the armature current. These losses are generally taken to be 1% of the rating of the machine. Another loss in the machine is due to the brush voltage drop. This is accounted for by increasing the armature resistance or assuming a constant-voltage drop across the brushes regardless of load. Repeat Problem 4, assuming the machine has 5% rotational losses, 1% stray-load losses, and a total voltage drop across the brushes of 2 V that should be added to that calculated in Problem 4*a*. Assume that the rotational and stray-load losses can be represented by a resistance connected across the terminals of the machine.

6 The parameters of a dc shunt machine are $r_a = 10\,\Omega$, $R_f = 50\,\Omega$, and $L_{AF} = 0.5\,\text{H}$. Neglect B_m and $V_a = V_f = 25\,\text{V}$. Calculate (*a*) the steady-state stall torque, (*b*) the no-load speed, and (*c*) the steady-state rotor speed with $T_L = 3.75 \times 10^{-3}\omega_r$.

7 A permanent-magnet dc motor is driven by a mechanical source at 3820 r/min. The measured open-circuit armature voltage is 7 V. The mechanical source is disconnected, and a 12-V electric source is connected to the armature. With zero load torque, $I_a = 0.1$ A and $\omega_r = 650\,\text{rad/s}$. Calculate k_v, B_m, and r_a.

8 Express the maximum steady-state power output of a dc shunt motor ($P_{\text{out}} = T_e\omega_r$) if the field current i_f and armature voltage v_a are held constant. Let $B_m = 0$. (*Hint:* First express the rotor speed for maximum power output.)

9 The parameters of a 5-hp dc shunt machine are $r_a = 0.6\,\Omega$, $L_{AA} = 0.012\,\text{H}$, $R_f = 120\,\Omega$, $L_{FF} = 120\,\text{H}$, $L_{AF} = 1.8\,\text{H}$, and $V_a = V_f = 240\,\text{V}$. Calculate the steady-state rotor speed ω_r for $I_t = 0$.

10 A dc series motor requires 100 W at full load. The full-load speed is 2000 r/min and the terminal voltage is 100 V, $r_a = 2\,\Omega$, and $r_{fs} = 1\,\Omega$. Calculate the stall torque of the motor (T_e with $\omega_r = 0$).

11 The torque load of a dc series motor is $T_L = 100\,\text{N}\cdot\text{m}$, $L_{AFs} = 0.6\,\text{H}$, $r_a = 2\,\Omega$, and $r_{fs} = 3\,\Omega$. The voltage applied to the motor is 200 V. Calculate the steady-state rotor speed. Assume $B_m = 0$.

12 A dc compound motor requires 100 W at full load. The full-load speed is 2000 r/min and the voltage applied to the terminals of the machine is 100 V, $r_a = 2\,\Omega$, $R_f = 200\,\Omega$, and $r_{fs} = 1\,\Omega$. The series field is differentially connected. Calculate the steady-state no-load speed if $L_{AFs} = 0.1 L_{AF}$.

13 Modify the state equations given by (2.6-14) for a permanent-magnet dc machine to include θ_r as a state variable.

14 Write the field voltage equation of a shunt dc machine in terms of k_v, and express the transfer function between k_v and v_f.

15 A separately excited dc machine is operating with no load ($T_L = 0$) and fixed field current. The armature resistance and inductance, r_a and L_{AA}, are small and can be neglected. Assume $B_m = 0$. Express the transfer function between i_a and v_a. Show that this motor appears as a capacitor to v_a.

16 Write the state equations in fundamental form for the coupled circuits considered in Section 1.2. Start with (1.2-34) and (1.2-35) and use λ_1 and λ_2' as state variables. Relate currents and flux linkages by (1.2-31) and (1.2-32).

17 Develop the time-domain block diagram for the coupled circuits considered in Problem 16.

18 Develop the time-domain block diagram for a series-connected dc machine.

19 Develop the time-domain block diagram for the compound machine with a long-shunt connection.

20 We see from Figs. 2.5-2 and 2.5-4 that the permanent-magnet dc motor and the 200-hp dc shunt motor both demonstrated a slightly oscillatory speed response to a step change in load torque. However, from Fig. 2.5-3, we note that the rotor speed response of the 5-hp dc shunt motor appears to be slightly overdamped. Use information in Example 2C to support these observations.

21 Construct the block diagram in terms of the Laplace operator for the permanent-magnet dc motor valid for load torque changes with the armature applied voltage held constant. Write the transfer function $\Delta\omega_r(s)/\Delta T_L(s)$.

22 Formulate the following transfer functions for a permanent-magnet dc machine:

(a) $\Delta i_a(s)/\Delta T_L(s)$ with $\Delta v_a(s) = 0$

(b) $\Delta i_a(s)/\Delta v_a(s)$ with $\Delta T_L(s) = 0$

(c) $\Delta T_e(s)/\Delta v_a(s)$ with $\Delta T_L(s) = 0$

23 The field current of a dc shunt motor is held constant. Express the transfer function $\Delta i_a(s)/\Delta v_a(s)$ for no-load conditions [$\Delta T_L(s) = 0$]. Show that if $B_m = 0$, this transfer function is identical in form to that of a series rLC circuit. Express r, L, and C in terms of machine parameters.

24 Consider the 200-hp dc shunt motor described in Section 2.5. Assume that the machine is initially operating at rated conditions. The armature voltage is suddenly stepped from rated to 110% rated value. The load torque and field voltage remain constant and $B_m = 0$. Express i_a.

25 Consider the 200-hp dc shunt motor described in Section 2.5. The machine is operating with no-load and rated terminal voltage; however, the field and armature circuits are supplied from separate sources. The armature is disconnected from its source, and a 0.5 Ω resistor is immediately connected across the armature. Neglect r_a, L_{AA}, and B_m. Assume $J = 20\,\mathrm{kg} \cdot \mathrm{m}^2$. Express ω_r. This method of slowing a dc machine is referred to as *dynamic braking*.

Chapter 3

REFERENCE-FRAME THEORY

3.1 INTRODUCTION

The voltage equations that describe the performance of induction and synchronous machines were established in Chapter 1. We found that some of the machine inductances are functions of the rotor speed, whereupon the coefficients of the differential equations (voltage equations) that describe the behavior of these machines are time-varying except when the rotor is stalled. A change of variables is often used to reduce the complexity of these differential equations. There are several changes of variables that are used, and it was originally thought that each change of variables was different and therefore they were treated separately [1–4]. It was later learned that all changes of variables used to transform real variables are contained in one [5,6]. This general transformation refers machine variables to a frame of reference that rotates at an arbitrary angular velocity. All known real transformations are obtained from this transformation by simply assigning the speed of the rotation of the reference frame.

In this chapter this transformation is set forth and, because many of its properties can be studied without the complexities of the machine equations, it is applied to the equations that describe resistive, inductive, and capacitive circuit elements. By this approach, many of the basic concepts and interpretations of this general transformation are readily and concisely established. Extending the material presented in this chapter to the analysis of ac machines is straightforward involving a minimum of trigonometric manipulations.

3.2 BACKGROUND

In the late 1920s, R. H. Park [1] introduced a new approach to electric machine analysis. He formulated a change of variables which, in effect, replaced the variables

109

(voltages, currents, and flux linkages) associated with the stator windings of a synchronous machine with variables associated with fictitious windings rotating with the rotor. In other words, he transformed, or referred, the stator variables to a frame of reference fixed in the rotor. Park's transformation, which revolutionized electric machine analysis, has the unique property of eliminating all time-varying inductances from the voltage equations of the synchronous machine which occur due to (1) electric circuits in relative motion and (2) electric circuits with varying magnetic reluctance.

In the late 1930s, H. C. Stanley [2] employed a change of variables in the analysis of induction machines. He showed that the time-varying inductances in the voltage equations of an induction machine due to electric circuits in relative motion could be eliminated by transforming the variables associated with the rotor windings (rotor variables) to variables associated with fictitious stationary windings. In this case the rotor variables are transformed to a frame reference fixed in the stator.

G. Kron [3] introduced a change of variables that eliminated the position or time-varying mutual inductances of a symmetrical induction machine by transforming both the stator variables and the rotor variables to a reference frame rotating in synchronism with the rotating magnetic field. This reference frame is commonly referred to as the synchronously rotating reference frame.

D. S. Brereton et al. [4] employed a change of variables that also eliminated the time-varying inductances of a symmetrical induction machine by transforming the stator variables to a reference frame fixed in the rotor. This is essentially Park's transformation applied to induction machines.

Park, Stanley, Kron, and Brereton et al. developed changes of variables, each of which appeared to be uniquely suited for a particular application. Consequently, each transformation was derived and treated separately in literature until it was noted in 1965 [5] that all known real transformations used in induction machine analysis are contained in one general transformation that eliminates all time-varying inductances by referring the stator and the rotor variables to a frame of reference that may rotate at any angular velocity or remain stationary. All known real transformations may then be obtained by simply assigning the appropriate speed of rotation, which may in fact be zero, to this so-called *arbitrary reference frame*. It is interesting to note that this transformation is sometimes referred to as the "generalized rotating real transformation," which may be somewhat misleading because the reference frame need not rotate. In any event, we will refer to it as the arbitrary reference frame as did the originators [5]. Later, it was noted that the stator variables of a synchronous machine could also be referred to the arbitrary reference frame [6]. However, we will find that the time-varying inductances of a synchronous machine are eliminated only if the reference frame is fixed in the rotor (Park's transformation); consequently the arbitrary reference frame does not offer the advantages in the analysis of the synchronous machines that it does in the case of induction machines.

3.3 EQUATIONS OF TRANSFORMATION: CHANGE OF VARIABLES

Although changes of variables are used in the analysis of ac machines to eliminate time-varying inductances, changes of variables are also employed in the analysis of various static, constant-parameter power-system components and control systems associated with electric drives. For example, in many of the computer programs used for transient and dynamic stability studies of large power systems, the variables of all power-system components except for the synchronous machines are represented in a reference frame rotating at synchronous speed. Hence, the variables associated with the transformers, transmission lines, loads, capacitor banks, and static var units, for example, must be transformed to the synchronous rotating reference frame by a change of variables. Similarly, the "average value" of the variables associated with the conversion process in electric drive systems and in high-voltage ac-dc systems are often expressed in the synchronously rotating reference frame.

Fortunately, all known real transformations for these components and controls are also contained in the transformation to the arbitrary reference frame, the same general transformation used for the stator variables of the induction and synchronous machines and for the rotor variables of induction machines. Although we could formulate one transformation to the arbitrary reference frame that could be applied to all variables, it is preferable to consider only the variables associated with stationary circuits in this chapter and then modify this analysis for the variables associated with the rotor windings of the induction machine at the time it is analyzed.

A change of variables that formulates a transformation of the 3-phase variables of stationary circuit elements to the arbitrary reference frame may by expressed as

$$\mathbf{f}_{qd0s} = \mathbf{K}_s \mathbf{f}_{abcs} \tag{3.3-1}$$

where

$$(\mathbf{f}_{qd0s})^T = [f_{qs} \quad f_{ds} \quad f_{0s}] \tag{3.3-2}$$

$$(\mathbf{f}_{abcs})^T = [f_{as} \quad f_{bs} \quad f_{cs}] \tag{3.3-3}$$

$$\mathbf{K}_s = \frac{2}{3} \begin{bmatrix} \cos\theta & \cos\left(\theta - \frac{2\pi}{3}\right) & \cos\left(\theta + \frac{2\pi}{3}\right) \\ \sin\theta & \sin\left(\theta - \frac{2\pi}{3}\right) & \sin\left(\theta + \frac{2\pi}{3}\right) \\ \frac{1}{2} & \frac{1}{2} & \frac{1}{2} \end{bmatrix} \tag{3.3-4}$$

$$\omega = \frac{d\theta}{dt} \tag{3.3-5}$$

It can be shown that for the inverse transformation we have

$$(\mathbf{K}_s)^{-1} = \begin{bmatrix} \cos\theta & \sin\theta & 1 \\ \cos\left(\theta - \frac{2\pi}{3}\right) & \sin\left(\theta - \frac{2\pi}{3}\right) & 1 \\ \cos\left(\theta + \frac{2\pi}{3}\right) & \sin\left(\theta + \frac{2\pi}{3}\right) & 1 \end{bmatrix} \tag{3.3-6}$$

The angular velocity, ω, and the angular displacement, θ, of the arbitrary reference frame are related by (3.3-5). Thus

$$\theta = \int \omega \, dt \qquad (3.3\text{-}7)$$

or in definite integral form

$$\theta = \int_0^t \omega(\xi) \, d\xi + \theta(0) \qquad (3.3\text{-}8)$$

where ξ is a dummy variable of integration. We will use one of the above relationships between θ and ω depending upon which is most convenient for analysis purposes.

In the above equations, f can represent either voltage, current, flux linkage, or electric charge. The superscript T denotes the transpose of a matrix. The s subscript indicates the variables, parameters, and transformation associated with stationary circuits. The angular displacement θ must be continuous; however, the angular velocity associated with the change of variables is unspecified. The frame of reference may rotate at any constant or varying angular velocity or it may remain stationary. The connotation of arbitrary stems from the fact that the angular velocity of the transformation is unspecified and can be selected arbitrarily to expedite the solution of the system equations or to satisfy the system constraints. The change of variables may be applied to variables of any waveform and time sequence; however, we will find that the transformation given above is particularly appropriate for an *abc* sequence.

Although the transformation to the arbitrary reference frame is a change of variables and needs no physical connotation, it is often convenient to visualize the transformation equations as trigonometric relationships between variables as shown in Fig. 3.3-1. In particular, the equations of transformation may be thought of as if the f_{qs} and f_{ds} variables are "directed" along paths orthogonal to each other and rotating at an angular velocity of ω, whereupon f_{as}, f_{bs}, and f_{cs} may be considered as variables directed along stationary paths, each displaced by $120°$. If f_{as}, f_{bs}, and f_{cs} are resolved into f_{qs}, the first row of (3.3-1) is obtained; and if f_{as}, f_{bs}, and f_{cs} are resolved into f_{ds}, the second row is obtained. It is important to note that f_{0s} variables are not associated with the arbitrary reference frame. Instead, the zero variables are related arithmetically to the *abc* variables, independent of θ. It is also important not to confuse f_{as}, f_{bs}, and f_{cs} with phasors. They are instantaneous quantities that may be any function of time. Portraying the transformation as shown in Fig. 3.3-1 is particularly convenient when applying it to ac machines where the direction of f_{as}, f_{bs}, and f_{cs} may also be thought of as the direction of the magnetic axes of the stator windings. We will find that the direction of f_{qs} and f_{ds} can be considered as the direction of the magnetic axes of the "new" windings created by the change of variables.

The total instantaneous power may be expressed in *abc* variables as

$$P_{abcs} = v_{as} i_{as} + v_{bs} i_{bs} + v_{cs} i_{cs} \qquad (3.3\text{-}9)$$

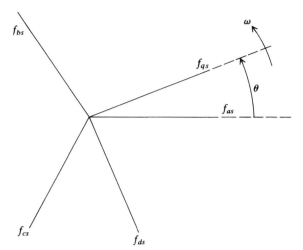

Figure 3.3-1 Transformation for stationary circuits portrayed by trigonometric relationships.

The total power expressed in the $qd0$ variables must equal the total power expressed in the abc variables; hence substituting (3.3-1) into (3.3-9) yields

$$P_{qd0s} = P_{abcs} = \frac{3}{2}(v_{qs}i_{qs} + v_{ds}i_{ds} + 2v_{0s}i_{0s}) \qquad (3.3\text{-}10)$$

The 3/2 factor comes about due to the choice of the constant used in the transformation. Although the waveforms of the qs and ds voltages, currents, flux linkages, and electric charges are dependent upon the angular velocity of the frame of reference, the waveform of total power is independent of the frame of reference. In other words, the waveform of the total power is the same regardless of the reference frame in which it is evaluated.

Example 3A   For the purpose of becoming familiar with this transformation let

$$f_{as} = \cos t \qquad f_{bs} = \frac{1}{2}t \qquad f_{cs} = -\sin t$$

Let us first determine expressions for f_{qs}, f_{ds}, and f_{0s}. From (3.3-1) we obtain

$$f_{qs} = \frac{2}{3}\left[\cos t \cos\theta + \frac{1}{2}t\cos\left(\theta - \frac{2\pi}{3}\right) - \sin t \cos\left(\theta + \frac{2\pi}{3}\right)\right] \qquad (3A\text{-}1)$$

$$f_{ds} = \frac{2}{3}\left[\cos t \sin\theta + \frac{1}{2}t\sin\left(\theta - \frac{2\pi}{3}\right) - \sin t \sin\left(\theta + \frac{2\pi}{3}\right)\right] \qquad (3A\text{-}2)$$

$$f_{0s} = \frac{1}{3}\left(\cos t + \frac{1}{2}t - \sin t\right) \qquad (3A\text{-}3)$$

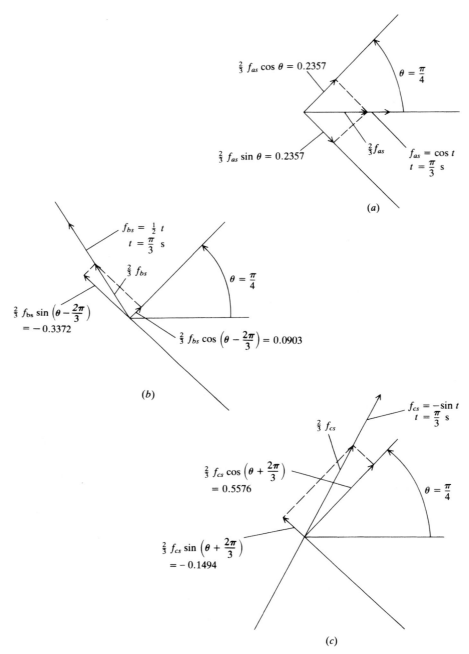

Figure 3A-1   Resolution of f_{as}, f_{bs}, and f_{cs} into the directions of f_{qs} and f_{ds} for $t = \pi/3$s and $\theta = \pi/4$ rad. (a) Resolution of f_{as}; (b) resolution of f_{bs}; (c) resolution of f_{cs}.

Now assume that $\theta(0) = -\pi/12$ and $\omega = 1$ rad/s. Let us evaluate (3A-1)–(3A-3) for $t = \pi/3$s and show the trigonometric relationship between the $f_{as}, f_{bs}, f_{cs}, f_{qs},$ and f_{ds} variables for this condition. From (3.3-8) we have

$$\theta_{(t=\pi/3)} = \frac{\pi}{3} - \frac{\pi}{12} = \frac{\pi}{4} \tag{3A-4}$$

Appropriate substitution into (3A-1)–(3A-3) yields

$$f_{qs} = \frac{2}{3}\left[\cos\frac{\pi}{3}\cos\frac{\pi}{4} + \left(\frac{1}{2}\right)\frac{\pi}{3}\cos\left(\frac{\pi}{4} - \frac{2\pi}{3}\right) - \sin\frac{\pi}{3}\cos\left(\frac{\pi}{4} + \frac{2\pi}{3}\right)\right]$$

$$= \frac{2}{3}\left[\left(\frac{1}{2}\right)\left(\frac{1}{\sqrt{2}}\right) + \left(\frac{1}{2}\right)\left(\frac{\pi}{3}\right)(0.2588) - \left(\frac{\sqrt{3}}{2}\right)(-0.9659)\right]$$

$$= 0.8836 \tag{3A-5}$$

$$f_{ds} = \frac{2}{3}\left[\cos\frac{\pi}{3}\sin\frac{\pi}{4} + \left(\frac{1}{2}\right)\frac{\pi}{3}\sin\left(\frac{\pi}{4} - \frac{2\pi}{3}\right) - \sin\frac{\pi}{3}\sin\left(\frac{\pi}{4} + \frac{2\pi}{3}\right)\right]$$

$$= \frac{2}{3}\left[\left(\frac{1}{2}\right)\left(\frac{1}{\sqrt{2}}\right) + \left(\frac{1}{2}\right)\left(\frac{\pi}{3}\right)(-0.9659) - \left(\frac{\sqrt{3}}{2}\right)(0.2588)\right]$$

$$= -0.2509 \tag{3A-6}$$

$$f_{0s} = \frac{1}{3}\left[\cos\frac{\pi}{3} + \left(\frac{1}{2}\right)\left(\frac{\pi}{3}\right) - \sin\frac{\pi}{3}\right]$$

$$= 0.0525 \tag{3A-7}$$

The trigonometric relationships between variables at $t = \pi/3$s and $\theta = \pi/4$ rad are shown in Figs. 3A-1 and 3A-2. In Fig. 3A-1a the resolution of f_{qs} and f_{ds} is depicted. The resulting component in the direction of f_{qs} is the first term of (3A-5). Likewise, the component in the direction of f_{ds} is the first term of (3A-6). Note that Figs. 3A-1b and 3A-1c the component in the f_{ds} direction is negative. The composite of all components is shown in Fig. 3A-2.

3.4 STATIONARY CIRCUIT VARIABLES TRANSFORMED TO THE ARBITRARY REFERENCE FRAME

It is convenient to treat resistive, inductive, and capacitive circuit elements separately.

Resistive Elements

For a 3-phase resistive circuit we have

$$\mathbf{v}_{abcs} = \mathbf{r}_s\mathbf{i}_{abcs} \tag{3.4-1}$$

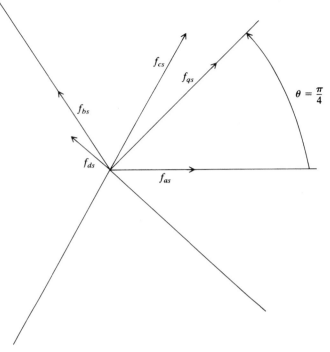

Figure 3A-2   Composite of Fig. 3A-1.

From (3.3-1) we obtain

$$\mathbf{v}_{qd0s} = \mathbf{K}_s \mathbf{r}_s (\mathbf{K}_s)^{-1} \mathbf{i}_{qd0s} \qquad (3.4\text{-}2)$$

It is necessary to specify the resistance matrix $\mathbf{r}_s$ before proceeding. All stator phase windings of either a synchronous or a symmetrical induction machine are designed to have the same resistance. Similarly, transformers, capacitor banks, transmission lines, and, in fact, all power-system components are designed so that all phases have equal resistances. Even power-system loads are distributed between phases so that all phases are loaded nearly equal. If the nonzero elements of the diagonal matrix $\mathbf{r}_s$ are equal, then

$$\mathbf{K}_s \mathbf{r}_s (\mathbf{K}_s)^{-1} = \mathbf{r}_s \qquad (3.4\text{-}3)$$

Thus, the resistance matrix associated with the arbitrary reference variables is equal to the resistance matrix associated with the actual variables if each phase of the actual circuit has the same resistance. If the phase resistances are unequal (unbalanced or unsymmetrical), then the resistance matrix associated with the arbitrary reference-frame variables contains sinusoidal functions of θ except when $\omega = 0$, whereupon $\mathbf{K}_s$ is algebraic. In other words, if the phase resistances are unbalanced, the transformation yields constant resistances only if the reference frame is fixed where the unbalance physically exists. This feature is quite easily illustrated by substituting $\mathbf{r}_s = \mathrm{diag}\,[r_{as}\ r_{bs}\ r_{cs}]$ into $\mathbf{K}_s \mathbf{r}_s (\mathbf{K}_s)^{-1}$.

Inductive Elements

For a 3-phase inductive circuit we have

$$\mathbf{v}_{abcs} = p\boldsymbol{\lambda}_{abcs} \tag{3.4-4}$$

where p is the operator d/dt. In the case of the magnetically linear system it has been customary to express the flux linkages as a product of inductance and current matrices before performing a change of variables. However, the transformation is valid for flux linkages, and an extensive amount of work can be avoided by transforming the flux linkages directly. This is especially true in the analysis of ac machines where the inductance matrix is a function of rotor position. Thus, in terms of the substitute variables, (3.4-4) becomes

$$\mathbf{v}_{qd0s} = \mathbf{K}_s p[(\mathbf{K}_s)^{-1}\boldsymbol{\lambda}_{qd0s}] \tag{3.4-5}$$

which can be written as

$$\mathbf{v}_{qd0s} = \mathbf{K}_s p[(\mathbf{K}_s)^{-1}\boldsymbol{\lambda}_{qd0s}] + \mathbf{K}_s(\mathbf{K}_s)^{-1} p\boldsymbol{\lambda}_{qd0s} \tag{3.4-6}$$

It is easy to show that

$$p[(\mathbf{K}_s)^{-1}] = \omega \begin{bmatrix} -\sin\theta & \cos\theta & 0 \\ -\sin\left(\theta - \frac{2\pi}{3}\right) & \cos\left(\theta - \frac{2\pi}{3}\right) & 0 \\ -\sin\left(\theta + \frac{2\pi}{3}\right) & \cos\left(\theta + \frac{2\pi}{3}\right) & 0 \end{bmatrix} \tag{3.4-7}$$

Therefore,

$$\mathbf{K}_s p[(\mathbf{K}_s)^{-1}] = \omega \begin{bmatrix} 0 & 1 & 0 \\ -1 & 0 & 0 \\ 0 & 0 & 0 \end{bmatrix} \tag{3.4-8}$$

Trigonometric identities given in Appendix A are helpful in obtaining (3.4-8). Equation (3.4-6) may now be expressed as

$$\mathbf{v}_{qd0s} = \omega\boldsymbol{\lambda}_{dqs} + p\boldsymbol{\lambda}_{qd0s} \tag{3.4-9}$$

where

$$(\boldsymbol{\lambda}_{dqs})^T = [\,\lambda_{ds}\ \ -\lambda_{qs}\ \ 0\,] \tag{3.4-10}$$

Equation (3.4-6) is often written in expanded form as

$$v_{qs} = \omega\lambda_{ds} + p\lambda_{qs} \tag{3.4-11}$$

$$v_{ds} = -\omega\lambda_{qs} + p\lambda_{ds} \tag{3.4-12}$$

$$v_{0s} = p\lambda_{0s} \tag{3.4-13}$$

The first term on the right side of (3.4-11) or (3.4-12) is referred to as a "speed voltage," with the speed being the angular velocity of the arbitrary reference frame. It is clear that the speed voltage terms are zero if ω is zero, which, of course, is when the reference frame is stationary. Clearly, the voltage equations for the 3-phase inductive circuit become the familiar time rate of change of flux linkages if the reference frame is fixed where the circuit physically exists. Also, because (3.4-4) is valid in general, it follows that (3.4-11)–(3.4-13) are valid regardless if the system is magnetically linear or nonlinear and regardless of the form of the inductance matrix if the system is magnetically linear.

For a linear magnetic system, the flux linkages may be expressed as

$$\lambda_{abcs} = \mathbf{L}_s\mathbf{i}_{abcs} \tag{3.4-14}$$

whereupon the flux linkages in the arbitrary reference frame may be written as

$$\lambda_{qd0s} = \mathbf{K}_s\mathbf{L}_s(\mathbf{K}_s)^{-1}\mathbf{i}_{qd0s} \tag{3.4-15}$$

As is the case of the resistive circuit, it is necessary to specify the inductance matrix before proceeding with the evaluation of (3.4-15). However, once the inductance matrix is specified, the procedure for expressing any 3-phase inductive circuit in the arbitrary reference frame reduces to one of evaluating (3.4-15) and substituting the resulting λ_{qs}, λ_{ds}, and λ_{0s} into the voltage equations (3.4-11)–(3.4-13). This procedure is straightforward with a minimum of matrix manipulations compared to the work involved if, for a linear system, the flux linkage matrix λ_{abcs} is replaced by $\mathbf{L}_s\mathbf{i}_{abcs}$ before performing the transformation.

If, for example, $\mathbf{L}_s$ is a diagonal matrix with all nonzero terms equal, then

$$\mathbf{K}_s\mathbf{L}_s(\mathbf{K}_s)^{-1} = \mathbf{L}_s \tag{3.4-16}$$

A matrix of this form could describe the inductance of a balanced 3-phase inductive load, a 3-phase set of line reactors used in high-voltage transmission systems or any symmetrical 3-phase inductive network without inductive coupling between phases. It is clear that the comments regarding unbalanced or unsymmetrical phase resistances also apply in the case of unsymmetrical inductances.

An inductance matrix that is common to synchronous and induction machines is of the form

$$\mathbf{L}_s = \begin{bmatrix} L_{ls} + L_{ms} & -\frac{1}{2}L_{ms} & -\frac{1}{2}L_{ms} \\ -\frac{1}{2}L_{ms} & L_{ls} + L_{ms} & -\frac{1}{2}L_{ms} \\ -\frac{1}{2}L_{ms} & -\frac{1}{2}L_{ms} & L_{ls} + L_{ms} \end{bmatrix} \tag{3.4-17}$$

where L_{ls} is a leakage inductance and L_{ms} is a magnetizing inductance. From our work in Chapter 1 we realize that this inductance matrix describes the self- and mutual inductance relationships of the stator phases of a symmetrical induction machine and the stator phases of a round rotor synchronous machine. It is left to the reader to show that for $\mathbf{L}_s$ given by (3.4-17) we have

$$
\mathbf{K}_s \mathbf{L}_s (\mathbf{K}_s)^{-1} = \begin{bmatrix} L_{ls} + \frac{3}{2} L_{ms} & 0 & 0 \\ 0 & L_{ls} + \frac{3}{2} L_{ms} & 0 \\ 0 & 0 & L_{ls} \end{bmatrix} \tag{3.4-18}
$$

Linear 3-phase coupled systems are magnetically symmetrical if the diagonal elements are equal and all off-diagonal elements of the inductance matrix are also equal. Equation (3.4-17) is of this form. We see from (3.4-18) that, for a symmetrical system, $\mathbf{K}_s \mathbf{L}_s (\mathbf{K}_s)^{-1}$ yields a diagonal matrix that, in effect, magnetically uncouples the substitute variables in all reference frames. On the other hand, we have seen in Chapter 1 that the self- and mutual inductances between the stator phases of the salient-pole synchronous machine form a magnetically unsymmetrical system. It will be shown that in this case there is only one reference frame wherein the substitute variables are not magnetically coupled.

Capacitive Elements

For a 3-phase capacitive circuit we have

$$
\mathbf{i}_{abcs} = p\mathbf{q}_{abcs} \tag{3.4-19}
$$

Incorporating the substitute variables yields

$$
\mathbf{i}_{qd0s} = \mathbf{K}_s p[(\mathbf{K}_s)^{-1} \mathbf{q}_{qd0s}] \tag{3.4-20}
$$

which can be written as

$$
\mathbf{i}_{qd0s} = \mathbf{K}_s p[(\mathbf{K}_s)^{-1}] \mathbf{q}_{qd0s} + \mathbf{K}_s (\mathbf{K}_s)^{-1} p\mathbf{q}_{qd0s} \tag{3.4-21}
$$

Utilizing (3.4-8) yields

$$
\mathbf{i}_{qd0s} = \omega \mathbf{q}_{dqs} + p\mathbf{q}_{qd0s} \tag{3.4-22}
$$

where

$$
(\mathbf{q}_{dqs})^T = [\, q_{ds} \quad -q_{qs} \quad 0 \,] \tag{3.4-23}
$$

In expanded form we have

$$
i_{qs} = \omega q_{ds} + pq_{qs} \tag{3.4-24}
$$
$$
i_{ds} = -\omega q_{qs} + pq_{ds} \tag{3.4-25}
$$
$$
i_{0s} = pq_{0s} \tag{3.4-26}
$$

Considering the terminology of "speed voltages" as used in the case of inductive circuits, it would seem appropriate to refer to the first term on the right side of either (3.4-24) or (3.4-25) as "speed currents." Also, as in the case of inductive circuits, the equations revert to the familiar form in the stationary reference frame ($\omega = 0$).

Equations (3.4-24)–(3.4-26) are valid regardless of the relationship between charge and voltage. For a linear capacitive system we have

$$\mathbf{q}_{abcs} = \mathbf{C}_s \mathbf{v}_{abcs} \tag{3.4-27}$$

Thus, in the arbitrary reference frame we have

$$\mathbf{q}_{qd0s} = \mathbf{K}_s \mathbf{C}_s (\mathbf{K}_s)^{-1} \mathbf{v}_{qd0s} \tag{3.4-28}$$

Once the capacitance matrix is specified, q_{qs}, q_{ds}, and q_{0s} can be determined and substituted into (3.4-24)–(3.4-26). The procedure and limitations are analogous to those in the case of the inductive circuits. A diagonal capacitance matrix with equal non-zero elements describes, for example, (i) a 3-phase capacitor bank used for power factor correction and (ii) the series capacitance used for transmission line compensation or any 3-phase electrostatic system without coupling between phases. A 3-phase transmission system is often approximated as a symmetrical system, whereupon the inductance and capacitance matrices may be written in a form similar to (3.4-17).

Example 3B   For the purpose of demonstrating the transformation of variables to the arbitrary reference frame, let us consider a 3-phase RL circuit defined by

$$\mathbf{r}_s = \mathrm{diag}\begin{bmatrix} r_s & r_s & r_s \end{bmatrix} \tag{3B-1}$$

$$\mathbf{L}_s = \begin{bmatrix} L_s & M & M \\ M & L_s & M \\ M & M & L_s \end{bmatrix} \tag{3B-2}$$

Two circuit configurations are shown in Fig. 3B-1. The circuit diagram shown in Fig. 3B-1a portrays the example resistance and inductance matrices as the series elements of a section of a symmetrical, 3-phase transmission line. The diagram in Fig. 3B-1b shows the circuit elements connected in a wye arrangement. If $L_s = L_{ls} + L_{ms}$ and $M = -\frac{1}{2}L_{ms}$, then the resistance and inductance matrices depict the stator phases of a symmetrical induction machine or a round-rotor synchronous machine. It must be pointed out, however, that this inductance matrix does not describe the complete magnetic system of a machine because the rotor windings are not considered. Nevertheless, this RL circuit can be used to advantage to illustrate the implementation of the change of variables.

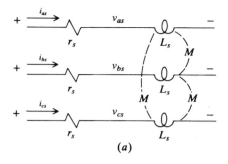

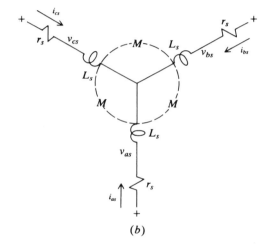

Figure 3B-1 Three-phase RL circuit. (*a*) Series elements of a symmetrical transmission line; (*b*) Stator windings of a symmetrical induction or round-rotor synchronous machine.

Each phase voltage can be expressed as the sum of the voltages across each element:

$$v_{as} = v_{asR} + v_{asL} \qquad (3B\text{-}3)$$

$$v_{bs} = v_{bsR} + v_{bsL} \qquad (3B\text{-}4)$$

$$v_{cs} = v_{csR} + v_{csL} \qquad (3B\text{-}5)$$

Similarly, in the arbitrary reference we have

$$v_{qs} = v_{qsR} + v_{qsL} \qquad (3B\text{-}6)$$

$$v_{ds} = v_{dsR} + v_{dsL} \qquad (3B\text{-}7)$$

$$v_{0s} = v_{0sR} + v_{0sL} \qquad (3B\text{-}8)$$

From (3.4-2) and (3.4-3) we obtain

$$v_{qsR} = r_s i_{qs} \tag{3B-9}$$

$$v_{dsR} = r_s i_{ds} \tag{3B-10}$$

$$v_{0sR} = r_s i_{0s} \tag{3B-11}$$

Equations (3.4-11)–(3.4-13) express the voltage equations for the inductive elements in the arbitrary reference. Because the example inductance matrix is in the same form as (3.4-17), we can use (3.4-18) as a guide to evaluate $\mathbf{K}_s \mathbf{L}_s (\mathbf{K}_s)^{-1}$. Thus

$$\mathbf{K}_s \mathbf{L}_s (\mathbf{K}_s)^{-1} = \begin{bmatrix} L_s - M & 0 & 0 \\ 0 & L_s - M & 0 \\ 0 & 0 & L_s + 2M \end{bmatrix} \tag{3B-12}$$

Therefore the voltage equations in the arbitrary reference frame may be expressed as

$$v_{qs} = r_s i_{qs} + \omega\lambda_{ds} + p\lambda_{qs} \tag{3B-13}$$

$$v_{ds} = r_s i_{ds} - \omega\lambda_{qs} + p\lambda_{ds} \tag{3B-14}$$

$$v_{0s} = r_s i_{0s} + p\lambda_{0s} \tag{3B-15}$$

where

$$\lambda_{qs} = (L_s - M)i_{qs} \tag{3B-16}$$

$$\lambda_{ds} = (L_s - M)i_{ds} \tag{3B-17}$$

$$\lambda_{0s} = (L_s - 2M)i_{0s} \tag{3B-18}$$

The equivalent circuit which portrays (3B-13)–(3B-18) is given in Fig. 3B-2.

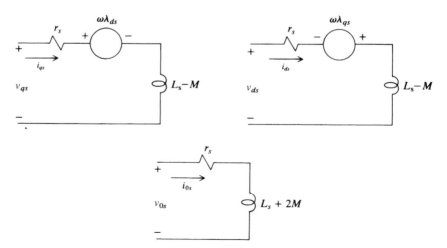

Figure 3B-2 Arbitrary reference-frame equivalent circuits for 3-phase RL circuit shown in Fig. 3B-1.

In this chapter we have chosen to introduce the transformation to the arbitrary reference frame by considering only stationary circuits. The complexities of the time-varying inductances are purposely omitted. Although the transformation diagonalizes the inductance matrix and thus uncouples the phases, one cannot see the advantages of transforming to any reference frame other than the stationary reference frame because it tends to complicate the voltage equations of the static circuits. In other words, the above voltage equations are most easily solved with $\omega = 0$. However, our purpose is to set forth the basic concepts and the interpretations of this general transformation; its advantages in machine analysis will be demonstrated in later chapters.

3.5 COMMONLY USED REFERENCE FRAMES

It is instructive to take a preliminary look at the reference frames commonly used in the analysis of electric machines and power system components—namely, the arbitrary, stationary, rotor, and synchronous reference frames. Information regarding each of these reference frames as applied to stationary circuits is given in the following table.

Reference Frame Speed	Interpretation	Notation	
		Variables	Transformation
ω (unspecified)	Stationary circuit variables referred to the arbitrary reference frame	$\mathbf{f}_{qd0s}$ or f_{qs}, f_{ds}, f_{0s}	$\mathbf{K}_s$
0	Stationary circuit variables referred to the stationary reference frame	$\mathbf{f}_{qd0s}^s$ or $f_{qs}^s, f_{ds}^s, f_{0s}$	$\mathbf{K}_s^s$
ω_r	Stationary circuit variables referred to a reference frame fixed in the rotor	$\mathbf{f}_{qd0s}^r$ or $f_{qs}^r, f_{ds}^r, f_{0s}$	$\mathbf{K}_s^r$
ω_e	Stationary circuit variables referred to the synchronously rotating reference frame	$\mathbf{f}_{qd0s}^e$ or $f_{qs}^e, f_{ds}^e, f_{0s}$	$\mathbf{K}_s^e$

For purposes at hand it is sufficient for us to define the synchronously rotating reference frame or the synchronous reference frame as the reference frame rotating at the electrical angular velocity corresponding to the fundamental frequency of the variables associated with stationary circuits, herein denoted as ω_e. In the case of ac machines, ω_e is the electrical angular velocity of the air-gap rotating magnetic field established by stator currents of fundamental frequency.

The notation requires some explanation. We have previously established that the s subscript denotes variables and transformations associated with circuits that are stationary in "real life" as opposed to rotor circuits that are free to rotate. Later we will use subscript r to denote variables and the transformation associated with rotor circuits. The raised index denotes the qs and ds variables and transformation associated with a specific reference frame except in the case of the arbitrary reference frame that carries no raised index. Because the $0s$ variables are independent of ω and therefore not associated with a reference frame, a raised index is not assigned to f_{0s}. The transformation of variables associated with stationary circuits to a stationary reference frame was developed by E. Clarke [7], who used the notation f_α, f_β, and f_0 rather than f_{qs}^s, f_{ds}^s, and f_{0s}. In Park's transformation to the rotor reference frame, he denoted the variables f_q, f_d, and f_0 rather than f_{qs}^r, f_{ds}^r, and f_{0s}. There appears to be no established notation for the variables in the synchronously rotating reference frame.

As mentioned previously, the voltage equations for all reference frames may be obtained from those in the arbitrary reference frame. The voltage equations for inductive or capacitive elements are obtained by assigning the appropriate value to ω in the speed voltages or speed currents. The transformation for a specific reference frame is obtained by substituting the appropriate reference-frame speed for ω into (3.3-8) to obtain the angular displacement. In most cases the initial or time-zero displacement can be selected equal to zero; however, there are situations where the initial displacement of the reference frame to which the variables are being transformed will not be zero.

3.6 TRANSFORMATION BETWEEN REFERENCE FRAMES

In some derivations and analyses it is convenient to relate variables in one reference frame to variables in another reference frame directly, without involving the abc variables in the transformation. In order to establish this transformation between any two frames of reference, let x denote the reference frame from which the variables are being transformed and let y denote the reference frame to which the variables are being transformed; then

$$\mathbf{f}_{qd0s}^y = {}^x\mathbf{K}^y\mathbf{f}_{qd0s}^x \tag{3.6-1}$$

From (3.3-1) we obtain

$$\mathbf{f}_{qd0s}^x = \mathbf{K}_s^x\mathbf{f}_{abcs} \tag{3.6-2}$$

Substituting (3.6-2) into (3.6-1) yields

$$\mathbf{f}_{qd0s}^y = {}^x\mathbf{K}^y\mathbf{K}_s^x\mathbf{f}_{abcs} \tag{3.6-3}$$

However, from (3.3-1) we obtain

$$\mathbf{f}_{qd0s}^y = \mathbf{K}_s^y\mathbf{f}_{abcs} \tag{3.6-4}$$

Thus

$$^x\mathbf{K}^y\mathbf{K}_s^x = \mathbf{K}_s^y \tag{3.6-5}$$

from which

$$^x\mathbf{K}^y = \mathbf{K}_s^y(\mathbf{K}_s^x)^{-1} \tag{3.6-6}$$

The desired transformation is obtained by substituting the appropriate transformations into (3.6-6). Hence

$$^x\mathbf{K}^y = \begin{bmatrix} \cos(\theta_y - \theta_x) & -\sin(\theta_y - \theta_x) & 0 \\ \sin(\theta_y - \theta_x) & \cos(\theta_y - \theta_x) & 0 \\ 0 & 0 & 1 \end{bmatrix} \tag{3.6-7}$$

Several of the trigonometric identities given in Appendix A are useful in obtaining (3.6-7). This transformation, which is sometimes referred to as a "vector rotator" or simply "rotator," can also be visualized from the trigonometric relationship between two sets of rotating, orthogonal quantities as shown in Fig. 3.6-1. Resolving f_{qs}^x and f_{ds}^x into f_{qs}^y yields the first row of (3.6-7), and resolving f_{qs}^x and f_{ds}^x into f_{ds}^y yields the second row. It is left for the reader to show that

$$(^x\mathbf{K}^y)^{-1} = (^x\mathbf{K}^y)^T \tag{3.6-8}$$

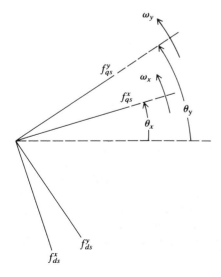

Figure 3.6-1 Transformation between two reference frames portrayed by trigonometric relationships.

3.7 TRANSFORMATION OF A BALANCED SET

Although the transformation equations are valid regardless of the waveform of the variables, it is instructive to consider the characteristics of the transformation when the 3-phase system is symmetrical and the voltages and currents form a balanced 3-phase set of *abc* sequence as given by (3.7-1)–(3.7-4). A balanced 3-phase set is generally defined as a set of equal-amplitude sinusoidal quantities that are displaced by 120°. Because the sum of this set is zero, the 0*s* variables are zero.

$$f_{as} = \sqrt{2}f_s \cos \theta_{ef} \tag{3.7-1}$$

$$f_{bs} = \sqrt{2}f_s \cos \left(\theta_{ef} - \frac{2\pi}{3} \right) \tag{3.7-2}$$

$$f_{cs} = \sqrt{2}f_s \cos \left(\theta_{ef} + \frac{2\pi}{3} \right) \tag{3.7-3}$$

where f_s may be a function of time and

$$\omega_e = \frac{d\theta_{ef}}{dt} \tag{3.7-4}$$

Substituting (3.7-1)–(3.7-3) into the transformation to the arbitrary reference frame (3.3-1) yields

$$f_{qs} = \sqrt{2}f_s \cos (\theta_{ef} - \theta) \tag{3.7-5}$$
$$f_{ds} = -\sqrt{2}f_s \sin (\theta_{ef} - \theta) \tag{3.7-6}$$
$$f_{0s} = 0 \tag{3.7-7}$$

With the 3-phase variables as given in (3.7-1)–(3.7-3), the *qs* and *ds* variables form a balanced 2-phase set in all reference frames except when $\omega = \omega_e$. In this, the synchronously rotating reference frame, the *qs* and *ds* quantities become

$$f_{qs}^e = \sqrt{2}f_s \cos (\theta_{ef} - \theta_e) \tag{3.7-8}$$
$$f_{ds}^e = -\sqrt{2}f_s \sin (\theta_{ef} - \theta_e) \tag{3.7-9}$$

Recall that θ_e is the angular position of the synchronously rotating reference frame and θ_e and θ_{ef} both have an angular velocity of ω_e. Hence, $\theta_{ef} - \theta_e$ is a constant depending upon the initial values of the variable being transformed and the initial position of the synchronously rotating reference frame. Equations (3.7-8) and (3.7-9) reveal a property that is noteworthy. There is one reference frame where a constant amplitude balanced set will appear as constants. In other words, if a constant amplitude balanced set appears in any reference frame, then there is another reference frame where this balanced set appears as constants.

3.8 BALANCED STEADY-STATE PHASOR RELATIONSHIPS

For balanced steady-state conditions, ω_e is constant and (3.7-1)–(3.7-3) may be expressed as

$$F_{as} = \sqrt{2}\,F_s \cos[\omega_e t + \theta_{ef}(0)]$$
$$= \mathrm{Re}[\sqrt{2}\,F_s\, e^{j\theta_{ef}(0)}\, e^{j\omega_e t}] \tag{3.8-1}$$

$$F_{bs} = \sqrt{2}\,F_s \cos\left[\omega_e t + \theta_{ef}(0) - \frac{2\pi}{3}\right]$$
$$= \mathrm{Re}[\sqrt{2}\,F_s\, e^{j[\theta_{ef}(0)-2\pi/3]}\, e^{j\omega_e t}] \tag{3.8-2}$$

$$F_{cs} = \sqrt{2}\,F_s \cos\left[\omega_e t + \theta_{ef}(0) + \frac{2\pi}{3}\right]$$
$$= \mathrm{Re}[\sqrt{2}\,F_s e^{j[\theta_{ef}(0)+2\pi/3]}\, e^{j\omega_e t}] \tag{3.8-3}$$

where $\theta_{ef}(0)$ corresponds to the time-zero value of the 3-phase variables. The upper-case letters are used to denote steady-state quantities. If the speed of the arbitrary reference frame is an unspecified constant, then for the balanced steady-state conditions we may express (3.7-5) and (3.7-6) as

$$F_{qs} = \sqrt{2}F_s \cos[(\omega_e - \omega)t + \theta_{ef}(0) - \theta(0)]$$
$$= \mathrm{Re}[\sqrt{2}F_s e^{j[\theta_{ef}(0)-\theta(0)]} e^{j(\omega_e-\omega)t}] \tag{3.8-4}$$

$$F_{ds} = -\sqrt{2}\,F_s \sin[(\omega_e - \omega)t + \theta_{ef}(0) - \theta(0)]$$
$$= \mathrm{Re}[j\sqrt{2}\,F_s\, e^{j[\theta_{ef}(0)-\theta(0)]}\, e^{j(\omega_e-\omega)t}] \tag{3.8-5}$$

From (3.8-1), the phasor representing the *as* variables is

$$\tilde{F}_{as} = F_s e^{j\theta_{ef}(0)} \tag{3.8-6}$$

If ω is not equal to ω_e, then F_{qs} and F_{ds} are sinusoidal quantities, and from (3.8-4) and (3.8-5) we obtain

$$\tilde{F}_{qs} = F_s e^{j[\theta_{ef}(0)-\theta(0)]} \tag{3.8-7}$$

$$\tilde{F}_{ds} = j\tilde{F}_{qs} \tag{3.8-8}$$

It is necessary to consider negative frequencies because ω can be greater than ω_e. The phasors rotate in the counterclockwise direction for $\omega < \omega_e$ and in the clockwise direction for $\omega > \omega_e$.

In the analysis of steady-state operation we are free to select time zero. It is often convenient to select it so that $\theta(0) = 0$; then from (3.8-6) and (3.8-7) we obtain

$$\tilde{F}_{as} = \tilde{F}_{qs} \tag{3.8-9}$$

Thus, in all asynchronously rotating reference frames ($\omega \neq \omega_e$) with $\theta(0) = 0$, the phasor representing the *as* is equal to the phasor representing the *qs* variables. For balanced steady-state conditions, the phasor representing the variables of one phase need only be shifted in order to represent the variables in the other phases.

In the synchronously rotating reference frame we have $\omega = \omega_e$ and $\theta(0) = \theta_e(0)$. If we continue to use uppercase letters to denote the constant steady-state variables in the synchronously rotating reference frame, then from (3.8-4) and (3.8-5) we obtain

$$F_{qs}^e = \text{Re}[\sqrt{2}F_s e^{j[\theta_{ef}(0)-\theta_e(0)]}] \tag{3.8-10}$$

$$F_{ds}^e = \text{Re}[j\sqrt{2}F_s e^{j[\theta_{ef}(0)-\theta_e(0)]}] \tag{3.8-11}$$

If we let the time-zero position of the reference frame be zero, then $\theta_e(0) = \theta(0) = 0$ and

$$F_{qs}^e = \sqrt{2}F_s \cos\theta_{ef}(0) \tag{3.8-12}$$

$$F_{ds}^e = -\sqrt{2}F_s \sin\theta_{ef}(0) \tag{3.8-13}$$

Thus, we see from a comparison of (3.8-6) with (3.8-12) and (3.8-13) that

$$\sqrt{2}\tilde{F}_{as} = F_{qs}^e - jF_{ds}^e \tag{3.8-14}$$

Because $\tilde{F}_{as} = \tilde{F}_{qs}$, (3.8-14) is important in that it relates the synchronously rotating reference-frame variables to a phasor in all other reference frames. $\tilde{F}_{as}$ is a phasor that represents a sinusoidal quantity; however, F_{qs}^e and F_{ds}^e are not phasors. They are real quantities representing the constant steady-state variables of the synchronously rotating reference frame.

Example 3C   It is helpful to discuss the difference between the directions of f_{as}, f_{bs}, and f_{cs} as shown in Fig. 3.3-1 and phasors. The relationships shown in Fig. 3.3-1 trigonometrically illustrate the transformation defined by (3.3-1)–(3.3-6). Figure 3.3-1 is not a phasor diagram and should not be interpreted as such. It simply depicts the relationships between the directions of f_{as}, f_{bs}, f_{cs}, f_{qs}, and f_{ds} as dictated by the equations of transformation regardless of the instantaneous values of these variables. On the other hand, phasors provide an analysis tool for steady-state sinusoidal variables. The magnitude and phase angle of the phasor are directly related to the amplitude of the sinusoidal variation and its phase position relative to a reference. The balanced set given by (3.7-1)–(3.7-3) may be written as (3.8-1)–(3.8-3) for steady-state conditions. The phasor representation for the *as* variables is given by (3.8-6). The phasor representation for the balanced set is

$$\tilde{F}_{as} = F_s e^{j\theta_{ef}(0)} \tag{3C-1}$$

$$\tilde{F}_{bs} = F_s e^{j[\theta_{ef}(0)-2\pi/3]} \tag{3C-2}$$

$$\tilde{F}_{cs} = F_s e^{j[\theta_{ef}(0)+2\pi/3]} \tag{3C-3}$$

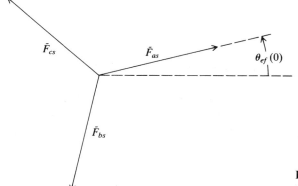

Figure 3C-1 Phasor representation for a 3-phase balanced set.

The phasor diagram is shown in Fig. 3C-1. For balanced conditions, the phasors that form an *abc* sequence are displaced from each other by 120° and each
with a phase angle of $\theta_{ef}(0)$. The directions of f_{as}, f_{bs}, and f_{cs} in Fig. 3.3-1,
which are fixed by the transformation, are such that f_{cs} is directed $-120°$
from f_{as}. However, $\tilde{F}_{cs}$ is $+120°$ from $\tilde{F}_{as}$ for balanced conditions
(Fig. 3C-1). Another important difference is that the phasor diagram must
be rotated at ω_e in the counterclockwise direction, and the real part of the phasors represents the instantaneous values of the 3-phase set. The diagram of f_{as},
f_{bs}, and f_{cs} shown in Fig. 3.3-1 is stationary for stationary circuits.

Phasors can also be used to analyze unbalanced steady-state sinusoidal variables. An example of a steady-state unbalanced 3-phase set is

$$F_{as} = \sqrt{2}\,F_s \cos\left(\omega_e t + \frac{\pi}{6}\right) \tag{3C-4}$$

$$F_{bs} = 2\sqrt{2}F_s \sin \omega_e t \tag{3C-5}$$

$$F_{cs} = \frac{\sqrt{2}}{2}F_s \cos\left(\omega_e t + \pi\right) \tag{3C-6}$$

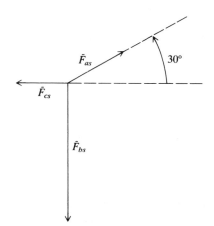

Figure 3C-2 Phasor representation for a 3-phase
unbalanced set.

The phasor diagram for this set is shown in Fig. 3C-2. It is clear that the positive directions of f_{as}, f_{bs}, and f_{cs} in Fig. 3.3-1 are unchanged.

3.9 BALANCED STEADY-STATE VOLTAGE EQUATIONS

If the 3-phase system is symmetrical and if the applied voltages form a balance set as given by (3.7-1)–(3.7-3), then the steady-state currents will also form a balanced set. For equal resistance in each phase, the steady-state voltage equation in terms of the *as* variables is

$$\tilde{V}_{as} = r_s \tilde{I}_{as} \qquad (3.9\text{-}1)$$

For linear, symmetrical inductive elements the steady-state voltage equation may be written as

$$\tilde{V}_{as} = j\omega_e \tilde{\Lambda}_{as} \qquad (3.9\text{-}2)$$

where $\tilde{\Lambda}_{as}$ is an inductance times $\tilde{I}_{as}$. For linear, symmetrical capacitive elements the steady-state current equation becomes

$$\tilde{I}_{as} = j\omega_e \tilde{Q}_{as} \qquad (3.9\text{-}3)$$

where $\tilde{Q}_{as}$ is a capacitance times $\tilde{V}_{as}$. It is clear that for any combination of linear symmetrical circuit elements the steady-state voltage equation may be expressed in phasor form as

$$\tilde{V}_{as} = Z_s \tilde{I}_{as} \qquad (3.9\text{-}4)$$

where Z_s is the impedance of each phase of the 3-phase system.

For equal resistance in each phase of the circuit, the balanced steady-state voltage equation for the *qs* variables in all asynchronously rotating reference frames can be written from (3.4-2) as

$$\tilde{V}_{qs} = r_s \tilde{I}_{qs} \qquad (3.9\text{-}5)$$

For linear symmetrical inductive elements, the steady-state *qs* voltage equation in all asynchronously rotating reference frames may be written from (3.4-11) as

$$\tilde{V}_{qs} = \omega \tilde{\Lambda}_{ds} + j(\omega_e - \omega)\tilde{\Lambda}_{qs} \qquad (3.9\text{-}6)$$

where the $(\omega_e - \omega)$ factor comes about due to the fact that the steady-state variables in all asynchronously rotating reference frames vary at the frequency corresponding to $(\omega_e - \omega)$. From (3.8-8), we obtain $\tilde{\Lambda}_{ds} = j\tilde{\Lambda}_{qs}$; thus (3.9-6) becomes

$$\tilde{V}_{qs} = j\omega_e \tilde{\Lambda}_{qs} \tag{3.9-7}$$

Similarly, for a linear symmetrical capacitive circuit the steady-state qs current phasor equation in all asynchronously rotating reference frames may be written from (3.4-24) as

$$\tilde{I}_{qs} = j\omega_e \tilde{Q}_{qs} \tag{3.9-8}$$

Thus, for any combination of linear symmetrical circuit elements the steady-state voltage equation in all asynchronously rotating reference frames may be expressed in phasor form as

$$\tilde{V}_{qs} = Z_s \tilde{I}_{qs} \tag{3.9-9}$$

where, for a given 3-phase system, Z_s is the same complex impedance as in (3.9-4).

The fact that the steady-state phasor voltage equations are identical for the as and qs variables was actually known beforehand because (3.8-9) tells us that for $\theta(0) = 0$ the phasors representing the as variables are equal to the phasors representing the qs variables in all asynchronously rotating reference frames; therefore, the as and qs circuits must have the same impedance.

Example 3D   It is instructive to derive the phasor voltage equation for the RL circuit used in Example 3B for balanced steady-state conditions. Three methods of deriving this equation are described in the previous section. We will use all three approaches to arrive at the same steady-state, phasor voltage equation. As the first approach, the as voltage equation may be written for the example RL circuit using steady-state notation as

$$V_{as} = r_s I_{as} + L_s p I_{as} + M p I_{bs} + M p I_{cs} \tag{3D-1}$$

For balanced conditions we have

$$F_{as} + F_{bs} + F_{cs} = 0 \tag{3D-2}$$

Thus, (3D-1) may be written as

$$V_{as} = r_s I_{as} + (L_s - M) p I_{as} \tag{3D-3}$$

For steady-state conditions, p is replaced by $j\omega_e$, whereupon (3D-3) can be written in phasor form as

$$\tilde{V}_{as} = [r_s + j\omega_e(L_s - M)]\tilde{I}_{as} \tag{3D-4}$$

Comparing (3D-4) with (3.9-2) and (3.9-4), we see that

$$\tilde{\Lambda}_{as} = (L_s - M)\tilde{I}_{as} \tag{3D-5}$$

and

$$Z = r_s + j\omega_e(L_s - M) \tag{3D-6}$$

In the two remaining derivations we will make use of the qs and ds voltage equations in the arbitrary reference frame. Thus, from (3B-13) and (3B-14) we obtain

$$v_{qs} = r_s i_{qs} + \omega\lambda_{ds} + p\lambda_{qs} \tag{3D-7}$$

$$v_{ds} = r_s i_{ds} - \omega\lambda_{qs} + p\lambda_{ds} \tag{3D-8}$$

where

$$\lambda_{qs} = (L_s - M)i_{qs} \tag{3D-9}$$

$$\lambda_{ds} = (L_s - M)i_{ds} \tag{3D-10}$$

For the second method, we will start with either the qs- or ds-voltage equation in the asynchronously rotating reference frame. Thus, using steady-state notation, (3D-7) may be written

$$V_{qs} = r_s I_{qs} + \omega\Lambda_{ds} + p\Lambda_{qs} \tag{3D-11}$$

For balanced steady-state conditions, p may be replaced by $j(\omega_e - \omega)$ and from (3.8-8), $\tilde{\Lambda}_{ds} = j\tilde{\Lambda}_{qs}$. Hence

$$\tilde{V}_{qs} = r_s \tilde{I}_{qs} + j\omega_e \tilde{\Lambda}_{qs} \tag{3D-12}$$

Clearly, (3.9-5) and (3.9-7) combine to give (3D-12). Substituting for $\tilde{\Lambda}_{qs}$ yields

$$\tilde{V}_{qs} = [r_s + j\omega_e(L_s - M)]\tilde{I}_{qs} \tag{3D-13}$$

Because in all asynchronously rotating reference frames with $\theta(0) = 0$

$$\tilde{F}_{as} = \tilde{F}_{qs} \tag{3D-14}$$

we have arrived at the same result as in the first case where we started with the as-voltage equation.

For the third approach, let us write the voltage equations in the synchronously rotating reference frame. Thus, using steady-state notation, (3D-7) and (3D-8) may be written in the synchronously rotating reference frame as

$$V_{qs}^e = r_s I_{qs}^e + \omega_e\Lambda_{ds}^e + p\Lambda_{qs}^e \tag{3D-15}$$

$$V_{ds}^e = r_s I_{ds}^e - \omega_e\Lambda_{qs}^e + p\Lambda_{ds}^e \tag{3D-16}$$

For balanced steady-state conditions the variables in the synchronously rotating reference frame are constants; therefore $p\Lambda_{qs}^e$ and $p\Lambda_{ds}^e$ are zero. Therefore,

(3D-15) and (3D-16) may be written as

$$V_{qs}^e = r_s I_{qs}^e + \omega_e(L_s - M)I_{ds}^e \tag{3D-17}$$

$$V_{ds}^e = r_s I_{ds}^e - \omega_e(L_s - M)I_{qs}^e \tag{3D-18}$$

wherein Λ_{qs}^e and Λ_{ds}^e have been written as a product of inductance and current. Now, (3.8-14) is

$$\sqrt{2}\tilde{F}_{as} = F_{qs}^e - jF_{ds}^e \tag{3D-19}$$

Thus

$$\sqrt{2}\tilde{V}_{as} = r_s I_{qs}^e + \omega_e(L_s - M)I_{ds}^e - j[r_s I_{ds}^e - \omega_e(L_s - M)I_{qs}^e] \tag{3D-20}$$

Now

$$\sqrt{2}\tilde{I}_{as} = I_{qs}^e - jI_{ds}^e \tag{3D-21}$$

and

$$j\sqrt{2}\tilde{I}_{as} = I_{ds}^e + jI_{qs}^e \tag{3D-22}$$

Substituting (3D-21) and (3D-22) into (3D-20) yields the desired equation:

$$\tilde{V}_{as} = [r_s + j\omega_e(L_s - M)]\tilde{I}_{as} \tag{3D-23}$$

3.10 VARIABLES OBSERVED FROM SEVERAL FRAMES OF REFERENCE

It is instructive to observe the waveform of the variables of a stationary 3-phase series RL circuit in the arbitrary reference frame and in commonly used reference frames. For this purpose we will assume that both $\mathbf{r}_s$ and $\mathbf{L}_s$ are diagonal matrices, each with equal nonzero elements, and the applied voltages are of the form

$$v_{as} = \sqrt{2}V_s \cos\omega_e t \tag{3.10-1}$$

$$v_{bs} = \sqrt{2}V_s \cos\left(\omega_e t - \frac{2\pi}{3}\right) \tag{3.10-2}$$

$$v_{cs} = \sqrt{2}V_s \cos\left(\omega_e t + \frac{2\pi}{3}\right) \tag{3.10-3}$$

where ω_e is an unspecified constant. The currents, which are assumed to be zero at $t = 0$, may be expressed as

$$i_{as} = \frac{\sqrt{2}V_s}{|Z_s|}[-e^{-t/\tau}\cos\alpha + \cos(\omega_e t - \alpha)] \tag{3.10-4}$$

$$i_{bs} = \frac{\sqrt{2}V_s}{|Z_s|}\left[-e^{-t/\tau}\cos\left(\alpha + \frac{2\pi}{3}\right) + \cos\left(\omega_e t - \alpha - \frac{2\pi}{3}\right)\right] \tag{3.10-5}$$

$$i_{cs} = \frac{\sqrt{2}V_s}{|Z_s|}\left[-e^{-t/\tau}\cos\left(\alpha - \frac{2\pi}{3}\right) + \cos\left(\omega_e t - \alpha + \frac{2\pi}{3}\right)\right] \tag{3.10-6}$$

where

$$Z_s = r_s + j\omega_e L_s \tag{3.10-7}$$

$$\tau = \frac{L_s}{r_s} \tag{3.10-8}$$

$$\alpha = \tan^{-1}\frac{\omega_e L_s}{r_s} \tag{3.10-9}$$

It may at first appear necessary to solve the voltage equations in the arbitrary reference frame in order to obtain the expression for the currents in the arbitrary reference frame. This is unnecessary because once the solution is known in one reference frame, it is known in all reference frames. In the example at hand, this may be accomplished by transforming (3.10-4)–(3.10-6) to the arbitrary reference frame. For illustrative purposes, let ω be an unspecified constant with $\theta(0) = 0$; then $\theta = \omega t$ and in the arbitrary reference frame we have

$$i_{qs} = \frac{\sqrt{2}V_s}{|Z_s|}\{-e^{-t/\tau}\cos(\omega t - \alpha) + \cos[(\omega_e - \omega)t - \alpha]\} \tag{3.10-10}$$

$$i_{ds} = \frac{\sqrt{2}V_s}{|Z_s|}\{-e^{-t/\tau}\sin(\omega t - \alpha) - \sin[(\omega_e - \omega)t - \alpha]\} \tag{3.10-11}$$

Clearly, the state of the electric system is independent of the frame of reference from which it is observed. Although the variables will appear differently in each reference frame, they will exhibit the same mode of operation (transient of steady state) regardless of the reference frame. In general, (3.10-10) and (3.10-11) contain two balanced sets. One, which represents the electric transient, decays exponentially at a frequency corresponding to the instantaneous angular velocity of the arbitrary reference frame. In this set, the qs variable leads the ds variable by 90° when $\omega > 0$ and lags by 90° when $\omega < 0$. The second balanced set, which represents the steady-state response, has a constant amplitude with a frequency corresponding to the difference in the angular velocity of the voltages applied to the stationary circuits and the angular velocity of the arbitrary reference frame. In this set, the qs variable lags the ds by 90° when $\omega < \omega_e$ and leads by 90° when $\omega > \omega_e$. This of course leads to the concept of negative frequency when relating phasors that represent qs and ds variables by (3.8-8).

There are two frames of reference that do not contain both balanced sets. In the stationary reference frame we have $\omega = 0$ and $i_{qs}^s = i_{as}$. The exponentially decaying balanced set becomes an exponential decay, and the constant amplitude balanced set varies at ω_e. In the synchronously rotating reference frame where $\omega = \omega_e$, the electric transients are represented by an exponentially decaying balanced set varying at ω_e and the constant amplitude set becomes constant.

The waveforms of the system variables in various reference frames are shown by computer simulation in Figs. 3.10-1–3.10-3 [8]. The voltages of the form given by

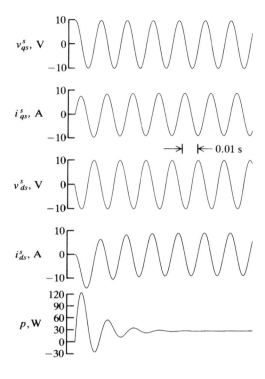

Figure 3.10-1   Variables of a stationary 3-phase system in the stationary reference frame.

(3.10-1)–(3.10-3) are applied to the 3-phase system with $V_s = 10/\sqrt{2}\ V$, $r_s = 0.216\ \Omega$, and $\omega_e L_s = 1.09\Omega$ with $\omega_e = 377$ rad/s. The response, for $t > 0$, of the electric system in the stationary reference frame is shown in Fig. 3.10-1. Because we have selected $\theta(0) = 0$, we obtain $f_{as} = f_{qs}^s$ and the plots of v_{qs}^s and i_{qs}^s are v_{as} and i_{as}, respectively. The variables for the same mode of operation are shown in the synchronously rotating reference frame in Fig. 3.10-2. Note, from (3.10-1)–(3.10-3), that we have selected $\theta_{ev}(0) = 0$; thus from (3.7-5) and (3.7-6) with $\theta(0) = 0$, we obtain $v_{qs}^e = 10$ V and $v_{ds}^e = 0$. In Fig. 3.10-3, with $\theta(0) = 0$ the speed of the reference frame is switched from its original value of -377 rad/s to zero and then ramped to 377 rad/s.

There are several features worthy of note. The waveform of the instantaneous electric power is the same in all cases. The electric transient is very evident in the waveforms of the instantaneous electric power and the currents in the synchronously rotating reference frame (Fig. 3.10-2); and because v_{ds}^e is zero, i_{qs}^e is related to the power by a constant $(3/2\ v_{qs}^e)$. In Fig. 3.10-3 we selected $\theta_{ev}(0) = 0$ and $\theta(0) = 0$. The voltages were applied and we observed the solution of the differential equations in the reference frame rotating clockwise at ω_e $(\omega = -\omega_e)$. The reference frame

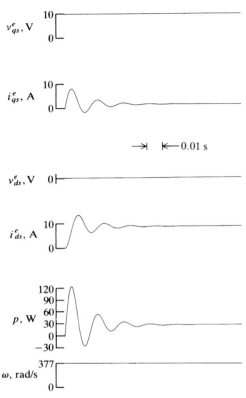

Figure 3.10-2   Variables of a stationary 3-phase system in synchronously rotating reference frame.

speed was then stepped from -377 rad/s to zero, whereupon the differential equations were solved in the stationary reference frame. However, when switching from one reference frame to another, the variables must be continuous. Therefore, after the switching occurs the solution continues using the stationary reference frame differential equations with the initial values determined by the instantaneous values of the variables in the previous reference frame ($\omega = -\omega_e$) at the time of switching. It is important to note the change in frequency of the variables as the reference frame speed is ramped from zero to ω_e. Here the differential equations being solved are continuously changing while the variables remain continuous. When the reference frame speed reaches synchronous speed, the variables have reached steady state; therefore they will be constant corresponding to their values at the instant ω becomes equal to ω_e. In essence we have applied a balanced 3-phase set of voltages to a symmetrical RL circuit, and in Fig. 3.10-3 we observed the actual variables from various reference frames by "jumping" or "running" from one reference frame to another.

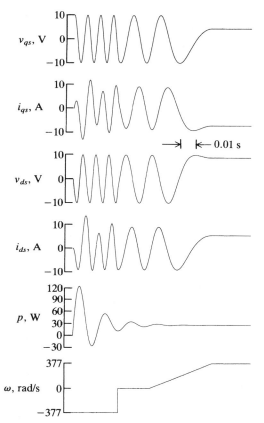

Figure 3.10-3 Variables of a stationary 3-phase system. First with $\omega = -\omega_e$, then ω is stepped to zero followed by a ramp change in reference frame speed to $\omega = \omega_e$.

REFERENCES

[1] R. H. Park, Two-Reaction Theory of Synchronous Machines—Generalized Method of Analysis, Part I, *AIEE Transactions*, Vol. 48, July 1929, pp. 716–727.

[2] H. C. Stanley, An Analysis of the Induction Motor, *AIEE Transactions*, Vol. 57 (Supplement), 1938, pp. 751–755.

[3] G. Kron, *Equivalent Circuits of Electric Machinery*, John Wiley and Sons, New York, 1951.

[4] D. S. Brereton, D. G. Lewis, and C. G. Young, Representation of Induction Motor Loads During Power System Stability Studies, *AIEE Transactions*, Vol. 76, August 1957, pp. 451–461.

[5] P. C. Krause and C. H. Thomas, Simulation of Symmetrical Induction Machinery, *IEEE Transactions on Power Apparatus and Systems*, Vol. 84, November 1965, pp. 1038–1053.

[6] P. C. Krause, F. Nozari, T. L. Skvarenina, and D. W. Olive, The Theory of Neglecting Stator Transients, *IEEE Transactions on Power Apparatus and Systems*, Vol. 98, January/February 1979, pp. 141–148.

[7] E. Clarke, *Circuit Analysis of A-C Power Systems*, Vol. I—*Symmetrical and Related Components*, John Wiley and Sons, New York, 1943.

[8] C. M. Ong, *Dynamic Simulation of Electric Machinery*, Prentice-Hall PTR, Upper Saddle River, NJ, 1998.

PROBLEMS

1 The transformation for a 2-phase set to the arbitrary reference frame is

$$\mathbf{f}_{qds} = \mathbf{K}_{2s}\mathbf{f}_{abs}$$

where

$$(\mathbf{f}_{qds})^T = [f_{qs} \quad f_{ds}]$$
$$(\mathbf{f}_{abs})^T = [f_{as} \quad f_{bs}]$$
$$\mathbf{K}_{2s} = \begin{bmatrix} \cos\theta & \sin\theta \\ \sin\theta & -\cos\theta \end{bmatrix}$$

where θ is defined by (3.3-5). Express the voltage equations in the arbitrary reference frame for a 2-phase resistive circuit if (a) $r_a = r_b = r_s$ and if (b) $r_a \neq r_b$.

2 Using the transformation given in Problem 1, express the voltage equations in the arbitrary reference frame for a 2-phase inductive circuit if (a) $L_a = L_b = L$ and if (b) $L_a \neq L_b$.

3 Using the transformation given in Problem 1, express the current equations in the arbitrary reference frame for a 2-phase capacitive circuit if (a) $C_a = C_b = C$ and if (b) $C_a \neq C_b$.

4 The phases of a 3-phase circuit consist of equal resistances, equal inductances, and equal capacitances connected in series. The phases are not coupled. Write the voltage equations in the arbitrary reference frame and draw the equivalent circuit. You should be able to use results from Problems 1, 2, and 3.

5 Repeat Problem 4 for the circuit elements in each phase connected in parallel.

6 Show that for a symmetrical 2-phase circuit the average power expressed in *as* and *bs* variables is equal to the average power expressed in *qs* and *ds* variables.

7 Clarke's transformation may be written as

$$\mathbf{f}_{\alpha\beta0} = \mathbf{C}\mathbf{f}_{abcs}$$

where

$$(\mathbf{f}_{\alpha\beta0})^T = [f_\alpha \quad f_\beta \quad f_0]$$
$$C = \frac{2}{3}\begin{bmatrix} 1 & -\frac{1}{2} & -\frac{1}{2} \\ 0 & -\frac{\sqrt{3}}{2} & \frac{\sqrt{3}}{2} \\ \frac{1}{2} & \frac{1}{2} & \frac{1}{2} \end{bmatrix}$$

Relate $f_{qs}^s, f_{ds}^s, f_\alpha, f_\beta$, and f_0.

8 A transformation that is sometimes used in the case of synchronous machines is one where f_{ds} leads f_{qs} in Fig. 3.3-1 by $90°$ with $\omega = \omega_r$.

(*a*) Express the transformation.

(*b*) Using this transformation, write the voltage equations for a 3-phase inductive circuit.

9 The inductance matrix that describes the self- and mutual inductances between the stator windings of a z-phase, salient-pole, synchronous machine is given below.

$$\mathbf{L}_s = \begin{bmatrix} L_{ls} + L_A - L_B \cos 2\theta_r & -L_B \sin 2\theta_r \\ -L_B \sin 2\theta_r & L_{ls} + L_A + L_B \cos 2\theta_r \end{bmatrix}$$

Evaluate $\mathbf{K}_{2s}^r \mathbf{L}_s (\mathbf{K}_{2s}^r)^{-1}$, where $\mathbf{K}_{2s}^r$ is given in Problem 1.

10 If A is one reference frame and B another, show that $(^A\mathbf{K}^B)^{-1} = {}^B\mathbf{K}^A$.

11 Equations (3.7-1)–(3.7-3) form an *abc* sequence. Express an *acb* sequence and transform this set to the arbitrary reference frame using (3.3-1). Express $\mathbf{f}_{qd0s}^e(\omega = \omega_e)$ and $\mathbf{f}_{qd0s}^{-e}(\omega = -\omega_e)$.

12 Devise a transformation that yields only constants when $\omega = \omega_e$ for a balanced 3-phase set with a phase sequence of *acb*.

13 Relate $\tilde{F}_{bs}$ and $\tilde{F}_{cs}$ to $\tilde{F}_{qs}$ and $\tilde{F}_{ds}$ for a balanced 3-phase set with a time sequence of *abc*.

14 For steady-state balanced conditions the total 3-phase power and reactive power may be expressed as

$$P_e = 3V_sI_s \cos[\theta_{ev}(0) - \theta_{ei}(0)]$$
$$Q_e = 3V_sI_s \sin[\theta_{ev}(0) - \theta_{ei}(0)]$$

Show that the following expressions are equal to those given above:

$$P_e = \frac{3}{2}(V_{qs}I_{qs} + V_{ds}I_{ds})$$

$$Q_e = \frac{3}{2}(V_{qs}I_{ds} - V_{ds}I_{qs})$$

15 Write the expressions for the currents in Figs. 3.10-1 and 3.10-2.

16 Assume the steady-state *abc* variables are of the form

$$F_{as} = \sqrt{2}F_a \cos \omega_e t$$

$$F_{bs} = \sqrt{2}F_b \cos\left(\omega_e t - \frac{2\pi}{3}\right)$$

$$F_{cs} = \sqrt{2}F_c \cos\left(\omega_e t + \frac{2\pi}{3}\right)$$

where F_a, F_b, and F_c are unequal constants. Show that this unbalanced set of *abc* variables forms 2-phase balanced sets of *qs* and *ds* variables in the arbitrary reference frame with the arguments of $(\omega_e t - \theta)$ and $(\omega_e t + \theta)$. Note the form of the qs and ds variables when $\omega = \omega_e$ and $\omega = -\omega_e$.

17 Repeat Problem 16 with

$$F_{as} = \sqrt{2}F_s \cos(\omega_e t + \phi_a)$$
$$F_{bs} = \sqrt{2}F_s \cos(\omega_e t + \phi_b)$$
$$F_{cs} = \sqrt{2}F_s \cos(\omega_e t + \phi_c)$$

where ϕ_a, ϕ_b, and ϕ_c are unequal constants.

18 It is often suggested that $\mathbf{K}_s$ should be changed so that $(\mathbf{K}_s)^T = (\mathbf{K}_s)^{-1}$. For example, if

$$\mathbf{K}_s = \sqrt{\frac{2}{3}} \begin{bmatrix} \cos\theta & \cos\left(\theta - \frac{2\pi}{3}\right) & \cos\left(\theta + \frac{2\pi}{3}\right) \\ \sin\theta & \sin\left(\theta - \frac{2\pi}{3}\right) & \sin\left(\theta + \frac{2\pi}{3}\right) \\ \frac{1}{\sqrt{2}} & \frac{1}{\sqrt{2}} & \frac{1}{\sqrt{2}} \end{bmatrix}$$

then $(\mathbf{K}_s)^T = (\mathbf{K}_s)^{-1}$. Show that this is true. Also, show that in this case

$$\mathbf{f}_{as}^2 + \mathbf{f}_{bs}^2 + \mathbf{f}_{cs}^2 = \mathbf{f}_{qs}^2 + \mathbf{f}_{ds}^2 + \mathbf{f}_{0s}^2$$

Chapter 4

SYMMETRICAL INDUCTION MACHINES

4.1 INTRODUCTION

The induction machine is used in a wide variety of applications as a means of converting electric power to mechanical work. It is without doubt the workhorse of the electric power industry. Pump, steel mill, and hoist drives are but a few applications of large multiphase induction motors. On a smaller scale, the 2-phase servomotor is used in position-follow-up control systems, and single-phase induction motors are widely used in household appliances as well as in hand and bench tools.

In the beginning of this chapter, classical techniques are used to establish the voltage and torque equations for a symmetrical induction machine expressed in terms of machine variables. Next, the transformation to the arbitrary reference frame presented in Chapter 3 is modified to accommodate rotating circuits. Once this groundwork has been laid, the machine voltage equations are written in the arbitrary reference frame, directly without a laborious exercise in trigonometry with which one is faced when starting from the substitution of the equations of transformations into the voltage equations expressed in machine variables. The equations may then be expressed in any reference frame by appropriate assignment of the reference-frame speed in the arbitrary reference-frame voltage equations rather than performing each transformation individually as in the past. Although the stationary reference frame, the reference frame fixed in the rotor, and the synchronously rotating reference frame are most frequently used, the arbitrary reference frame offers a direct means of obtaining the voltage equations in these and all other reference frames.

The steady-state voltage equations for an induction machine are obtained from the voltage equations in the arbitrary reference frame by direct application of the material presented in Chapter 3. Computer solutions are used to illustrate the

141

dynamic performance of typical induction machines and to depict the variables in various reference frames during free acceleration. Finally, the equations for an induction machine are arranged appropriately for computer simulation. The material presented in this chapter forms the basis for solution of more advanced problems. In particular, these basic concepts are fundamental to the analysis of induction machines in most power system and variable frequency drive applications.

4.2 VOLTAGE EQUATIONS IN MACHINE VARIABLES

The winding arrangement for a 2-pole, 3-phase, wye-connected, symmetrical induction machine is shown in Fig. 1.5-2. Figure 4.2-1 is Fig. 1.5-2 repeated here for convenience. The stator windings are identical, sinusoidally distributed windings, displaced 120°, with N_s equivalent turns and resistance r_s[1]. For the purpose at hand, the rotor windings will also be considered as three identical sinusoidally distributed windings, displaced 120°, with N_r equivalent turns and resistance r_r. The positive direction of the magnetic axis of each winding is shown in Fig. 4.2-1. It is important to note that the positive direction of the magnetic axes of the stator windings coincides with the direction of f_{as}, f_{bs}, and f_{cs} as specified by the equations of transformation and shown in Fig. 3.3-1.

The voltage equations in machine variables may be expressed as

$$\mathbf{v}_{abcs} = \mathbf{r}_s \mathbf{i}_{abcs} + p\boldsymbol{\lambda}_{abcs} \tag{4.2-1}$$

$$\mathbf{v}_{abcr} = \mathbf{r}_r \mathbf{i}_{abcr} + p\boldsymbol{\lambda}_{abcr} \tag{4.2-2}$$

where

$$\left(\mathbf{f}_{abcs}\right)^T = \begin{bmatrix} f_{as} & f_{bs} & f_{cs} \end{bmatrix} \tag{4.2-3}$$

$$\left(\mathbf{f}_{abcr}\right)^T = \begin{bmatrix} f_{ar} & f_{br} & f_{cr} \end{bmatrix} \tag{4.2-4}$$

In the above equations the s subscript denotes variables and parameters associated with the stator circuits, and the r subscript denotes variables and parameters associated with the rotor circuits. Both $\mathbf{r}_s$ and $\mathbf{r}_r$ are diagonal matrices, each with equal nonzero elements. For a magnetically linear system, the flux linkages may be expressed as

$$\begin{bmatrix} \boldsymbol{\lambda}_{abcs} \\ \boldsymbol{\lambda}_{abcr} \end{bmatrix} = \begin{bmatrix} \mathbf{L}_s & \mathbf{L}_{sr} \\ (\mathbf{L}_{sr})^T & \mathbf{L}_r \end{bmatrix} \begin{bmatrix} \mathbf{i}_{abcs} \\ \mathbf{i}_{abcr} \end{bmatrix} \tag{4.2-5}$$

The winding inductances are derived in Section 1.5. In particular,

$$\mathbf{L}_s = \begin{bmatrix} L_{ls} + L_{ms} & -\frac{1}{2}L_{ms} & -\frac{1}{2}L_{ms} \\ -\frac{1}{2}L_{ms} & L_{ls} + L_{ms} & -\frac{1}{2}L_{ms} \\ -\frac{1}{2}L_{ms} & -\frac{1}{2}L_{ms} & L_{ls} + L_{ms} \end{bmatrix} \tag{4.2-6}$$

$$\mathbf{L}_r = \begin{bmatrix} L_{lr} + L_{mr} & -\frac{1}{2}L_{mr} & -\frac{1}{2}L_{mr} \\ -\frac{1}{2}L_{mr} & L_{lr} + L_{mr} & -\frac{1}{2}L_{mr} \\ -\frac{1}{2}L_{mr} & -\frac{1}{2}L_{mr} & L_{lr} + L_{mr} \end{bmatrix} \tag{4.2-7}$$

$$\mathbf{L}_{sr} = L_{sr} \begin{bmatrix} \cos\theta_r & \cos\left(\theta_r + \frac{2\pi}{3}\right) & \cos\left(\theta_r - \frac{2\pi}{3}\right) \\ \cos\left(\theta_r - \frac{2\pi}{3}\right) & \cos\theta_r & \cos\left(\theta_r + \frac{2\pi}{3}\right) \\ \cos\left(\theta_r + \frac{2\pi}{3}\right) & \cos\left(\theta_r - \frac{2\pi}{3}\right) & \cos\theta_r \end{bmatrix} \tag{4.2-8}$$

In the above inductance equations, L_{ls} and L_{ms} are, respectively, the leakage and magnetizing inductances of the stator windings; L_{lr} and L_{mr} are for the rotor

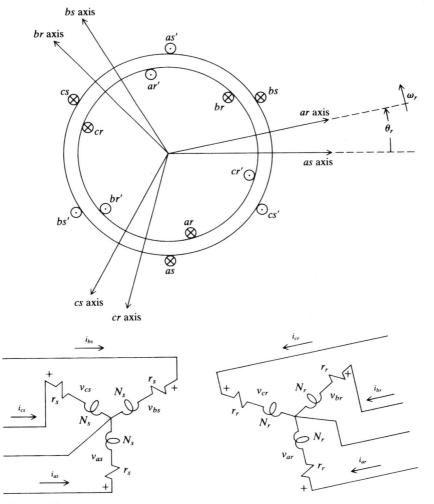

Figure 4.2-1   Two-pole, 3-phase, wye-connected symmetrical induction machine.

windings. The inductance L_{sr} is the amplitude of the mutual inductances between stator and rotor windings.

A majority of induction machines are not equipped with coil-wound rotor windings; instead, the current flows in copper or aluminum bars which are uniformly distributed and which are embedded in a ferromagnetic material with all bars terminated in a common ring at each end of the rotor. This type of rotor configuration is referred to as a squirrel-cage rotor. It may at first appear that the mutual inductance between a uniformly distributed rotor winding and a sinusoidally distributed stator winding would not be of the form given by (4.2-8). However, in most cases a uniformly distributed winding is adequately described by its fundamental sinusoidal component and is represented by an equivalent 3-phase winding. Generally, this representation consists of one equivalent winding per phase; however, the rotor construction of some machines is such that its performance is more accurately described by representing each phase with two equivalent windings connected in parallel. This type of machine is commonly referred to as a double-cage rotor machine.

Another consideration is that in a practical machine, the rotor conductors are often skewed. That is, the conductors are not placed in the plane of the axis of rotation of the rotor. Instead, the conductors are skewed slightly with the axis of rotation. This type of conductor arrangement helps to reduce the magnitude of harmonic torques due to the harmonic content of the MMF waves. Such design features are not considered here. Instead, it is assumed that all effects upon the amplitude of the fundamental component of the MMF waveform due to skewing and uniformly distributed rotor windings are accounted for in the value of N_r. The assumption that the induction machine is a linear (no saturation) and MMF harmonic-free device is an oversimplification that cannot describe the behavior of induction machines in all modes of operation. However, in a majority of applications, its behavior can be adequately predicted with this simplified representation.

When expressing the voltage equations in machine variable form, it is convenient to refer all rotor variables to the stator windings by appropriate turns ratios:

$$\mathbf{i}'_{abcr} = \frac{N_r}{N_s} \mathbf{i}_{abcr} \tag{4.2-9}$$

$$\mathbf{v}'_{abcr} = \frac{N_s}{N_r} \mathbf{v}_{abcr} \tag{4.2-10}$$

$$\boldsymbol{\lambda}'_{abcr} = \frac{N_s}{N_r} \boldsymbol{\lambda}_{abcr} \tag{4.2-11}$$

The magnetizing and mutual inductances are associated with the same magnetic flux path; therefore L_{ms}, L_{mr}, and L_{sr} are related as set forth by (1.5-44), (1.5-47), and (1.5-52). In particular,

$$L_{ms} = \frac{N_s}{N_r} L_{sr} \tag{4.2-12}$$

Thus, we will define

$$
\begin{aligned}
\mathbf{L}'_{sr} &= \frac{N_s}{N_r} \mathbf{L}_{sr} \\
&= L_{ms} \begin{bmatrix}
\cos\theta_r & \cos\left(\theta_r + \frac{2\pi}{3}\right) & \cos\left(\theta_r - \frac{2\pi}{3}\right) \\
\cos\left(\theta_r - \frac{2\pi}{3}\right) & \cos\theta_r & \cos\left(\theta_r + \frac{2\pi}{3}\right) \\
\cos\left(\theta_r + \frac{2\pi}{3}\right) & \cos\left(\theta_r - \frac{2\pi}{3}\right) & \cos\theta_r
\end{bmatrix}
\end{aligned}
\tag{4.2-13}
$$

Also, from (1.5-44) and (1.5-47), L_{mr} may be expressed as

$$
L_{mr} = \left(\frac{N_r}{N_s}\right)^2 L_{ms}
\tag{4.2-14}
$$

and if we let

$$
\mathbf{L}'_r = \left(\frac{N_s}{N_r}\right)^2 \mathbf{L}_r
\tag{4.2-15}
$$

then from (4.2-7) we obtain

$$
\mathbf{L}'_r = \begin{bmatrix}
L'_{lr} + L_{ms} & -\frac{1}{2}L_{ms} & -\frac{1}{2}L_{ms} \\
-\frac{1}{2}L_{ms} & L'_{lr} + L_{ms} & -\frac{1}{2}L_{ms} \\
-\frac{1}{2}L_{ms} & -\frac{1}{2}L_{ms} & L'_{lr} + L_{ms}
\end{bmatrix}
\tag{4.2-16}
$$

where

$$
L'_{lr} = \left(\frac{N_s}{N_r}\right)^2 L_{lr}
\tag{4.2-17}
$$

The flux linkages may now be expressed as

$$
\begin{bmatrix} \boldsymbol{\lambda}_{abcs} \\ \boldsymbol{\lambda}'_{abcr} \end{bmatrix} = \begin{bmatrix} \mathbf{L}_s & \mathbf{L}'_{sr} \\ (\mathbf{L}'_{sr})^T & \mathbf{L}'_r \end{bmatrix} \begin{bmatrix} \mathbf{i}_{abcs} \\ \mathbf{i}'_{abcr} \end{bmatrix}
\tag{4.2-18}
$$

The voltage equations expressed in terms of machine variables referred to the stator windings may now be written as

$$
\begin{bmatrix} \mathbf{v}_{abcs} \\ \mathbf{v}'_{abcr} \end{bmatrix} = \begin{bmatrix} \mathbf{r}_s + p\mathbf{L}_s & p\mathbf{L}'_{sr} \\ p(\mathbf{L}'_{sr})^T & \mathbf{r}'_r + p\mathbf{L}'_r \end{bmatrix} \begin{bmatrix} \mathbf{i}_{abcs} \\ \mathbf{i}'_{abcr} \end{bmatrix}
\tag{4.2-19}
$$

where

$$
\mathbf{r}'_r = \left(\frac{N_s}{N_r}\right)^2 \mathbf{r}_r
\tag{4.2-20}
$$

4.3 TORQUE EQUATION IN MACHINE VARIABLES

Evaluation of the energy stored in the coupling field by (1.3-51) yields the familiar expression for energy stored in a magnetically linear system. In particular, the stored energy is the sum of the self-inductance of each winding times one-half the square of its current and all mutual inductances, each times the currents in the two winding coupled by the mutual inductance. It is clear that the energy stored in the leakage inductances is not a part of the energy stored in the coupling field. Thus, the energy stored in the coupling field may be written as

$$W_f = \frac{1}{2}(\mathbf{i}_{abcs})^T (\mathbf{L}_s - L_{ls}\mathbf{I})\mathbf{i}_{abcs} + (\mathbf{i}_{abcs})^T \mathbf{L}'_{sr}\mathbf{i}'_{abcr}$$

$$+ \frac{1}{2}(\mathbf{i}'_{abcr})^T (\mathbf{L}'_r - L'_{lr}\mathbf{I})\mathbf{i}'_{abcr} \tag{4.3-1}$$

where $\mathbf{I}$ is the identity matrix. Because the machine is assumed to be magnetically linear, the field energy W_f is equal to the coenergy W_c.

Before using the second entry of Table 1.3-1 to express the electromagnetic torque, it is necessary to modify the expressions given in Table 1.3-1 to account for a P-pole machine. The change of mechanical energy in a rotational system with one mechanical input may be written from (1.3-71) as

$$dW_m = -T_e d\theta_{rm} \tag{4.3-2}$$

where T_e is the electromagnetic torque positive for motor action (torque output) and θ_{rm} is the actual angular displacement of the rotor. The flux linkages, currents, W_f, and W_c are all expressed as functions of the electrical angular displacement θ_r. Because

$$\theta_r = \left(\frac{P}{2}\right)\theta_{rm} \tag{4.3-3}$$

where P is the number of poles in the machine, then

$$dW_m = -T_e \left(\frac{2}{P}\right) d\theta_r \tag{4.3-4}$$

Therefore to account for a P-pole machine all terms on the right-hand side of Table 1.3-1 should be multiplied by $P/2$. Because $W_f = W_c$, the electromagnetic torque may be evaluated from

$$T_e(i_j, \theta_r) = \left(\frac{P}{2}\right)\frac{\partial W_c(i_j, \theta_r)}{\partial \theta_r} \tag{4.3-5}$$

The abbreviated functional notation, as used in Table 1.3-1, is also used here for the currents. Because $\mathbf{L}_s$ and $\mathbf{L}_r'$ are not functions of θ_r, substituting W_f from (4.3-1) into (4.3-5) yields the electromagnetic torque in Newton · meters $(\text{N} \cdot \text{m})$:

$$T_e = \left(\frac{P}{2}\right)(\mathbf{i}_{abcs})^T \frac{\partial}{\partial \theta_r}[\mathbf{L}_{sr}']\mathbf{i}_{abcr}' \tag{4.3-6}$$

In expanded form, (4.3-6) becomes

$$
\begin{aligned}
T_e = -\left(\frac{P}{2}\right)L_{ms}\Bigg\{ &\left[i_{as}\left(i_{ar}' - \frac{1}{2}i_{br}' - \frac{1}{2}i_{cr}'\right) + i_{bs}\left(i_{br}' - \frac{1}{2}i_{ar}' - \frac{1}{2}i_{cr}'\right)\right.\\
&\left. + i_{cs}\left(i_{cr}' - \frac{1}{2}i_{br}' - \frac{1}{2}i_{ar}'\right)\right]\sin\theta_r + \frac{\sqrt{3}}{2}\left[i_{as}(i_{br}' - i_{cr}') + i_{bs}(i_{cr}' - i_{ar}')\right.\\
&\left. + i_{cs}(i_{ar}' - i_{br}')\right]\cos\theta_r \Bigg\}
\end{aligned}
\tag{4.3-7}
$$

The torque and rotor speed are related by

$$T_e = J\left(\frac{2}{P}\right)p\omega_r + T_L \tag{4.3-8}$$

where J is the inertia of the rotor and in some cases the connected load. The first term on the right-hand side is the inertial torque. In (4.3-8) the units of J are kilogram · meter^2 (kg · m^2) or joules · second^2 (J · s^2). Often the inertia is given as a quantity called WR^2 expressed in units of pound mass · feet^2 (lbm · ft^2). The load torque T_L is positive for a torque load on the shaft of the induction machine.

4.4   EQUATIONS OF TRANSFORMATION FOR ROTOR CIRCUITS

In Chapter 3 the concept of the arbitrary reference frame was introduced and applied to stationary circuits. However, in the analysis of induction machines it is also desirable to transform the variables associated with the symmetrical rotor windings to the arbitrary reference frame. A change of variables which formulates a transformation of the 3-phase variables of the rotor circuits to the arbitrary reference frame is

$$\mathbf{f}_{qd0r}' = \mathbf{K}_r\mathbf{f}_{abcr}' \tag{4.4-1}$$

where

$$\left(\mathbf{f}_{qd0r}'\right)^T = [f_{qr}' \quad f_{dr}' \quad f_{0r}'] \tag{4.4-2}$$

$$\left(\mathbf{f}_{abcr}'\right)^T = [f_{ar}' \quad f_{br}' \quad f_{cr}'] \tag{4.4-3}$$

$$\mathbf{K}_r = \frac{2}{3}\begin{bmatrix} \cos\beta & \cos\left(\beta - \frac{2\pi}{3}\right) & \cos\left(\beta + \frac{2\pi}{2}\right) \\ \sin\beta & \sin\left(\beta - \frac{2\pi}{3}\right) & \sin\left(\beta + \frac{2\pi}{3}\right) \\ \frac{1}{2} & \frac{1}{2} & \frac{1}{2} \end{bmatrix} \tag{4.4-4}$$

$$\beta = \theta - \theta_r \tag{4.4-5}$$

The angular displacement θ is defined by (3.3-5)–(3.3-8) and θ_r is defined by

$$\omega_r = \frac{d\theta_r}{dt} \tag{4.4-6}$$

and we realize that (4.4-6) may be solved for θ_r and written in indefinite or definite integral form as (3.3-7) and (3.3-8), respectively. The inverse is

$$(\mathbf{K}_r)^{-1} = \begin{bmatrix} \cos\beta & \sin\beta & 1 \\ \cos\left(\beta - \frac{2\pi}{3}\right) & \sin\left(\beta - \frac{2\pi}{3}\right) & 1 \\ \cos\left(\beta + \frac{2\pi}{3}\right) & \sin\left(\beta + \frac{2\pi}{3}\right) & 1 \end{bmatrix} \tag{4.4-7}$$

The r subscript indicates the variables, parameters, and transformation associated with rotating circuits. Although this change of variables needs no physical interpretation, it is convenient, as in the case of stationary circuits, to visualize these transformation equations as trigonometric relationships between vector quantities as shown in Fig. 4.4-1.

It is clear that the above transformation equations for rotor circuits are the transformation equations for stationary circuits with β used as the angular displacement rather than θ. In fact, the equations of transformation for stationary and rotor circuits are special cases of a transformation for all circuits, stationary or rotating. In particular, if in β, θ_r is replaced by θ_c where

$$\omega_c = \frac{d\theta_c}{dt} \tag{4.4-8}$$

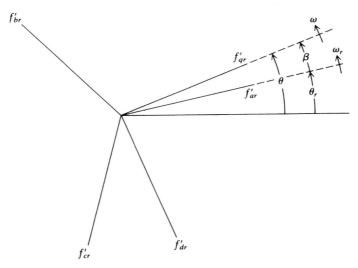

Figure 4.4-1   Transformation for rotating circuits portrayed by trigonometric relationships.

then ω_c, the angular velocity of the circuits, may be selected to correspond to the circuits being transformed; that is, $\omega_c = 0$ for stationary circuits and $\omega_c = \omega_r$ for rotor circuits. Although this more general approach could have been used in Chapter 3, it does add to the complexity of the transformation, making it somewhat more difficult to follow without deriving any advantage from the generality of the approach because only two types of circuits, stationary or fixed in the rotor, are considered in this text.

It follows that all equations for stationary circuits in Sections 3.3 and 3.4 are valid for rotor circuits if θ is replaced by β and ω by $\omega - \omega_r$. The phasor and steady-state relations for stationary circuits, given in Sections 3.7–3.9, also apply to rotor circuits of an induction machine if we realize that the rotor variables, during balanced, steady-state operation, are of the form

$$F'_{ar} = \sqrt{2}F'_r \cos\left[(\omega_e - \omega_r)t + \theta_{erf}(0)\right] \tag{4.4-9}$$

$$F'_{br} = \sqrt{2}F'_r \cos\left[(\omega_e - \omega_r)t + \theta_{erf}(0) - \frac{2\pi}{3}\right] \tag{4.4-10}$$

$$F'_{cr} = \sqrt{2}F'_r \cos\left[(\omega_e - \omega_r)t + \theta_{erf}(0) + \frac{2\pi}{3}\right] \tag{4.4-11}$$

where $\theta_{erf}(0)$ is the phase angle of F'_{ar} at time zero.

4.5 VOLTAGE EQUATIONS IN ARBITRARY REFERENCE-FRAME VARIABLES

Using the information set forth in Chapter 3 and in the previous section, we know the form of the voltage equations in the arbitrary reference frame without any further analysis [2]. In particular,

$$\mathbf{v}_{qd0s} = \mathbf{r}_s\mathbf{i}_{qd0s} + \omega\boldsymbol{\lambda}_{dqs} + p\boldsymbol{\lambda}_{qd0s} \tag{4.5-1}$$

$$\mathbf{v}'_{qd0r} = \mathbf{r}'_r\mathbf{i}'_{qd0r} + (\omega - \omega_r)\boldsymbol{\lambda}'_{dqr} + p\boldsymbol{\lambda}'_{qd0r} \tag{4.5-2}$$

where

$$(\boldsymbol{\lambda}_{dqs})^T = [\lambda_{ds} \quad -\lambda_{qs} \quad 0] \tag{4.5-3}$$

$$(\boldsymbol{\lambda}'_{dqr})^T = [\lambda'_{dr} \quad -\lambda'_{qr} \quad 0] \tag{4.5-4}$$

The set of equations is complete once the expressions for the flux linkages are determined. Substituting the equations of transformation, (3.3-1) and (4.4-1), into the flux linkage equations expressed in *abc* variables (4.2-18) yields the flux linkage equations for a magnetically linear system:

$$\begin{bmatrix} \boldsymbol{\lambda}_{qd0s} \\ \boldsymbol{\lambda}'_{qd0r} \end{bmatrix} = \begin{bmatrix} \mathbf{K}_s\mathbf{L}_s(\mathbf{K}_s)^{-1} & \mathbf{K}_s\mathbf{L}'_{sr}(\mathbf{K}_r)^{-1} \\ \mathbf{K}_r(\mathbf{L}'_{sr})^T(\mathbf{K}_s)^{-1} & \mathbf{K}_r\mathbf{L}'_r(\mathbf{K}_r)^{-1} \end{bmatrix} \begin{bmatrix} \mathbf{i}_{qd0s} \\ \mathbf{i}'_{qd0r} \end{bmatrix} \tag{4.5-5}$$

We know from Chapter 3 that for $\mathbf{L}_s$ of the form given by (4.2-6) we have

$$\mathbf{K}_s \mathbf{L}_s (\mathbf{K}_s)^{-1} = \begin{bmatrix} L_{ls} + L_M & 0 & 0 \\ 0 & L_{ls} + L_M & 0 \\ 0 & 0 & L_{ls} \end{bmatrix} \qquad (4.5\text{-}6)$$

where

$$L_M = \frac{3}{2} L_{ms} \qquad (4.5\text{-}7)$$

Because $\mathbf{L}'_r$ is similar in form to $\mathbf{L}_s$, it follows that

$$\mathbf{K}_r \mathbf{L}'_r (\mathbf{K}_r)^{-1} = \begin{bmatrix} L'_{lr} + L_M & 0 & 0 \\ 0 & L'_{lr} + L_M & 0 \\ 0 & 0 & L'_{lr} \end{bmatrix} \qquad (4.5\text{-}8)$$

It can be shown that

$$\mathbf{K}_s \mathbf{L}'_{sr} (\mathbf{K}_r)^{-1} = \mathbf{K}_r (\mathbf{L}'_{sr})^T (\mathbf{K}_s)^{-1} = \begin{bmatrix} L_M & 0 & 0 \\ 0 & L_M & 0 \\ 0 & 0 & 0 \end{bmatrix} \qquad (4.5\text{-}9)$$

The voltage equations are often written in expanded form. From (4.5-1) and (4.5-2) we obtain

$$v_{qs} = r_s i_{qs} + \omega \lambda_{ds} + p\lambda_{qs} \qquad (4.5\text{-}10)$$

$$v_{ds} = r_s i_{ds} - \omega \lambda_{qs} + p\lambda_{ds} \qquad (4.5\text{-}11)$$

$$v_{0s} = r_s i_{0s} + p\lambda_{0s} \qquad (4.5\text{-}12)$$

$$v'_{qr} = r'_r i'_{qr} + (\omega - \omega_r)\lambda'_{dr} + p\lambda'_{qr} \qquad (4.5\text{-}13)$$

$$v'_{dr} = r'_r i'_{dr} - (\omega - \omega_r)\lambda'_{qr} + p\lambda'_{dr} \qquad (4.5\text{-}14)$$

$$v'_{0r} = r'_r i'_{0r} + p\lambda'_{0r} \qquad (4.5\text{-}15)$$

Substituting (4.5-6), (4.5-8), and (4.5-9) into (4.5-5) yields the expressions for the flux linkages. In expanded form we have

$$\lambda_{qs} = L_{ls} i_{qs} + L_M (i_{qs} + i'_{qr}) \qquad (4.5\text{-}16)$$

$$\lambda_{ds} = L_{ls} i_{ds} + L_M (i_{ds} + i'_{dr}) \qquad (4.5\text{-}17)$$

$$\lambda_{0s} = L_{ls} i_{0s} \qquad (4.5\text{-}18)$$

$$\lambda'_{qr} = L'_{lr} i'_{qr} + L_M (i_{qs} + i'_{qr}) \qquad (4.5\text{-}19)$$

$$\lambda'_{dr} = L'_{lr} i'_{dr} + L_M (i_{ds} + i'_{dr}) \qquad (4.5\text{-}20)$$

$$\lambda'_{0r} = L'_{lr} i'_{0r} \qquad (4.5\text{-}21)$$

The voltage and flux linkage equations suggest the equivalent circuits shown in Fig. 4.5-1.

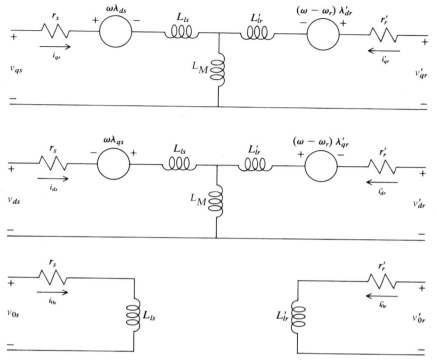

Figure 4.5-1 Arbitrary reference-frame equivalent circuits for a 3-phase, symmetrical induction machine.

Because machine and power system parameters are nearly always given in ohms or percent or per unit of a base impedance, it is convenient to express the voltage and flux linkage equations in terms of reactances rather than inductances. Hence, (4.5-10)–(4.5-15) are often written as

$$v_{qs} = r_s i_{qs} + \frac{\omega}{\omega_b}\psi_{ds} + \frac{p}{\omega_b}\psi_{qs} \tag{4.5-22}$$

$$v_{ds} = r_s i_{ds} - \frac{\omega}{\omega_b}\psi_{qs} + \frac{p}{\omega_b}\psi_{ds} \tag{4.5-23}$$

$$v_{0s} = r_s i_{0s} + \frac{p}{\omega_b}\psi_{0s} \tag{4.5-24}$$

$$v'_{qr} = r'_r i'_{qr} + \left(\frac{\omega - \omega_r}{\omega_b}\right)\psi'_{dr} + \frac{p}{\omega_b}\psi'_{qr} \tag{4.5-25}$$

$$v'_{dr} = r'_r i'_{dr} - \left(\frac{\omega - \omega_r}{\omega_b}\right)\psi'_{qr} + \frac{p}{\omega_b}\psi'_{dr} \tag{4.5-26}$$

$$v'_{0r} = r'_r i'_{0r} + \frac{p}{\omega_b}\psi'_{0r} \tag{4.5-27}$$

where ω_b is the base electrical angular velocity used to calculate the inductive reactances. Flux linkages (4.5-16)–(4.5-21) now become flux linkages per second with the units of volts:

$$\psi_{qs} = X_{ls}i_{qs} + X_M(i_{qs} + i'_{qr}) \tag{4.5-28}$$

$$\psi_{ds} = X_{ls}i_{ds} + X_M(i_{ds} + i'_{dr}) \tag{4.5-29}$$

$$\psi_{0s} = X_{ls}i_{0s} \tag{4.5-30}$$

$$\psi'_{qr} = X'_{lr}i'_{qr} + X_M(i_{qs} + i'_{qr}) \tag{4.5-31}$$

$$\psi'_{dr} = X'_{lr}i'_{dr} + X_M(i_{ds} + i'_{dr}) \tag{4.5-32}$$

$$\psi'_{0r} = X'_{lr}i'_{0r} \tag{4.5-33}$$

In the above equations the inductive reactances are obtained by multiplying ω_b times inductance. It is left to the reader to modify the equivalent circuits shown in Fig. 4.5-1 to accommodate the use of reactances rather than inductances in the voltage equations.

The voltage equations (4.5-10)–(4.5-15) or (4.5-22)–(4.5-27) are written in terms of currents and flux linkages (flux linkages per second). Clearly, the currents and flux linkages are related and both cannot be independent or state variables. In transfer function formulation and computer simulation of induction machines we will find it desirable to express the voltage equations in terms of either currents or flux linkages (flux linkages per second).

If the currents are selected as independent variables and the flux linkages or flux linkages per second replaced by the currents, the voltage equations become

$$
\begin{bmatrix} v_{qs} \\ v_{ds} \\ v_{0s} \\ v'_{qr} \\ v'_{dr} \\ v'_{0r} \end{bmatrix} =
\begin{bmatrix}
r_s + \frac{p}{\omega_b}X_{ss} & \frac{\omega}{\omega_b}X_{ss} & 0 & \frac{p}{\omega_b}X_M & \frac{\omega}{\omega_b}X_M & 0 \\
-\frac{\omega}{\omega_b}X_{ss} & r_s + \frac{p}{\omega_b}X_{ss} & 0 & -\frac{\omega}{\omega_b}X_M & \frac{p}{\omega_b}X_M & 0 \\
0 & 0 & r_s + \frac{p}{\omega_b}X_{ls} & 0 & 0 & 0 \\
\frac{p}{\omega_b}X_M & \left(\frac{\omega-\omega_r}{\omega_b}\right)X_M & 0 & r'_r + \frac{p}{\omega_b}X'_{rr} & \left(\frac{\omega-\omega_r}{\omega_b}\right)X'_{rr} & 0 \\
-\left(\frac{\omega-\omega_r}{\omega_b}\right)X_M & \frac{p}{\omega_b}X_M & 0 & -\left(\frac{\omega-\omega_r}{\omega_b}\right)X'_{rr} & r'_r + \frac{p}{\omega_b}X'_{rr} & 0 \\
0 & 0 & 0 & 0 & 0 & r'_r + \frac{p}{\omega_b}X'_{lr}
\end{bmatrix}
\begin{bmatrix} i_{qs} \\ i_{ds} \\ i_{0s} \\ i'_{qr} \\ i'_{dr} \\ i'_{0r} \end{bmatrix}
$$

$$\tag{4.5-34}$$

where

$$X_{ss} = X_{ls} + X_M \tag{4.5-35}$$

$$X'_{rr} = X'_{lr} + X_M \tag{4.5-36}$$

The flux linkages per second may be expressed from (4.5-28)–(4.5-33) as

$$
\begin{bmatrix} \psi_{qs} \\ \psi_{ds} \\ \psi_{0s} \\ \psi'_{qr} \\ \psi'_{dr} \\ \psi'_{0r} \end{bmatrix} =
\begin{bmatrix}
X_{ss} & 0 & 0 & X_M & 0 & 0 \\
0 & X_{ss} & 0 & 0 & X_M & 0 \\
0 & 0 & X_{ls} & 0 & 0 & 0 \\
X_M & 0 & 0 & X'_{rr} & 0 & 0 \\
0 & X_M & 0 & 0 & X'_{rr} & 0 \\
0 & 0 & 0 & 0 & 0 & X'_{lr}
\end{bmatrix}
\begin{bmatrix} i_{qs} \\ i_{ds} \\ i_{0s} \\ i'_{qr} \\ i'_{dr} \\ i'_{0r} \end{bmatrix}
\tag{4.5-37}
$$

If flux linkages or flux linkages per second are selected as independent variables then (4.5-37) may be solved for currents and written as

$$
\begin{bmatrix} i_{qs} \\ i_{ds} \\ i_{0s} \\ i'_{qr} \\ i'_{dr} \\ i'_{0r} \end{bmatrix} = \frac{1}{D} \begin{bmatrix} X'_{rr} & 0 & 0 & -X_M & 0 & 0 \\ 0 & X'_{rr} & 0 & 0 & -X_M & 0 \\ 0 & 0 & \frac{D}{X_{ls}} & 0 & 0 & 0 \\ -X_M & 0 & 0 & X_{ss} & 0 & 0 \\ 0 & -X_M & 0 & 0 & X_{ss} & 0 \\ 0 & 0 & 0 & 0 & 0 & \frac{D}{X'_{lr}} \end{bmatrix} \begin{bmatrix} \psi_{qs} \\ \psi_{ds} \\ \psi_{0s} \\ \psi'_{qr} \\ \psi'_{dr} \\ \psi'_{0r} \end{bmatrix}
\tag{4.5-38}
$$

where

$$
D = X_{ss}X'_{rr} - X_M^2
\tag{4.5-39}
$$

Substituting (4.5-38) for the currents into (4.5-22)-(4.5-27) yields the voltage equations in terms of flux linkages per second as given by (4.5-40).

$$
\begin{bmatrix} v_{qs} \\ v_{ds} \\ v_{0s} \\ v'_{qr} \\ v'_{dr} \\ v'_{0r} \end{bmatrix} = \begin{bmatrix} \frac{r_s X'_{rr}}{D} + \frac{p}{\omega_b} & \frac{\omega}{\omega_b} & 0 & -\frac{r_s X_M}{D} & 0 & 0 \\ -\frac{\omega}{\omega_b} & \frac{r_s X'_{rr}}{D} + \frac{p}{\omega_b} & 0 & 0 & -\frac{r_s X_M}{D} & 0 \\ 0 & 0 & \frac{r_s}{X_{ls}} + \frac{p}{\omega_b} & 0 & 0 & 0 \\ -\frac{r'_r X_M}{D} & 0 & 0 & \frac{r'_r X_{ss}}{D} + \frac{p}{\omega_b} & \frac{\omega-\omega_r}{\omega_b} & 0 \\ 0 & -\frac{r'_r X_M}{D} & 0 & -\frac{\omega-\omega_r}{\omega_b} & \frac{r'_r X_{ss}}{D} + \frac{p}{\omega_b} & 0 \\ 0 & 0 & 0 & 0 & 0 & \frac{r'_r}{X'_{lr}} + \frac{p}{\omega_b} \end{bmatrix} \begin{bmatrix} \psi_{qs} \\ \psi_{ds} \\ \psi_{0s} \\ \psi'_{qr} \\ \psi'_{dr} \\ \psi'_{0r} \end{bmatrix}
$$

$$
\tag{4.5-40}
$$

It is interesting to note that each q- and d-voltage equation contains two derivatives of current when currents are selected as independent or state variables, (4.5-34). When flux linkages are selected as independent variables, (4.5-40), each q- and d-voltage equation contains only one derivative of flux linkage. We will see that this property makes it more convenient to implement a computer simulation of an induction machine with flux linkages as state variables rather than with currents.

4.6 TORQUE EQUATION IN ARBITRARY REFERENCE-FRAME VARIABLES

The expression for the electromagnetic torque in terms of arbitrary reference-frame variables may be obtained by substituting the equations of transformation into (4.3-6). Thus

$$
T_e = \left(\frac{P}{2}\right)[(\mathbf{K}_s)^{-1}\mathbf{i}_{qd0s}]^T \frac{\partial}{\partial\theta_r}[\mathbf{L}'_{sr}](\mathbf{K}_r)^{-1}\mathbf{i}'_{qd0r}
\tag{4.6-1}
$$

This expression yields the torque expressed in terms of currents as

$$T_e = \left(\frac{3}{2}\right)\left(\frac{P}{2}\right) L_M (i_{qs} i'_{dr} - i_{ds} i'_{qr}) \qquad (4.6\text{-}2)$$

where T_e is positive for motor action. Other equivalent expressions for the electromagnetic torque of an induction machine are

$$T_e = \left(\frac{3}{2}\right)\left(\frac{P}{2}\right) (\lambda'_{qr} i'_{dr} - \lambda'_{dr} i'_{qr}) \qquad (4.6\text{-}3)$$

$$T_e = \left(\frac{3}{2}\right)\left(\frac{P}{2}\right) (\lambda_{ds} i_{qs} - \lambda_{qs} i_{ds}) \qquad (4.6\text{-}4)$$

Equations (4.6-3) and (4.6-4) may be somewhat misleading because they seem to imply that the leakage inductances are involved in the energy conversion process. This, however, is not the case. Even though the flux linkages in (4.6-3) and (4.6-4) contain the leakage inductances, they are eliminated by the algebra within the parentheses. The above expressions for torque are also often written in terms of flux linkages per second and currents. For example, (4.6-3) can be written as

$$T_e = \left(\frac{3}{2}\right)\left(\frac{P}{2}\right)\left(\frac{1}{\omega_b}\right) (\psi'_{qr} i'_{dr} - \psi'_{dr} i'_{qr}) \qquad (4.6\text{-}5)$$

It is left to the reader to show that in terms of flux linkages per second the electromagnetic torque may be expressed as

$$T_e = \left(\frac{3}{2}\right)\left(\frac{P}{2}\right)\left(\frac{X_M}{D\omega_b}\right) (\psi_{qs} \psi'_{dr} - \psi'_{qr} \psi_{ds}) \qquad (4.6\text{-}6)$$

where D is defined by (4.5-39).

4.7 COMMONLY USED REFERENCE FRAMES

Although the behavior of a symmetrical induction machine may be described in any frame of reference, there are three that are commonly employed: the stationary reference frame first employed by H. C. Stanley [3], the rotor reference frame which is Park's transformation [4] applied to induction machines (Brereton et al. [5]), and the synchronously rotating reference frame [6]. The voltage equations for each of these reference frames may be obtained from the voltage equations in the arbitrary reference frame by assigning the appropriate speed to ω. That is, $\omega = 0$ for the stationary, $\omega = \omega_r$ for the rotor, and $\omega = \omega_e$ for the synchronously rotating reference frame.

Generally, the conditions of operation will determine the most convenient reference frame for analysis and/or simulation purposes [2]. If, for example, the stator voltages are unbalanced or discontinuous and the rotor applied voltages are balanced or zero, the stationary reference frame should be used to simulate the performance of the induction machine. If, on the other hand, the external rotor circuits are unbalanced but the applied stator voltages are balanced, then the reference frame fixed in the rotor is most convenient. Either the stationary or synchronously rotating reference frame is generally used to analyze balanced or symmetrical conditions. Linearized machine equations that are used to determine eigenvalues and to express linearized transfer functions for use in control system analysis are obtained from the voltage equations expressed in the synchronously rotating reference frame. The synchronously rotating reference frame is also particularly convenient when incorporating the dynamic characteristics of an induction machine into a digital computer program used to study the transient and dynamic stability of large power systems. The synchronously rotating reference frame may also be useful in variable frequency applications if it is permissible to assume that the stator voltages are a balanced sinusoidal set. In this case, variable frequency operation may be analyzed by varying the speed of the arbitrary reference frame to coincide with the electrical angular velocity of the applied stator voltages. A word of caution is perhaps appropriate. Regardless of the reference frame being used, the stator and rotor voltages and currents must be properly transformed to and from this reference frame. In most cases these transformations are straightforward and can be accomplished implicitly. However, it may be necessary to actually include or implement a transformation in the analysis or computer simulation of an induction machine, whereupon special care must be taken.

4.8 PER UNIT SYSTEM

It is often convenient to express machine parameters and variables as per unit quantities. Base power and base voltage are selected, and all parameters and variables are normalized using these base quantities. When the machine is being considered separately, the base power is generally selected as the horsepower rating of the machine in volt-amperes (i.e., horsepower times 746). If, on the other hand, the machine is a part of a power system and if it is desirable to convert the entire system to per unit quantities, then only one power base (VA base) is selected which would most likely be different from the rating of any machine in the system. Here we will consider the machine separately with the rating of the machine taken as base power.

Although we will violate this convention from time to time when dealing with instantaneous quantities, the rms value of the rated phase voltage is generally selected as base voltage for the *abc* variables while the peak value is generally selected as base voltage for the $qd0$ variables. That is, $V_{B(abc)}$ is the rms voltage selected as base voltage for the *abc* variables then $V_{B(qd0)} = \sqrt{2}V_{B(abc)}$. The base power may be expressed as

$$P_B = 3V_{B(abc)}I_{B(abc)} \qquad (4.8\text{-}1)$$

or

$$P_B = \left(\frac{3}{2}\right) V_{B(qd0)} I_{B(qd0)} \qquad (4.8\text{-}2)$$

Therefore, because base voltage and base power are selected, base current can be calculated from either (4.8-1) or (4.8-2). It follows that the base impedance may be expressed as

$$Z_B = \frac{V_{B(abc)}}{I_{B(abc)}} = \frac{3 V_{B(abc)}^2}{P_B} \qquad (4.8\text{-}3)$$

or

$$Z_B = \frac{V_{B(qd0)}}{I_{B(qd0)}} = \left(\frac{3}{2}\right) \frac{V_{B(qd0)}^2}{P_B} \qquad (4.8\text{-}4)$$

The $qd0$ equations written in terms of reactances, (4.5-22)–(4.5-33), can be readily converted to per unit by dividing the voltages by $V_{B(qd0)}$, the currents by $I_{B(qd0)}$, and the resistances and reactances by Z_B. Note that because a flux linkage per second is a volt, it is per unitized by dividing by base voltage.

Although the voltage and flux linkage per second equations do not change form when per unitized, the torque equation is modified by the per unitizing process. For this purpose the base torque may be expressed as

$$T_B = \frac{P_B}{(2/P)\omega_b} \qquad (4.8\text{-}5)$$

where ω_b corresponds to rated or base frequency of the machine. A word of caution is appropriate. If, in (4.8-5), P_B is the rated power output of the machine, then base torque T_B will not be rated torque. We will find that, in the case of an induction machine, rated power output generally occurs at a speed (rated speed) slightly less than synchronous. Hence, T_B will be less than rated torque by the ratio of rated speed to synchronous speed.

If the torque expression given by (4.6-5) is divided by (4.8-5), with (4.8-2) substituted for P_B, the multiplier $(\frac{3}{2})(P/2)(1/\omega_b)$ is eliminated, and with all quantities expressed in per unit, the per unit torque becomes

$$T_e = \psi'_{qr} i'_{dr} - \psi'_{dr} i'_{qr} \qquad (4.8\text{-}6)$$

If the electrical variables are expressed in volts, amperes, and watts, then the inertia of the rotor is expressed in mks units. If, however, the per unit system is used, the inertia is expressed in seconds. This can be shown by first recalling from (4.3-8) that the inertial torque T_{IT} for a P-pole machine may be expressed as

$$T_{IT} = J\left(\frac{2}{P}\right) p\omega_r \qquad (4.8\text{-}7)$$

where ω_r is the electrical angular velocity of the rotor and J is the inertia of the rotor and connected mechanical load expressed in kg · m^2. In order to express (4.8-7) in per unit, it is divided by base torque and the rotor speed is normalized to base speed. Thus

$$T_{IT} = \frac{J(2/P)\omega_b}{T_B} p \frac{\omega_r}{\omega_b} \qquad (4.8\text{-}8)$$

By definition, the inertia constant expressed in seconds is

$$H = \left(\frac{1}{2}\right)\left(\frac{2}{P}\right)\frac{J\omega_b}{T_B} = \left(\frac{1}{2}\right)\left(\frac{2}{P}\right)^2\frac{J\omega_b^2}{P_B} \qquad (4.8\text{-}9)$$

Thus, in per unit, (4.3-8) becomes

$$T_e = 2Hp\frac{\omega_r}{\omega_b} + T_L \qquad (4.8\text{-}10)$$

It is important to become familiar with both systems of units and to be able to convert readily from one to the other. We will use both systems interchangeably throughout the text.

4.9 ANALYSIS OF STEADY-STATE OPERATION

The voltage equations that describe the balanced steady-state operation of an induction machine may be obtained in several ways. For balanced conditions the zero quantities of the stator and rotor are zero, and from our work in Chapter 3 we know that for balanced steady-state conditions the q and d variables are sinusoidal in all reference frames except the synchronously rotating reference frame wherein they are constant. Hence, one method of obtaining the steady-state voltage equations for balanced conditions is to first recall that in an asynchronously rotating reference frame the steady-state voltages are related by

$$\tilde{F}_{ds} = j\tilde{F}_{qs} \qquad (4.9\text{-}1)$$

and with $\theta(0) = 0$

$$\tilde{F}_{qs} = \tilde{F}_{as} \qquad (4.9\text{-}2)$$

Because the induction machine is a symmetrical device, (4.9-1) and (4.9-2) also apply to the stator currents and flux linkages. Likewise, the steady-state rotor variables are related by

$$\tilde{F}'_{dr} = j\tilde{F}'_{qr} \qquad (4.9\text{-}3)$$

and with $\theta(0)$ and $\theta_r(0)$ both selected equal to zero we obtain

$$\tilde{F}'_{qr} = \tilde{F}'_{ar} \tag{4.9-4}$$

Appropriate substitution of these equations into either (4.5-22) and (4.5-25) or (4.5-23) and (4.5-26) yields the standard steady-state voltage equations in phasor form. A second method is to express (4.5-22), (4.5-23), (4.5-25), and (4.5-26) in the synchronously rotating reference frame and neglect the time rate of change of all flux linkages and then employ the relationships

$$\sqrt{2}\tilde{F}_{as} = F^e_{qs} - jF^e_{ds} \tag{4.9-5}$$

$$\sqrt{2}\tilde{F}'_{ar} = \tilde{F}'^e_{qr} - j\tilde{F}'^e_{dr} \tag{4.9-6}$$

We will proceed using the first approach and leave the latter as an exercise for the reader.

If, in (4.5-22) and (4.5-25), p is replaced by $j(\omega_e - \omega)$, the equations may be written in phasor form as

$$\tilde{V}_{qs} = r_s\tilde{I}_{qs} + \frac{\omega}{\omega_b}\tilde{\Psi}_{ds} + j\left(\frac{\omega_e - \omega}{\omega_b}\right)\tilde{\Psi}_{qs} \tag{4.9-7}$$

$$\tilde{V}'_{qr} = r'_r\tilde{I}'_{qr} + \left(\frac{\omega - \omega_r}{\omega_b}\right)\tilde{\Psi}'_{dr} + j\left(\frac{\omega_e - \omega}{\omega_b}\right)\tilde{\Psi}'_{qr} \tag{4.9-8}$$

Substituting (4.9-1) and (4.9-3) into the above equations yields

$$\tilde{V}_{qs} = r_s\tilde{I}_{qs} + j\frac{\omega_e}{\omega_b}\tilde{\Psi}_{qs} \tag{4.9-9}$$

$$\tilde{V}'_{qr} = r'_r\tilde{I}'_{qr} + j\left(\frac{\omega_e - \omega_r}{\omega_b}\right)\tilde{\Psi}'_{qr} \tag{4.9-10}$$

The well-known steady-state phasor voltage equations, which are actually valid in all asynchronously rotating reference frames, are obtained by substituting the phasor form of (4.5-28) and (4.5-31) for $\tilde{\psi}_{qs}$ and $\tilde{\psi}'_{qr}$, respectively, into (4.9-9) and (4.9-10) and employing (4.9-2) and (4.9-4) to replace qs and qr variables with as and ar variables, respectively:

$$\tilde{V}_{as} = \left(r_s + j\frac{\omega_e}{\omega_b}X_{ls}\right)\tilde{I}_{as} + j\frac{\omega_e}{\omega_b}X_M(\tilde{I}_{as} + \tilde{I}'_{ar}) \tag{4.9-11}$$

$$\frac{\tilde{V}'_{ar}}{s} = \left(\frac{r'_r}{s} + j\frac{\omega_e}{\omega_b}X'_{lr}\right)\tilde{I}'_{ar} + j\frac{\omega_e}{\omega_b}X_M(\tilde{I}_{as} + \tilde{I}'_{ar}) \tag{4.9-12}$$

where the slip s is defined as

$$s = \frac{\omega_e - \omega_r}{\omega_e} \tag{4.9-13}$$

Equations (4.9-11) and (4.9-12) suggest the equivalent circuit shown in Fig. 4.9-1.

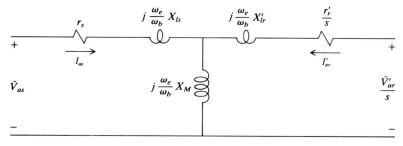

Figure 4.9-1 Equivalent circuit for steady-state operation of a symmetrical induction machine.

All voltage equations given in this chapter are valid regardless of the frequency of operation in that ω_e and ω_b are both given explicitly. Because ω_b corresponds to rated frequency, it is generally used to calculate and per unitize the reactances. Although the ratio of ω_e to ω_b is generally not included in the steady-state voltage equations, this ratio makes (4.9-11) and (4.9-12) valid for applied voltages of any constant frequency. It should be clear that all products of the ratio of ω_e to ω_b times a reactance may be replaced by ω_e times the appropriate inductance.

The steady-state electromagnetic torque can be expressed in terms of currents in phasor form by first writing the torque in terms of currents in the synchronously rotating reference frame and then utilizing (4.9-5) and (4.9-6) to relate synchronously rotating reference frame and phasor quantities. The torque expression becomes

$$T_e = 3\left(\frac{P}{2}\right)\left(\frac{X_M}{\omega_b}\right)\mathrm{Re}\left[j\tilde{I}_{as}^*\tilde{I}_{ar}'\right] \tag{4.9-14}$$

where $\tilde{I}_{as}^*$ is the conjugate of $\tilde{I}_{as}$.

The balanced steady-state torque-speed or torque-slip characteristics of a singly excited or singly fed induction machine warrants discussion. The vast majority of induction machines in use today are singly excited wherein electric power is transferred to or from the induction machine via the stator circuits with the rotor windings short-circuited. Moreover, a vast majority of the singly excited induction machines are of the squirrel-cage rotor type. For singly fed machines, $\tilde{V}_{ar}'$ is zero whereupon

$$\tilde{I}_{ar}' = -\frac{j(\omega_e/\omega_b)X_M}{r_r'/s + j(\omega_e/\omega_b)X_{rr}'}\tilde{I}_{as} \tag{4.9-15}$$

where X_{rr}' is defined from (4.5-36). Substituting (4.9-15) into (4.9-14) yields

$$T_e = \frac{3\left(\frac{P}{2}\right)\left(\frac{\omega_e}{\omega_b}\right)\left(\frac{X_M^2}{\omega_b}\right)\left(\frac{r_r'}{s}\right)|\tilde{I}_{as}|^2}{\left(\frac{r_r'}{s}\right)^2 + \left(\frac{\omega_e}{\omega_b}\right)^2 X_{rr}'^2} \tag{4.9-16}$$

Now, the input impedance of the equivalent circuit shown in Fig. 4.9-1, with $\tilde{V}'_{ar}$ equal to zero, is

$$Z = \frac{\frac{r_s r'_r}{s} + \left(\frac{\omega_e}{\omega_b}\right)^2 \left(X_M^2 - X_{ss} X'_{rr}\right) + j\frac{\omega_e}{\omega_b}\left(\frac{r'_r}{s} X_{ss} + r_s X'_{rr}\right)}{\frac{r'_r}{s} + j\frac{\omega_e}{\omega_b} X'_{rr}}$$

(4.9-17)

where X_{ss} is defined from (4.5-35). Because

$$|\tilde{I}_{as}| = \frac{|\tilde{V}_{as}|}{|Z|}$$

(4.9-18)

the torque for a singly fed induction machine may be expressed as

$$T_e = \frac{3\left(\frac{P}{2}\right)\frac{\omega_e}{\omega_b}\left(\frac{X_M^2}{\omega_b}\right) r'_r s |\tilde{V}_{as}|^2}{\left[r_s r'_r + s\left(\frac{\omega_e}{\omega_b}\right)^2 \left(X_M^2 - X_{ss} X'_{rr}\right)\right]^2 + \left(\frac{\omega_e}{\omega_b}\right)^2 \left(r'_r X_{ss} + s r_s X'_{rr}\right)^2}$$

(4.9-19)

In per unit, (4.9-19) becomes

$$T_e = \frac{\frac{\omega_e}{\omega_b} X_M^2 r'_r s |\tilde{V}_{as}|^2}{\left[r_s r'_r + s\left(\frac{\omega_e}{\omega_b}\right)^2 \left(X_M^2 - X_{ss} X'_{rr}\right)\right]^2 + \left(\frac{\omega_e}{\omega_b}\right)^2 \left(r'_r X_{ss} + s r_s X'_{rr}\right)^2}$$

(4.9-20)

The steady-state torque–speed or torque–slip characteristics typical of many singly excited multiphase induction machines are shown in Fig. 4.9-2, where $\omega_e = \omega_b$. Generally, the parameters of the machine are selected so that maximum torque occurs near synchronous speed and the maximum torque output (motor action) is two or three times the rated torque of the machine.

An expression for the slip at maximum torque may be obtained by taking the derivative of (4.9-19) with respect to slip and setting the result equal to zero. Thus

$$s_m = r'_r G$$

(4.9-21)

where

$$G = \pm \left[\frac{(\omega_e/\omega_b)^{-2} r_s^2 + X_{ss}^2}{(X_M^2 - X_{ss} X'_{rr})^2 \left(\frac{\omega_e}{\omega_b}\right)^2 + r_s^2 X'^2_{rr}}\right]$$

(4.9-22)

Two values of slip at maximum torque are obtained: one for motor action and one for generator action. It is important to note that G is not a function of r'_r; thus the slip at

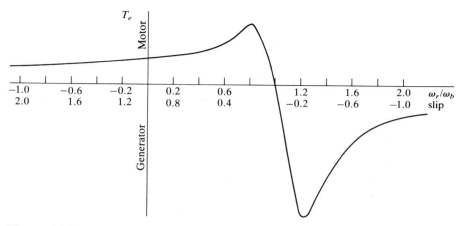

Figure 4.9-2 Steady-state torque–speed characteristics of a singly excited induction machine.

maximum torque is directly proportional to r'_r. Consequently, because all other machine parameters are constant, the speed at which maximum steady-state torque occurs may be varied by inserting external rotor resistance. This feature is often used when starting large motors that have coil-wound rotors with slip rings. In this application, balanced external rotor resistances are inserted so that maximum torque occurs near stall. As the machine speeds up, the external resistors are reduced to zero. On the other hand, 2-phase induction machines used as servomotors are designed with high-resistance rotor windings. Because it is used in positioning devices, the rotor is generally stationary and rotates only when an error signal causes the applied voltage(s) to increase from zero. Consequently, maximum torque must be available at or near stall in order to provide fast response of the position-follow-up system.

It may at first appear that the magnitude of the maximum torque would be influenced by r'_r. However, if (4.9-21) is substituted in (4.9-19), the maximum torque is expressed as

$$T_{e(max)} = \frac{3\left(\frac{P}{2}\right)\left(\frac{\omega_e}{\omega_b}\right)\left(\frac{X_M^2}{\omega_b}\right)G|\tilde{V}_{as}|^2}{\left[r_s + G\left(\frac{\omega_e}{\omega_b}\right)^2(X_M^2 - X_{ss}X'_{rr})\right]^2 + \left(\frac{\omega_e}{\omega_b}\right)^2(X_{ss} + Gr_sX'_{rr})^2} \qquad (4.9\text{-}23)$$

Equation (4.9-23) is independent of r'_r. Thus the maximum torque remains constant if only r'_r is varied. The effect of changing r'_r is illustrated in Fig. 4.9-3, where $r'_{r3} > r'_{r2}$ and $r'_{r2} > r'_{r1}$.

In variable frequency drive systems the operating speed of the electric machine (reluctance, synchronous, or induction) is changed by changing the frequency of the applied voltages by either an inverter or a cycloconverter arrangement. As

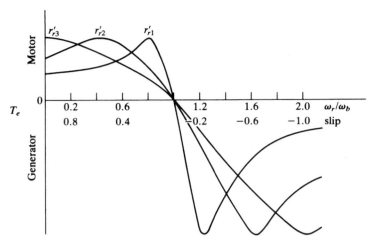

Figure 4.9-3 Steady-state torque–speed characteristics of a singly excited induction machine for various values of rotor resistance.

mentioned previously, the phasor voltage equations are applicable regardless of the frequency of operation. It is only necessary to keep in mind that the reactances given in the steady-state equivalent circuits (Fig. 4.9-1) are defined by the product of ω_b and the inductances. As frequency is decreased, the time rate of change of the steady-state variables is decreased proportionally. Thus, the reactances, or $(\omega_e/\omega_b)X$ in the equivalent circuit, decrease linearly with frequency. If the amplitude of the applied voltages is maintained at the rated value, the current will become excessive. In order to prevent large currents, the magnitude of the stator voltages is decreased as the frequency is decreased. In many applications the voltage magnitude is reduced linearly with frequency until a low frequency is reached, whereupon the decrease in voltage is programmed in a manner to compensate for the effects of the stator resistance.

The influence of frequency upon the steady-state torque–speed characteristics is illustrated in Fig. 4.9-4. These characteristics are for a linear relationship between the magnitude of the applied voltages and frequency. The machine is designed to operate at $\omega_e = \omega_b$, where ω_b corresponds to the base or rated frequency. Rated voltage is applied at rated frequency; that is, when $\omega_e = \omega_b$ we have $|\tilde{V}_{as}| = V_B$, where V_B is the rated or base voltage. The maximum torque is reduced markedly at $\omega_e/\omega_b = 0.1$. At this frequency the voltage would probably be increased somewhat so as to obtain a higher torque. Perhaps a voltage of say $0.15V_B$ or $0.2V_B$ would be used rather than $0.1V_B$. Saturation may, however, cause the current to be excessive at this higher voltage. These practical considerations of variable frequency drives are of major importance but are beyond the scope of this chapter.

Example 4A   The parameters for the equivalent circuit shown in Fig. 4.9-1 may be calculated using electric field theory or determined from tests [7,8].

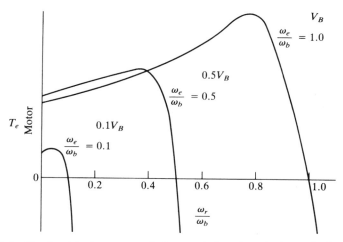

Figure 4.9-4 Steady-state torque–speed characteristics of a singly excited symmetrical induction machine for various operating frequencies.

The tests generally performed are a dc test, a no-load test, and a blocked-rotor test. The following test data are given for a 5-hp, 4-pole, 220-volt, 3-phase, 60-Hz induction machine [8] where all ac voltages and currents are rms values.

dc test	No-load test	Blocked-rotor test
$V_{dc} = 13.8$ V	$V_{nl} = 220$ V	$V_{br} = 23.5$ V
$I_{dc} = 13.0$ A	$I_{nl} = 3.86$ A	$I_{br} = 12.9$ A
	$P_{nl} = 200$ W	$P_{br} = 469$ W
	$f = 60$ Hz	$f = 15$ Hz

During the dc test a dc voltage is applied across two terminals while the machine is at standstill. Thus

$$r_s = \frac{1}{2}\frac{V_{dc}}{I_{dc}} = \frac{13.8}{2 \times 13} = 0.531 \ \Omega \tag{4A-1}$$

The no-load test, which is analogous to the transformer open-circuit test, is performed with balanced 3-phase, 60-Hz voltages applied to the stator windings without mechanical load on the machine (no load). The power input during this test is the sum of the stator ohmic losses, the core losses due to hysteresis and eddy current losses, and rotational losses due to friction and windage. The stator ohmic losses are (I_{nl} is a phase current)

$$P_{I^2 r_s} = 3I_{nl}^2 r_s = 3 \times (3.86)^2 \times 0.531 = 23.7 \ \text{W} \tag{4A-2}$$

Therefore the power loss due to friction and windage losses and core losses is

$$P_{fWC} = P_{nl} - P_{I^2 r_s} = 200 - 23.7 = 176.3 \text{ W} \qquad (4A\text{-}3)$$

In the equivalent circuit shown in Fig. 4.9-1, this loss is neglected. It is generally small, and in most cases little error is introduced by neglecting it. It can be taken into account by placing a resistor in shunt with the magnetizing reactance X_M or by applying a small mechanical load (torque) to the shaft of the machine.

It is noted from the no-load test data that the power factor is very small because the apparent power is (V_{nl} is a line-to-line voltage)

$$|S_{nl}| = \sqrt{3} V_{nl} I_{nl} = \sqrt{3} \times 220 \times 3.86 = 1470.9 \text{ V} \qquad (4A\text{-}4)$$

Therefore, the no-load impedance is highly inductive and its magnitude is assumed to be the sum of the stator leakage reactance and the magnetizing reactance because the rotor speed is essentially synchronous, whereupon r'_r/s is much larger than X_M. Thus

$$X_{ls} + X_M \approx \frac{V_{nl}}{\sqrt{3} I_{nl}}$$
$$= \frac{220}{\sqrt{3} \times 3.86} = 32.9 \qquad (4A\text{-}5)$$

During the blocked-rotor test, which is analogous to the transformer short-circuit test, the rotor is locked by some external means and balanced 3-phase stator voltages are applied. The frequency of the applied voltage is often less than rated in order to obtain a representative value of r'_r because, during normal operation, the frequency of the rotor currents is low and the rotor resistances of some induction machines vary considerably with frequency. During stall the impedance $r'_r + jX'_{lr}$ is much smaller in magnitude than X_M, whereupon the current flowing in the magnetizing reactance may be neglected. Hence

$$P_{br} = 3I_{br}^2 (r_s + r'_r) \qquad (4A\text{-}6)$$

from which

$$r'_r = \frac{P_{br}}{3I_{br}^2} - r_s = \frac{469}{3 \times (12.9)^2} - 0.531 = 0.408 \ \Omega \qquad (4A\text{-}7)$$

The magnitude of the blocked-rotor input impedance is

$$|Z_{br}| = \frac{V_{br}}{\sqrt{3} I_{br}} = \frac{23.5}{\sqrt{3} \times 12.9} = 1.052 \ \Omega \qquad (4A\text{-}8)$$

Now

$$\left| (r_s + r'_r) + j\frac{15}{60}(X_{ls} + X'_{lr}) \right| = 1.052\,\Omega \qquad (4A\text{-}9)$$

from which

$$\begin{aligned}
\left[\frac{15}{16}(X_{ls} + X'_{lr})\right]^2 &= (1.052)^2 - (r_s + r'_r)^2 \\
&= (1.052)^2 - (0.531 + 0.408)^2 \\
&= 0.225\,\Omega \qquad (4A\text{-}10)
\end{aligned}$$

Thus

$$X_{ls} + X'_{lr} = 1.9\,\Omega \qquad (4A\text{-}11)$$

Generally, X_{ls} and X'_{lr} are assumed to be equal; however, in some types of induction machines a different ratio is suggested. We will assume $X_{ls} = X'_{lr}$, whereupon we have determined the machine parameters. In particular, for $\omega_b = 377$ rad/s the parameters are

$$\begin{aligned}
r_s &= 0.531\,\Omega & X_M &= 31.95\,\Omega & r'_r &= 0.408\,\Omega \\
X_{ls} &= 0.95\,\Omega & & & X'_{lr} &= 0.95\,\Omega
\end{aligned}$$

4.10 FREE ACCELERATION CHARACTERISTICS

It is instructive to observe the variables of several induction machines during free (no-load) acceleration from stall. For this purpose, the nonlinear differential equations that describe the induction machine were simulated on a computer, and studies were performed. The parameters of the machines are from [9] and given here in Table 4.10-1. Each machine is a 4-pole, 60-Hz, 3-phase induction motor. The parameters are expressed in ohms using the 60-Hz value of the reactances. In Table 4.10-1 the

Table 4.10-1 Induction Machine Parameters

Machine Rating			T_B	$I_{B(abc)}$	r_s	X_{ls}	X_M	X'_{lr}	r'_r	J
hp	Volts	rpm	(N·m)	(amps)	(ohms)	(ohms)	(ohms)	(ohms)	(ohms)	(kg·m²)
3	220	1710	11.9	5.8	0.435	0.754	26.13	0.754	0.816	0.089
50	460	1705	198	46.8	0.087	0.302	13.08	0.302	0.228	1.662
500	2300	1773	1.98×10^3	93.6	0.262	1.206	54.02	1.206	0.187	11.06
2250	2300	1786	8.9×10^3	421.2	0.029	0.226	13.04	0.226	0.022	63.87

voltage is the rated rms line-to-line voltage, the speed is rated speed, and J includes the inertia of the load which is assumed to be equal to the inertia of the rotor. Base torque, as calculated from (4.8-5), and base or rated current (rms) are also given.

The torque-versus-speed characteristics during free acceleration are shown for each machine in Figs. 4.10-1 through 4.10-4. In each case, the machine is initially stalled when rated balanced voltage is applied with $v_{as} = \sqrt{2}V_s \cos \omega_e t$. The machine currents along with the electromagnetic torque and speed for the 3- and 2250-hp machines during free acceleration are shown in Figs. 4.10-5 and 4.10-6. Because friction and windage losses are not represented, the machines accelerate to synchronous speed. In all figures, the scales of the currents are given in multiples of rated peak values. The scale of the torque is given in multiples of base torque.

At stall the input impedance of the induction machine is essentially the stator resistance and leakage reactance in series with the rotor resistance and leakage reactance. Consequently, with rated voltage applied, the starting current is large, in some cases on the order of 10 times the rated value. Therefore, in practice a compensator (transformer) is generally used to start large-horsepower machines with reduced

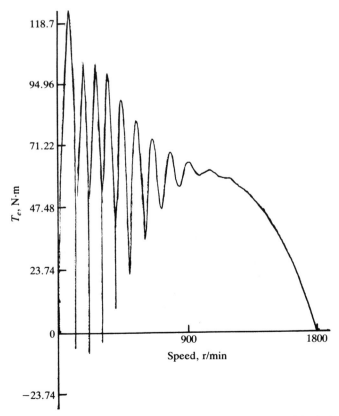

Figure 4.10-1 Torque–speed characteristics during free acceleration: 3-hp induction motor.

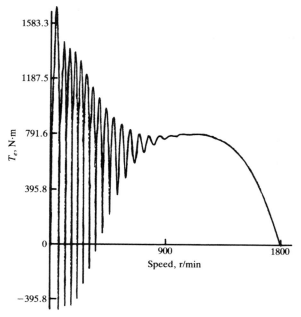

Figure 4.10-2 Torque–speed characteristics during free acceleration: 50-hp induction motor.

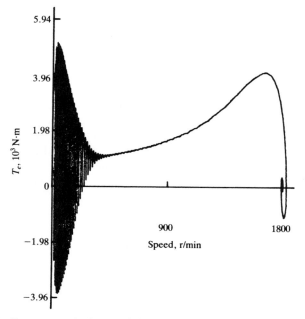

Figure 4.10-3 Torque–speed characteristics during free acceleration: 500-hp induction motor.

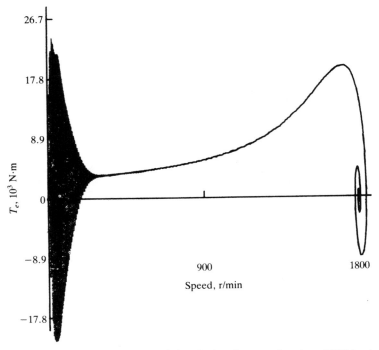

Figure 4.10-4 Torque–speed characteristics during free acceleration: 2250-hp induction motor.

voltage until the machine has reached 60% to 80% of synchronous speed whereupon full voltage is applied.

The 3- and 50-hp machines are relatively high-slip machines; that is, rated torque is developed at a speed considerably less than synchronous speed. On the other hand, the 500- and 2250-hp machines are low-slip machines. These characteristics are evident in the torque–speed characteristics shown in Figs. 4.10-1 through 4.10-4.

The transient torque–speed characteristics are different from the steady-state torque–speed characteristics in several respects. The instantaneous electromagnetic torque, immediately following the application of the stator voltages, varies at 60 Hz about an average positive value. This decaying, 60-Hz variation in the instantaneous torque is due to the transient offset in the stator currents. Although the offset in each of the stator currents depends upon the values of the source voltages at the time of application, the instantaneous torque is independent of the initial values of balanced source voltages because the machine is symmetrical. We also note from the current traces in Figs. 4.10-5 and 4.10-6 that the envelope of the machine currents varies during the transient period. It is shown in a subsequent chapter that this is due to the interaction of the stator and rotor electric transients.

Another noticeable difference between the dynamic and steady-state torque–speed characteristics occurs in the case of the 500- and 2250-hp machines. In

particular, the rotor speed overshoots synchronous speed and the instantaneous torque and speed demonstrate decayed oscillations about the final operating point. This
characteristic is especially evident in the larger-horsepower machines; however, in
the case of the 3- and 50-hp machines the rotor speed is highly damped and the final
operating condition is attained without oscillations. It is noted from Table 4.10-1 that

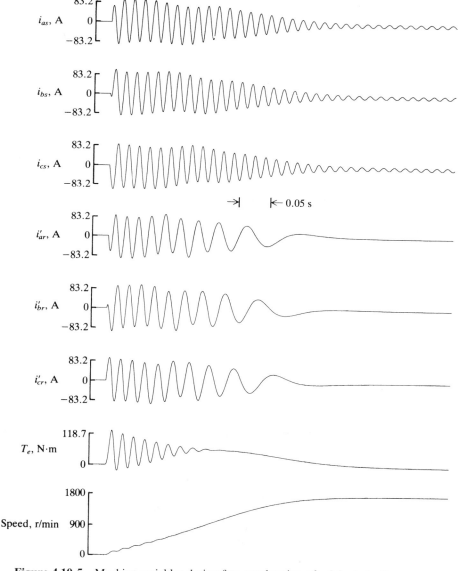

Figure 4.10-5   Machine variables during free acceleration of a 3-hp induction motor.

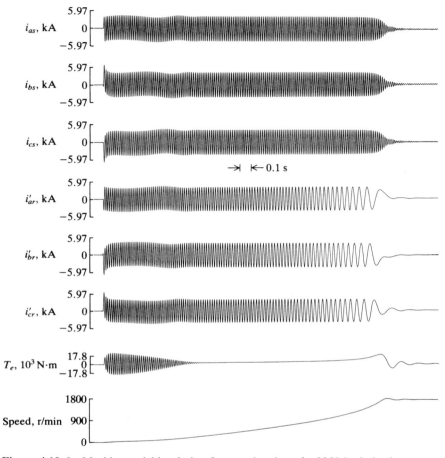

Figure 4.10-6   Machine variables during free acceleration of a 2250-hp induction motor.

the ratio of rotor leakage reactance to rotor resistance is much higher for the larger-horsepower machines than for the smaller-horsepower machines. In a later chapter we find that the complex conjugate pair of eigenvalues, associated with the rotor circuits, are much less damped in the case of the 500- and 2250-hp machines than in the case of the 3- and 50-hp machines.

If we were to plot the steady-state torque–speed characteristics of the 3- and 50-hp machines upon the free acceleration torque–speed characteristics, we would find that the steady-state torque corresponds very closely to the average of the transient torque. This, however, is not the case for the 500- and 2250-hp machines, where the steady-state value of maximum torque is much larger than that observed from the free acceleration characteristics. This is illustrated in Figs. 4.10-7 and 4.10-8, where the steady-state torque–speed characteristic is superimposed upon the free acceleration characteristic for the 500- and 2250-hp machines. In a later

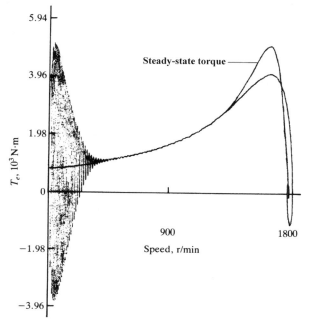

Figure 4.10-7 Comparison of steady-state and free acceleration torque–speed characteristics: 500-hp induction motor.

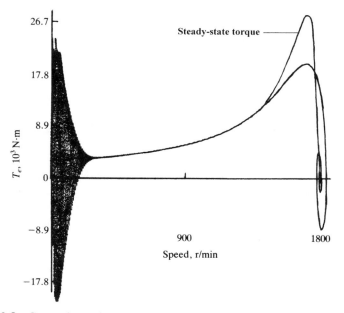

Figure 4.10-8 Comparison of steady-state and free acceleration torque–speed characteristics: 2250-hp induction motor.

chapter this difference is shown to be due primarily to the electric transients in the rotor circuits.

Example 4B   Let us calculate the steady-state torque and current at stall for the 3-hp machine given in Table 4.10-1 and compare these values to those shown in Figs. 4.10-1 and 4.10-5. From (4.9-19) and Table 4.10-1 with $s = 1$ we obtain

$$T_e = \frac{(3)(4/2)(1)[(26.13)^2/377](0.816)(1)(220/\sqrt{3})^2}{[(0.435)(0.816) + (1)(1)^2(26.13^2 - 26.884 \times 26.884)]^2 + (1)^2(0.816 \times 26.884 + 1 \times 0.435 \times 26.884)^2}$$

$$= 51.9\,\text{N} \cdot \text{m} \tag{4B-1}$$

This is approximately the average of the pulsating torque at $\omega_r = 0$ depicted in Figs. 4.10-1 and 4.10-5.

The stall, steady-state current may be calculated from

$$\tilde{I}_{as} = \frac{\tilde{V}_{as}}{(r_s + r_r') + j(X_{ls} + X_{lr}')}$$

$$= \frac{(220/\sqrt{3})\underline{/0^\circ}}{(0.435 + 0.816) + j(0.754 + 0.754)}$$

$$= 64.8\underline{/-50.3^\circ}A \tag{4B-2}$$

This value is the steady-state current that would occur if the rotor is locked and after all electric transients have subsided. It is somewhat difficult to compare this value with that shown in Fig. 4.10-5 because the electric transients cause the currents to be offset in Fig. 4.10-5. However, i_{bs} in Fig. 4.10-5 contains the least offset, and it compares quite well. In particular, the rms value of the first cycle of i_{bs} is approximately 69 A, which is on the order of 12 times rated current.

4.11   FREE ACCELERATION CHARACTERISTICS VIEWED FROM VARIOUS REFERENCE FRAMES

It is also instructive to observe the variables of an induction machine in various reference frames during free acceleration from stall. The machine simulated on the computer, for this purpose, is a singly excited, 6-pole, 3-phase, 220-V (line-to-line), 10-hp, 60-Hz induction motor with the following parameters expressed in per unit [10].

$$r_s = 0.0453 \qquad X_M = 2.042 \qquad r_r' = 0.0222$$
$$X_{ls} = 0.0775 \qquad\qquad\qquad X_{lr}' = 0.0322$$

The inertia of the rotor is $H = 0.5$ s.

The machine variables during free acceleration are shown in Fig. 4.11-1. All variables are plotted in per unit with the peak value of the base sinusoidal quantities

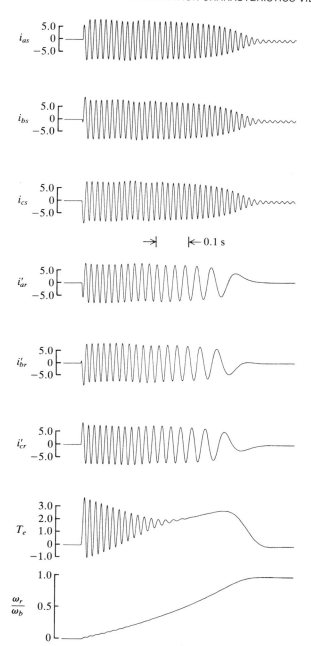

Figure 4.11-1 Free acceleration characteristics of a 10-hp induction motor in machine variables.

given as 1.0 pu. If we were to follow the convention set forth in Section 4.8, we would use the rms value as 1.0 pu. However, the selection of peak values as 1.0 pu allows a more direct comparison with the $qd0s$ variables shown later. Also, base torque rather than rated torque is taken as one per unit torque. At $t = 0$, rated voltage, with v_{as} a cosine, is applied to the machine. As in the studies reported in the previous section, the rotor accelerates from stall with zero load torque and, because friction and windage losses are not taken into account, the machine accelerates to synchronous speed.

The same free acceleration characteristics are shown in different reference frames in Figs. 4.11-2 through 4.11-5. The stationary reference-frame variables during free acceleration are shown in Fig. 4.11-2. With the reference frame fixed in the stator the qs and ds variables are arithmetically related to the abc variables. In particular, the zero position of the reference frame is zero, therefore $f_{as} = f_{qs}^s$. Thus, v_{qs}^s and i_{qs}^s are identical to v_{as} and i_{as} of Fig. 4.11-1. The rotor variables are referred to the stationary reference frame and vary therein at 60 Hz.

The free acceleration characteristics with the reference frame fixed in the rotor is given in Fig. 4.11-3. Here the zero position of the rotor and the reference frame are both zero, therefore $f_{ar}' = f_{qr}'^r$. Hence, $i_{qr}'^r$ in Fig. 4.11-3 is identical to i_{ar}' of Fig. 4.11-1 and, because the stator variables are referred to the rotor, they vary at slip frequency. At stall the rotor reference frame coincides with the stationary reference frame. At synchronous speed the rotor reference frame becomes the synchronously rotating reference frame. It is important to note that because the machine essentially operates in the steady-state mode upon reaching synchronous speed, the variables become constant corresponding to their instantaneous values at the time the rotor speed becomes equal to synchronous speed.

Free acceleration with the reference frame rotating in synchronism with the electrical angular velocity of the applied voltages is shown in Fig. 4.11-4. Here, the zero position of the reference frame is selected so that v_{qs}^e is the amplitude of the stator applied phase voltages and $v_{ds}^e = 0$.

The fact that the reference frame may rotate at any speed, constant or varying, is depicted in Fig. 4.11-5. During free acceleration from stall the reference-frame speed is initially rotating at synchronous speed opposite to the direction of rotation of the rotating magnetic field established by the stator currents. The reference-frame speed is then stepped to zero, whereupon it becomes a stationary reference frame and the variables correspond to those in Fig. 4.11-2. The speed of the reference frame is then varied sinusoidally at base frequency, and finally it is held fixed at synchronous speed. It is interesting to note that the waveform of the torque is the same regardless of the speed or the change in the speed of the reference frame.

4.12 DYNAMIC PERFORMANCE DURING SUDDEN CHANGES IN LOAD TORQUE

The dynamic behavior of the 3- and 2250-hp induction motors during step changes in load torque is shown in Figs. 4.12-1 and 4.12-2, respectively. Initially

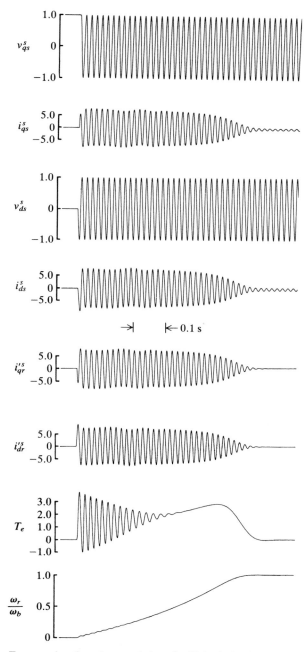

Figure 4.11-2 Free acceleration characteristics of a 10-hp induction motor in the stationary reference frame.

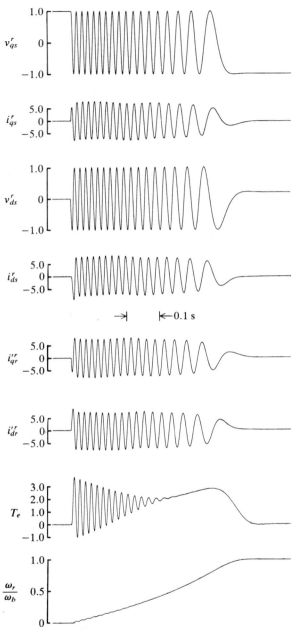

Figure 4.11-3 Free acceleration characteristics of a 10-hp induction motor in a reference frame fixed in rotor.

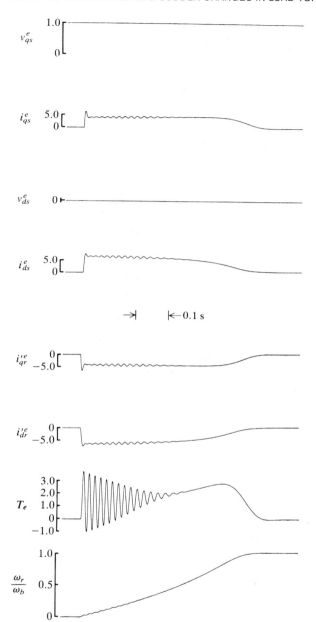

Figure 4.11-4 Free acceleration characteristics of a 10-hp induction motor in the synchronously rotating reference frame.

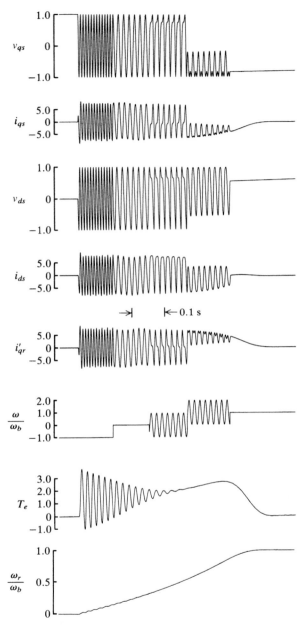

Figure 4.11-5 Free acceleration characteristics of a 10-hp induction motor in the arbitrary reference frame. First $\omega = -\omega_e$, then $\omega = 0$, followed by a sinusoidal variation about zero of ω with amplitude ω_e and frequency ω_e, then the same variation about ω_e. Finally ω is set equal to ω_e.

each machine is operating at synchronous speed. The load torque is first stepped from zero to base torque (slightly less than rated) and the machine allowed to establish this new operating point. Next, the load torque is stepped from base torque back to zero whereupon the machine reestablishes its original operating condition.

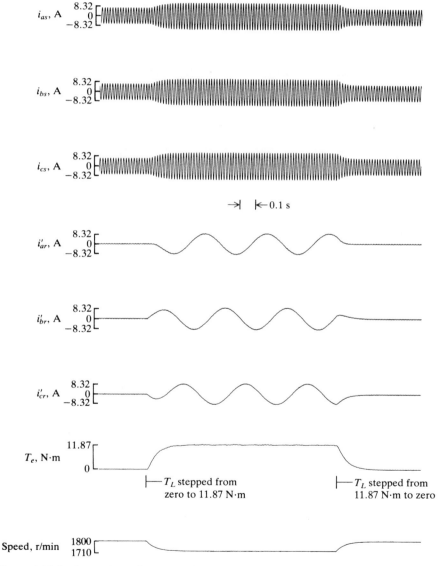

Figure 4.12-1 Dynamic performance of a 3-hp induction motor during step changes in load torque from zero to 11.87 N · m to zero.

The variables of the 3-hp machine approach each new operating condition in an overdamped manner. This is characteristic of the 3- and 50-hp machines given in Table 4.10-1. We previously found that for these machines the steady-state torque–speed characteristic nearly duplicates the free acceleration characteristic

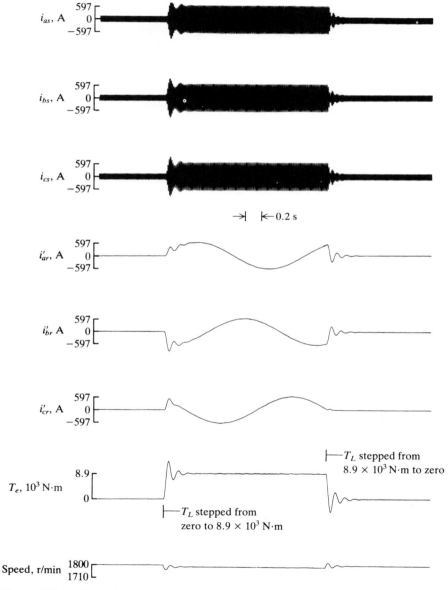

Figure 4.12-2 Dynamic performance of a 2250-hp induction motor during step changes in load torque from zero to 8.9×10^3 N $\cdot$ m to zero.

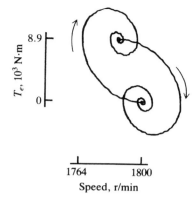

Figure 4.12-3 Torque versus speed for 2250-hp induction motor during load torque changes shown in Fig. 4.12-2.

once the electrical transient associated with the stator circuits has subsided; therefore we are not surprised to find that the dynamics during load torque changes can be predicted adequately by the steady-state torque–speed characteristics. Indeed, this is the case; the plot of torque versus speed during the load torque changes depicted in Fig. 4.12-1 follow nearly exactly the steady-state torque–speed curve. Therefore, the dynamic behavior of most smaller induction machines during normal load torque changes can be predicted by using the steady-state voltage and torque equations to calculate the currents and torque and using (4.3-8) or (4.8-10) to relate torque to the inertia and rotor speed.

The dynamic performance of the 2250-hp machine during load torque changes is strongly influenced by the rotor electric transients. This influence of the rotor circuit transients, which is illustrated more clearly in a later chapter, causes the 2250-hp machine to exhibit damped oscillations about the new operating point. At best, the steady-state torque–speed characteristics could approximate the average of this dynamic response; it could not predict the complete dynamics during normal load torque changes for the larger machines. This fact is further emphasized by the plot of torque versus speed for the 2250-hp machine in Fig. 4.12-3. The steady-state torque–speed characteristic would be nearly a straight line drawn between the two operating points. We, of course, expected this from the previous comparison of the steady-state torque–speed curve with the free acceleration characteristics (Fig. 4.10-8).

4.13 DYNAMIC PERFORMANCE DURING A 3-PHASE FAULT AT THE MACHINE TERMINALS

The dynamic performance of the 3- and 2250-hp induction machines is shown, respectively, in Figs. 4.13-1 and 4.13-2 during and following a 3-phase fault at the terminals. Initially, each motor is operating at essentially rated conditions with a load torque equal to base torque. The 3-phase fault at the terminals is simulated by setting v_{as}, v_{bs}, and v_{cs} to zero at the instant v_{as} passes through zero going positive. After six cycles the source voltages are reapplied.

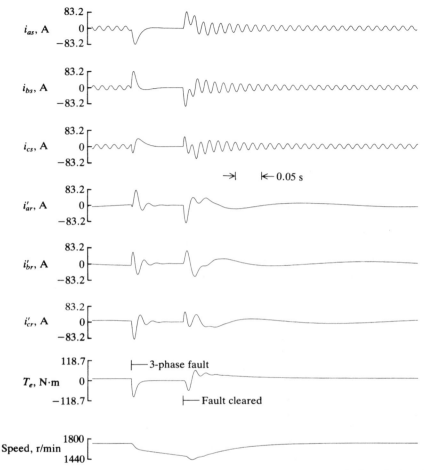

Figure 4.13-1 Dynamic performance of a 3-hp induction motor during a 3-phase fault at the terminals.

The stepping of the terminal voltages to zero at the time of the fault gives rise to decaying offsets in both the stator and rotor currents. These transient offsets in the stator currents appear in the rotor circuits as decaying oscillations of near 60 Hz (because the rotor speed is slightly less than synchronous) which are superimposed upon the transients of the rotor circuits. Similarly, the transient offsets in the rotor currents appear as decaying oscillations in the stator currents at a frequency corresponding to the rotor speed. In the case of the 3-hp machine, both the stator and rotor transients are highly damped and subside before the fault is removed and the voltages reapplied. With all machine currents equal to zero, the electromagnetic torque is, of course, zero; therefore full load torque decelerates the machine.

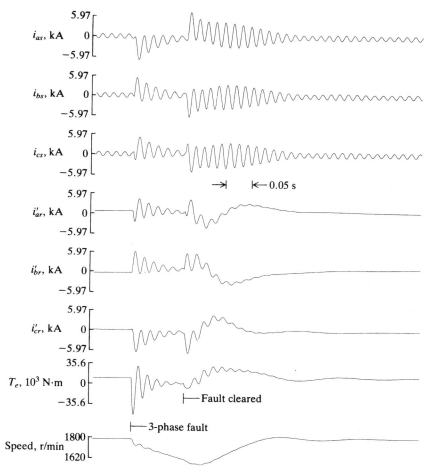

Figure 4.13-2 Dynamic performance of a 2250-hp induction motor during a 3-phase fault at the terminals.

In the case of the 2250-hp machine, which has a higher leakage reactance to resistance ratio in both the stator and rotor than the 3-hp machine, the stator and rotor transients are still present when the fault is removed and the voltages reapplied. Hence, T_e has an average value throughout the fault period corresponding to the ohmic losses.

When the voltages are reapplied the offsets again occur in the stator currents. The machine reestablishes the original operating condition, in a well-damped manner in the case of 3-hp machine and in an oscillatory manner in the case of the 2250-hp machine. The decaying oscillations in the stator and rotor circuits, due to the transient offsets, are very apparent in Fig. 4.13-2. We will discuss the performance of an induction motor during a 3-phase fault at the terminals later when calculating eigen-values and when considering reduced-order equations.

4.14 COMPUTER SIMULATION IN THE ARBITRARY REFERENCE FRAME

There are numerous ways of formulating the equations of an induction machine for the purposes of computer simulation. The form, that has become a standard and that we will use, is reported in reference 2. Also see references 10 and 11. In particular, the computer representation of the symmetrical induction machine in the arbitrary reference frame will be used as the basis from which various modes of operation are represented. This simulation is quite convenient not only from the standpoint of representing all practical modes of operation but also because it permits the effects of saturation to be readily simulated. The simulation of saturation is covered in Section 5.16 for synchronous machines. This method may also be used to simulate the saturation of an induction machine.

The equations convenient for simulating the symmetrical induction machine in the arbitrary reference frame may be established by first solving the flux linkage equations or flux linkage per second equations for the currents. Thus, from (4.5-28)–(4.5-33) we can write

$$i_{qs} = \frac{1}{X_{ls}}(\psi_{qs} - \psi_{mq}) \tag{4.14-1}$$

$$i_{ds} = \frac{1}{X_{ls}}(\psi_{ds} - \psi_{md}) \tag{4.14-2}$$

$$i_{0s} = \frac{1}{X_{ls}}\psi_{0s} \tag{4.14-3}$$

$$i'_{qr} = \frac{1}{X'_{lr}}(\psi'_{qr} - \psi_{mq}) \tag{4.14-4}$$

$$i'_{dr} = \frac{1}{X'_{lr}}(\psi'_{dr} - \psi_{md}) \tag{4.14-5}$$

$$i'_{0r} = \frac{1}{X'_{lr}}\psi'_{0r} \tag{4.14-6}$$

where ψ_{mq} and ψ_{md}, which are useful when representing saturation, are defined as

$$\psi_{mq} = X_M(i_{qs} + i'_{qr}) \tag{4.14-7}$$
$$\psi_{md} = X_M(i_{ds} + i'_{dr}) \tag{4.14-8}$$

If (4.14-1)–(4.14-6) are used to eliminate the currents in (4.14-7) and (4.14-8) as well as in the voltage equations in the arbitrary reference frame given by

(4.5-22)–(4.5-27) and if the resulting voltage equations are solved for the flux linkages per second, we can write the following integral equations:

$$\psi_{qs} = \frac{\omega_b}{p}\left[v_{qs} - \frac{\omega}{\omega_b}\psi_{ds} + \frac{r_s}{X_{ls}}(\psi_{mq} - \psi_{qs})\right] \tag{4.14-9}$$

$$\psi_{ds} = \frac{\omega_b}{p}\left[v_{ds} + \frac{\omega}{\omega_b}\psi_{qs} + \frac{r_s}{X_{ls}}(\psi_{md} - \psi_{ds})\right] \tag{4.14-10}$$

$$\psi_{0s} = \frac{\omega_b}{p}\left[v_{0s} - \frac{r_s}{X_{ls}}\psi_{0s}\right] \tag{4.14-11}$$

$$\psi'_{qr} = \frac{\omega_b}{p}\left[v'_{qr} - \left(\frac{\omega - \omega_r}{\omega_b}\right)\psi'_{dr} + \frac{r'_r}{X'_{lr}}(\psi_{mq} - \psi'_{qr})\right] \tag{4.14-12}$$

$$\psi'_{dr} = \frac{\omega_b}{p}\left[v'_{dr} + \left(\frac{\omega - \omega_r}{\omega_b}\right)\psi'_{qr} + \frac{r'_r}{X'_{lr}}(\psi_{md} - \psi'_{dr})\right] \tag{4.14-13}$$

$$\psi'_{0r} = \frac{\omega_b}{p}\left[v'_{0r} - \frac{r'_r}{X'_{lr}}\psi'_{0r}\right] \tag{4.14-14}$$

Equations (4.14-7) and (4.14-8) are now expressed as

$$\psi_{mq} = X_{aq}\left(\frac{\psi_{qs}}{X_{ls}} + \frac{\psi'_{qr}}{X'_{lr}}\right) \tag{4.14-15}$$

$$\psi_{md} = X_{ad}\left(\frac{\psi_{ds}}{X_{ls}} + \frac{\psi'_{dr}}{X'_{lr}}\right) \tag{4.14-16}$$

in which

$$X_{aq} = X_{ad} = \left(\frac{1}{X_M} + \frac{1}{X_{ls}} + \frac{1}{X'_{lr}}\right)^{-1} \tag{4.14-17}$$

In the computer simulation, (4.14-9)–(4.14-16) are used to solve for the flux linkages per second and (4.14-1)–(4.14-8) are used to obtain the currents from the flux linkages per second. It is clear that in the case of the zero quantities, i_{0s} and i'_{0r} may be solved for directly from (4.14-11) and (4.14-14), respectively, by substituting (4.14-3) for ψ_{0s} and (4.14-6) for ψ'_{0r} into these integral equations.

With saturation, the electromagnetic torque must be expressed by either (4.6-3) or (4.6-4) because X_M is not constant. The flux linkage per second version of (4.6-4) is often the most convenient. Expressed in per unit this relationship is

$$T_e = \psi_{ds}i_{qs} - \psi_{qs}i_{ds} \tag{4.14-18}$$

which is positive for motor action. The expression used to compute the rotor speed is written from (4.8-10) as

$$\omega_r = \frac{\omega_b}{2Hp} (T_e - T_L) \tag{4.14-19}$$

Often per unit rotor speed ω_r/ω_b is used rather than ω_r, because it is conveniently incorporated into the integral flux linkage equations.

A block diagram illustrating the computer representation of a symmetrical 3-phase induction machine in the arbitrary reference frame is given in Fig. 4.14-1. The machine equations that are used in the computer simulation are indicated by number in the block diagram. The transformations are defined by (3.3-1) for the stator variables and (4.4-1) for the rotor variables. The sinusoidal functions involved in the transformations are generated from ω in the case of $\mathbf{K}_s$ and $\omega - \omega_r$ for $\mathbf{K}_r$. Although saturation may be included in the flux linkage equations, the method of representing saturation is deferred until the computer simulation of the synchronous machine is established (Section 5.16). The programming of the equations is left to the reader. Our purpose is to set forth the computer equations and to discuss the procedures for representing various modes of operation.

The representation of the symmetrical induction machine in the arbitrary reference frame forms a general simulation from which the simulation for any practical mode of operation may be derived. Although the representation in the arbitrary reference frame may be used as depicted in Fig. 4.14-1 for all modes of operation, the simulation in this form is more involved than is generally necessary. Depending upon the mode of operation, the computer simulation is implemented in either the stationary, rotor, or synchronous reference frames.

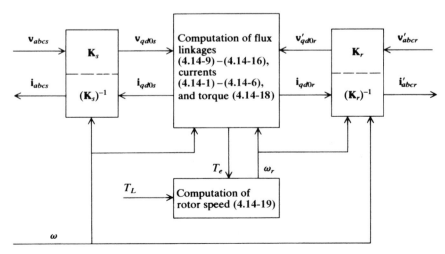

Figure 4.14-1   Simulation of a symmetrical 3-phase induction machine in the arbitrary reference frame shown in block diagram form.

Often there is little or no advantage in using one reference frame over another when simulating balanced conditions and disturbances such as a change in load torque, a symmetrical change in voltage, or a symmetrical 3-phase fault. However, there are situations where it may be desirable to represent the stator phase variables (*abc* variables). In this case, the stationary reference frame ($\omega = 0$) would offer advantage especially when simulating a singly fed induction machine with symmetrical rotor circuits, whereupon it would not be necessary to implement the transformations $\mathbf{K}_r^s$ and $(\mathbf{K}_r^s)^{-1}$ in the computer simulation. On the other hand, if the machine is doubly fed, it may be advantageous to use the rotor reference frame ($\omega = \omega_r$). If, in this case, the stator is supplied from a balanced source and if it is not necessary to have the stator currents available in the form of *abc* variables, then it is possible to reduce the transformation of the stator variables from that shown in Fig. 4.14-1. In particular, for balanced stator conditions we can write v_{qs}^r and v_{ds}^r directly from (3.7-5) and (3.7-6) as

$$v_{qs}^r = \sqrt{2}v_s \cos\left(\theta_{ev} - \theta_r\right) \tag{4.14-20}$$

$$v_{ds}^r = -\sqrt{2}v_s \sin\left(\theta_{ev} - \theta_r\right) \tag{4.14-21}$$

Perhaps the most widely used reference frame for the simulation of balanced operation of symmetrical induction machine is the synchronously rotating reference frame ($\omega = \omega_e$). This reference frame is particularly convenient for simulation purposes because the variables are constant during steady-state operation and vary only when a system disturbance occurs. Moreover, if the system to which the stator of the induction machine is connected is symmetrical and if it can also be represented in the synchronously rotating reference frame, then it is unnecessary to program the transformation of stator variables on the computer because both the network and machine are represented in the same reference frame.

Unbalanced conditions may be simulated by appropriate modification of the simulation block diagram shown in Fig. 4.14-1. Unbalanced conditions such as unbalanced stator voltages, unsymmetrical stator impedances, opening of a stator phase, and unsymmetrical rotor resistors are all analyzed and the procedures for simulation are set forth in Chapters 9 and 12 of reference 12.

REFERENCES

[1] D. C. White and H. H. Woodson, *Electromechanical Energy Conversion*, John Wiley and Sons, New York, 1959.

[2] P. C. Krause and C. H. Thomas, Simulation of Symmetrical Induction Machinery, *IEEE Transaction on Power Apparatus and Systems*, Vol. 84, November 1965, pp. 1038–1053.

[3] H. C. Stanley, An Analysis of the Induction Motor, *AIEE Transactions*, Vol. 57 (Supplement), 1938, pp. 751–755.

[4] R. H. Park, Two-Reaction Theory of Synchronous Machines—Generalized Method of Analysis, Part I, *AIEE Transactions*, Vol. 48, July 1929, pp. 716–727.

[5] D. S. Brereton, D. G. Lewis, and C. G. Young, Representation of Induction Motor Loads During Power System Stability Studies, *AIEE Transactions*, Vol. 76, August 1957, pp. 451–461.

[6] G. Kron, *Equivalent Circuits of Electric Machinery*, John Wiley and Sons, New York, 1951.

[7] A. E. Fitzgerald, C. Kingsley, Jr., and S. D. Umans, *Electric Machinery*, 5th ed., McGraw-Hill, New York, 1990.

[8] G. McPherson and R. D. Laramore, *An Introduction to Electrical Machines and Transformers*, 2rd ed., John Wiley and Sons, New York, 1990.

[9] J. J. Cathey, R. K. Calvin, III, and A. K. Ayoub, Transient Load Model of an Induction Machine, *IEEE Transactions on Power Apparatus and Systems*, Vol. 92, July/August 1973, pp. 1399–1406.

[10] C. M. Ong, *Dynamic Simulation of Electric Machinery*, Prentice-Hall PTR, Upper Saddle River, NJ, 1998.

[11] O. Wasynczuk and S. D. Sudhoff, Automated State Model Generation Algorithm for Power Circuits and Systems, *IEEE Transactions on Power Systems*, Vol. 11, No. 4, November 1996, pp. 1951–1956.

[12] P. C. Krause, O. Wasynczuk, and S. Sudhoff, *Analysis of Electric Machinery*, IEEE Press, Piscataway, NJ, 1995.

PROBLEMS

1 Consider the symmetrical 2-pole, 2-phase symmetrical induction machine shown in Fig. 1.P-8. Derive the voltage equations in machine variables.

2 Repeat Problem 1 for the 4-pole, 2-phase symmetrical induction machine shown in Fig. 1.P-9. Extend this development to show that if the electrical displacement and angular velocity of the rotor are defined as $\theta_r = (P/2)\theta_{rm}$ and $\omega_r = (P/2)\omega_{rm}$, then the voltage equations are identical to those obtained in Problem 1.

3 Prove that for a 3-wire, wye-connected, stator winding we have $\lambda_{as} + \lambda_{bs} + \lambda_{cs} = 0$. For this to be true, is it necessary for the rotor windings to also be connected in a 3-wire wye arrangement? Is $\lambda_{as} + \lambda_{bs} = 0$ for a 2-phase stator winding? Why?

4 Consider the 2-phase induction machine shown in Fig. 1.P-8. $I_{as} = \sqrt{2}I_s \cos \omega_e t$ and $I_{bs} = -\sqrt{2}I_s \sin \omega_e t$. If the rotor windings are open-circuited, express V_{ar}.

5 Derive an expression for the torque between the *as* and *bs* windings of the 2-pole, 2-phase induction machine shown in Fig. 1.P-8 with all other windings open-circuited. Repeat for a 4-pole, 2-phase induction machine.

6 Derive the expression for electromagnetic torque similar in form to (4.3-7) for a P-pole, 2-phase induction machine.

7 Using the equations of transformation given in Problem 1 of Chapter 3, derive the voltage equations of a 2-phase symmetrical induction machine in the arbitrary reference frame. Compare to those for a 3-phase machine given by (4.5-10)–(4.5-21).

8 Write the voltage equations derived in Problem 7 with (*a*) currents as state variables and (*b*) flux linkages as state variables. (*c*) Draw the time-domain block diagram for (*a*) and (*b*). Why is it advantageous to use flux linkages as state variables when implementing a computer simulation?

9 A 2-phase induction machine has two sets of 2-phase windings on the rotor. The *a* phases of the 2-phase sets are tightly coupled and have the same magnetic axis; similarly for the *b* phases. The parameters are r'_{r1} and X'_{lr1} for one set and are r'_{r2} and X'_{lr2} for the other. Draw the equivalent circuits in the arbitrary reference frame. Express all flux linkages per second and label all components and variables necessary to completely define the equivalent circuit of this double-cage rotor machine.

10 An induction machine has a 3-phase stator winding as shown in Fig. 4.2-1 and has a 2-phase rotor winding as shown in Fig. 1.P-8. Develop the equivalent circuits for this machine in the arbitrary reference frame.

11 The stator windings of a 3-phase induction machine are connected in delta. By equations of transformation, relate the line currents and line-to-line voltages to the $qd0$ variables in the arbitrary reference frame.

12 Derive an expression for electromagnetic torque in arbitrary reference frame variables for a 2-phase machine similar in form to (*a*) (4.6-2) and (*b*) (4.6-5).

13 Derive an expression for the electromagnetic torque in the arbitrary reference-frame variables for (*a*) the machine in Problem 9 and for (*b*) the machine in Problem 10.

14 For the device shown in Fig. 1.P-8, let $r_s = r'_r = 0.5\ \Omega$, $L_{ss} = L'_{rr} = 0.09$ H, and $L_{ms} = 0.08$ H. With $V_{as} = \sqrt{2}\cos t$, $\omega_r = 0$, $\theta_r = \frac{\pi}{3}$. The *bs* and *br* windings are open-circuited and the *ar* winding is short-circuited. Calculate $\tilde{I}_{as}$.

15 Show that the inertia constant H is equivalent to the stored energy of the rotor at synchronous speed normalized to the base power.

16 Devise a relationship that can be used to convert a per unit impedance from one VA base to another.

17 Per unitize the machine parameters given in Table 4.10-1.

18 Convert the per unit parameters given for the 10-hp machine in Section 2.11 to ohms and henrys and with the inertia in kg · m².

19 Verify (4.9-9) and (4.9-10).

20 In Section 4.9 two methods of deriving the steady-state voltage equations for an induction machine are discussed. The first method involves relating asynchronously rotating reference-frame variables to phasors; the second involves relating synchronously rotating reference-frame variables to phasors. The first method was used to obtain the steady-state voltage equations. Obtain these same equations using the second method.

21 Derive (4.9-14). What would this expression be in per unit?

22 A 4-pole, 7.5-hp, 3-phase induction motor has the following parameters:

$$r_s = 0.3\ \Omega \qquad L_{ms} = 0.035\ \text{H} \qquad r'_r = 0.15\ \Omega$$

$$L_{ls} = 0.0015\ \text{H} \qquad\qquad\qquad L'_{lr} = 0.0007\ \text{H}$$

The machine is supplied from a 110-V line to neutral 60-Hz source. (*a*) Assume ω_r is held at zero. Calculate T_e and express $\tilde{I}_{as}$, I_{bs} (instantaneous), and $\tilde{I}_{cs}$ assuming an *acb* sequence. (*b*) Repeat (*a*) if $\omega_r = \omega_e$ and friction and windage losses are neglected.

23 Repeat Problem 22 with the machine supplied from an 11-V line to neutral 6-Hz source.

24 Calculate the maximum steady-state torque (motor and generator action) for the 500-hp induction machine given in Table 4.10-1. Express I_{as} (instantaneous) for each condition.

25 Calculate the speed at maximum torque (motor action) for the 50-hp machine given in Table 4.10-1 when connected to a source of (a) 120 Hz, (b) 60 Hz, (c) 30 Hz, and (d) 6 Hz.

26 The 3-hp induction machine given in Table 4.10-1 is operating at no-load. The sequence of the applied voltages is suddenly reversed. Assume the electrical system establishes steady-state operation before the speed of the rotor has changed appreciably. Calculate the torque.

27 Select three identical capacitors so that when they are connected in parallel with the 500-hp induction machine given in Table 4.10-1, the capacitor–induction machine combination operates at a 0.95 lagging power factor at rated power output.

28 Per unitize the variables given in Figs. 4.10-5 and 4.10-6.

29 Use the steady-state torque-speed characteristics and graphical integration to determine the time required to reach synchronous speed from stall for the 3-and 2250-hp machines. Compare with the acceleration time shown in Figs. 4.10-5 and 4.10-6.

30 Use the steady-state torque-speed characteristics and graphically integration to determine the time required for the 3- and 2250-hp machines to slow from synchronous speed to the new operating speed when the base load torque is applied. Compare with the time required in Figs. 4.12-1 and 4.12-2.

31 The block diagram in Fig. 4.P-1 illustrates a simulation technique which is appropriate for computer simulation of 3-phase induction machine if saturation is neglected. We may write the operation performed in block 1 as

$$\boldsymbol{\psi}_{qd0s} = \frac{1}{p} \{ (\mathbf{W}_p)^{-1} [(\mathbf{W}_k + \mathbf{W}_\omega + \mathbf{W}_{\omega_r}) \boldsymbol{\psi}_{qd0s} + \mathbf{B} \mathbf{v}_{qd0s}] \}$$

where $\mathbf{W}_p$ are the terms associated with the derivatives, $\mathbf{W}_\omega$ are the terms containing ω, $\mathbf{W}_{\omega_r}$ containing ω_r, and $\mathbf{W}_k$ contains the remaining terms. Identify $\mathbf{W}_p, \mathbf{W}_k, \mathbf{W}_\omega, \mathbf{W}_{\omega_r}, \mathbf{B}$, and the equations in the remaining blocks.

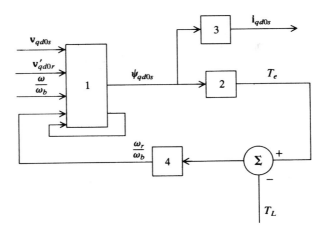

Figure 4.P-1 Block diagram for simulation of an induction machine with saturation neglected.

Chapter 5

SYNCHRONOUS MACHINES

5.1 INTRODUCTION

Nearly all of the electric power used throughout the world is generated by synchronous machines driven either by hydro or steam turbines or by combustion engines. Just as the induction machine is the workhorse when it comes to converting energy from electrical to mechanical, the synchronous machine is the principal means of converting energy from mechanical to electrical. The electrical and electromechanical behavior of most synchronous machines can be predicted from the equations that describe the 3-phase salient-pole synchronous machine. In particular, these equations can be used directly to predict the performance of hydro and steam turbine synchronous generators, synchronous motors, and—with only slight modifications—reluctance motors.

The rotor of a synchronous machine is equipped with a field winding and one or more damper windings; and, in general, all rotor windings have different electrical characteristics. Moreover, the rotor of a salient-pole synchronous machine is magnetically unsymmetrical. As a result of these rotor asymmetries, a change of variables for the rotor variables offers no advantage. However, a change of variables is beneficial for the stator variables. In most cases, the stator variables are transformed to a reference frame fixed in the rotor (Park's equations) [1]; however, the stator variables may also be expressed in the arbitrary reference frame, which is convenient for some computer simulations.

In this chapter, the voltage and electromagnetic torque equations are first established in machine variables. Reference-frame theory set forth in Chapter 3 is then used to establish the machine equations with the stator variables in the arbitrary reference frame and in the rotor reference frame (Park's equations). The equations that describe the steady-state behavior are then derived from Park's equations using the theory established in Chapter 3. Computer traces are given to illustrate the dynamic behavior of typical hydro and steam turbine generators during sudden changes in input torque and during and following a 3-phase fault at the terminals. These dynamic responses, which are calculated using the detailed set of nonlinear differential equations, are compared to those predicted by an approximate method of calculating the transient torque-angle characteristics which was widely used before the advent of modern computers and which still offer an unequalled means of visualizing the transient behavior of synchronous machines.

   The analysis given in this chapter is valid for a linear magnetic system; saturation is not considered. In some cases, saturation has only a secondary effect upon the overall performance of the machine; in other cases it is very important and it must be taken into account when predicting the performance. Analytically incorporating saturation into the voltage equations that describe its dynamic behavior is quite involved. Nevertheless, saturation can be taking into account in a computer simulation. In the final section of this chapter the machine equations are arranged appropriately for computer simulation. Therein, a method of accounting for saturation is set forth.

5.2   VOLTAGE EQUATIONS IN MACHINE VARIABLES

A 2-pole, 3-phase, wye-connected, salient-pole synchronous machine is shown in Fig. 5.2-1. The stator windings are identical sinusoidally distributed windings, displaced $120°$, with N_s equivalent turns and resistance r_s. The rotor is equipped with a field winding and three damper windings. The field winding (fd winding) has N_{fd} equivalent turns with resistance r_{fd}. One damper winding has the same magnetic axis as the field winding. This winding, the kd winding, has N_{kd} equivalent turns with resistance r_{kd}. The magnetic axis of the second and third damper windings, the $kq1$ and $kq2$ windings, is displaced $90°$ ahead of the magnetic axis of the fd and kd windings. The $kq1$ and $kq2$ windings have N_{kq1} and N_{kq2} equivalent turns, respectively, with resistances r_{kq1} and r_{kq2}. It is assumed that all rotor windings are sinusoidally distributed. The synchronous machine depicted in Fig. 5.2-1 differs from the one shown in Fig. 1.5-1 in that damper windings are included in Fig. 5.2-1. Also, the assumed direction of positive stator currents is out of the terminals convenient to describe generator action. Hence, with the assumed positive direction of the magnetic axes, negative flux linkages result due to positive stator currents.

   In Fig. 5.2-1 the magnetic axes of the stator windings are denoted by the *as, bs,* and *cs* axes. This notation was also used for the stator windings of the induction machine. The quadrature axis (*q* axis) and direct axis (*d* axis) are introduced in Fig. 5.2-1. The *q* axis is the magnetic axis of the *kq1* and *kq2* windings while the *d* axis is the magnetic axis of the *fd* and *kd* windings. The use of the *q* and *d* axes was in existence prior to Park's work [1]; and as mentioned in Chapter 3, Park used the notation of f_q, f_d, and f_0 in his transformation. Perhaps he made this choice of

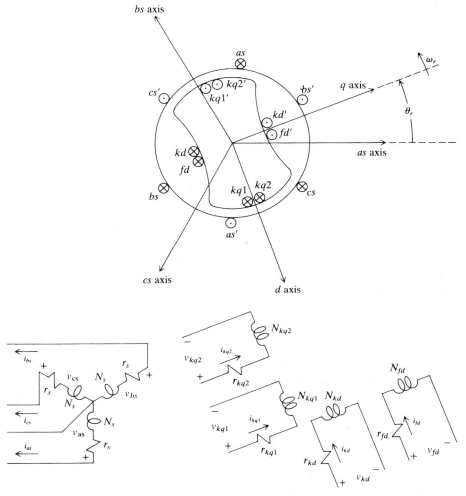

Figure 5.2-1   Two-pole, 3-phase, wye-connected, salient-pole synchronous machine.

notation because, in effect, this transformation referred the stator variables to the rotor where the traditional q and d axes are located.

We have used f_{qs}, f_{ds}, and f_{0s}, as well as f'_{qr}, f'_{dr}, and f'_{0r}, to denote transformed induction machine variables without introducing the connotation of a q or d axis. Instead, the q and d axes have been reserved to denote the rotor magnetic axes of the synchronous machine where they have an established physical meaning quite independent of any transformation. For this reason, one may argue that the q and d subscripts should not be used to denote the transformation to the arbitrary reference frame. Indeed, this line of reasoning has merit, however, because the transformation to the arbitrary reference frame is in essence a generalization of Park's transformation; the q and d subscripts have been selected for use in the transformation to the arbitrary reference primarily out of respect for Park's work, which is the basis of it all.

Although the damper windings are shown with provisions to apply a voltage, they are, in fact, short-circuited windings that represent the paths for induced rotor currents. Currents may flow in either cage-type windings similar to the squirrel-cage windings of induction machines or in the actual iron of the rotor. In salient-pole machines, at least the rotor surface is laminated and the damping winding currents are confined, for the most part, to the cage windings embedded in the rotor surface. In the high-speed, 2- or 4-pole machines the rotor is cylindrical, made of solid iron with a cage-type winding embedded in the rotor surface. Here currents can flow either in the cage winding or in the solid iron.

The performance of nearly all types of synchronous machines may be adequately described by straightforward modifications of the equations describing the performance of the machine shown in Fig. 5.2-1. For example, the behavior of low-speed hydro turbine generators, which are always salient-pole machines, is generally predicted sufficiently by one equivalent damper winding in the q axis. Hence, the performance of this type of machine may be described from the equations derived for the machine shown in Fig. 5.2-1 by eliminating all terms involving one of the kq windings. The reluctance machine, which has no field winding and generally only one damper winding in the q axis, may be described by eliminating the terms involving the fd winding and one of the kq windings. In solid iron rotor, steam turbine generators the magnetic characteristics of the q and d axes are identical, or nearly so; hence the inductances associated with the two axes are essentially the same. Also, it is necessary, in most cases, to include all three damper windings in order to portray adequately the transient characteristics of the stator variables and the electromagnetic torque of solid iron rotor machines [2].

Because the synchronous machine is generally operated as a generator, it is convenient to assume that the direction of positive stator current is out of the terminals as shown in Fig. 5.2-1. With this convention the voltage equations in machine variables may be expressed in matrix form as

$$\mathbf{v}_{abcs} = -\mathbf{r}_s \mathbf{i}_{abcs} + p\boldsymbol{\lambda}_{abcs} \qquad (5.2\text{-}1)$$

$$\mathbf{v}_{qdr} = \mathbf{r}_r \mathbf{i}_{qdr} + p\boldsymbol{\lambda}_{qdr} \qquad (5.2\text{-}2)$$

where

$$\left(\mathbf{f}_{abcs}\right)^T = \begin{bmatrix} f_{as} & f_{bs} & f_{cs} \end{bmatrix} \qquad (5.2\text{-}3)$$

$$\left(\mathbf{f}_{qdr}\right)^T = \begin{bmatrix} f_{kq1} & f_{kq2} & f_{fd} & f_{kd} \end{bmatrix} \qquad (5.2\text{-}4)$$

In the above equations the s and r subscripts denote variables associated with the stator and rotor windings, respectively. Both $\mathbf{r}_s$ and $\mathbf{r}_r$ are diagonal matrices;

in particular,

$$\mathbf{r}_s = \text{diag}[\, r_s \quad r_s \quad r_s \,] \tag{5.2-5}$$

$$\mathbf{r}_r = \text{diag}[\, r_{kq1} \quad r_{kq2} \quad r_{fd} \quad r_{kd} \,] \tag{5.2-6}$$

In Fig. 5.2-1 the positive *as, bs,* and *cs* axes are drawn in the direction of negative flux linkages relative to the assumed positive direction of the stator currents. In this case, the flux linkage equations become

$$
\begin{bmatrix} \boldsymbol{\lambda}_{abcs} \\ \boldsymbol{\lambda}_{qdr} \end{bmatrix} = \begin{bmatrix} \mathbf{L}_s & \mathbf{L}_{sr} \\ (\mathbf{L}_{sr})^T & \mathbf{L}_r \end{bmatrix} \begin{bmatrix} -\mathbf{i}_{abcs} \\ \mathbf{i}_{qdr} \end{bmatrix} \tag{5.2-7}
$$

From the work in Section 1.5 we can write $\mathbf{L}_s$ as

$$
\mathbf{L}_s = \begin{bmatrix} L_{ls} + L_A - L_B \cos 2\theta_r & -\tfrac{1}{2}L_A - L_B \cos 2\left(\theta_r - \tfrac{\pi}{3}\right) & -\tfrac{1}{2}L_A - L_B \cos 2\left(\theta_r + \tfrac{\pi}{3}\right) \\ -\tfrac{1}{2}L_A - L_B \cos 2\left(\theta_r - \tfrac{\pi}{3}\right) & L_{ls} + L_A - L_B \cos 2\left(\theta_r - \tfrac{2\pi}{3}\right) & -\tfrac{1}{2}L_A - L_B \cos 2(\theta_r + \pi) \\ -\tfrac{1}{2}L_A - L_B \cos 2\left(\theta_r + \tfrac{\pi}{3}\right) & -\tfrac{1}{2}L_A - L_B \cos 2(\theta_r + \pi) & L_{ls} + L_A - L_B \cos 2\left(\theta_r + \tfrac{2\pi}{3}\right) \end{bmatrix} \tag{5.2-8}
$$

By a straightforward extension of the work in Section 1.5 we can express the self- and mutual inductances of the damper windings. The inductance matrices $\mathbf{L}_{sr}$ and $\mathbf{L}_r$ may then be expressed as

$$
\mathbf{L}_{sr} = \begin{bmatrix} L_{skq1} \cos \theta_r & L_{skq2} \cos \theta_r & L_{sfd} \sin \theta_r & L_{skd} \sin \theta_r \\ L_{skq1} \cos \left(\theta_r - \tfrac{2\pi}{3}\right) & L_{skq2} \cos \left(\theta_r - \tfrac{2\pi}{3}\right) & L_{sfd} \sin \left(\theta_r - \tfrac{2\pi}{3}\right) & L_{skd} \sin \left(\theta_r - \tfrac{2\pi}{3}\right) \\ L_{skq1} \cos \left(\theta_r + \tfrac{2\pi}{3}\right) & L_{skq2} \cos \left(\theta_r + \tfrac{2\pi}{3}\right) & L_{sfd} \sin \left(\theta_r + \tfrac{2\pi}{3}\right) & L_{skd} \sin \left(\theta_r + \tfrac{2\pi}{3}\right) \end{bmatrix} \tag{5.2-9}
$$

$$
\mathbf{L}_r = \begin{bmatrix} L_{lkq1} + L_{mkq1} & L_{kq1kq2} & 0 & 0 \\ L_{kq1kq2} & L_{lkq2} + L_{mkq2} & 0 & 0 \\ 0 & 0 & L_{lfd} + L_{mfd} & L_{fdkd} \\ 0 & 0 & L_{fdkd} & L_{lkd} + L_{mkd} \end{bmatrix} \tag{5.2-10}
$$

In (5.2-8), $L_A > L_B$ and L_B is zero for a round rotor machine. Also in (5.2-8) and (5.2-10), the leakage inductances are denoted with l in the subscript. The subscripts *skq*1, *skq*2, *sfd*, and *skd* in (5.2-9) denote mutual inductances between stator and rotor windings.

The magnetizing inductances are defined as

$$L_{mq} = \tfrac{3}{2}(L_A - L_B) \tag{5.2-11}$$

$$L_{md} = \tfrac{3}{2}(L_A + L_B) \tag{5.2-12}$$

whereupon it is easy to show that (Section 1.5)

$$L_{skq1} = \left(\frac{N_{kq1}}{N_s}\right)\left(\frac{2}{3}\right)L_{mq} \tag{5.2-13}$$

$$L_{skq2} = \left(\frac{N_{kq2}}{N_s}\right)\left(\frac{2}{3}\right)L_{mq} \tag{5.2-14}$$

$$L_{sfd} = \left(\frac{N_{fd}}{N_s}\right)\left(\frac{2}{3}\right)L_{md} \tag{5.2-15}$$

$$L_{skd} = \left(\frac{N_{kd}}{N_s}\right)\left(\frac{2}{3}\right)L_{md} \tag{5.2-16}$$

$$L_{mkq1} = \left(\frac{N_{kq1}}{N_s}\right)^2\left(\frac{2}{3}\right)L_{mq} \tag{5.2-17}$$

$$L_{mkq2} = \left(\frac{N_{kq2}}{N_s}\right)^2\left(\frac{2}{3}\right)L_{mq} \tag{5.2-18}$$

$$L_{mfd} = \left(\frac{N_{fd}}{N_s}\right)^2\left(\frac{2}{3}\right)L_{md} \tag{5.2-19}$$

$$L_{mkd} = \left(\frac{N_{kd}}{N_s}\right)^2\left(\frac{2}{3}\right)L_{md} \tag{5.2-20}$$

$$L_{kq1kq2} = \left(\frac{N_{kq2}}{N_{kq1}}\right)L_{mkq1}$$
$$= \left(\frac{N_{kq1}}{N_{kq2}}\right)L_{mkq2} \tag{5.2-21}$$

$$L_{fdkd} = \left(\frac{N_{kd}}{N_{fd}}\right)L_{mfd}$$
$$= \left(\frac{N_{fd}}{N_{kd}}\right)L_{mkd} \tag{5.2-22}$$

It is convenient to incorporate the following substitute variables that refer the rotor variables to the stator windings:

$$i'_j = \left(\frac{2}{3}\right)\left(\frac{N_j}{N_s}\right)i_j \tag{5.2-23}$$

$$v'_j = \left(\frac{N_s}{N_j}\right)v_j \tag{5.2-24}$$

$$\lambda'_j = \left(\frac{N_s}{N_j}\right)\lambda_j \tag{5.2-25}$$

where j may be $kq1$, $kq2$, fd, or kd.

The flux linkages may now be written as

$$\begin{bmatrix} \boldsymbol{\lambda}_{abcs} \\ \boldsymbol{\lambda}'_{qdr} \end{bmatrix} = \begin{bmatrix} \mathbf{L}_s & \mathbf{L}'_{sr} \\ \frac{2}{3}(\mathbf{L}'_{sr})^T & \mathbf{L}'_r \end{bmatrix} \begin{bmatrix} -\mathbf{i}_{abcs} \\ \mathbf{i}'_{qdr} \end{bmatrix} \tag{5.2-26}$$

where $\mathbf{L}_s$ is defined by (5.2-8) and

$$\mathbf{L}'_{sr} = \begin{bmatrix} L_{mq}\cos\theta_r & L_{mq}\cos\theta_r & L_{md}\sin\theta_r & L_{md}\sin\theta_r \\ L_{mq}\cos\left(\theta_r - \frac{2\pi}{3}\right) & L_{mq}\cos\left(\theta_r - \frac{2\pi}{3}\right) & L_{md}\sin\left(\theta_r - \frac{2\pi}{3}\right) & L_{md}\sin\left(\theta_r - \frac{2\pi}{3}\right) \\ L_{mq}\cos\left(\theta_r + \frac{2\pi}{3}\right) & L_{mq}\cos\left(\theta_r + \frac{2\pi}{3}\right) & L_{md}\sin\left(\theta_r + \frac{2\pi}{3}\right) & L_{md}\sin\left(\theta_r + \frac{2\pi}{3}\right) \end{bmatrix} \tag{5.2-27}$$

$$\mathbf{L}'_r = \begin{bmatrix} L'_{lkq1} + L_{mq} & L_{mq} & 0 & 0 \\ L_{mq} & L'_{lkq2} + L_{mq} & 0 & 0 \\ 0 & 0 & L'_{lfd} + L_{md} & L_{md} \\ 0 & 0 & L_{md} & L'_{lkd} + L_{md} \end{bmatrix} \tag{5.2-28}$$

The voltage equations expressed in terms of machine variables referred to the stator windings are

$$\begin{bmatrix} \mathbf{v}_{abcs} \\ \mathbf{v}'_{qdr} \end{bmatrix} = \begin{bmatrix} \mathbf{r}_s + p\mathbf{L}_s & p\mathbf{L}'_{sr} \\ \frac{2}{3}p(\mathbf{L}'_{sr})^T & \mathbf{r}'_r + p\mathbf{L}'_r \end{bmatrix} \begin{bmatrix} -\mathbf{i}_{abcs} \\ \mathbf{i}'_{qdr} \end{bmatrix} \tag{5.2-29}$$

In (5.2-28) and (5.2-29) we have

$$r'_j = \left(\frac{3}{2}\right)\left(\frac{N_s}{N_j}\right)^2 r_j \tag{5.2-30}$$

$$L'_{lj} = \left(\frac{3}{2}\right)\left(\frac{N_s}{N_j}\right)^2 L_{lj} \tag{5.2-31}$$

where, again, j may be $kq1$, $kq2$, fd, or kd.

The voltage equations given in (5.2-29) are valid for the positive direction of stator current assumed out of the stator terminals. If the positive direction of stator current is reversed, making analysis of motor action somewhat more convenient, the only modification to (5.2-29) is a change in the sign preceding $\mathbf{i}_{abcs}$.

5.3 TORQUE EQUATION IN MACHINE VARIABLES

The energy stored in the coupling field of a synchronous machine may be expressed as

$$W_f = \frac{1}{2}(\mathbf{i}_{abcs})^T (\mathbf{L}_s - L_{ls}\mathbf{I})\mathbf{i}_{abcs} - (\mathbf{i}_{abcs})^T \mathbf{L}'_{sr}\mathbf{i}'_{qdr} + \frac{1}{2}\left(\frac{3}{2}\right)(\mathbf{i}'_{qdr})^T (\mathbf{L}'_r - L'_{lr}\mathbf{I})\mathbf{i}'_{qdr} \tag{5.3-1}$$

where **I** is the identity matrix and

$$\mathbf{L}'_{lr} = \text{diag}[L'_{lkq1} \quad L'_{lkq2} \quad L'_{lfd} \quad L'_{lkd}] \tag{5.3-2}$$

Because the magnetic system is assumed to be linear, $W_f = W_c$ and the second entry of Table 1.3-1 may be used with the factor $P/2$ included to account for a P-pole machine (Section 4.3) and a negative sign to make T_e positive for generator action. Thus

$$T_e = \left(\frac{P}{2}\right)\left\{ -\frac{1}{2}(\mathbf{i}_{abcs})^T \frac{\partial}{\partial\theta_r}[\mathbf{L}_s - L_{ls}\mathbf{I}]\mathbf{i}_{abcs} + (\mathbf{i}_{abcs})^T \frac{\partial}{\partial\theta_r}[\mathbf{L}'_{sr}]\mathbf{i}'_{qdr}\right\} \tag{5.3-3}$$

In expanded form, (5.3-3) becomes

$$T_e = \left(\frac{P}{2}\right)\left\{ -\frac{(L_{md} - L_{mq})}{3}\left[\left(i_{as}^2 - \frac{1}{2}i_{bs}^2 - \frac{1}{2}i_{cs}^2 - i_{as}i_{bs} - i_{as}i_{cs} + 2i_{bs}i_{cs}\right)\sin 2\theta_r \right.\right.$$
$$\left. + \frac{\sqrt{3}}{2}(i_{bs}^2 + i_{cs}^2 - 2i_{as}i_{bs} + 2i_{as}i_{cs})\cos 2\theta_r\right]$$
$$- L_{mq}(i'_{kq1} + i'_{kq2})\left[\left(i_{as} - \frac{1}{2}i_{bs} - \frac{1}{2}i_{cs}\right)\sin\theta_r - \frac{\sqrt{3}}{2}(i_{bs} - i_{cs})\cos\theta_r\right]$$
$$\left. + L_{md}(i'_{fd} + i'_{kd})\left[\left(i_{as} - \frac{1}{2}i_{bs} - \frac{1}{2}i_{cs}\right)\cos\theta_r + \frac{\sqrt{3}}{2}(i_{bs} - i_{cs})\sin\theta_r\right]\right\} \tag{5.3-4}$$

The above expression for torque is positive for generator action with the positive direction of stator current assumed out of the stator terminals. If the positive direction of stator current is assumed into the stator terminals, then the above expression for torque is positive for motor action if the sign of the coefficients of $\sin 2\theta_r$ and $\cos 2\theta_r$ are changed. This, of course, can be accomplished by changing the sign preceding the multiplier $(L_{md} - L_{mq})$.

The torque and rotor speed are related by

$$T_e = -J\left(\frac{2}{P}\right)p\omega_r + T_I \tag{5.3-5}$$

where J is the inertia expressed in kilogram meters^2 (kg·m^2) or joule seconds^2 (J·s^2). Often the inertia is given as WR^2 in units of pound mass feet^2 (lbm·ft^2). The input torque T_I is positive for a torque input to the shaft of the synchronous machine.

5.4 STATOR VOLTAGE EQUATIONS IN ARBITRARY REFERENCE-FRAME VARIABLES

The voltage equations of the stator windings of a synchronous machine can be expressed in the arbitrary reference frame. In particular, by using the results presented in Chapter 3, the voltage equations for the stator windings may be written in the arbitrary reference frame as [3]

$$\mathbf{v}_{qd0s} = -\mathbf{r}_s \mathbf{i}_{qd0s} + \omega \boldsymbol{\lambda}_{dqs} + p\boldsymbol{\lambda}_{qd0s} \tag{5.4-1}$$

where

$$(\boldsymbol{\lambda}_{dqs})^T = [\lambda_{ds} \quad -\lambda_{qs} \quad 0] \tag{5.4-2}$$

The only restriction on (5.4-1) is that the resistance of each phase is the same; otherwise, the first term on the right-hand side of (5.4-1) would be written as in (3.4-2).

The rotor windings of a synchronous machine are different; therefore, the change of variables set forth in Section 4.4 offers no advantage in the analysis of the rotor circuits. Because the rotor variables are not transformed, the rotor voltage equations are expressed only in the rotor reference frame. Hence, from (5.2-2) with the appropriate turns ratios included and raised index r used to denote the rotor reference frame, the rotor voltage equations are

$$\mathbf{v}_{qdr}^{\prime r} = \mathbf{r}_r^\prime \mathbf{i}_{qdr}^{\prime r} + p\boldsymbol{\lambda}_{qdr}^{\prime r} \tag{5.4-3}$$

For linear magnetic system, the flux linkage equations may be expressed from (5.2-7) with the transformation of the stator variables to the arbitrary reference frame incorporated:

$$\begin{bmatrix} \boldsymbol{\lambda}_{qd0s} \\ \boldsymbol{\lambda}_{qdr}^{\prime r} \end{bmatrix} = \begin{bmatrix} \mathbf{K}_s \mathbf{L}_s (\mathbf{K}_s)^{-1} & \mathbf{K}_s \mathbf{L}_{sr}^\prime \\ \frac{2}{3}(\mathbf{L}_{sr}^\prime)^T (\mathbf{K}_s)^{-1} & \mathbf{L}_r^\prime \end{bmatrix} \begin{bmatrix} -\mathbf{i}_{qd0s} \\ \mathbf{i}_{qdr}^{\prime r} \end{bmatrix} \tag{5.4-4}$$

It can be shown that all terms of the inductance matrix of (5.4-4) are sinusoidal in nature except $\mathbf{L}_r^\prime$. For example, by using trigonometric identities given in Appendix A we obtain

$$\mathbf{K}_s \mathbf{L}_{sr}^\prime =$$

$$\begin{bmatrix} L_{mq}\cos(\theta - \theta_r) & L_{mq}\cos(\theta - \theta_r) & -L_{md}\sin(\theta - \theta_r) & -L_{md}\sin(\theta - \theta_r) \\ L_{mq}\sin(\theta - \theta_r) & L_{mq}\sin(\theta - \theta_r) & L_{md}\cos(\theta - \theta_r) & L_{md}\cos(\theta - \theta_r) \\ 0 & 0 & 0 & 0 \end{bmatrix}$$

$$\tag{5.4-5}$$

The sinusoidal terms of (5.4-5) are constant, independent of ω and ω_r only if $\omega = \omega_r$. Similarly, $\mathbf{K}_s \mathbf{L}_s (\mathbf{K}_s)^{-1}$ and $\frac{2}{3}(\mathbf{L}_{sr}^\prime)^T (\mathbf{K}_s)^{-1}$ are constant only if $\omega = \omega_r$. Therefore, the time-varying inductances are eliminated from the voltage equations only if the reference frame is fixed in the rotor. Hence, it would appear that only the rotor reference frame is useful in the analysis of synchronous machines. Although this is essentially the case, there are situations, especially in computer simulations, where it is convenient to express the stator voltage equations in a reference frame other than the one fixed in the rotor. For these applications it is necessary to relate the arbitrary reference-frame variables to the variables in the rotor reference frame. This may be accomplished by using (3.6-1), from which

$$\mathbf{f}_{qd0s}^r = \mathbf{K}^r \mathbf{f}_{qd0s} \tag{5.4-6}$$

From (3.6-7) we obtain

$$\mathbf{K}^r = \begin{bmatrix} \cos(\theta_r - \theta) & -\sin(\theta_r - \theta) & 0 \\ \sin(\theta_r - \theta) & \cos(\theta_r - \theta) & 0 \\ 0 & 0 & 1 \end{bmatrix} \tag{5.4-7}$$

Here we must again recall that the arbitrary reference does not carry a raised index.

5.5 VOLTAGE EQUATIONS IN ROTOR REFERENCE-FRAME VARIABLES: PARK'S EQUATIONS

R. H. Park was the first to incorporate a change of variables in the analysis of synchronous machines [1]. He transformed the stator variables to the rotor reference frame that eliminates the time-varying inductances in the voltage equations. Park's equations are obtained from (5.4-1) and (5.4-3) by setting the speed of the arbitrary reference frame equal to the rotor speed ($\omega = \omega_r$). Thus

$$\mathbf{v}_{qd0s}^r = -\mathbf{r}_s \mathbf{i}_{qd0s}^r + \omega_r \lambda_{dqs}^r + p\lambda_{qd0s}^r \tag{5.5-1}$$

$$\mathbf{v}_{qdr}^{\prime r} = \mathbf{r}_r^\prime \mathbf{i}_{qdr}^{\prime r} + p\lambda_{qdr}^{\prime r} \tag{5.5-2}$$

where

$$(\lambda_{qds}^r)^T = [\lambda_{ds}^r \quad -\lambda_{qs}^r \quad 0] \tag{5.5-3}$$

For a magnetically linear system, the flux linkages may be expressed in the rotor reference frame from (5.4-4) by setting $\theta = \theta_r$, whereupon $\mathbf{K}_s$ becomes $\mathbf{K}_s^r$. Thus

$$\begin{bmatrix} \lambda_{qd0s}^r \\ \lambda_{qdr}^{\prime r} \end{bmatrix} = \begin{bmatrix} \mathbf{K}_s^r \mathbf{L}_s (\mathbf{K}_s^r)^{-1} & \mathbf{K}_s^r \mathbf{L}_{sr}^\prime \\ \frac{2}{3}(\mathbf{L}_{sr}^\prime)^T (\mathbf{K}_s^r)^{-1} & \mathbf{L}_r^\prime \end{bmatrix} \begin{bmatrix} -\mathbf{i}_{qd0s}^r \\ \mathbf{i}_{qdr}^{\prime r} \end{bmatrix} \tag{5.5-4}$$

Using trigonometric identities from Appendix A, it can be shown that

$$\mathbf{K}_s^r \mathbf{L}_s (\mathbf{K}_s^r)^{-1} = \begin{bmatrix} L_{ls} + L_{mq} & 0 & 0 \\ 0 & L_{ls} + L_{md} & 0 \\ 0 & 0 & L_{ls} \end{bmatrix} \tag{5.5-5}$$

$$\mathbf{K}_s^r \mathbf{L}_{sr}' = \begin{bmatrix} L_{mq} & L_{mq} & 0 & 0 \\ 0 & 0 & L_{md} & L_{md} \\ 0 & 0 & 0 & 0 \end{bmatrix} \tag{5.5-6}$$

$$\frac{2}{3}(\mathbf{L}_{sr}')^T (\mathbf{K}_s^r)^{-1} = \begin{bmatrix} L_{mq} & 0 & 0 \\ L_{mq} & 0 & 0 \\ 0 & L_{md} & 0 \\ 0 & L_{md} & 0 \end{bmatrix} \tag{5.5-7}$$

Park's voltage equations are often written in expanded form; thus from (5.5-1) and (5.5-2) we have

$$v_{qs}^r = -r_s i_{qs}^r + \omega_r \lambda_{ds}^r + p \lambda_{qs}^r \tag{5.5-8}$$

$$v_{ds}^r = -r_s i_{ds}^r - \omega_r \lambda_{qs}^r + p \lambda_{ds}^r \tag{5.5-9}$$

$$v_{0s} = -r_s i_{0s} + p \lambda_{0s} \tag{5.5-10}$$

$$v_{kq1}^{\prime r} = r_{kq1}' i_{kq1}^{\prime r} + p \lambda_{kq1}^{\prime r} \tag{5.5-11}$$

$$v_{kq2}^{\prime r} = r_{kq2}' i_{kq2}^{\prime r} + p \lambda_{kq2}^{\prime r} \tag{5.5-12}$$

$$v_{fd}^{\prime r} = r_{fd}' i_{fd}^{\prime r} + p \lambda_{fd}^{\prime r} \tag{5.5-13}$$

$$v_{kd}^{\prime r} = r_{kd}' i_{kd}^{\prime r} + p \lambda_{kd}^{\prime r} \tag{5.5-14}$$

Substituting (5.5-5)–(5.5-7) and (5.2-28) into (5.5-4) yields the expressions for the flux linkages. In expanded form we have

$$\lambda_{qs}^r = -L_{ls} i_{qs}^r + L_{mq}(-i_{qs}^r + i_{kq1}^{\prime r} + i_{kq2}^{\prime r}) \tag{5.5-15}$$

$$\lambda_{ds}^r = -L_{ls} i_{ds}^r + L_{md}(-i_{ds}^r + i_{fd}^{\prime r} + i_{kd}^{\prime r}) \tag{5.5-16}$$

$$\lambda_{0s} = -L_{ls} i_{0s} \tag{5.5-17}$$

$$\lambda_{kq1}^{\prime r} = L_{lkq1}' i_{kq1}^{\prime r} + L_{mq}(-i_{qs}^r + i_{kq1}^{\prime r} + i_{kq2}^{\prime r}) \tag{5.5-18}$$

$$\lambda_{kq2}^{\prime r} = L_{lkq2}' i_{kq2}^{\prime r} + L_{mq}(-i_{qs}^r + i_{kq1}^{\prime r} + i_{kq2}^{\prime r}) \tag{5.5-19}$$

$$\lambda_{fd}^{\prime r} = L_{lfd}' i_{fd}^{\prime r} + L_{md}(-i_{ds}^r + i_{fd}^{\prime r} + i_{kd}^{\prime r}) \tag{5.5-20}$$

$$\lambda_{kd}^{\prime r} = L_{lkd}' i_{kd}^{\prime r} + L_{md}(-i_{ds}^r + i_{fd}^{\prime r} + i_{kd}^{\prime r}) \tag{5.5-21}$$

The voltage and flux linkage equations suggest the equivalent circuits shown in Fig. 5.5-1.

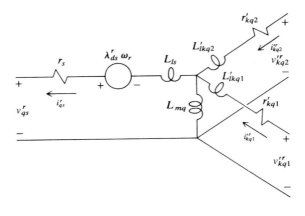

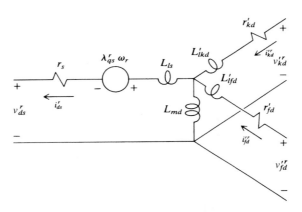

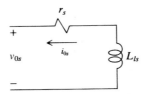

Figure 5.5-1   Equivalent circuits of a 3-phase synchronous machine with the reference frame fixed in a rotor: Park's equations.

As in the case of the induction machine, it is often convenient to express the voltage and flux linkage equations in terms of reactances rather than inductances. Hence, (5.5-8)–(5.5-14) are often written as

$$v_{qs}^r = -r_s i_{qs}^r + \frac{\omega_r}{\omega_b}\psi_{ds}^r + \frac{p}{\omega_b}\psi_{qs}^r \tag{5.5-22}$$

$$v_{ds}^r = -r_s i_{ds}^r - \frac{\omega_r}{\omega_b}\psi_{qs}^r + \frac{p}{\omega_b}\psi_{ds}^r \tag{5.5-23}$$

$$v_{0s} = -r_s i_{0s} + \frac{p}{\omega_b}\psi_{0s} \tag{5.5-24}$$

$$v_{kq1}^{\prime r} = r_{kq1}^{\prime} i_{kq1}^{\prime r} + \frac{p}{\omega_b}\psi_{kq1}^{\prime r} \tag{5.5-25}$$

$$v_{kq2}^{\prime r} = r_{kq2}^{\prime} i_{kq2}^{\prime r} + \frac{p}{\omega_b}\psi_{kq2}^{\prime r} \tag{5.5-26}$$

$$v_{fd}^{\prime r} = r_{fd}^{\prime} i_{fd}^{\prime r} + \frac{p}{\omega_b}\psi_{fd}^{\prime r} \tag{5.5-27}$$

$$v_{kd}^{\prime r} = r_{kd}^{\prime} i_{kd}^{\prime r} + \frac{p}{\omega_b}\psi_{kd}^{\prime r} \tag{5.5-28}$$

where ω_b is the base electrical angular velocity used to calculate the inductive reactances. The flux linkages per second are

$$\psi_{qs}^r = -X_{ls}i_{qs}^r + X_{mq}(-i_{qs}^r + i_{kq1}^{\prime r} + i_{kq2}^{\prime r}) \tag{5.5-29}$$

$$\psi_{ds}^r = -X_{ls}i_{ds}^r + X_{md}(-i_{ds}^r + i_{fd}^{\prime r} + i_{kd}^{\prime r}) \tag{5.5-30}$$

$$\psi_{0s} = -X_{ls}i_{0s} \tag{5.5-31}$$

$$\psi_{kq1}^{\prime r} = X_{lkq1}^{\prime} i_{kq1}^{\prime r} + X_{mq}(-i_{qs}^r + i_{kq1}^{\prime r} + i_{kq2}^{\prime r}) \tag{5.5-32}$$

$$\psi_{kq2}^{\prime r} = X_{lkq2}^{\prime} i_{kq2}^{\prime r} + X_{mq}(-i_{qs}^r + i_{kq1}^{\prime r} + i_{kq2}^{\prime r}) \tag{5.5-33}$$

$$\psi_{fd}^{\prime r} = X_{lfd}^{\prime} i_{fd}^{\prime r} + X_{md}(-i_{ds}^r + i_{fd}^{\prime r} + i_{kd}^{\prime r}) \tag{5.5-34}$$

$$\psi_{kd}^{\prime r} = X_{lkd}^{\prime} i_{kd}^{\prime r} + X_{md}(-i_{ds}^r + i_{fd}^{\prime r} + i_{kd}^{\prime r}) \tag{5.5-35}$$

Park's equations are generally written without the superscript r, the subscript s, and the primes that denote referred quantities. Also, we will later find that it is convenient to define

$$e_{xfd}^{\prime r} = v_{fd}^{\prime r} \frac{X_{md}}{r_{fd}^{\prime}} \tag{5.5-36}$$

and to substitute this relationship into the expression for field voltage so that (5.5-27) becomes

$$e_{xfd}^{\prime r} = \frac{X_{md}}{r_{fd}^{\prime}}\left(r_{fd}^{\prime} i_{fd}^{\prime r} + \frac{p}{\omega_b}\psi_{fd}^{\prime r}\right) \tag{5.5-37}$$

The voltage equations for the synchronous machine given thus far are in terms of currents and flux linkages per second. As we have pointed out earlier, the current and flux linkages are related, and both cannot be independent or state variables. As in the case of the induction machine, we will need to express the voltage equations in terms of either currents or flux linkages (flux linkages per second) when formulating transfer functions and implementing a computer simulation.

If we select the currents as independent variables, the flux linkages (flux linkages per second) are replaced by currents and the voltage equations given by (5.5-22)–(5.5-28) with (5.5-37) used instead of (5.5-27) become

$$
\begin{bmatrix} v_{qs}^r \\ v_{ds}^r \\ v_{0s} \\ v_{kq1}^{\prime r} \\ v_{kq2}^{\prime r} \\ e_{xfd}^{\prime r} \\ v_{kd}^{\prime r} \end{bmatrix}
=
\begin{bmatrix}
-r_s - \frac{p}{\omega_b}X_q & -\frac{\omega_r}{\omega_b}X_d & 0 & \frac{p}{\omega_b}X_{mq} & \frac{p}{\omega_b}X_{mq} & \frac{\omega_r}{\omega_b}X_{md} & \frac{\omega_r}{\omega_b}X_{md} \\
\frac{\omega_r}{\omega_b}X_q & -r_s - \frac{p}{\omega_b}X_d & 0 & -\frac{\omega_r}{\omega_b}X_{mq} & -\frac{\omega_r}{\omega_b}X_{mq} & \frac{p}{\omega_b}X_{md} & \frac{p}{\omega_b}X_{md} \\
0 & 0 & -r_s - \frac{p}{\omega_b}X_{ls} & 0 & 0 & 0 & 0 \\
-\frac{p}{\omega_b}X_{mq} & 0 & 0 & r_{kq1}' + \frac{p}{\omega_b}X_{kq1}' & \frac{p}{\omega_b}X_{mq} & 0 & 0 \\
-\frac{p}{\omega_b}X_{mq} & 0 & 0 & \frac{p}{\omega_b}X_{mq} & r_{kq2}' + \frac{p}{\omega_b}X_{kq2}' & 0 & 0 \\
0 & -\frac{X_{md}}{r_{fd}'}\times & 0 & 0 & 0 & \frac{X_{md}}{r_{fd}'}\left(r_{fd}'+ & \frac{X_{md}}{r_{fd}'}\times \\
 & \left(\frac{p}{\omega_b}X_{md}\right) & & & & \frac{p}{\omega_b}X_{fd}'\right) & \left(\frac{p}{\omega_b}X_{md}\right) \\
0 & -\frac{p}{\omega_b}X_{md} & 0 & 0 & 0 & \frac{p}{\omega_b}X_{md} & r_{kd}' + \frac{p}{\omega_b}X_{kd}'
\end{bmatrix}
\begin{bmatrix} i_{qs}^r \\ i_{ds}^r \\ i_{0s} \\ i_{kq1}^{\prime r} \\ i_{kq2}^{\prime r} \\ i_{fd}^{\prime r} \\ i_{kd}^{\prime r} \end{bmatrix}
$$

$$(5.5\text{-}38)$$

$$X_q = X_{ls} + X_{mq} \tag{5.5-39}$$
$$X_d = X_{ls} + X_{md} \tag{5.5-40}$$
$$X_{kq1}' = X_{lkq1}' + X_{mq} \tag{5.5-41}$$
$$X_{kq2}' = X_{lkq2}' + X_{mq} \tag{5.5-42}$$
$$X_{fd}' = X_{lfd}' + X_{md} \tag{5.5-43}$$
$$X_{kd}' = X_{lkd}' + X_{md} \tag{5.5-44}$$

The reactances X_q and X_d are generally referred to as q- and d-axis reactances, respectively. The flux linkages per second may be expressed from (5.5-29)–(5.5-35) as

$$
\begin{bmatrix} \psi_{qs}^r \\ \psi_{ds}^r \\ \psi_{0s} \\ \psi_{kq1}^{\prime r} \\ \psi_{kq2}^{\prime r} \\ \psi_{fd}^{\prime r} \\ \psi_{kd}^{\prime r} \end{bmatrix}
=
\begin{bmatrix}
-X_q & 0 & 0 & X_{mq} & X_{mq} & 0 & 0 \\
0 & -X_d & 0 & 0 & 0 & X_{md} & X_{md} \\
0 & 0 & X_{ls} & 0 & 0 & 0 & 0 \\
-X_{mq} & 0 & 0 & X_{kq1}' & X_{mq} & 0 & 0 \\
-X_{mq} & 0 & 0 & X_{mq} & X_{kq2}' & 0 & 0 \\
0 & -X_{md} & 0 & 0 & 0 & X_{fd}' & X_{md} \\
0 & -X_{md} & 0 & 0 & 0 & X_{md} & X_{kd}'
\end{bmatrix}
\begin{bmatrix} i_{qs}^r \\ i_{ds}^r \\ i_{0s} \\ i_{kq1}^{\prime r} \\ i_{kq2}^{\prime r} \\ i_{fd}^{\prime r} \\ i_{kd}^{\prime r} \end{bmatrix}
\tag{5.5-45}
$$

If the flux linkages or flux linkages per second are selected as independent variables it is convenient to first express (5.5-45) as

$$
\begin{bmatrix} \psi_{qs}^r \\ \psi_{kq1}^{\prime r} \\ \psi_{kq2}^{\prime r} \end{bmatrix} = \begin{bmatrix} -X_q & X_{mq} & X_{mq} \\ -X_{mq} & X_{kq1}^{\prime} & X_{mq} \\ -X_{mq} & X_{mq} & X_{kq2}^{\prime} \end{bmatrix} \begin{bmatrix} i_{qs}^r \\ i_{kq1}^{\prime r} \\ i_{kq2}^{\prime r} \end{bmatrix}
\tag{5.5-46}
$$

$$
\begin{bmatrix} \psi_{ds}^r \\ \psi_{fd}^{\prime r} \\ \psi_{kd}^{\prime r} \end{bmatrix} = \begin{bmatrix} -X_d & X_{md} & X_{md} \\ -X_{md} & X_{fd}^{\prime} & X_{md} \\ -X_{md} & X_{md} & X_{kd}^{\prime} \end{bmatrix} \begin{bmatrix} i_{ds}^r \\ i_{fd}^{\prime r} \\ i_{kd}^{\prime r} \end{bmatrix}
\tag{5.5-47}
$$

$$
\psi_{0s} = X_{ls} i_{0s}
\tag{5.5-48}
$$

Solving the above equations for currents yields

$$
\begin{bmatrix} i_{qs}^r \\ i_{kq1}^{\prime r} \\ i_{kq2}^{\prime r} \end{bmatrix} = \frac{1}{D_q} \begin{bmatrix} X_{kq1}^{\prime}X_{kq2}^{\prime} - X_{mq}^2 & -X_{mq}X_{kq2}^{\prime} + X_{mq}^2 & -X_{mq}X_{kq1}^{\prime} + X_{mq}^2 \\ X_{mq}X_{kq2}^{\prime} - X_{mq}^2 & -X_q X_{kq2}^{\prime} + X_{mq}^2 & X_q X_{mq} - X_{mq}^2 \\ X_{mq}X_{kq1}^{\prime} - X_{mq}^2 & X_q X_{mq} - X_{mq}^2 & -X_q X_{kq1}^{\prime} + X_{mq}^2 \end{bmatrix} \begin{bmatrix} \psi_{qs}^r \\ \psi_{kq1}^{\prime r} \\ \psi_{kq2}^{\prime r} \end{bmatrix}
\tag{5.5-49}
$$

$$
\begin{bmatrix} i_{ds}^r \\ i_{fd}^{\prime r} \\ i_{kd}^{\prime r} \end{bmatrix} = \frac{1}{D_d} \begin{bmatrix} X_{fd}^{\prime}X_{kd}^{\prime} - X_{md}^2 & -X_{md}X_{kd}^{\prime} + X_{md}^2 & -X_{md}X_{fd}^{\prime} + X_{md}^2 \\ X_{md}X_{kd}^{\prime} - X_{md}^2 & -X_d X_{kd}^{\prime} + X_{md}^2 & X_d X_{md} - X_{md}^2 \\ X_{md}X_{fd}^{\prime} - X_{md}^2 & X_d X_{md} - X_{md}^2 & -X_d X_{fd}^{\prime} + X_{md}^2 \end{bmatrix} \begin{bmatrix} \psi_{ds}^r \\ \psi_{fd}^{\prime r} \\ \psi_{kd}^{\prime r} \end{bmatrix}
\tag{5.5-50}
$$

$$
i_{0s} = \frac{1}{X_{ls}} \psi_{0s}
\tag{5.5-51}
$$

where

$$
D_q = X_{mq}^2 (X_q - 2X_{mq} + X_{kq1}^{\prime} + X_{kq2}^{\prime}) - X_q X_{kq1}^{\prime} X_{kq2}^{\prime}
\tag{5.5-52}
$$

$$
D_d = X_{md}^2 (X_d - 2X_{md} + X_{fd}^{\prime} + X_{kd}^{\prime}) - X_d X_{fd}^{\prime} X_{kd}^{\prime}
\tag{5.5-53}
$$

Substituting (5.5-49)–(5.5-51) for the currents into the voltage equations (5.5-22)–(5.5-26), (5.5-37) and (5.5-38) yields

$$
\begin{bmatrix} v_{qs}^r \\ v_{ds}^r \\ v_{0s} \\ v_{kq1}^{\prime r} \\ v_{kq2}^{\prime r} \\ e_{xfd}^{\prime r} \\ v_{kd}^{\prime r} \end{bmatrix} =
\begin{bmatrix}
-r_s a_{11} + \frac{p}{\omega_b} & \frac{\omega_r}{\omega_b} & 0 & -r_s a_{12} & -r_s a_{13} & 0 & 0 \\
-\frac{\omega_r}{\omega_b} & -r_s b_{11} + \frac{p}{\omega_b} & 0 & 0 & 0 & -r_s b_{12} & -r_s b_{13} \\
0 & 0 & -\frac{r_s}{X_{ls}} + \frac{p}{\omega_b} & 0 & 0 & 0 & 0 \\
r_{kq1}^{\prime} a_{21} & 0 & 0 & r_{kq1}^{\prime} a_{22} + \frac{p}{\omega_b} & r_{kq1}^{\prime} a_{23} & 0 & 0 \\
r_{kq2}^{\prime} a_{31} & 0 & 0 & r_{kq2}^{\prime} a_{32} & r_{kq2}^{\prime} a_{33} + \frac{p}{\omega_b} & 0 & 0 \\
0 & X_{md} b_{21} & 0 & 0 & 0 & X_{md} b_{22} + \frac{X_{md}}{r_{fd}^{\prime}}\frac{p}{\omega_b} & X_{md} b_{23} \\
0 & r_{kd}^{\prime} b_{31} & 0 & 0 & 0 & r_{kd}^{\prime} b_{32} & r_{kd}^{\prime} b_{33} + \frac{p}{\omega_b}
\end{bmatrix}
\begin{bmatrix} \psi_{qs}^r \\ \psi_{ds}^r \\ \psi_{0s} \\ \psi_{kq1}^{\prime r} \\ \psi_{kq2}^{\prime r} \\ \psi_{fd}^{\prime r} \\ \psi_{kd}^{\prime r} \end{bmatrix}
\tag{5.5-54}
$$

In (5.5-54) a_{ij} and b_{ij} are the elements of the 3×3 matrices given in (5.5-49) and (5.5-50), respectively. The statements at the end of Section 4.5 regarding the computer simulation of induction machines also apply in the case of synchronous machines.

In some applications, especially in variable-frequency drive systems, the synchronous machine is operated as a motor. When analyzing motor operation it may be desirable to write the equations with the direction of positive stator current into the stator terminals. The equations are modified to accommodate this change in assumed direction of positive current by simply changing the sign of i^r_{qs}, i^r_{ds}, and i_{0s} in all voltage and flux linkage equations.

5.6 TORQUE EQUATIONS IN SUBSTITUTE VARIABLES

The expression for the electromagnetic torque in terms of rotor reference-frame variables may be obtained by substituting the equations of transformation into (5.3-3). Hence

$$
T_e = \left(\frac{P}{2}\right) [(\mathbf{K}^r_s)^{-1} \mathbf{i}^r_{qd0s}]^T \left\{ -\frac{1}{2} \frac{\partial}{\partial \theta_r} [\mathbf{L}_s - L_{ls}\mathbf{I}](\mathbf{K}^r_s)^{-1} \mathbf{i}_{qd0s} + \frac{\partial}{\partial \theta_r} [\mathbf{L}'_{sr}] \mathbf{i}'^r_{qdr} \right\}
$$

$$(5.6\text{-}1)$$

After considerable work the above equation reduces to

$$
T_e = \left(\frac{3}{2}\right) \left(\frac{P}{2}\right) [L_{md}(-i^r_{ds} + i'^r_{fd} + i'^r_{kd}) i^r_{qs} - L_{mq}(-i^r_{qs} + i'^r_{kq1} + i'^r_{kq2}) i^r_{ds}] \qquad (5.6\text{-}2)
$$

Equation (5.6-2) is equivalent to

$$
T_e = \left(\frac{3}{2}\right) \left(\frac{P}{2}\right) (\lambda^r_{ds} i^r_{qs} - \lambda^r_{qs} i^r_{ds}) \qquad (5.6\text{-}3)
$$

In terms of flux linkages per second and currents, we have

$$
T_e = \left(\frac{3}{2}\right) \left(\frac{P}{2}\right) \left(\frac{1}{\omega_b}\right) (\psi^r_{ds} i^r_{qs} - \psi^r_{qs} i^r_{ds}) \qquad (5.6\text{-}4)
$$

It is left to the reader to show that in terms of flux linkages per second the electromagnetic torque may be expressed as

$$
T_e = \left(\frac{3}{2}\right) \left(\frac{P}{2}\right) \left(\frac{1}{\omega_b}\right) [(a_{11} - b_{11}) \psi^r_{qs} \psi^r_{ds} + \psi^r_{ds}(a_{12} \psi'^r_{kq1}
$$
$$
+ a_{13} \psi'^r_{kq2}) - \psi^r_{qs}(b_{12} \psi'^r_{fd} + b_{13} \psi'^r_{kd})] \qquad (5.6\text{-}5)
$$

where a_{ij} and b_{ij} are the elements of the 3×3 matrices given in (5.5-49) and (5.5-50), respectively.

The above equations yield positive torque for generator action with the positive direction of stator current assumed out of the stator terminals. If, for the analysis of motor action, the positive direction of stator current is assumed into the stator terminals, then the voltage and flux linkage equations are modified as discussed previously. The expression for torque given by (5.6-3), (5.6-4), and (5.6-5) is then positive for motor action without change; however, (5.6-2) must be modified by changing the sign of i_{ds}^r in the first term on the right-hand side and of i_{qs}^r in the second term.

The electromagnetic torque expressed with the stator variables in the arbitrary reference frame may be obtained by employing the transformation of the stator variables from the rotor reference frame to the arbitrary reference frame. From (3.6-7) we obtain

$$^r\mathbf{K} = \begin{bmatrix} \cos(\theta - \theta_r) & -\sin(\theta - \theta_r) & 0 \\ \sin(\theta - \theta_r) & \cos(\theta - \theta_r) & 0 \\ 0 & 0 & 1 \end{bmatrix} \quad (5.6\text{-}6)$$

Because only the qs and ds variables are involved in the transformation, we have

$$\begin{bmatrix} f_{qs} \\ f_{ds} \end{bmatrix} = \begin{bmatrix} \cos(\theta - \theta_r) & -\sin(\theta - \theta_r) \\ \sin(\theta - \theta_r) & \cos(\theta - \theta_r) \end{bmatrix} \begin{bmatrix} f_{qs}^r \\ f_{ds}^r \end{bmatrix} \quad (5.6\text{-}7)$$

The inverse is

$$\begin{bmatrix} f_{qs}^r \\ f_{ds}^r \end{bmatrix} = \begin{bmatrix} \cos(\theta - \theta_r) & \sin(\theta - \theta_r) \\ -\sin(\theta - \theta_r) & \cos(\theta - \theta_r) \end{bmatrix} \begin{bmatrix} f_{qs} \\ f_{ds} \end{bmatrix} \quad (5.6\text{-}8)$$

Appropriate substitution of (5.6-8) into the above torque equations yields the torque expressed with the stator variables in the arbitrary reference frame. For example, (5.6-3) becomes

$$T_e = \left(\frac{3}{2}\right)\left(\frac{P}{2}\right)(\lambda_{ds}i_{qs} - \lambda_{qs}i_{ds}) \quad (5.6\text{-}9)$$

5.7 ROTOR ANGLE AND ANGLE BETWEEN ROTORS

Except for isolated operation, it is convenient for analysis and interpretation purposes to relate the position of the rotor of a synchronous machine to a voltage or to the rotor of another machine. The electrical angular displacement of the rotor relative to its terminal voltage is defined as the rotor angle, while in a multiple-machine power system it is customary to express the angle between machine rotors.

The rotor angle is the displacement of the rotor generally referenced to the maximum positive value of the fundamental component of the terminal voltage of phase a. Therefore, the rotor angle expressed in radians is

$$\delta = \theta_r - \theta_{ev} \tag{5.7-1}$$

The electrical angular velocity of the rotor is ω_r; ω_e is the electrical angular velocity of the terminal voltages. The above definition of δ is valid regardless of the mode of operation (either or both ω_r and ω_e may vary). Because a physical interpretation is most easily visualized during balanced steady-state operation, we will defer this explanation until the steady-state voltage and torque equations have been written in terms of δ.

It is important to note that the rotor angle is often used as the argument in the transformation between the rotor and synchronously rotating reference frames because ω_e is the speed of the synchronously rotating reference frame and it is also the angular velocity of θ_{ev}. From (3.6-1) we obtain

$$\mathbf{f}^r_{qd0s} = {}^e\mathbf{K}^r \mathbf{f}^e_{qd0s} \tag{5.7-2}$$

where

$$
{}^e\mathbf{K}^r = \begin{bmatrix} \cos\delta & -\sin\delta & 0 \\ \sin\delta & \cos\delta & 0 \\ 0 & 0 & 1 \end{bmatrix} \tag{5.7-3}
$$

The rotor angle is often used in relating torque and rotor speed. In particular, if ω_e is constant, then (5.3-5) may be written as

$$T_e = -J\left(\frac{2}{P}\right)p^2\delta + T_l \tag{5.7-4}$$

where δ is expressed in electrical radians.

The angular displacement between rotors of machines in a power system may be expressed as

$$\delta_{21} = \theta_{r2} - \theta_{r1} \tag{5.7-5}$$

Here, δ_{21} is the angular displacement between the q axis of the rotor of machine 2 and the q axis of the rotor of machine 1. It follows that δ_{21} would be the argument in the transformation between these reference frames, that is,

$$
{}^{r1}\mathbf{K}^{r2} = \begin{bmatrix} \cos\delta_{21} & -\sin\delta_{21} & 0 \\ \sin\delta_{21} & \cos\delta_{21} & 0 \\ 0 & 0 & 1 \end{bmatrix} \tag{5.7-6}
$$

5.8 PER UNIT SYSTEM

The equations for a synchronous machine may be written in per unit by following the same procedure as in the case of the induction machine. Base voltage is generally selected as the rms value of the rated phase voltage for the *abc* variables and the peak value for the *qd*0 variables. However, we will often use the same base value when comparing *abc* and *qd*0 variables. When considering the machine separately, the power base is selected as its volt-ampere rating. When considering power systems, a system power base (system base) is selected which is generally different from the power base of the machine (machine base).

Once the base quantities are established, the corresponding base current and base impedance may be calculated. Park's equations written in terms of flux linkages per second and reactances are readily per unitized by dividing each term by the peak of the base voltage (or the peak value of the base current times base impedance). The form of these equations remains unchanged as a result of per unitizing. When per unitizing the voltage equation of the field winding (*fd* winding) it is convenient to use the form given by (5.5-37) involving e''_{xfd}. The reason for this choice is established later.

Base torque is the base power divided by the synchronous speed of the rotor. Thus

$$T_B = \frac{P_B}{(2/P)\omega_b} = \frac{(\frac{3}{2})V_{B(qd0)}I_{B(qd0)}}{(2/P)\omega_b} \tag{5.8-1}$$

where ω_b corresponds to rated or base frequency, P_B is the base power, $V_{B(qd0)}$ is the peak value of the base phase voltage and $I_{B(qd0)}$ is the peak value of the base phase current. Dividing the torque equations by (5.8-1) yields the torque expressed in per unit. For example, (5.6-4) with all quantities expressed in per unit becomes

$$T_e = (\psi^r_{ds}i^r_{qs} - \psi^r_{qs}i^r_{ds}) \tag{5.8-2}$$

Equation (5.3-5), which relates torque and speed, is expressed in per unit as

$$T_e = -2Hp\frac{\omega_r}{\omega_b} + T_I \tag{5.8-3}$$

If ω_e is constant, then this relationship becomes

$$T_e = -\frac{2H}{\omega_b}p^2\delta + T_I \tag{5.8-4}$$

where δ is in electrical radians. The inertia constant H is in seconds. It is defined as

$$H = \left(\frac{1}{2}\right)\left(\frac{2}{P}\right)\frac{J\omega_b}{T_B} = \left(\frac{1}{2}\right)\left(\frac{2}{P}\right)^2\frac{J\omega_b^2}{P_B} \tag{5.8-5}$$

where J is often the combined inertia of the rotor and prime mover expressed in $kg \cdot m^2$ or given as the quantity WR^2 in $lb \cdot ft^2$.

5.9 ANALYSIS OF STEADY-STATE OPERATION

Although the voltage equations that describe balanced steady-state operation of synchronous machines may be derived using several approaches, it is convenient to use Park's equations in this derivation. For balanced conditions the 0s quantities are zero. For balanced steady-state conditions the electrical angular velocity of the rotor is constant and equal to ω_e, whereupon the electrical angular velocity of the rotor reference frame becomes the electrical angular velocity of the synchronously rotating reference frame. In this mode of operation the rotor windings do not experience a change of flux linkages; hence current is not flowing in the short-circuited damper windings. Thus, with ω_r set equal to ω_e and the time rate of change of all flux linkages neglected, the steady-state versions of (5.5-22), (5.5-23), and (5.5-27) become

$$V_{qs}^r = -r_s I_{qs}^r - \frac{\omega_e}{\omega_b} X_d I_{ds}^r + \frac{\omega_e}{\omega_b} X_{md} I_{fd}^{\prime r} \tag{5.9-1}$$

$$V_{ds}^r = -r_s I_{ds}^r + \frac{\omega_e}{\omega_b} X_q I_{qs}^r \tag{5.9-2}$$

$$V_{fd}^{\prime r} = r_{fd}^{\prime} I_{fd}^{\prime r} \tag{5.9-3}$$

Here the ω_e to ω_b ratio is again included to accommodate analysis when the operating frequency is other than rated. It is recalled that all reactances used in this text are calculated using base or rated frequency.

The reactances X_q and X_d are defined by (5.5-39) and (5.5-40); that is, $X_q = X_{ls} + X_{mq}$ and $X_d = X_{ls} + X_{md}$. As mentioned previously, Park's equations are generally written with the primes and the s and r indexes omitted. The uppercase letters are used here to denote steady-state quantities.

Equations (3.7-5) and (3.7-6) express the instantaneous variables in the arbitrary reference frame for balanced conditions. In the rotor reference frame these expressions become

$$f_{qs}^r = \sqrt{2} f_s \cos(\theta_{ef} - \theta_r) \tag{5.9-4}$$

$$f_{ds}^r = -\sqrt{2} f_s \sin(\theta_{ef} - \theta_r) \tag{5.9-5}$$

For steady-state balanced conditions, (5.9-4) and (5.9-5) may be expressed as

$$F_{qs}^r = \text{Re}[\sqrt{2} F_s e^{j(\theta_{ef} - \theta_r)}] \tag{5.9-6}$$

$$F_{ds}^r = \text{Re}[j\sqrt{2} F_s e^{j(\theta_{ef} - \theta_r)}] \tag{5.9-7}$$

It is to our advantage to express (5.9-6) and (5.9-7) in terms of δ [see (5.7-1)]. Hence, if we multiply each equation by $e^{j\theta_{ev}(1-1)}$ and because θ_{ef} and θ_{ev} are both functions of ω_e, the above equations may be written as

$$F^r_{qs} = \mathrm{Re}[\sqrt{2}F_s e^{j[\theta_{ef}(0)-\theta_{ev}(0)]}e^{-j\delta}]$$ (5.9-8)

$$F^r_{ds} = \mathrm{Re}[j\sqrt{2}F_s e^{j[\theta_{ef}(0)-\theta_{ev}(0)]}e^{-j\delta}]$$ (5.9-9)

It is important to note that

$$\tilde{F}_{as} = F_s e^{j[\theta_{ef}(0)-\theta_{ev}(0)]}$$ (5.9-10)

is a phasor that represents the *as* variables referenced to the time zero position of θ_{ev} which we will select so that maximum v_{as} occurs at $t = 0$.

From (5.9-8) and (5.9-9) we obtain

$$F^r_{qs} = \sqrt{2}F_s \cos[\theta_{ef}(0) - \theta_{ev}(0) - \delta]$$ (5.9-11)

$$F^r_{ds} = -\sqrt{2}F_s \sin[\theta_{ef}(0) - \theta_{ev}(0) - \delta]$$ (5.9-12)

from which

$$\sqrt{2}\tilde{F}_{as}e^{-j\delta} = F^r_{qs} - jF^r_{ds}$$ (5.9-13)

where $\tilde{F}_{as}$ is defined by (5.9-10). Hence

$$\sqrt{2}\tilde{V}_{as}e^{-j\delta} = V^r_{qs} - jV^r_{ds}$$ (5.9-14)

Substituting (5.9-1) and (5.9-2) into (5.9-14) yields

$$\sqrt{2}\tilde{V}_{as}e^{-j\delta} = -r_s I^r_{qs} - \frac{\omega_e}{\omega_b}X_d I^r_{ds} + \frac{\omega_e}{\omega_b}X_{md}I'^r_{fd} + j\left(r_s I^r_{ds} - \frac{\omega_e}{\omega_b}X_q I^r_{qs}\right)$$ (5.9-15)

If $(\omega_e/\omega_b)X_q I^r_{ds}$ is added to and subtracted from the right-hand side of (5.9-15) and if it is noted that

$$j\sqrt{2}\tilde{I}_{as}e^{-j\delta} = I^r_{ds} - jI^r_{qs}$$ (5.9-16)

then (5.9-15) may be written as

$$\tilde{V}_{as} = -\left(r_s + j\frac{\omega_e}{\omega_b}X_q\right)\tilde{I}_{as} + \frac{1}{\sqrt{2}}\left[-\frac{\omega_e}{\omega_b}(X_d - X_q)I^r_{ds} + \frac{\omega_e}{\omega_b}X_{md}I'^r_{fd}\right]e^{j\delta}$$ (5.9-17)

It is convenient to define the last term on the right-hand side of (5.9-17) as

$$\tilde{E}_a = \frac{1}{\sqrt{2}}\left[-\left(\frac{\omega_e}{\omega_b}\right)(X_d - X_q)\,I_{ds}^r + \left(\frac{\omega_e}{\omega_b}\right)X_{md}\,I_{fd}^{\prime r}\right]e^{j\delta} \qquad (5.9\text{-}18)$$

which is sometimes referred to as the excitation voltage. Thus, (5.9-17) becomes

$$\tilde{V}_{as} = -\left(r_s + j\frac{\omega_e}{\omega_b}X_q\right)\tilde{I}_{as} + \tilde{E}_a \qquad (5.9\text{-}19)$$

Equations (5.9-18) and (5.9-19) are written for the positive direction of stator current assumed out of the machine, convenient for generator action. The ω_e to ω_b ratio is included so that the equations are valid for the analysis of balanced steady-state operation at a frequency other than rated.

When the positive direction of stator current is assumed into the machine, convenient for motor action, then

$$\tilde{V}_{as} = \left(r_s + j\frac{\omega_e}{\omega_b}X_q\right)\tilde{I}_{as} + \tilde{E}_a \qquad (5.9\text{-}20)$$

where

$$\tilde{E}_a = \frac{1}{\sqrt{2}}\left[\frac{\omega_e}{\omega_b}(X_d - X_q)I_{ds}^r + \frac{\omega_e}{\omega_b}X_{md}I_{fd}^{\prime r}\right]e^{j\delta} \qquad (5.9\text{-}21)$$

If (5.9-1) and (5.9-2) are solved for I_{qs}^r and I_{ds}^r and the results substituted into (5.6-2), the expression for the balanced steady-state electromagnetic torque can be written as

$$
\begin{aligned}
T_e = & -\left(\frac{3}{2}\right)\left(\frac{P}{2}\right)\left(\frac{1}{\omega_b}\right) \\
& \times \left\{ \frac{r_s X_{md}I_{fd}^{\prime r}}{r_s^2 + (\omega_e/\omega_b)^2 X_q X_d}\left(V_{qs}^r - \frac{\omega_e}{\omega_b}X_{md}I_{fd}^{\prime r} - \frac{\omega_e}{\omega_b}\frac{X_d}{r_s}V_{ds}^r\right) \right. \\
& + \frac{X_d - X_q}{[r_s^2 + (\omega_e/\omega_b)^2 X_q X_d]^2}\left[r_s\frac{\omega_e}{\omega_b}X_q\left(V_{qs}^r - \frac{\omega_e}{\omega_b}X_{md}I_{fd}^{\prime r}\right)\right]^2 \\
& + \left[r_s^2 - \left(\frac{\omega_e}{\omega_b}\right)^2 X_q X_d\right]V_{ds}^r\left(V_{qs}^r - \frac{\omega_e}{\omega_b}X_{md}I_{fd}^{\prime r}\right) - r_s\frac{\omega_e}{\omega_b}X_d(V_{ds}^r)^2 \right\} \qquad (5.9\text{-}22)
\end{aligned}
$$

where P is the number of poles, ω_b is the base electrical angular velocity used to calculate the reactances, and ω_e corresponds to the operating frequency.

For balanced operation the stator voltages may be expressed in the form given by (3.7-1)–(3.7-3). Thus

$$v_{as} = \sqrt{2}v_s \cos\theta_{ev} \tag{5.9-23}$$

$$v_{bs} = \sqrt{2}v_s \cos\left(\theta_{ev} - \frac{2\pi}{3}\right) \tag{5.9-24}$$

$$v_{cs} = \sqrt{2}v_s \cos\left(\theta_{ev} + \frac{2\pi}{3}\right) \tag{5.9-25}$$

where

$$\theta_{ev} = \int_0^t \omega_e(\xi)\,d\xi + \theta_{ev}(0) \tag{5.9-26}$$

where ξ is a dummy variable of integration. These voltages may be expressed in the rotor reference frame by replacing θ with θ_r in (3.7-5) and (3.7-6).

$$v_{qs}^r = \sqrt{2}v_s \cos(\theta_{ev} - \theta_r) \tag{5.9-27}$$

$$v_{ds}^r = -\sqrt{2}v_s \sin(\theta_{ev} - \theta_r) \tag{5.9-28}$$

If the rotor angle from (5.7-1) is substituted into (5.9-27) and (5.9-28), we obtain

$$v_{qs}^r = \sqrt{2}v_s \cos\delta \tag{5.9-29}$$

$$v_{ds}^r = \sqrt{2}v_s \sin\delta \tag{5.9-30}$$

The only restrictions on (5.9-29) and (5.9-30) are that the stator voltages form a balanced set. These equations are valid for transient and steady-state operation; that is, v_s and δ may both be functions of time with $\theta_{ev}(0)$ generally set equal to zero.

The torque given by (5.9-22) is for balanced steady-state conditions. In this mode of operation, (5.9-29) and (5.9-30) are constants because v_s and δ are both constants. Before proceeding, it is noted that from (5.5-36) that for balanced steady-state operation we obtain

$$E_{xfd}^{\prime r} = X_{md}I_{fd}^{\prime r} \tag{5.9-31}$$

Although this expression is sometimes substituted into the above steady-state voltage equations, it is most often used in the expression for torque. In particular, if (5.9-31) and the steady-state versions of (5.9-29) and (5.9-30) are substituted in (5.9-22) and if r_s is neglected, the torque may be expressed as

$$T_e = \left(\frac{3}{2}\right)\left(\frac{P}{2}\right)\left(\frac{1}{\omega_b}\right)\left[\frac{E_{xfd}^{\prime r}\sqrt{2}V_s}{(\omega_e/\omega_b)X_d}\sin\delta + \left(\frac{1}{2}\right)\left(\frac{\omega_e}{\omega_b}\right)^{-2}\left(\frac{1}{X_q}-\frac{1}{X_d}\right)(\sqrt{2}V_s)^2\sin 2\delta\right]$$

$$\tag{5.9-32}$$

In per unit, (5.9-32) becomes

$$T_e = \frac{E_{xfd}^{'r} V_s}{(\omega_e/\omega_b) X_d} \sin\delta + \left(\frac{1}{2}\right)\left(\frac{\omega_e}{\omega_b}\right)^{-2}\left(\frac{1}{X_q} - \frac{1}{X_d}\right) V_s^2 \sin 2\delta \qquad (5.9\text{-}33)$$

Neglecting r_s is justified if r_s is small relative to the reactances of the machine. In variable-frequency drive systems, this may not be the case at low frequencies, whereupon (5.9-22) must be used to calculate torque rather than (5.9-32). With the stator resistance neglected, steady-state power and torque are related by rotor speed; and if torque and power are expressed in per unit, they are equal during steady-state operation. The above expressions are positive for generator action. If one wishes positive torque for motor action, then (5.9-22), (5.9-32), and (5.9-33) must be multiplied by -1.

Although (5.9-32) is valid only for balanced steady-state operation and if the stator resistance is small relative to the magnetizing reactances (X_{mq} and X_{md}) of the machine, it permits a quantitative description of the nature of the steady-state electromagnetic torque of a synchronous machine. The first term on the right-hand side of (5.9-32) is due to the interaction of the magnetic system produced by the currents flowing in the stator windings and the magnetic system produced by the current flowing in the field winding. The second term is due to the saliency of the rotor. This component is commonly referred to as the reluctance torque. The predominate torque is the torque due to the interaction of the stator and field currents. The amplitude of this component is proportional to the magnitudes of the stator voltage, V_s, and the voltage applied to the field, $E_{xfd}^{'r}$. In power systems it is desirable to maintain the stator voltage near rated. This is achieved by automatically adjusting the voltage applied to the field winding. Hence, the amplitude of this torque component varies as $E_{xfd}^{'r}$ is varied to maintain the terminal voltage at or near rated and/or to control reactive power flow. The reluctance torque component is generally a relatively small part of the total torque of a synchronous generator. In power systems where the terminal voltage is maintained nearly constant, the amplitude of the reluctance torque would also be nearly constant, a function of only the parameters of the machine. A steady-state reluctance torque does not exist in round or cylindrical rotor synchronous machines because $X_q = X_d$. On the other hand, a reluctance machine is a device that is not equipped with a field winding; hence, the only torque produced is reluctance torque. Reluctance machines are used widely as motors, especially in variable-frequency drive systems.

Let us return for a moment to the steady-state voltage equation for generator action, (5.9-19). With $\theta_{ev}(0) = 0$, $\tilde{V}_{as}$ lies along the positive real axis of a phasor diagram. Because δ is the angle associated with $\tilde{E}_a$ [see (5.9-18)], its position relative to $\tilde{V}_{as}$ is also the position of the q axis of the machine relative to $\tilde{V}_{as}$. Therefore, we can superimpose the q and d axes of the synchronous machine upon the phasor diagram.

If T_I is assumed zero and if we neglect friction and windage losses along with the stator resistance, then T_e and δ are also zero and the machine will theoretically run at synchronous speed without absorbing energy from either the electrical system or the mechanical system. Although this mode of operation is not feasible in practice because the machine will actually absorb some small amount of energy to satisfy the ohmic and friction and windage losses, it is convenient for purposes of explanation. With the machine "floating on the line" the field voltage can be adjusted to establish the desired terminal conditions. Three situations may exist: (1) $|\tilde{E}_a| = |\tilde{V}_{as}|$, whereupon $\tilde{I}_{as} = 0$; (2) $|\tilde{E}_a| > |\tilde{V}_{as}|$, whereupon $\tilde{I}_{as}$ lags $\tilde{V}_{as}$, and because $\tilde{I}_{as}$ is positive out of the machine the synchronous machine appears as a capacitor supplying reactive power to the system; or (3) $|\tilde{E}_a| < |\tilde{V}_{as}|$ with $\tilde{I}_{as}$ leading $\tilde{V}_{as}$, whereupon the machine is absorbing reactive power appearing as an inductor to the system.

In order to maintain the voltage in a power system at rated value the synchronous generators are normally operated in the overexcited mode with $|\tilde{E}_a| > |\tilde{V}_{as}|$ because they are the main source of reactive power for the inductive loads throughout the system. In fact, some synchronous machines are placed in the power system for the sole purpose of supplying reactive power without any provision to provide real power. During peak load conditions when the system voltage is depressed, these so-called "synchronous condensers" are brought on line and the field voltage is adjusted to help increase the system voltage. In this mode of operation the synchronous machine behaves like an adjustable capacitor. On the other hand, it may be necessary for a generator to absorb reactive power in order to regulate voltage in a high-voltage transmission system during light load conditions. This mode of operation is, however, not desirable and should be avoided because machine oscillations become less damped as the reactive power required is decreased. This will be shown in a later chapter when we calculate eigenvalues.

The manner in which torque is produced in a synchronous machine may now be further explained with a somewhat more detailed consideration of the interaction of the resulting air-gap magnetomotive force (MMF) established by the stator currents and the field current with (1) the MMF established by the field current and (2) the minimum reluctance path of the rotor. With the machine operating with T_I equal zero and $|\tilde{E}_a| > |\tilde{V}_{as}|$ the stator currents are

$$i_{as} = \sqrt{2} I_s \cos\left(\omega_e t - \frac{\pi}{2}\right), \quad \text{etc.} \tag{5.9-34}$$

The rotor angle δ is zero and the q axis of the machine coincides with the real axis of a phasor diagram and the d axis with the negative imaginary axis as shown in Fig. 5.9-1. Electromagnetic torque is developed so as to align the poles or the MMF created by the field current with the resultant air-gap MMF produced by the field and stator currents. In this mode of operation, the MMF due to the field current is downward in the direction of the positive d axis at the instant v_{as} is maximum. At this time, i_{as} is zero while i_{bs} and i_{cs} are equal and opposite. Hence, the MMF produced by the stator currents is directed upward in the direction of the negative d axis. The resultant of these two MMFs must be in the direction of the positive d axis because it was the increasing of the field MMF, by increasing the field current, which caused the stator current to lag the voltage (positive current out of machine), thereby causing the MMF produced by the stator currents to oppose the MMF

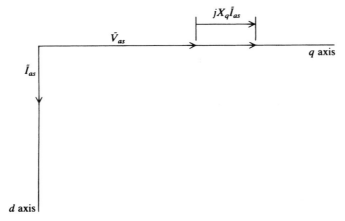

Figure 5.9-1 Phasor diagram with all losses neglected and zero power transfer.

produced by the field current. Therefore the resultant air-gap MMF and the field MMF are aligned. Moreover, the resultant air-gap MMF and the minimum reluctance path of the rotor (d axis) are also aligned. It follows that zero torque is produced and the rotor and MMFs will rotate while maintaining this alignment. If, however, the rotor tries to move from this alignment by either speeding up or slowing down ever so slightly, there will be both a torque due to the interaction of stator and field currents and a reluctance torque to bring the rotor back into alignment.

Let us now consider the procedure by which generator action is established. A prime mover is mechanically connected to the shaft of the synchronous generator. This prime mover can be either a steam turbine, a hydro turbine, or a combustion engine. If, initially, the torque input on the shaft due to the prime mover is zero, then T_e is very slightly negative due to losses. The synchronous machine is essentially floating on the line. If now the input torque is increased to some positive value by supplying steam to the turbine blades, for example, a torque imbalance occurs because T_e must remain at its original value until δ changes. Hence the rotor will temporarily accelerate slightly above synchronous speed, whereupon δ will increase in accordance with (5.7-1). Thus, T_e increases and a new operating point will be established with a positive δ where T_I is equal to T_e plus the losses. The rotor will again rotate at synchronous speed with a torque exerted on it in an attempt to align the field MMF with the resultant air-gap MMF. The actual dynamic response of the electrical and mechanical systems during this loading process is illustrated by computer traces in the following section. If, during generator operation, the torque input from the prime mover is increased to a value greater than the maximum possible value of T_e, the machine will be unable to maintain steady-state operation because it cannot transmit the power supplied to the shaft. In this case, the device will accelerate above synchronous speed theoretically without bound; however, protection is normally provided which disconnects the machine from the system

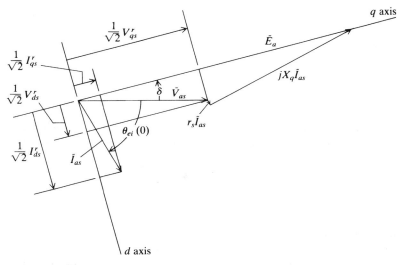

Figure 5.9-2   Phasor diagram for generator operation.

and reduces the input torque to zero by closing the steam valves of the steam turbine, for example, when it exceeds synchronous speed by generally 3% to 5%.

Normal steady-state generator operation is depicted by the phasor diagram shown in Fig. 5.9-2. Here, $\theta_{ei}(0)$ is the angle between the voltage and the current because the time zero position is $\theta_{ev}(0) = 0$ after steady-state operation is established. Because the phasor diagram and the q and d axes of the machine may be superimposed, the rotor reference-frame voltages and currents are also shown in Fig. 5.9-2. For example, V_{qs}^r and I_{qs}^r are shown directed along the positive q axis. If we wish to show each component of V_{qs}^r, it can be broken up according to (5.9-1) and each term cane be added algebraically along the q axis. Care must be taken, however, when interpreting the diagram. $\tilde{V}_{as}$, $\tilde{I}_{as}$, and $\tilde{E}_a$ are phasors representing sinusoidal quantities. On the other hand, all rotor reference-frame quantities are constants. They do not represent phasors in the rotor reference frame even though they are displayed on the phasor diagram.

There is one last detail to clear up. In (5.5-36) we defined $e_{xfd}^{\prime r}$ ($E_{xfd}^{\prime r}$ for steady-state operation) and indicated we would later find a convenient use for this term. If we assume that the stator of the synchronous machine is open-circuited and the rotor is being driven at synchronous speed, then, from (5.9-19) we obtain

$$\tilde{V}_{as} = \tilde{E}_a \tag{5.9-35}$$

Substituting (5.9-18) for $\tilde{E}_a$ with I_{ds}^r equal to zero yields

$$\sqrt{2}\left|\tilde{V}_{as}\right| = \frac{\omega_e}{\omega_b} X_{md} I_{fd}^{\prime r} \tag{5.9-36}$$

However, as given by (5.9-31) for balanced steady-state operation, we have

$$E_{xfd}^{\prime r} = X_{md} I_{fd}^{\prime r} \tag{5.9-37}$$

hence

$$\sqrt{2}|\tilde{V}_{as}| = \frac{\omega_e}{\omega_b} E_{xfd}^{\prime r} \tag{5.9-38}$$

Now let us per unitize the above equation. To do so, we must divide each side of (5.9-38) by $V_{B(qd0)}$ or $\sqrt{2} V_{B(abc)}$ because $E_{xfd}^{\prime r}$ is a rotor reference frame quantity. Thus

$$\frac{\sqrt{2}|\tilde{V}_{as}|}{\sqrt{2} V_{B(abc)}} = \frac{(\omega_e/\omega_b) E_{xfd}^{\prime r}}{V_{B(qd0)}} \tag{5.9-39}$$

Therefore, when $|\tilde{V}_{as}|$ is one per unit, $(\omega_e/\omega_b) E_{xfd}^{\prime r}$ is one per unit. During steady-state rated speed operation, (ω_e/ω_b) is unity and therefore one per unit $E_{xfd}^{\prime r}$ produces one per unit open-circuit terminal voltage. Because this provides a convenient relationship, $E_{xfd}^{\prime r}$ is used extensively to define the field voltage rather than the actual voltage applied to the field winding.

> **Example 5A**   A 3-phase, 64-pole, hydro turbine generator is rated at 325 MVA, with 20-kV line-to-line voltage and a power factor of 0.85. The machine parameters in ohms at 60 Hz are: $r_s = 0.00234$, $X_q = 0.5911$, and $X_d = 1.0467$. For balanced, steady-state rated conditions calculate (a) $\tilde{E}_a$, (b) $E_{xfd}^{\prime r}$, and (c) T_e.
>   The apparent power $|S|$ is
>
> $$|S| = 3|\tilde{V}_{as}||\tilde{I}_{as}| \tag{5A-1}$$

Thus

$$|\tilde{I}_{as}| = \frac{|S|}{3|\tilde{V}_{as}|} = \frac{325 \times 10^6}{(3 \times 20 \times 10^3)/\sqrt{3}} = 9.37\,\text{kA} \tag{5A-2}$$

The power factor angle is $\cos^{-1} 0.85 = 31.8°$. Because current is positive out of the terminals of the generator, reactive power is delivered by the generator when the current is lagging the terminal voltage. Thus, $\tilde{I}_{as} = 9.37\underline{/-31.8°}$ kA because the machine would be delivering reactive power during normal operating conditions. Therefore from (5.9-19) we can obtain the answer to part a:

$$\tilde{E}_a = \tilde{V}_{as} + \left(r_s + j\frac{\omega_e}{\omega_b} X_q \right) \tilde{I}_{as}$$

$$= \frac{20 \times 10^3}{\sqrt{3}} \underline{/0°} + [0.00234 + j(1)(0.5911)]9.37 \times 10^3 \underline{/-31.8°}$$

$$= 15.2\underline{/18°}\,\text{kV} \tag{5A-3}$$

Hence $\delta = 18°$.

We can solve for E''^r_{xfd} by first substituting (5.9-31) into (5.9-18), however, I^r_{ds} is required before E''^r_{xfd} can be evaluated. Thus from (5.9-12) we obtain

$$
\begin{aligned}
I^r_{ds} &= -\sqrt{2}I_s \sin\left[\theta_{ei}(0) - \theta_{ev}(0) - \delta\right] \\
&= -\sqrt{2}\left|\tilde{I}_{as}\right| \sin\left[-31.8° - 0 - 18°\right] \\
&= -\sqrt{2}(9.37 \times 10^3)\sin(-49.8°) \\
&= 10.12\,\text{kA}
\end{aligned}
\tag{5A-4}
$$

From (5.9-18) and (5.9-31) we have

$$
\begin{aligned}
E''^r_{xfd} &= \frac{\omega_e}{\omega_b}\left[\sqrt{2}\left|\tilde{E}_a\right| + \frac{\omega_e}{\omega_b}(X_d - X_q)I^r_{ds}\right] \\
&= \sqrt{2}(15.2 \times 10^3) + (1.0467 - 0.5911)10.12 \times 10^3 \\
&= 26.1\,\text{kV}
\end{aligned}
\tag{5A-5}
$$

Because r_s is small, T_e may be calculated by substitution into (5.9-32):

$$
\begin{aligned}
T_e &= \left(\frac{3}{2}\right)\left(\frac{P}{2}\right)\left(\frac{1}{\omega_b}\right)\left[\frac{E''^r_{xfd}\sqrt{2}\left|\tilde{V}_{as}\right|}{(\omega_e/\omega_b)X_d}\sin\delta\right. \\
&\quad \left. + \left(\frac{1}{2}\right)\left(\frac{\omega_e}{\omega_b}\right)^{-2}\left(\frac{1}{X_q} - \frac{1}{X_d}\right)(\sqrt{2}\left|\tilde{V}_{as}\right|)^2\sin 2\delta\right] \\
&= \left(\frac{3}{2}\right)\left(\frac{64}{2}\right)\left(\frac{1}{377}\right)\left\{\frac{(26.1 \times 10^3)(\sqrt{2})[(20 \times 10^3)/(\sqrt{3})]}{1.0467}\sin 18°\right. \\
&\quad \left. + \left(\frac{1}{2}\right)\left(\frac{1}{0.5911} - \frac{1}{1.0467}\right)\left[\sqrt{2}\left(\frac{20 \times 10^3}{\sqrt{3}}\right)^2\right]\sin 36°\right\} \\
&= 23.4 \times 10^6\,\text{N}\cdot\text{m}
\end{aligned}
\tag{5A-6}
$$

5.10   DYNAMIC PERFORMANCE DURING A SUDDEN CHANGE IN INPUT TORQUE

It is instructive to observe the dynamic performance of a synchronous machine during a step change in input torque. For this purpose, the differential equations that describe the synchronous machine were programmed on a computer and a study was performed [4]. Two large machines are considered: a low-speed hydro turbine generator and a high-speed steam turbine generator. Information regarding each machine is given in Tables 5.10-1 and 5-10.2. In the case of hydro turbine generator, parameters are given for only one damper winding in the q axis. The reason for

Table 5.10-1 Hydro Turbine Generator

Rating: 325 MVA
Line-to-line voltage: 20 kV
Power factor: 0.85
Poles: 64
Speed: 112.5 r/min
Combined inertia of generator and turbine:
$J = 35.1 \times 10^6 \text{J} \cdot \text{s}^2$, or $WR^2 = 833.1 \times 10^6 \text{ lbm} \cdot \text{ft}^2$ $H = 7.5$ s
Parameters in ohms and per unit:
$r_s = 0.00234\,\Omega,\ 0.0019$ pu
$X_{ls} = 0.1478\,\Omega,\ 0.120$ pu

$X_q = 0.5911\,\Omega, 0.480$ pu	$X_d = 1.0467\,\Omega, 0.850$ pu
	$r'_{fd} = 0.00050\,\Omega, 0.00041$ pu
	$X'_{lfd} = 0.2523\,\Omega,\ 0.2049$ pu
$r'_{kq2} = 0.01675\,\Omega,\ 0.0136$ pu	$r'_{kd} = 0.01736\,\Omega,\ 0.0141$ pu
$X'_{lkq2} = 0.1267\,\Omega,\ 0.1029$ pu	$X'_{lkd} = 0.1970\,\Omega,\ 0.160$ pu

denoting this winding as the $kq2$ winding rather than the $kq1$ winding will become clear in Chapter 7.

The computer traces shown in Figs. 5.10-1 and 5.10-2 illustrate the dynamic behavior of the hydro turbine generator following a step change in input torque from zero to 27.6×10^6 N·m (rated for unity power factor). The dynamic behavior

Table 5.10-2 Steam Turbine Generator

Rating: 835 MVA
Line-to-line voltage: 26 kV
Power factor: 0.85
Poles: 2
Speed: 3600 r/min
Combined inertia of generator and turbine:
$J = 0.0658 \times 10^6 \text{ J} \cdot \text{s}^2$, or $WR^2 = 1.56 \times 10^6 \text{ lbm} \cdot \text{ft}^2$ $H = 5.6$ s
Parameters in ohms and per unit:

$r_s = 0.00243\,\Omega, 0.003$ pu
$X_{ls} = 0.1538\,\Omega, 0.19$ pu

$X_q = 1.457\,\Omega, 1.8$ pu	$X_d = 1.457\,\Omega, 1.8$ pu
$r'_{kq1} = 0.00144\,\Omega, 0.00178$ pu	$r'_{fd} = 0.00075\,\Omega, 0.000929$ pu
$X'_{lkq1} = 0.6578\,\Omega$, pu 0.8125 pu	$X'_{lfd} = 0.1145\,\Omega, 0.1414$ pu
$r'_{kq2} = 0.00681\,\Omega, 0.00841$ pu	$r'_{kd} = 0.01080\,\Omega, 0.01334$ pu
$X'_{lkq2} = 0.07602\,\Omega, 0.0939$ pu	$X'_{lkd} = 0.06577\,\Omega, 0.08125$ pu

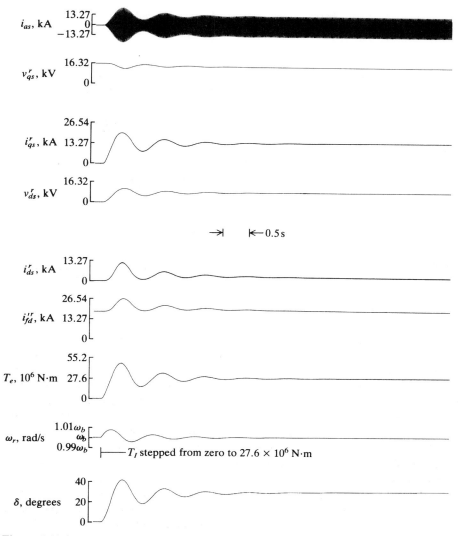

Figure 5.10-1 Dynamic performance of a hydro turbine generator during a step increase in input torque from zero to rated.

of the steam turbine generator is depicted in Figs. 5.10-3 and 5.10-4. In this case the step change in input torque is from zero to 1.11×10^6 N·m (50% rated). In Figs. 5.10-1 and 5.10-3 the following variables are plotted: i_{as}, v_{qs}^r, i_{qs}^r, v_{ds}^r, i_{ds}^r, $i_{fd}^{\prime r}$, T_e, ω_r, and δ, where ω_r is in electrical radians per second and δ is in electrical degrees. Figures 5.10-2 and 5.10-4 illustrate the dynamic torque versus rotor angle characteristics. In all figures, the scales of the voltages and currents are given in multiples of peak rated values.

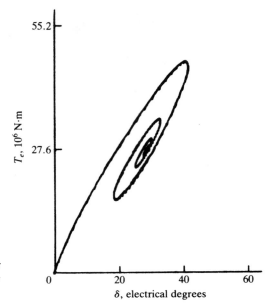

Figure 5.10-2 Torque versus rotor angle characteristics for the study shown in Fig. 5.10-1.

In each study, it is assumed that the machine is connected to a bus whose voltage and frequency remain constant, at the rated values, regardless of the stator current. This is commonly referred to as an infinite bus, because its characteristics do not change regardless of the power supplied or consumed by any device connected to it. Although an infinite bus cannot be realized in practice, its characteristics are approached if the power delivery capability of the system, at the point where the machine is connected, is much larger than the rating of the machine.

Initially each machine is operating with zero input torque with the excitation held fixed at the value that gives rated open-circuit terminal voltage at synchronous speed. It is instructive to observe the plots of T_e, ω_r, and δ following the step change input torque. In particular, consider the response of the hydro turbine generator (Fig. 5.10-1) where the machine is subjected to a step increase in input torque from zero to 27.6×10^6 N · m. The rotor speed begins to increase immediately following the step increase in input torque as predicted by (5.8-3), whereupon the rotor angle increases in accordance with (5.7-1). The rotor speeds up until the accelerating torque on the rotor is zero. As noted in Fig. 5.10-1, the speed increases to approximately 380 electrical radians per second, at which time T_e is equal to T_I because the change of ω_r is zero and hence the inertial torque (T_{IT}) is zero. Even though the accelerating torque is zero at this time, the rotor is running above synchronous speed; hence δ, and thus T_e, will continue to increase. The increase in T_e, which is an increase in the power output of the machine, causes the rotor to decelerate toward synchronous speed. However, when synchronous speed is reached, the magnitude of δ has become larger than necessary to satisfy the input torque. Note that at the first synchronous speed crossing of ω_r after the change in input torque, δ is

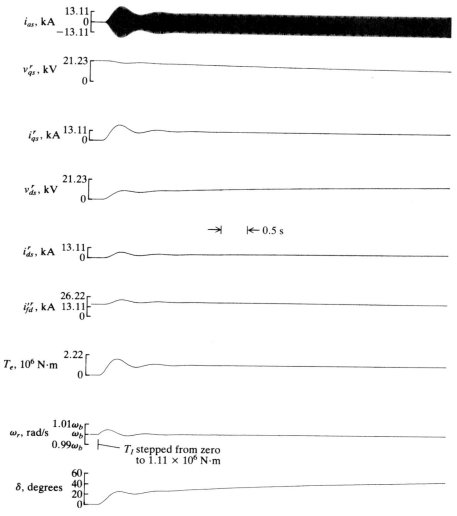

Figure 5.10-3 Dynamic performance of a steam turbine generator during a step increase in input torque from zero to 50% rated.

approximately 42 electrical degrees and T_e is approximately 47×10^6 N·m. Hence, the rotor continues to decelerate below synchronous speed and consequently δ begins to decrease, which in turn decreases T_e. Damped oscillations of the machine variables continue, and a new steady-state operating point is finally attained.

In the case of the hydro turbine generator (Fig. 5.10-1) the oscillations in machine variables subside in a matter of 2 or 3 s and the machine establishes the new steady-state operating point within 8 or 10 s. In the case of the steam turbine generator (Fig. 5.10-3) the oscillations subside rapidly, but the new steady-state operating

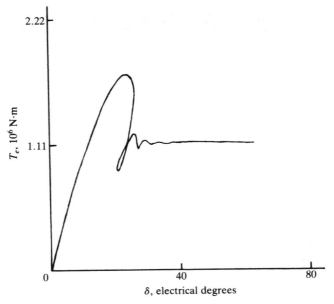

Figure 5.10-4   Torque versus rotor angle characteristics for the study shown in Fig. 5.10-3.

point is slowly approached. The damping is of course a function of the damper windings and can be determined from an eigenvalue analysis that will be discussed later. The point of interest here is the time required for the machine variables to reestablish steady-state operation after the torque disturbance. This rather slow approach to the new steady-state operating point in the case of the steam turbine generator is also apparent from the plot of T_e versus δ (Fig. 5.10-4).

Let us consider, for a moment, the expression for steady-state torque, (5.9-32). For the hydro turbine generator with $E''^r_{xfd} = \sqrt{\frac{2}{3}}\,20\,\text{kV}$ we have

$$T_e = (32.5\sin\delta + 12.5\sin 2\delta) \times 10^6\,\text{N}\cdot\text{m} \tag{5.10-1}$$

and for the steam turbine generator with $E''^r_{xfd} = \sqrt{\frac{2}{3}}\,26\,\text{kV}$ we obtain

$$T_e = 1.23 \times 10^6 \sin\delta\,\text{N}\cdot\text{m} \tag{5.10-2}$$

If (5.10-1) and (5.10-2) are plotted on Figs. 5.10-2 and 5.10-4, respectively, the steady-state T_e versus δ curves will pass through the final value of the dynamic T_e versus δ plots. However, the dynamic torque–angle characteristics immediately following the input torque disturbance yields a much larger T_e for a given value of δ than does the steady-state characteristic. In other words, the dynamic or transient torque–angle characteristic is considerably different from the steady-state characteristic and the steady-state T_e versus δ curve applies only after all transients have subsided. Although the computation of the transient torque during speed variations requires the solution of nonlinear differential equations, it can be approximated quite simply. This is the subject of a following section.

The studies shown in this section are for generator action. Motor action, wherein a load torque is applied to the machine, would essentially yield the mirror image of the T_e versus δ plots differing only by the ohmic losses.

5.11 DYNAMIC PERFORMANCE DURING A 3-PHASE FAULT AT THE MACHINE TERMINALS

The stability of synchronous machines throughout a power system following a fault is of major concern. A 3-phase fault or short-circuit rarely occurs, and a 3-phase fault at the machine terminals is even more uncommon; nevertheless, it is instructive to observe the dynamic performance of a synchronous machine during this type of a fault.

The computer traces shown in Figs. 5.11-1 and 5.11-2 illustrate the dynamic behavior of the hydro turbine generator during and following a 3-phase fault at the terminals. The dynamic behavior of the steam turbine generator as a result of a 3-phase terminal fault is shown in Figs. 5.11-3 and 5.11-4. The parameters of the machines are those given in the previous section. In Figs. 5.11-1 and 5.11-3 the following variables are plotted: i_{as}, v_{qs}^r, i_{qs}^r, v_{ds}^r, i_{ds}^r, $i_{fd}^{\prime r}$, T_e, ω_r, and δ. Figures 5.11-2 and 5.11-4 illustrate the dynamic torque–angle characteristics during and following the 3-phase fault.

In each case the machine is initially connected to an infinite bus delivering rated MVA at rated power factor. In the case of the hydro turbine generator the input torque is held constant at $(0.85)27.6 \times 10^6$ N·m with $E_{xfd}^{\prime r}$ fixed at $(1.6)\sqrt{\frac{2}{3}}20$ kV; for the steam turbine generator $T_I = (0.85)2.22 \times 10^6$ N·m and $E_{xfd}^{\prime r} = (2.48)\sqrt{\frac{2}{3}}26$ kV. (Rated operating conditions for the hydro turbine generator are calculated in Example 5A.) With the machines operating in this steady-state condition, a 3-phase terminal fault is simulated by setting v_{as}, v_{bs}, and v_{cs} to zero, in the simulation, at the instant v_{as} passes through zero going positive. The transient offset in the phase currents is reflected into the rotor reference-frame variables and the instantaneous torque as a decaying 60-Hz pulsation. Because the terminal voltage is zero during the 3-phase fault, the machine is unable to transmit power to the system. Hence, all of the input torque, with the exception of the ohmic losses, accelerates the rotor.

In the case of the hydro turbine generator the fault is removed in 0.466 s; 0.362 s in the case of the steam turbine generator. If the fault had been allowed to remain on the system slightly longer, the machines would have become unstable; that is, they would either not have returned to synchronous speed after removal of the fault or slipped poles before returning to synchronous speed. Asynchronous operation (pole slipping) is discussed in Chapter 10 of reference 5.

When the fault is cleared, the system voltages are reapplied to the machine; offsets again occur in the phase currents, giving rise to the decaying 60-Hz oscillations in the rotor reference-frame variables and the instantaneous torque. The dynamic torque–angle characteristics shown in Figs. 5.11-2 and 5.11-4 yield a very lucid illustration of the fault and switching sequence and the return of the machine to its original operating condition after the fault is cleared. These torque–angle plots

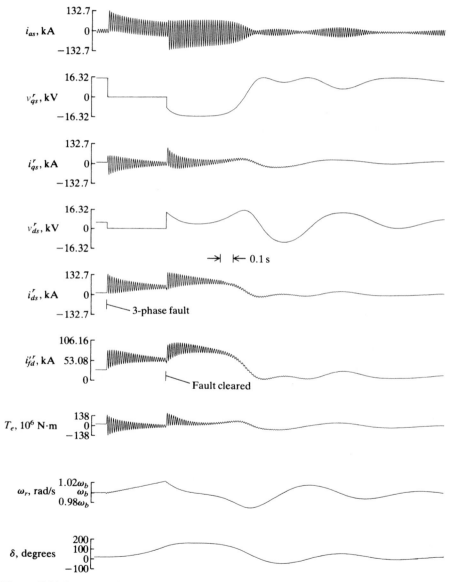

Figure 5.11-1 Dynamic performance of a hydro turbine generator during a 3-phase fault at the terminals.

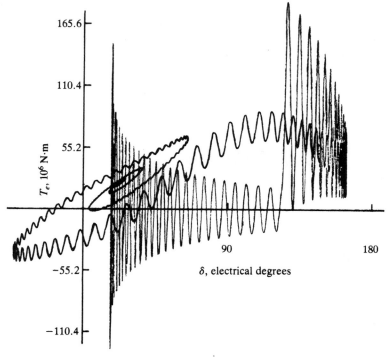

Figure 5.11-2 Torque versus rotor angle characteristics for the study shown in Fig. 5.11-1.

are shown later in Figs. 9.4-2 and 9.4-4, respectively, with the stator electric transients neglected, which eliminates the 60-Hz pulsating electromagnetic torque and permits the average torque to be more clearly depicted.

The expression for the steady-state torque–angle characteristic for the hydro turbine generator is

$$T_e = (52.1 \sin \delta + 12.5 \sin 2\delta) \times 10^6 \, \text{N} \cdot \text{m} \qquad (5.11\text{-}1)$$

For the steam turbine generator

$$T_e = 3.05 \times 10^6 \sin \delta \, \text{N} \cdot \text{m} \qquad (5.11\text{-}2)$$

If these steady-state torque–angle characteristics are plotted on Figs. 5.11-2 and 5.11-4, respectively, they would pass through only the initial (or final) steady-state operating point. As in the case of a sudden change in input torque, the instantaneous and/or average value of the dynamic or transient torque–angle characteristic differs markedly from the steady-state torque–angle characteristics.

It is perhaps appropriate to mention that this example is somewhat impractical. In the case of a 3-phase fault close to a fully loaded machine, the circuit breakers would probably remove the machine from the system and reclosing would be prohibited because the machine would accelerate beyond speed limits before it would be physically possible to reclose the circuit breakers. A practical situation that is approximated by the example might be a 3-phase fault on a large radial transmission line close to the machine terminals. Clearing or "switching out" of this line would then remove the fault from the system.

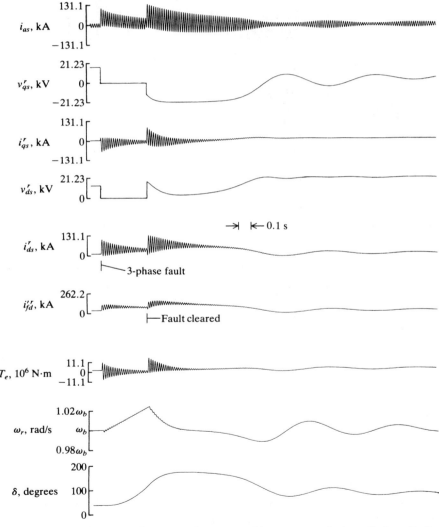

Figure 5.11-3 Dynamic performance of a steam turbine generator during a 3-phase fault at the terminals.

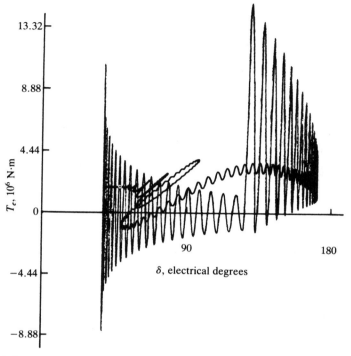

Figure 5.11-4   Torque versus rotor angle characteristics for the study shown in Fig. 5.11-3.

5.12 APPROXIMATE TRANSIENT TORQUE VERSUS ROTOR ANGLE CHARACTERISTICS

As pointed out in the previous sections, the transient and steady-state torque versus rotor angle characteristics are quite different. Because the transient characteristics will determine if the machine remains in synchronism after a disturbance, it is necessary to calculate these characteristics accurately whenever determining the transient stability of a synchronous machine. With present-day computers the calculation of the transient electromagnetic torque is a straightforward procedure. Consequently, it is difficult to appreciate the complex computational problems faced by machine and power system analysts before the advent of the computer and the techniques that they devised to simplify these problems.

In the late 1920s, R. E. Doherty and C. A. Nickle [6] described a simple method of approximating the transient torque–angle characteristics. This method, combined with the concept of equal-area criterion, which is discussed in a following section, formed the basis for transient stability studies of power systems until the 1960s. Although it is not the purpose here to dwell on techniques that have long been replaced, it is interesting to look back for a moment, not only to gain some appreciation of this early work, but to become acquainted with an approximate method that still remains invaluable in visualizing machine stability.

The method of approximating the transient torque–angle characteristics set forth by Doherty and Nickle [6] is based on the fact that the flux linkages will tend to remain constant in circuits that are largely inductive with a relative small resistance. Therefore, because the field winding has a small resistance with a large self-inductance, it is generally assumed that the field flux linkages will remain constant during the early part of the transient period. Moreover, it is assumed that all electric transients can be neglected and the action of the damper windings ignored, whereupon the steady-state voltage and flux linkage equations apply. With these assumptions, the steady-state versions of (5.5-30) and (5.5-34) may be written as

$$\Psi_{ds}^r = -X_d I_{ds}^r + X_{md} I_{fd}^{\prime r} \tag{5.12-1}$$

$$\Psi_{fd}^{\prime r} = X_{fd}^{\prime} I_{fd}^{\prime r} - X_{md} I_{ds}^r \tag{5.12-2}$$

where X_d is defined by (5.5-40) and $X_{fd}^{\prime}$ is

$$X_{fd}^{\prime} = X_{lfd}^{\prime} + X_{md} \tag{5.12-3}$$

Solving (5.12-1) for $I_{fd}^{\prime r}$ and substituting the result into (5.12-2) yields

$$\frac{X_{md}}{X_{fd}^{\prime}} \Psi_{fd}^{\prime r} = \Psi_{ds}^r + \left(X_d - \frac{X_{md}^2}{X_{fd}^{\prime}} \right) I_{ds}^r \tag{5.12-4}$$

Let us now define

$$X_d^{\prime} = X_d - \frac{X_{md}^2}{X_{fd}^{\prime}} \tag{5.12-5}$$

and

$$E_q^{\prime} = \frac{X_{md}}{X_{fd}^{\prime}} \Psi_{fd}^{\prime r} \tag{5.12-6}$$

Equation (5.12-4) may now be written as

$$\Psi_{ds}^r = -X_d^{\prime} I_{ds}^r + E_q^{\prime} \tag{5.12-7}$$

The reactance $X_d^{\prime}$ is referred to as the d-axis transient reactance. If the field flux linkages are assumed constant, then $E_q^{\prime}$, which is commonly referred to as the voltage behind transient reactance, is also constant.

The quantities $X_d^{\prime}$ and $E_q^{\prime}$, which are specifically related to the transient period, are each denoted by a prime. Heretofore, we have used the prime to denote rotor variables and rotor parameters referred to the stator windings by a turns ratio. As mentioned, the prime or any other distinguishing notation is seldom used in literature to denote referred quantities; on the other hand, the primes are always used to denote transient quantities. We will use the prime to denote both; the double meaning should not be confusing because the primed quantities that pertain to the transient period are few in number and readily recognized.

Let us now return to the method used to obtain the steady-state voltage and torque equations. The steady-state voltage equations in the rotor reference frame, (5.9-1) and (5.9-2), were obtained from (5.5-22) and (5.5-23) with the time rate of change of all flux linkages neglected and ω_r set equal to ω_e. These equations could have also been written in the form

$$V_{qs}^r = -r_s I_{qs}^r + \frac{\omega_e}{\omega_b} \Psi_{ds}^r \tag{5.12-8}$$

$$V_{ds}^r = -r_s I_{ds}^r - \frac{\omega_e}{\omega_b} \Psi_{qs}^r \tag{5.12-9}$$

where

$$\Psi_{ds}^r = -X_d I_{ds}^r + X_{md} I_{fd}^{'r} \tag{5.12-10}$$

$$\Psi_{qs}^r = -X_q I_{qs}^r \tag{5.12-11}$$

If we compare (5.12-10) and (5.12-7), we see that the two equations have the same form. Therefore if, in our previous derivation, we replace X_d with X_d' and $X_{md} I_{fd}^{'r}$ or $E_{xfd}^{'r}$ with E_q', we will obtain voltage and torque expressions that should approximate the behavior of the synchronous machine during the early part of the transient period, assuming the field flux linkages remain constant. In particular, the so-called transient torque–angle characteristic is expressed as

$$T_e = \left(\frac{3}{2}\right)\left(\frac{P}{2}\right)\left(\frac{1}{\omega_b}\right)\left[\frac{E_q' \sqrt{2} V_s}{(\omega_e/\omega_b)X_d'}\sin\delta + \left(\frac{1}{2}\right)\left(\frac{\omega_e}{\omega_b}\right)^{-2}\left(\frac{1}{X_q} - \frac{1}{X_d'}\right)(\sqrt{2}V_s)^2 \sin 2\delta\right]$$

$$\tag{5.12-12}$$

In per unit we obtain

$$T_e = \frac{E_q' V_s}{(\omega_e/\omega_b)X_d'}\sin\delta + \left(\frac{V_s^2}{2}\right)\left(\frac{\omega_e}{\omega_b}\right)^{-2}\left(\frac{1}{X_q} - \frac{1}{X_d'}\right)\sin 2\delta \tag{5.12-13}$$

As in the case of (5.9-32) and (5.9-33), (5.12-12) and (5.12-13) are valid only if r_s can be neglected. Also, it is interesting to note that the coefficient of $\sin 2\delta$ is zero if $X_q = X_d'$, which is seldom, if ever, the case. Actually $X_q > X_d'$ and the coefficient of $\sin 2\delta$ is negative. In other words, the transient electromagnetic torque–angle curve is a function of $\sin 2\delta$ even if $X_q = X_d$.

We have yet to determine E'_q from a readily available quantity. If in the expression for $\tilde{E}_a$, (5.9-18), X_d is replaced with X'_d and $X_{md}I'_{fd}$ with E'_q, then

$$\tilde{E}_a = \frac{1}{\sqrt{2}}\left(\frac{\omega_e}{\omega_b}\right)[-(X'_d - X_q)I^r_{ds} + E'_q]e^{j\delta} \tag{5.12-14}$$

Hence, the familiar phasor voltage equation given by (5.9-19) can be used to calculate the predisturbance $\tilde{E}_a$, whereupon we can use (5.12-14) to determine E'_q, which is assumed to remain constant during the early part of the transient period.

5.13 COMPARISON OF ACTUAL AND APPROXIMATE TRANSIENT TORQUE–ANGLE CHARACTERISTICS DURING A SUDDEN CHANGE IN INPUT TORQUE: FIRST SWING TRANSIENT STABILITY LIMIT

In the studies involving a step increase in input torque, which were reported in a previous section, the machines were initially operating with essentially zero stator current and zero input torque. Hence, $\tilde{E}_a = \tilde{V}_{as}$ and $E'_q = \sqrt{2}V_s$. The steady-state torque–angle curve is given by (5.10-1) for the hydro unit and by (5.10-2) for the steam unit. For the hydro turbine generator, $X'_d = 0.3448$ ohms and the transient torque–angle characteristic is

$$T_e = (98.5\sin\delta - 20.5\sin 2\delta) \times 10^6\,\text{N}\cdot\text{m} \tag{5.13-1}$$

For the steam turbine generator where $X'_d = 0.2591$ ohms, we obtain

$$T_e = (6.92\sin\delta - 2.84\sin 2\delta) \times 10^6\,\text{N}\cdot\text{m} \tag{5.13-2}$$

Figures 5.13-1 and 5.13-2 show the approximate transient and steady-state torque–angle curves plotted along with the actual dynamic torque–angle characteristics obtained from the studies involving a step increase in input torque (Figs. 5.10-2 and 5.10-4). During the initial swing of the rotor the dynamic torque–angle characteristic follows the approximate transient torque–angle curve more closely than the steady-state curve even though the approximation is rather crude, especially in the case of the steam turbine generator. As the transients subside, the actual torque–angle characteristic moves toward the steady-state torque–angle curve.

The approximate transient torque–angle characteristic is most often used along with the equal-area criterion to predict the maximum change in input torque possible without the machine becoming unstable (transient stability limit) rather than the dynamic performance during a relatively small step increase in input torque as portrayed in Figs. 5.13-1 and 5.13-2. In order to compare the results obtained here with the transient stability limit that we will calculate in a later section using the equal-area criterion, it is necessary to define the "first swing" transient stability limit as the maximum value of input torque which can be suddenly applied, and the rotor just

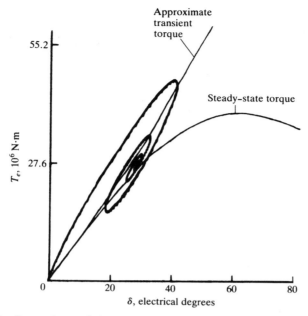

Figure 5.13-1 Comparison of the dynamic torque–angle characteristic during a step increase in input torque from zero to rated with the calculated steady-state and approximate transient torque–angle characteristics: hydro turbine generator.

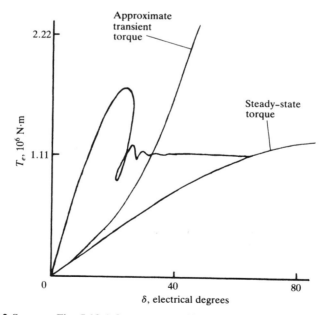

Figure 5.13-2 Same as Fig. 5.13-1 for a steam turbine generator with input torque stepped from zero to 50% rated.

returns to synchronous speed at the end of the first acceleration above synchronous speed. By trial and error, the transient stability limit for the hydro turbine generator was found to be 76.7×10^6 N·m and 5.2×10^6 N·m for the steam turbine generator. The computer traces shown in Figs. 5.13-3 through 5.13-6 show the transient response of the machine variables with the step input torque equal to the value at

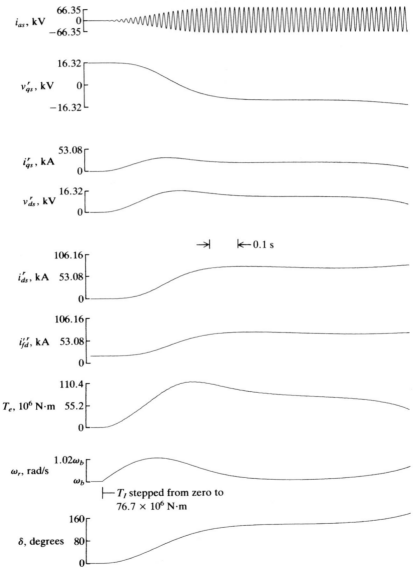

Figure 5.13-3 Dynamic performance of a hydro turbine generator at the "first swing" transient stability limit.

the "first swing" transient stability limit. Figures 5.13-3 and 5.13-4 are for the hydro turbine generator and steam turbine generator, respectively, with the same variables shown as in Figs. 5.10-1 and 5.10-3. The machine flux linkages per second are shown in Fig. 5.13-5 for the hydro turbine generator and in Fig. 5.13-6 for the steam turbine generator.

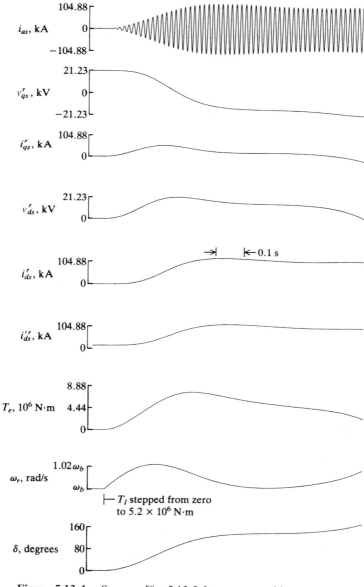

Figure 5.13-4   Same as Fig. 5.13-3 for a steam turbine generator.

In case of the hydro unit the field flux linkages are relatively constant during the first swing of the machine varying approximately 17% from the original value (Fig. 5.13-5); however, the field flux linkages for the steam unit (Fig. 5.13-6) vary approximately 40% during the first swing. This is due to the fact that the field circuit of the hydro unit has a higher reactance-to-resistance ratio than the field circuit of the steam unit. This observation casts doubt on the accuracy of the approximate transient torque–angle characteristics, especially for the steam unit, as we have already noted in Fig. 5.13-2. It is also important to note from Fig. 5.13-6 that the change in $\psi_{kq1}^{\prime r}$ is much less than that in the case of $\psi_{fd}^{\prime r}$. The reactance-to-resistance ratios of the two circuits gives an indication of this result. In particular, the reactance-to-

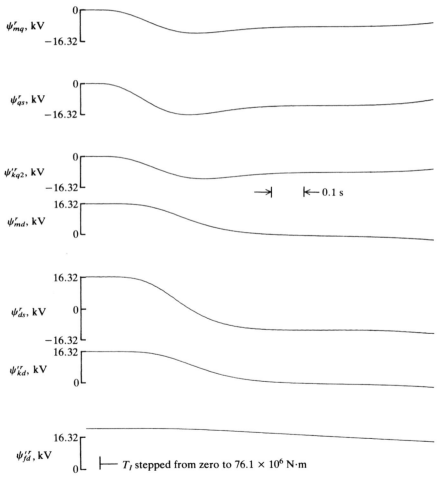

Figure 5.13-5   Traces of flux linkages per second for a hydro turbine generator at the "first swing" transient stability limit.

resistance ratio of the field circuit of the steam unit is 1333 whereas this ratio for the $kq1$ winding is 1468.

The actual dynamic torque–angle characteristics obtained from the computer study are shown in Figs. 5.13-7 and 5.13-8 with the approximate transient torque–angle curve calculated using constant flux linkages superimposed thereon. In the computer study the machines went unstable at the transient stability limit. Nevertheless, the transient stability limit has been defined here as the "first swing" transient stability limit in an attempt to duplicate the conditions for which the equal-area

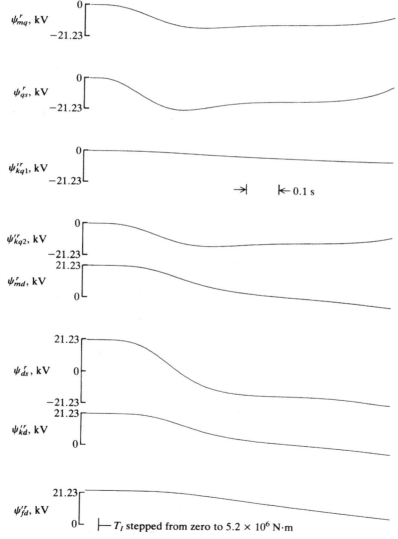

Figure 5.13-6 Same as Fig. 5.13-5 for a steam turbine generator.

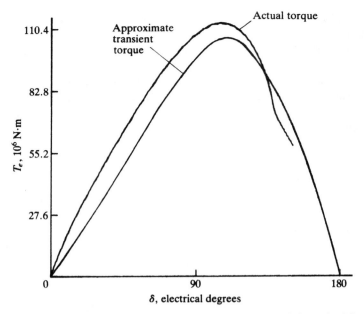

Figure 5.13-7   Comparison of the dynamic torque–angle characteristic at the "first swing" transient stability limit with the approximate transient torque–angle curve: hydro turbine generator.

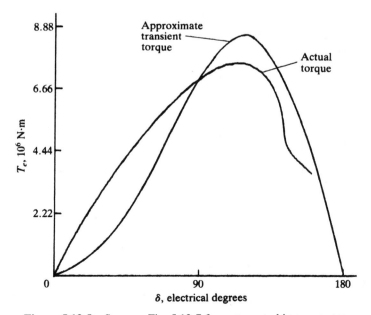

Figure 5.13-8   Same as Fig. 5.13-7 for a steam turbine generator.

criterion is to predict. The facility of the approximate transient torque–angle curve in predicting this transient stability must be deferred until we have described this criterion.

5.14 COMPARISON OF ACTUAL AND APPROXIMATE TRANSIENT TORQUE–ANGLE CHARACTERISTICS DURING A 3-PHASE FAULT AT THE TERMINALS: CRITICAL CLEARING TIME

In the studies involving a 3-phase fault at the terminals (Figs. 5.11-1 through 5.11-4), each machine is initially operating at rated conditions. In the case of the hydro turbine generator the steady-state torque–angle curve is given by (5.11-1), and for the steam turbine generator it is given by (5.11-2). For rated conditions, $E'_q = (1.16)\sqrt{\frac{2}{3}}20$ kV for the hydro unit and the approximate transient torque–angle characteristic is

$$T_e = (114.3\sin\delta - 20.5\sin 2\delta) \times 10^6\,\text{N}\cdot\text{m} \qquad (5.14\text{-}1)$$

For the steam turbine generator, $E'_q = (1.09)\sqrt{\frac{2}{3}}26$ kV and the approximate transient torque–angle characteristic is

$$T_e = (7.53\sin\delta - 2.84\sin 2\delta) \times 10^6\,\text{N}\cdot\text{m} \qquad (5.14\text{-}2)$$

Figures 5.14-1 and 5.14-2 show the approximate transient and steady-state torque–angle curves for the hydro and steam units, respectively, plotted along with the actual dynamic torque–angle characteristics shown previously in Figs. 5.11-2 and 5.11-4. It is important to note that the approximate transient and steady-state torque–angle curves both pass through the steady-state operating point. The flux linkages per second during a 3-phase fault and subsequent clearing are shown in Fig. 5.14-3 for the hydro turbine generator and in Fig. 5.14-4 for the steam turbine generator. The corresponding plots of voltages, currents, torque, speed, and rotor angle are shown in Figs. 5.11-1 and 5.11-3.

As mentioned earlier, the situation portrayed in this study (Figs. 5.11-1 through 5.11-4 and Figs. 5.14-3 and 5.14-4) is one where only a slight increase in the fault time would cause the machines to slip poles before returning to synchronous speed. This limiting condition is commonly referred to as the *critical clearing time*, and the corresponding maximum rotor angle attained is called the *critical clearing angle*. In the case of the hydro turbine generator the critical clearing time and angle were found by trial and error to be 0.466 s and 123°. For the steam turbine generator these values were found to be 0.362 s and 128°.

During the fault, the average value of the electromagnetic torque is essentially zero because the ohmic losses are small. The approximate transient torque–angle curve during this period is zero because the ohmic losses are neglected. The

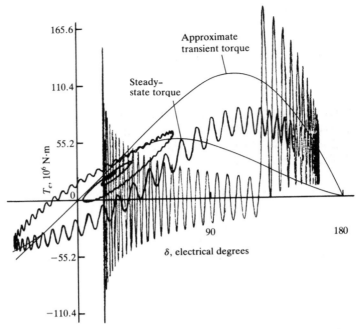

Figure 5.14-1 Comparison of the dynamic torque–angle characteristic during a 3-phase fault with the calculated steady-state and approximate transient torque–angle characteristics: hydro turbine generator.

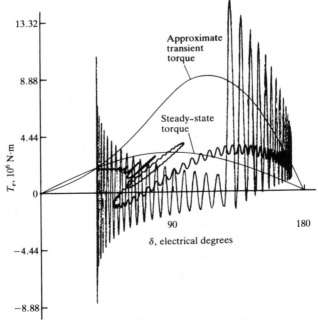

Figure 5.14-2 Same as Fig. 5.14-1 for a steam turbine generator.

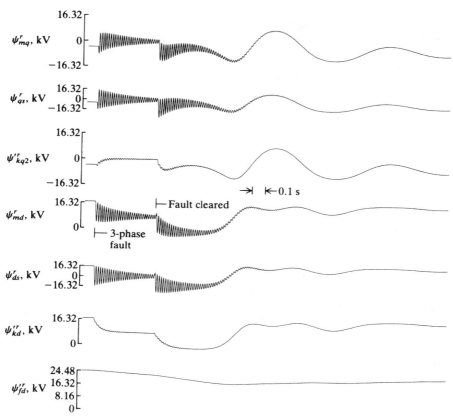

Figure 5.14-3 Traces of flux linkages per second for a hydro turbine generator during a 3-phase fault at the terminals.

approximate transient torque–angle curve plotted in Figs. 5.14-1 and 5.14-2 applies only before the 3-phase fault and after it is cleared. If the approximate transient torque–angle curve accurately portrayed the electromagnetic torque following the fault, the average of the instantaneous torque would traverse back and forth on this torque–angle characteristic until the initial operating condition was reestablished. The approximate characteristic appears to be adequate immediately following the fault when the rotor has reached the maximum angle. Thereafter, the approximation is quite inadequate especially in the case of the steam turbine geneator (Fig. 5.14-2). This inaccuracy can of course be contributed primarily to the fact that the average value of the field flux linkages does not remain constant during and following the fault. The change in field flux linkages is less in the case of the hydro unit, and the approximation is more accurate than that in the case of the steam unit.

As in the case of a step increase in input torque, the damper winding flux linkages, ψ'^r_{kq1}, of the steam unit remains more nearly constant than any of the other

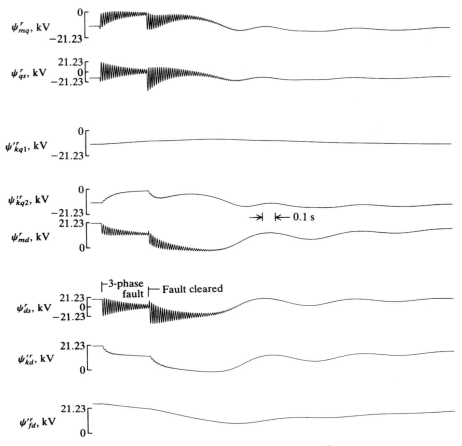

Figure 5.14-4 Same as Fig. 5.14-3 for a steam turbine generator.

machine flux linkages. Although the assumption of constant field flux linkages is by far the most common, there are refinements that can be made to yield the approximate transient torque–angle characteristic more accurate. In particular, it is sometimes assumed that the flux linkages of a q-axis damper winding also remains constant along with the field flux linkages. A voltage, different from E'_q, is then calculated behind a transient impedance. We will not consider this or other refinements in this text.

5.15 EQUAL-AREA CRITERION

As mentioned previously, the approximate transient torque–angle characteristics along with the equal-area method were used extensively to predict the transient response of synchronous machines. In most cases, these concepts were used to determine the transient stability limit and the critical clearing time. The theory underlying the equal-area method as applied to an input torque disturbance or a system fault can be readily established. Regardless of discrepancies or questions that might arise as to the validity of the results, the approximate method of determining the transient torque and the application of the equal-area criterion to predict stability are very useful in understanding the overall dynamic behavior of synchronous machines.

Input Torque Change

For this development, let us consider the approximate transient torque–angle curve for the hydro turbine generator shown in Fig. 5.13-7 and given again in Fig. 5.15-1. Consider a sudden step increase in input torque of T_I from an initial value of zero so as to correspond with our earlier work. This torque level of T_I is identified by a horizontal line in Fig. 5.15-1. At the instant the input torque is applied, the accelerating torque is T_I because T_e is initially zero and the losses are neglected. We see that the accelerating torque on the rotor is positive when $T_I > T_e$, where here T_e is the approximate transient torque–angle curve.

Work or energy is the integral of force times a differential distance or, in the case of a rotational system, the integral of torque times a differential angular displacement. Hence, the energy stored in the rotor during the initial acceleration is

$$\int_{\delta_0}^{\delta_1} (T_I - T_e)\, d\delta = \text{area } OABO \qquad (5.15\text{-}1)$$

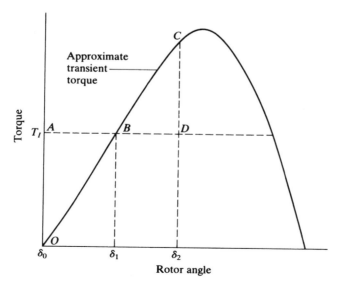

Figure 5.15-1   Equal-area criterion for sudden change in input torque.

The energy given up by the rotor as it decelerates back to synchronous speed is

$$\int_{\delta_1}^{\delta_2} (T_I - T_e)\,d\delta = \text{area } BDCB \tag{5.15-2}$$

The maximum angle is reached and the rotor will return to synchronous speed after the application of the input torque when

$$|\text{area } OABO| = |\text{area } BDCB| \tag{5.15-3}$$

It follows that the rotor angle of the synchronous machine would then oscillate back and forth between δ_0 and δ_2. We know, however, that the action of the damper windings will cause these oscillations to subside. One might then be led to believe that the new steady-state operating point would be established at point B. We, of course, know that this is not the case; instead, the new operating point is reestablished on the steady-state torque–angle curve as illustrated in Figs. 5.10-2 and 5.10-4 or Figs. 5.13-1 and 5.13-2. We are aware of the limitations of the approximate transient torque–angle curve, and thus any method utilizing this approximation will, at best, be adequate only for the first swing of the rotor from synchronous speed. It is left to the reader to show that the application of the equal-area criterion, which involves a graphical solution, gives a transient stability limit of $T_I = 68.1 \times 10^6$ N·m for the hydro turbine generator and $T_I = 4.65 \times 10^6$ N·m for the steam turbine generator.

It is instructive to compare the "first swing" transient stability limit obtained by means of a computer study for the two machines under consideration with the limits obtained by applying equal-area criterion. It is recalled that the actual dynamic torque–angle characteristics, which are shown in Figs. 5.13-7 and 5.13-8 with the transient torque–angle curve calculated using constant flux linkages superimposed thereon, gave a "first swing" transient stability limit for the hydro turbine generator of 76.7×10^6 N·m and 5.2×10^6 N·m for the steam turbine generator. The maximum torques obtained using the equal-area criterion are approximately 10% less than those obtained from the computer study.

It should also be mentioned that the term involving the $\sin 2\delta$ is often ignored and that only the $\sin \delta$ term of (5.12-12) is used to calculate the approximate torque–angle curve. This enables a trial-and-error analytical solution, and in this case it is a closer approximation. In particular, this method yields a transient stability limit of 71.2×10^6 N·m for the hydro turbine generator and 5.03×10^6 N·m for the steam turbine generator.

3-Phase Fault

The approximate transient torque–angle curve along with the equal-area criterion is most often used to predict the large excursion dynamic behavior of a synchronous machine during a system fault. The application of the equal-area method during a 3-phase system fault can be described by considering the approximate transient torque–angle curve of the hydro unit given in Fig. 5.15-2. Assume that the input torque T_I is constant and the machine is operating steadily, delivering power to the system with a rotor angle δ_0. When the 3-phase fault occurs at the terminals, the power output drops to zero and thus the approximate T_e is zero because the resistances are neglected. The machine accelerates with the total input torque as the accelerating torque. The fault is cleared at δ_1, and in this case the torque immediately becomes the value of the approximate transient torque (point D in Fig. 5.15-2). The energy stored in the rotor during the acceleration or advance in angle from δ_0 to δ_1 is

$$\int_{\delta_0}^{\delta_1} (T_I - T_e)\, d\delta = \text{area } OABCO \tag{5.15-4}$$

where T_e is zero and T_I is constant.

After the clearing of the fault the rotor decelerates back to synchronous speed. The energy given up by the rotor during this time is

$$\int_{\delta_1}^{\delta_2} (T_I - T_e)\, d\delta = \text{area } CDEFC \tag{5.15-5}$$

The maximum rotor angle is reached when

$$|\text{area } OABCO| = |\text{area} = CDEFC| \tag{5.15-6}$$

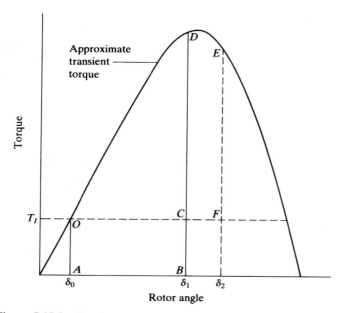

Figure 5.15-2 Equal-area criterion for a 3-phase fault at the terminals.

The critical clearing angle is reached when any future increase in δ_1 causes the total area representing decelerating energy to become less than the area representing the accelerating energy. This occurs when δ_2, or point F, is at the intersection of T_I and T_e.

The clearing time may be calculated from the clearing angle by using (5.8-4), which may be written as

$$\delta_1 - \delta_0 = \frac{\omega_b}{2H} \int_{t_0}^{t_1} \int_{t_0}^{\tau} (T_I - T_e)\, dt\, d\tau \qquad (5.15\text{-}7)$$

If, as in this example, $T_e = 0$ and T_I is a constant during the fault and with $t_0 = 0$, the clearing time T_1 becomes

$$t_1 = \sqrt{\frac{(\delta_1 - \delta_2)4H}{\omega_b T_I}} \qquad (5.15\text{-}8)$$

It is recalled that the critical clearing times and angles obtained by computer study were 0.466 s and 123° for the hydro unit and 0.362 s and 128° for the steam unit. These values compare to 0.41 s and 122° for the hydro unit and 0.33 s and 128° for the steam unit, obtained by a graphical application of the method of equal area. Because the critical clearing angles were found to be essentially the same by both methods, the larger critical clearing times can be attributed to the power lost in the resistance during the fault. The accuracy of the approximate method is indeed surprising, a result that should not be taken as an indication that the approximate method is highly accurate in general.

As mentioned previously, the $\sin 2\delta$ terms are often ignored in the approximate transient torque–angle characteristics. With the machine operating under load, ignoring the $\sin 2\delta$ term yields a fictitious initial rotor angle. Although this leads to a simplified method of solution and even though it was used widely in the past, we will not consider it in this text. Actually, sufficient background has been established so that the reader can readily develop this technique if the need arises.

5.16   COMPUTER SIMULATION

There are two general types of computer simulations of a synchronous machine which we will consider. The simulation used most widely is derived from the voltage equations expressed in the rotor reference frame with the equations arranged in the same form as in the case of the induction machine. This simulation was first developed by C. H. Thomas [7]. The second type of simulation that we will develop is one wherein the stator flux linkages per second are calculated in the arbitrary reference frame with the rotor flux linkages per second computed in the rotor reference frame. For other approaches see references 4 and 8–10.

Simulation in Rotor Reference Frame

The voltage equations expressed in the rotor reference frame are given by (5.5-22)–(5.5-28) with the equations defining the flux linkages per second given by

(5.5-29)–(5.5-35). These equations may be manipulated in the same way as the corresponding equations for the symmetrical induction machine in Chapter 4, and a similar computer simulation will be obtained. It is sufficient to give only the resulting integral equations [7]:

$$\psi_{qs}^r = \frac{\omega_b}{p}\left[v_{qs}^r - \frac{\omega_r}{\omega_b}\psi_{ds}^r + \frac{r_s}{X_{ls}}(\psi_{mq}^r - \psi_{qs}^r)\right] \tag{5.16-1}$$

$$\psi_{ds}^r = \frac{\omega_b}{p}\left[v_{ds}^r + \frac{\omega_r}{\omega_b}\psi_{qs}^r + \frac{r_s}{X_{ls}}(\psi_{md}^r - \psi_{ds}^r)\right] \tag{5.16-2}$$

$$\psi_{0s} = \frac{\omega_b}{p}\left[v_{0s} - \frac{r_s}{X_{ls}}\psi_{0s}\right] \tag{5.16-3}$$

$$\psi_{kq1}^{\prime r} = \frac{\omega_b}{p}\left[v_{kq1}^{\prime r} + \frac{r_{kq1}^{\prime}}{X_{lkq1}^{\prime}}(\psi_{mq}^r - \psi_{kq1}^{\prime r})\right] \tag{5.16-4}$$

$$\psi_{kq2}^{\prime r} = \frac{\omega_b}{p}\left[v_{kq2}^{\prime r} + \frac{r_{kq2}^{\prime}}{X_{lkq2}^{\prime}}(\psi_{mq}^r - \psi_{kq2}^{\prime r})\right] \tag{5.16-5}$$

$$\psi_{fd}^{\prime r} = \frac{\omega_b}{p}\left[\frac{r_{fd}^{\prime}}{X_{md}}e_{xfd}^{\prime r} + \frac{r_{fd}^{\prime}}{X_{lfd}^{\prime}}(\psi_{md}^r - \psi_{fd}^{\prime r})\right] \tag{5.16-6}$$

$$\psi_{kd}^{\prime r} = \frac{\omega_b}{p}\left[v_{kd}^{\prime r} + \frac{r_{kd}^{\prime}}{X_{lkd}^{\prime}}(\psi_{md}^r - \psi_{kd}^{\prime r})\right] \tag{5.16-7}$$

where

$$i_{qs}^r = -\frac{1}{X_{ls}}(\psi_{qs}^r - \psi_{mq}^r) \tag{5.16-8}$$

$$i_{ds}^r = -\frac{1}{X_{ls}}(\psi_{ds}^r - \psi_{md}^r) \tag{5.16-9}$$

$$i_{0s} = -\frac{1}{X_{ls}}\psi_{0s} \tag{5.16-10}$$

$$i_{kq1}^{\prime r} = \frac{1}{X_{lkq1}^{\prime}}(\psi_{kq1}^{\prime r} - \psi_{mq}^r) \tag{5.16-11}$$

$$i_{kq2}^{\prime r} = \frac{1}{X_{lkq2}^{\prime}}(\psi_{kq2}^{\prime r} - \psi_{mq}^r) \tag{5.16-12}$$

$$i_{fd}^{\prime r} = \frac{1}{X_{lfd}^{\prime}}(\psi_{fd}^{\prime r} - \psi_{md}^r) \tag{5.16-13}$$

$$i_{kd}^{\prime r} = \frac{1}{X_{lkd}^{\prime}}(\psi_{kd}^{\prime r} - \psi_{md}^r) \tag{5.16-14}$$

In the above equations

$$\psi^r_{mq} = X_{aq}\left(\frac{\psi^r_{qs}}{X_{ls}} + \frac{\psi'^r_{kq1}}{X'_{lkq1}} + \frac{\psi'^r_{kq2}}{X'_{lkq2}}\right) \tag{5.16-15}$$

$$\psi^r_{md} = X_{ad}\left(\frac{\psi^r_{ds}}{X_{ls}} + \frac{\psi'^r_{fd}}{X'_{lfd}} + \frac{\psi'^r_{kd}}{X'_{lkd}}\right) \tag{5.16-16}$$

$$X_{aq} = \left(\frac{1}{X_{mq}} + \frac{1}{X_{ls}} + \frac{1}{X'_{lkq1}} + \frac{1}{X'_{lkq2}}\right)^{-1} \tag{5.16-17}$$

$$X_{ad} = \left(\frac{1}{X_{md}} + \frac{1}{X_{ls}} + \frac{1}{X'_{lfd}} + \frac{1}{X'_{lkd}}\right)^{-1} \tag{5.16-18}$$

If saturation is to be taken into account, the torque equation that may be used in the simulation is (5.6-4) or the per unitized version given by (5.8-2). The rotor speed is obtained from (5.8-3) as

$$\omega_r = -\frac{\omega_b}{2Hp}(T_e - T_I) \tag{5.16-19}$$

Block diagrams showing the computer simulation of a synchronous machine in the rotor reference frame are shown in Fig. 5.16-1. The equations used to perform the computations are indicated by number. The voltages applied to the damper windings are not shown because these windings are always short-circuited and the voltages are zero. The two simulations shown in Fig. 5.16-1 differ only in the way that $\mathbf{v}^r_{qd0s}$ is obtained. In Fig. 5.16-1a, $\mathbf{v}_{abcs}$ is transformed directly to $\mathbf{v}^r_{qd0s}$; in Fig. 5.16-1b, $\mathbf{v}_{abcs}$ is first transformed to the arbitrary reference frame and then to the rotor reference frame.

Simulation with Stator Voltage Equations in Arbitrary Reference Frame

A synchronous machine simulation that is beginning to find use is one where the stator voltage equations are simulated in a reference frame other than the rotor reference frame. Either the stationary or synchronous reference frame is used; however, we will develop the simulation of stator equations in the arbitrary reference frame which, of course, encompasses all reference frames. The stator voltage equations in the arbitrary reference frame are given by (5.4-1). The integral equations in terms of flux linkages per second may be written as

$$\psi_{qs} = \frac{\omega_b}{p}\left[v_{qs} - \frac{\omega}{\omega_b}\psi_{ds} + r_s i_{qs}\right] \tag{5.16-20}$$

$$\psi_{ds} = \frac{\omega_b}{p}\left[v_{ds} + \frac{\omega}{\omega_b}\psi_{qs} + r_s i_{ds}\right] \tag{5.16-21}$$

$$\psi_{0s} = \frac{\omega_b}{p}(v_{0s} + r_s i_{0s}) \tag{5.16-22}$$

Here we have not substituted for the currents as in the previous formulation. Instead, ψ_{qs} and ψ_{ds} are computed and transformed to the rotor reference frame, whereupon i_{qs}^r and i_{ds}^r are calculated and transformed back to arbitrary reference. It is clear that

$$i_{0s} = -\frac{\psi_{0s}}{X_{ls}} \tag{5.16-23}$$

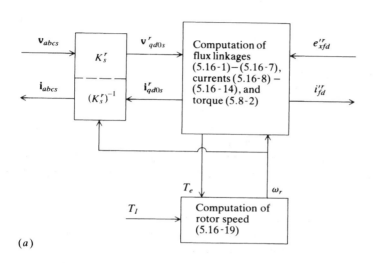

(a)

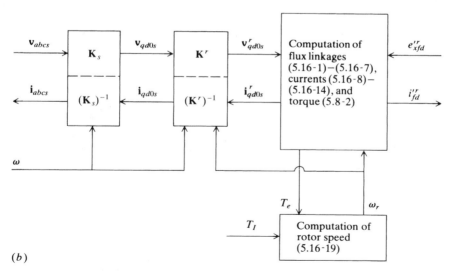

(b)

Figure 5.16-1 Simulation of a synchronous machine in the rotor reference frame shown in block diagram form. (a) abc variables transformed directly to the rotor reference frame; (b) abc variables transformed to arbitrary reference frame then to rotor reference frame.

250   SYNCHRONOUS MACHINES

This type of computer simulation is depicted in Fig. 5.16-2. Only the rotor flux linkage equations are calculated in the rotor reference frame; (5.16-1) and (5.16-3) are not used. The transformation from the arbitrary reference frame to the rotor reference frame is performed by implementing (3.6-7) with the zero quantities omitted. As mentioned, the stator equations are generally simulated either in the stationary reference frame or in the synchronously rotating reference frame. The transformation between the synchronous reference frame and the rotor reference frame may be written in terms of the rotor angle δ. This transformation may be obtained from (3.6-7) or it is given by (5.7-3).

Simulation of Saturation

A method of simulating the effects of saturation in the direct axis of a synchronous machine was developed by C. H. Thomas [7]. In the case of salient pole machines, it is sufficient to represent saturation only in the direct axis; however, in the case of round rotor machines (induction or synchronous), saturation should be represented in both axes. The method of Thomas will be presented and then extended to include saturation in both axes of a round rotor machine. For another approach of simulating saturation see reference 11.

The open-circuit test of a synchronous machine gives us a plot of the open-circuit terminal voltage versus field current as shown in Fig. 5.16-3a. Therein the straight-line relation between the stator terminal voltage and field current is referred to as the air-gap line. Because the terminals are open-circuited, the stator currents are zero; therefore $\psi_{qs}^{\prime r}$ and v_{ds}^r are zero. Hence v_{qs}^r is equal to the peak value of the line-to-neutral open-circuit stator voltage. Moreover, v_{qs}^r is equal to ψ_{ds}^r, which is ψ_{md}^r because the stator currents are zero. Thus, V_t in Fig. 5.16-3a may be replaced by ψ_{md}^r. Now, the current along the x axis of the open-circuit characteristics is i_{fd}^r;

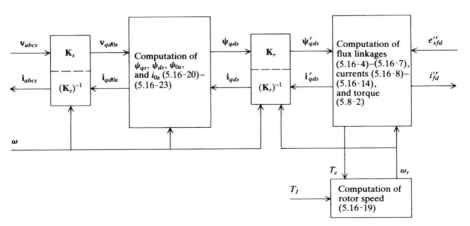

Figure 5.16-2   Block diagram of the simulation of a synchronous machine with stator voltage equations in the arbitrary reference frame.

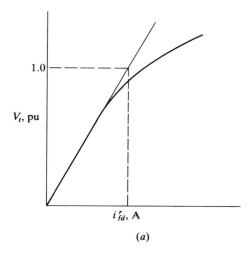

(a)

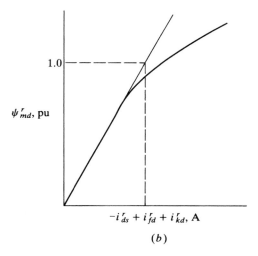

(b)

Figure 5.16-3 Saturation characteristics of a synchronous machine. (a) Open-circuited terminal voltage versus field current; (b) equivalent to characteristics shown in (a).

however, this same characteristic would occur regardless of the current flowing in the magnetizing reactance. Hence, i^r_{fd} may be replaced by the sum of $-i^r_{ds} + i'^r_{fd} + i'^r_{kd}$, which are the d-axis currents. It follows that Figs. 5.16-3a and 5.16-3b are identical.

It is convenient to plot the saturation curve shown in Fig. 5.16-3b with the scale in the x axis selected so that the air-gap line is at a 45° angle as shown in Fig. 5.16-4. The point that 1.0 per unit ψ^r_{md} intersects the air-gap line is 1.0 per unit $X_{md}(-i_{ds} + i'^r_{fd} + i'^r_{kd})$. Note from Fig. 5.16-4 that we can write

$$\psi^r_{md} = X_{md}(-i^r_{ds} + i'^r_{fd} + i'^r_{kd}) - f(\psi^r_{md}) \qquad (5.16\text{-}24)$$

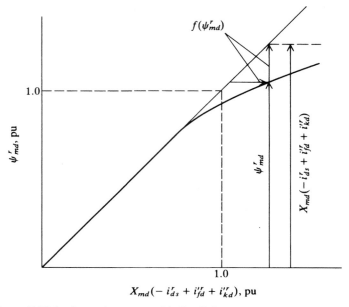

Figure 5.16-4 Saturation curve with flux linkages per second as variables.

where $f(\psi_{md}^r)$ is defined in Fig. 5.16-4. Therefore, saturation may be simulated by replacing (5.16-16) with (5.16-24) in the computer simulation. However, before this can be done we must first obtain $f(\psi_{md}^r)$. This of course can be accomplished directly from Fig. 5.16-4, and it is shown in Fig. 5.16-5. Next, the currents in (5.16-24) must be replaced by flux linkages per second. This leads to

$$\psi_{md}^r = X_{ad}\left(\frac{\psi_{ds}^r}{X_{ls}} + \frac{\psi_{fd}'^r}{X_{lfd}'} + \frac{\psi_{kd}'^r}{X_{lkd}'}\right) - \frac{X_{ad}}{X_{md}}f(\psi_{md}^r) \tag{5.16-25}$$

where X_{ad} is defined by (5.16-18). Therefore, saturation in the d-axis axis of a synchronous machine is simulated by generating the function $f(\psi_{md}^r)$ and replacing (5.16-16) with (5.16-25).

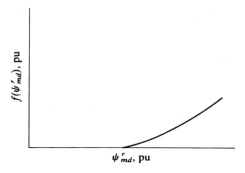

Figure 5.16-5 Plot of $f(\psi_{md}^r)$ obtained from Fig. 5.16-4.

A computer simulation of saturation in both axes of a round rotor machine, synchronous or induction, has been set forth by D. M. Triezenberg [12]. For this development let us consider the diagram shown in Fig. 5.16-6. The values of the variables shown are for a given instant of time. From Fig. 5.16-6 we have

$$\psi_m^r = [(\psi_{mq}^r)^2 + (\psi_{md}^r)^2]^{1/2} \qquad (5.16\text{-}26)$$

$$f(\psi_{mq}^r) = \frac{\psi_{mq}^r}{\psi_m^r} f(\psi_m^r) \qquad (5.16\text{-}27)$$

$$f(\psi_{md}^r) = \frac{\psi_{md}^r}{\psi_m^r} f(\psi_m^r) \qquad (5.16\text{-}28)$$

If, in the case of a round rotor synchronous machine, the open-circuit saturation curve approximates the magnetic characteristics of the rotor regardless if the MMF is directed along the q or d axis or anywhere in between, then Fig. 5.16-5 also describes the relationship between $f(\psi_m^r)$ and ψ_m^r. Actually it is convenient to use $f(\psi_m^r)/\psi_m^r$ rather than $f(\psi_m^r)$ as the ordinate when representing saturation in both axes. Also, it is convenient to plot $f(\psi_m^r)/\psi_m^r$ versus $(\psi_m^r)^2$, which eliminates the need to perform the square root of (5.16-26) in order to obtain ψ_m^r. A block diagram depicting this method of representing saturation in both axes is shown in Fig. 5.16-7.

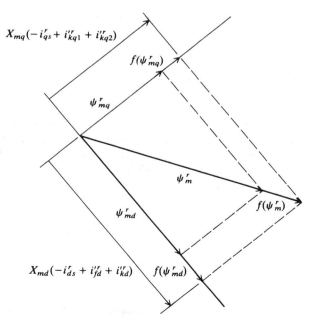

Figure 5.16-6 Saturation in both axes of a synchronous machine.

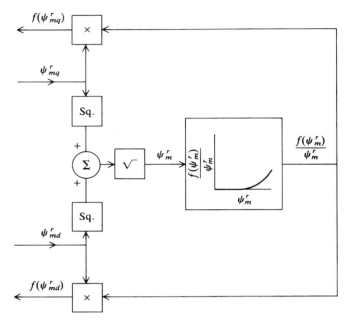

Figure 5.16-7 Simulation of saturation in both axes of a round rotor synchronous machine.

The expression for ψ_{md}^r is given by (5.16-25); ψ_{mq}^r now becomes

$$\psi_{mq}^r = X_{aq}\left(\frac{\psi_{qs}^r}{X_{ls}} + \frac{\psi_{kq1}^{'r}}{X'_{lkq1}} + \frac{\psi_{kq2}^{'r}}{X'_{lkq2}}\right) - \frac{X_{aq}}{X_{mq}}f(\psi_{mq}^r) \qquad (5.16\text{-}29)$$

where X_{aq} is defined by (5.16-17).

It is apparent that saturation of an induction machine may be represented in the same way. In this case, the saturation curve is generally obtained from the terminal voltage versus no-load stator current. The above equations for ψ_{mq}^r and ψ_{md}^r may be readily modified to express ψ_{mq} and ψ_{md} in the arbitrary reference frame. In particular, for an induction machine

$$\psi_{mq} = X_{aq}\left(\frac{\psi_{qs}}{X_{ls}} + \frac{\psi_{qr}'}{X'_{lr}}\right) - \frac{X_{aq}}{X_M}f(\psi_{mq}) \qquad (5.16\text{-}30)$$

$$\psi_{md} = X_{ad}\left(\frac{\psi_{ds}}{X_{ls}} + \frac{\psi_{dr}'}{X'_{lr}}\right) - \frac{X_{ad}}{X_M}f(\psi_{md}) \qquad (5.16\text{-}31)$$

where X_{aq} and X_{ad} are equal and expressed by (5.16-17). It is clear that $f(\psi_{mq})$ and $f(\psi_{md})$ are obtained from ψ_{mq} and ψ_{md} as shown in Fig. 5.16-7.

Simulation of Balanced Conditions

Any of the simulations given in Figs. 5.16-1 and 5.16-2 may be used to represent balanced or symmetrical conditions. However, a simplified version of Fig. 5.16-1b is probably most widely used. If it is unnecessary to have the stator phase variables available, then the synchronously rotating reference frame $(\omega - \omega_e)$ is convenient, especially in the simulation of large power systems. The transformation of the variables from the synchronously rotating reference frame to the rotor reference frame is performed using the rotor angle δ in the transformation given by (3.6-7) or (5.7-3). It is clear that the simulation given in Fig. 5.16-2 may also be used in a similar manner; the only difference is that ψ_{qs}^e and ψ_{ds}^e would be computed in the synchronously rotating reference frame and transformed to the rotor reference frame rather than transforming voltages as in Fig. 5.16-1b.

Simulation of Unbalanced Stator Conditions

Because the rotor flux linkages are computed in the rotor reference frame the simulation of various modes of operation of a synchronous machine is somewhat more difficult than in the case of the induction machine which may be simulated in any reference frame. This is especially true when it is desirable to represent unbalanced or discontinuous stator operation, and in some cases approximations are helpful in order to make the simulation manageable.

Perhaps the simulations shown in Fig. 5.16-1a and Fig. 5.16-2 offer the most convenient means of representing a synchronous machine during unbalanced stator conditions. It is generally advantageous to represent unbalanced or discontinuous stator conditions in *abc* variables. If the simulation given in Fig. 5.16-1a is used, the stator voltages and currents are transformed between the stationary and rotor reference frames. If the simulation given in Fig. 5.16-2 is used, ω is set equal to zero and the flux linkages per second ψ_{qs}^s and ψ_{ds}^s are computed in the stationary reference frame and then transformed to the rotor reference frame. Asynchronous and unbalanced operation including motor starting and pole slipping as well as line-to-neutral, line-to-line, and line-to-line-to-neutral faults are all demonstrated and simulated in Chapters 10 and 12 of reference 5.

REFERENCES

[1] R. H. Park, Two-Reaction Theory of Synchronous Machines—Generalized Method of Analysis, Part I, *AIEE Transactions*, Vol. 48, July 1929, pp. 716–727.

[2] D. R. Brown and P. C. Krause, Modeling of Transient Electrical Torques in Solid Iron Rotor Turbogenerators, *IEEE Transactions on Power Apparatus and Systems*, Vol. 98, September/October 1979, pp. 1502–1508.

[3] P. C. Krause, F. Nozari, T. L. Skvarenina, and D. W. Olive, The Theory of Neglecting Stator Transients, *IEEE Transactions on Power Apparatus and Systems*, Vol. 98, January/February 1979, pp. 141–148.

[4] C. M. Ong, *Dynamic Simulation of Electric Machinery*, Prentice-Hall PTR, Upper Saddle River, NJ, 1998.

[5] P. C. Krause, O. Wasynczuk, and S. D. Sudhoff, *Analysis of Electric Machinery*, IEEE Press, Piscataway, NJ, 1995.

[6] R. E. Doherty and C. A. Nickle, Synchronous Machines—III, Torque–Angle Characteristics Under Transient Conditions, *AIEE Transactions*, Vol. 46, January 1927, pp. 1–8.

[7] C. H. Thomas, Discussion of Analogue Computer Representations of Synchronous Generators in Voltage-Regulation Studies, *Transactions of AIEE* (Power Apparatus and Systems), Vol. 75, December 1956, pp. 1182–1184.

[8] O. Wasynczuk and S. D. Sudhoff, Automated State Model Generation Algorithm for Power Circuits and Systems, *IEEE Transactions on Power Systems*, Vol. 11, No. 9, November 1996, pp. 1951–1956.

[9] J. S. Mayer and O. Wasynczuk, An Efficient Method of Simulating Stiffly Connected Power Systems with Stator and Network Transients Included, *IEEE Transactions on Power Systems*, Vol. 6, No. 3, August 1991, pp. 922–929.

[10] S. D. Pekarek and O. Wasynczuk, An Efficient and Accurate Voltage Behind-Reactance Model of Synchronous Machines for Simulation and Analysis of Machine-Converter Systems, *IEEE Transactions on Energy Conversion*, Vol. 13, No. 4, March 1998, pp. 42–48.

[11] K. A. Corzine, B. Kuhn, and S. D. Sudhoff, An Improved Method for Incorporating Magnetic Saturation in the QD Synchronous Machine Model, *IEEE Transactions on Energy Conversion*, Vol. 13, No. 3, September 1998, pp. 270–275.

[12] D. M. Triezenberg, Private communications with authors, 1977.

PROBLEMS

1   A 2-pole, 2-phase, salient-pole synchronous machine is shown in Fig. 5.P-1. In the case of a 2-phase synchronous machine the magnetizing inductances are defined:

$$L_{mq} = L_A - L_B$$
$$L_{md} = L_A + L_B$$

Derive the voltage equations similar in form to (5.2-29) and the expression for electro-magnetic torque in machine variables. Express all resistance and inductance matrices.

2   Consider the torque expression derived in Problem 1 above. Identify the terms associated with (*a*) the reluctance torque, (*b*) the induction motor (damping) torque, and (*c*) the torque due to the interaction of MMF$_s$ and the field winding.

3   Modify the voltage equations (5.2-29) to describe a 3-phase round rotor synchronous machine.

4   Modify the voltage equations (5.2-29) to describe a 3-phase reluctance machine. Assume that the *fd* and *kq*1 windings are not present and denote the *kq*2 winding as simply the *kq* winding. Also, assume positive stator currents flow into the machine.

5   Justify the statement following (5.3-4) regarding the fact that this torque expression will be positive for positive currents assumed into the machine if the sign of the coefficients of $\sin 2\theta_r$ and $\cos 2\theta_r$ are changed.

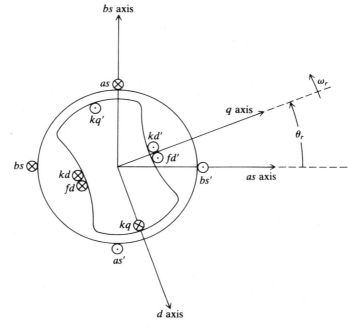

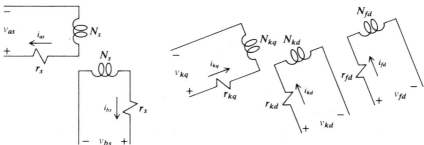

Figure 5.P-1   Two-pole, 2-phase, salient-pole synchronous machine.

6 Calculate the field current necessary to produce rated open-circuited stator voltages when driven at rated speed for (*a*) the hydro turbine generator given in Table 5.10-1 and (*b*) the steam turbine generator given in Table 5.10-2.

7 Determine $\mathbf{K}_s \mathbf{L}_s (\mathbf{K}_s)^{-1}$ given in (5.4-4). Show that it is independent of θ and θ_r only if $\theta = \theta_r$.

8 Repeat Problem 7 for $(\mathbf{L}_{sr})^{\mathrm{T}}(\mathbf{K}_s)^{-1}$.

9 Show that if the arbitrary reference-frame variables given in (5.4-4) are transformed to the rotor reference frame, (5.5-4) results if $\theta = \theta_r$. Is this also true if only $\omega = \omega_r$? Why?

10 Verify (5.5-5)–(5.5-7).

11 Modify Park's equations, (5.5-22)–(5.5-35), to describe a 3-phase reluctance machine with the rotor winding arrangement as indicated in Problem 4. Assume positive stator currents flow into the machine.

12 Derive (5.6-1).

13 Starting with the transformation given in Problem 1 of Chapter 3, derive Park's equations for the 2-phase synchronous machine shown in Fig. 5.P-1 in the form similar to (5.5-22)–(5.5-35). Also derive the express for electromagnetic torque similar to (5.6-4).

14 Consider the hydro turbine generator given in Section 5.10. Select the rated MVA as the base power. Determine the base voltage, base current, and base impedance. Check the parameters given in per unit.

15 Repeat Problem 14 for the steam turbine generator given in Section 5.10.

16 Starting with the results of Problem 13, derive the steady-state voltage equation for a 2-phase synchronous machine in the form similar to (5.9-19).

17 Derive the steady-state voltage equation for the synchronous machine (5.9-19) using the stator voltage equations expressed in the arbitrary reference.

18 A 4-pole, 3-phase, salient-pole synchronous machine is supplied from a 440-V (rms) line-to-line, 60-Hz source. The machine is operated as a motor with the total input power of 40 kW at the terminals. The parameters are

$$r_s = 0.3\,\Omega \qquad L_{md} = 0.015\,\text{H}$$
$$L_{ls} = 0.001\,\text{H} \qquad L_{mq} = 0.008\,\text{H}$$

Assume the positive direction of current is into the stator terminals.

(a) The excitation is adjusted so that $\tilde{I}_{as}$ lags $\tilde{V}_{as}$ by $30°$. Calculate $\tilde{E}_a$ and the reactive power Q. Draw the phasor diagram.

(b) Repeat a with the excitation adjusted so that $\tilde{I}_{as}$ is in phase with $\tilde{V}_{as}$.

(c) Repeat a with the excitation adjusted so that $\tilde{I}_{as}$ leads $\tilde{V}_{as}$ by $30°$.

19 A 2-pole, 220-V (rms) line-to-line, 5-hp, 3-phase reluctance machine has the following parameters

$$r_s = 1\,\Omega \qquad L_{md} = 0.10\,\text{H}$$
$$L_{ls} = 0.005\,\text{H} \qquad L_{mq} = 0.02\,\text{H}$$

The motor is supplied from a 60-Hz, 220-V source and it is operating at rated torque output. Calculate δ and $\tilde{I}_{as}$.

20 A 4-pole, 2-hp, 2-phase, round-rotor synchronous machine is connected to a 110-V, 60-Hz source. It delivers 1 kW at the terminals. $\tilde{I}_{as}$ lags $\tilde{V}_{as}$ by $20°$ (positive current out of machine). $r_s = 0.5\,\Omega$, $L_{ls} = 0.005\,\text{H}$, $L_{mq} = L_{md} = 0.05\,\text{H}$. Calculate $\tilde{E}_a$ and draw the phasor diagram showing $\tilde{V}_{as}$, $\tilde{I}_{as}$, $\tilde{E}_a$, and $(r_s + jX_q)\tilde{I}_{as}$.

21 The stator terminals of two synchronous machines are connected in a phase-to-phase arrangement. Let $\tilde{E}_{a1} = E_{a1}/\underline{\delta_1}$ and $\tilde{E}_{a2} = E_{a2}/\underline{\delta_2}$. Derive an expression for the steady-state power flowing between the two machines in terms of E_{a1}, E_{a2}, δ_1, and δ_2. Neglect the stator resistance of both machines.

22 Calculate the steady-state $\tilde{I}_{as}$ and δ for the hydro turbine generator for the final operating condition depicted in Figs. 5.10-1 and 5.10-2.

23 Repeat Problem 22 for the final operating conditions for the steam turbine generator depicted in Figs. 5.10-3 and 5.10-4.

24 Calculate the maximum steady-state torque that the steam turbine generator can deliver to an infinite bus with the excitation adjusted to produce rated open-circuit terminal voltage. Sketch the torque versus rotor angle response if a torque slightly larger than this maximum value is suddenly applied to the unloaded machine when connected to an infinite bus.

25 Calculate the initial, prefault values of all variables of the hydro turbine generator for the condition shown in Fig. 5.11-1.

26 Calculate the initial, prefault values of all variables of the steam turbine generator for the condition shown in Fig. 5.11-3.

27 Show that the rotor angle at which maximum steady-state torque occurs is

$$\cos \delta = -\frac{X_q E'^r_{xfd}}{4(X_d - X_q)V_s} \pm \sqrt{\left[\frac{X_q E'^r_{xfd}}{4(X_d - X_q)V_s}\right]^2 + \frac{1}{2}}$$

where all values are in per unit and generally only the positive value of the radical is used.

28 Derive an expression similar to that given in Problem 27 for the rotor angle at which maximum transient torque occurs as predicted by the approximate transient torque–angle characteristic. Check the values of rotor angle at maximum transient torque for the hydro turbine generator shown in Fig. 5.13-7 and for the steam turbine generator shown in Fig. 5.13-8.

29 Calculate the prefault values of the flux linkages per second for the hydro turbine generator shown in Fig. 5.14-3.

30 Calculate the prefault values of the flux linkages per second for the steam turbine generator shown in Fig. 5.14-4.

31 Check the values given in Section 5.15 for the "first swing" transient stability limit of the hydro and steam units using both the $\sin \delta$ and $\sin 2\delta$ terms of the approximate transient torque–angle curve.

32 Repeat Problem 31 with the $\sin 2\delta$ terms omitted.

33 Check the values of critical clearing time and critical clearing angle given in Section 5.15 for the hydro and steam units using the approximate transient torque–angle curve.

34 Determine the critical clearing time and critical clearing angle for the hydro and steam units using only the $\sin \delta$ terms of the approximate transient torque–angle curve.

Chapter 6

THEORY OF BRUSHLESS dc MACHINES

6.1 INTRODUCTION

The brushless dc motor is becoming widely used as a small horsepower control motor. This device has the physical appearance of a 3-phase permanent magnet synchronous machine that is supplied from an inverter that converts a dc voltage to 3-phase alternating-current (ac) voltages with frequency corresponding instantaneously to the rotor speed. The inverter–machine combination has the terminal and output (T_e vs. ω_r) characteristics resembling those of a dc shunt machine during motor operation; hence the name *brushless dc motor*. In this brief chapter, equations are derived which describe the operation of a brushless dc machine. The operation of brushless dc motor drives is considered in Chapter 15. In this chapter we look at the performance of the brushless dc motor with balanced 3-phase applied stator voltage with frequency corresponding to the rotor speed. This allows us to become familiar with the salient operating features of the inverter–motor combination without becoming involved with the harmonics of the phase voltages that occur due to the switching of the inverter.

6.2 VOLTAGE AND TORQUE EQUATIONS IN MACHINE VARIABLES

A 2-pole, brushless dc machine is depicted in Fig. 6.2-1. It has 3-phase, wye-connected stator windings and a permanent magnet rotor. It is a synchronous machine. The stator windings are identical windings displaced 120°, each with N_s equivalent turns and resistance r_s. For our analysis we will assume that the stator windings are sinusoidally distributed. The three sensors shown in Fig. 6.2-1 are Hall effect devices. When the north pole is under a sensor, its output is nonzero; with a south pole under the sensor, its output is zero. In most applications the stator is supplied from an inverter that is switched at a frequency corresponding to the rotor speed. The states of the three sensors are used to determine the switching logic for the inverter.

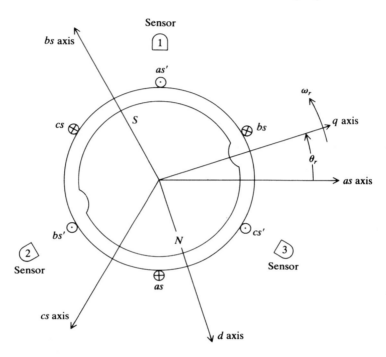

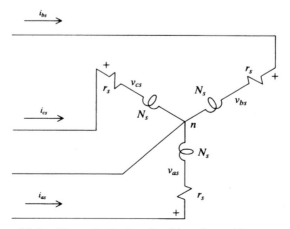

Figure 6.2-1 Two-pole, 3-phase brushless dc machine.

In the actual machine the sensors are not positioned over the rotor as shown in Fig. 6.2-1. Instead they are placed over a ring that is mounted on the shaft external to the stator windings and magnetized as the rotor. We will return to these sensors and the role they play later in this analysis. It is first necessary to establish the voltage and torque equations that can be used to describe the behavior of the permanent magnet synchronous machine.

The voltage equations in machine variables are

$$\mathbf{v}_{abcs} = \mathbf{r}_s \mathbf{i}_{abcs} + p\boldsymbol{\lambda}_{abcs} \tag{6.2-1}$$

where

$$(\mathbf{f}_{abcs})^T = [f_{as} \quad f_{bs} \quad f_{cs}] \tag{6.2-2}$$

$$\mathbf{r}_s = \text{diag}[r_s \quad r_s \quad r_s] \tag{6.2-3}$$

The flux linkages may be written

$$\boldsymbol{\lambda}_{abcs} = \mathbf{L}_s \mathbf{i}_{abcs} + \boldsymbol{\lambda}'_m \tag{6.2-4}$$

where $\mathbf{L}_s$ may be written from (1.5-25)–(1.5-27) and (1.5-29)–(1.5-31) or directly from (5.2-8). Also,

$$\boldsymbol{\lambda}'_m = \lambda'_m \begin{bmatrix} \sin\theta_r \\ \sin\left(\theta_r - \frac{2\pi}{3}\right) \\ \sin\left(\theta_r + \frac{2\pi}{3}\right) \end{bmatrix} \tag{6.2-5}$$

where λ'_m is the amplitude of the flux linkages established by the permanent magnet as viewed from the stator phase windings. In other words the magnitude of $p\lambda'_m$ would be the open-circuit voltage induced in each stator phase winding. Damper windings are neglected because the permanent magnet is a poor electrical conductor and the eddy currents that flow in the nonmagnetic materials securing the magnets are small. Hence, large armature currents can be tolerated without significant demagnetization. We have assumed by (6.2-5) that the voltages induced in the stator windings by the permanent magnet are constant-amplitude sinusoidal voltages. For nonsinusoidal induced voltages see references 1 and 2.

The expression for the electromagnetic torque may be written in machine variables from (5.3-4) by letting $\lambda'_m = L_{md} i'_{fd}$. Thus

$$\begin{aligned}
T_e = \left(\frac{P}{2}\right) &\left\{ \frac{(L_{md} - L_{mq})}{3} \left[\left(i_{as}^2 - \frac{1}{2}i_{bs}^2 - \frac{1}{2}i_{cs}^2 - i_{as}i_{bs} - i_{as}i_{cs} + 2i_{bs}i_{cs} \right) \sin 2\theta_r \right.\right. \\
&\left. + \frac{\sqrt{3}}{2}(i_{bs}^2 i_{cs}^2 - 2i_{as}i_{bs} + 2i_{as}i_{cs})\cos 2\theta_r \right] \\
&\left. + \lambda'_m \left[\left(i_{as} - \frac{1}{2}i_{bs} - \frac{1}{2}i_{cs} \right)\cos\theta_r + \frac{\sqrt{3}}{2}(i_{bs} - i_{cs})\sin\theta_r \right] \right\} \tag{6.2-6}
\end{aligned}$$

where L_{mq} and L_{md} are defined by (5.2-11) and (5.2-12), respectively. The above expression for torque is positive for motor action. The torque and speed may be related as

$$T_e = J\left(\frac{2}{P}\right)p\omega_r + B_m\left(\frac{2}{P}\right)\omega_r + T_L \tag{6.2-7}$$

where J is $\text{kg}\cdot\text{m}^2$; it is the inertia of the rotor and the connected load. Because we will be concerned primarily with motor action, the torque T_L is positive for a torque load. The constant B_m is a damping coefficient associated with the rotational system of the machine and the mechanical load. It has the units $\text{N}\cdot\text{m}\cdot\text{s}$ per radian of mechanical rotation, and it is generally small and often neglected.

6.3 VOLTAGE AND TORQUE EQUATIONS IN ROTOR REFERENCE-FRAME VARIABLES

The voltage equations in the rotor reference frame may be written directly from Section 3.4 with $\omega = \omega_r$ or from (5.5-1) with the direction of positive stator currents into the machine.

$$\mathbf{v}^r_{qd0s} = \mathbf{r}_s\mathbf{i}^r_{qd0s} + \omega_r\boldsymbol{\lambda}^r_{dqs} + p\boldsymbol{\lambda}^r_{qd0s} \tag{6.3-1}$$

where

$$(\boldsymbol{\lambda}^r_{dqs})^T = [\,\lambda^r_{ds} \quad -\lambda^r_{qs} \quad 0\,] \tag{6.3-2}$$

$$\boldsymbol{\lambda}^r_{qd0s} = \begin{bmatrix} L_{ls} + L_{mq} & 0 & 0 \\ 0 & L_{ls} + L_{md} & 0 \\ 0 & 0 & L_{ls} \end{bmatrix} \begin{bmatrix} i^r_{qs} \\ i^r_{ds} \\ i_{0s} \end{bmatrix} + \lambda'^r_m \begin{bmatrix} 0 \\ 1 \\ 0 \end{bmatrix} \tag{6.3-3}$$

To be consistent with our previous notation, we have added the superscript r to λ'_m. In expanded form we have

$$v^r_{qs} = r_s i^r_{qs} + \omega_r \lambda^r_{ds} + p\lambda^r_{qs} \tag{6.3-4}$$

$$v^r_{ds} = r_s i^r_{ds} - \omega_r \lambda^r_{qs} + p\lambda^r_{ds} \tag{6.3-5}$$

$$v_{0s} = r_s i_{0s} + p\lambda_{0s} \tag{6.3-6}$$

where

$$\lambda^r_{qs} = L_q i^r_{qs} \tag{6.3-7}$$

$$\lambda^r_{ds} = L_d i^r_{ds} + \lambda'^r_m \tag{6.3-8}$$

$$\lambda_{0s} = L_{ls} i_{0s} \tag{6.3-9}$$

where $L_q = L_{ls} + L_{mq}$ and $L_d = L_{ls} + L_{md}$.

Substituting (6.3-7)–(6.3-9) into (6.3-4)–(6.3-6) and because $p\lambda_m'' = 0$, we can write

$$v_{qs}^r = (r_s + pL_q)i_{qs}^r + \omega_r L_d i_{ds}^r + \omega_r \lambda_m''$$ (6.3-10)

$$v_{ds}^r = (r_s + pL_d)i_{ds}^r - \omega_r L_q i_{qs}^r$$ (6.3-11)

$$v_{0s} = (r_s + pL_{ls})i_{0s}$$ (6.3-12)

The expression for electromagnetic torque may be written from (5.6-9) as

$$T_e = \left(\frac{3}{2}\right)\left(\frac{P}{2}\right)(\lambda_{ds}^r i_{qs}^r - \lambda_{qs}^r i_{ds}^r)$$ (6.3-13)

Substituting (6.3-7) and (6.3-8) into (6.3-13) yields

$$T_e = \left(\frac{3}{2}\right)\left(\frac{P}{2}\right)[\lambda_m'' i_{qs}^r + (L_d - L_q)i_{qs}^r i_{ds}^r]$$ (6.3-14)

The electromagnetic torque is positive for motor action.

As pointed out in the previous section, the state of the sensors provides us with information regarding the position of the poles and thus the position of the q and d axes. In other words, when the machine is supplied from an inverter, it is possible, by controlling the firing of the inverter, to change the values of v_{qs}^r and v_{ds}^r. Recall that θ_r in the transformation equation to the rotor reference frame can be written as

$$\omega_r = \frac{d\theta_r}{dt}$$ (6.3-15)

For purposes of discussion, let us assume that the applied stator voltages are sinusoidal so that

$$v_{as} = \sqrt{2}v_s \cos\theta_{ev}$$ (6.3-16)

$$v_{bs} = \sqrt{2}v_s \cos\left(\theta_{ev} - \frac{2\pi}{3}\right)$$ (6.3-17)

$$v_{cs} = \sqrt{2}v_s \cos\left(\theta_{ev} + \frac{2\pi}{3}\right)$$ (6.3-18)

When the machine is supplied from an inverter, the stator voltages will have a stepped waveform. Nevertheless, (6.3-16)–(6.3-18) may be considered as the fundamental components of these stepped phase voltages. Also,

$$\omega_e = \frac{d\theta_{ev}}{dt}$$ (6.3-19)

The brushless dc machine is, by definition, a device where the frequency of the fundamental component of the applied stator voltages corresponds to the speed of the rotor, and we understand that this is accomplished by appropriately firing the inverter supplying (driving) the machine. Hence, in a brushless dc machine, ω_e in (6.3-19) is ω_r; and if (6.3-16)–(6.3-18) are substituted into $\mathbf{K}_s^r$, we obtain

$$v_{qs}^r = \sqrt{2}v_s \cos\phi_v \tag{6.3-20}$$

$$v_{ds}^r = -\sqrt{2}v_s \sin\phi_v \tag{6.3-21}$$

where

$$\phi_v = \theta_{ev} - \theta_r \tag{6.3-22}$$

Now, because $\omega_r = \omega_e$ at all times, ϕ_v is a constant during steady-state operation or is changed by advancing or retarding the firing of the inverter relative to the rotor position. In other words, we can, at any time, instantaneously adjust ϕ_v by appropriate firing of the inverter, thereby changing the phase relationship between the fundamental component of the 3-phase stator voltages and the rotor (permanent magnet). In most applications, however, ϕ_v is fixed at zero so that as far as the fundamental component is concerned, v_{ds}^r is always zero and $v_{qs}^r = \sqrt{2}v_s$.

In Section 2.6 the time-domain block diagrams and the state equations were derived for a dc machine. By following the same procedure, the time-domain diagram and state equations can be established for the brushless dc machine directly from the equations given above. Because this essentially is a repeat of the work in Section 2.6, the development is left as an exercise for the readers.

6.4 ANALYSIS OF STEADY-STATE OPERATION

For steady-state operation with balanced, sinusoidal applied stator voltages, (6.3-10) and (6.3-11) may be written as

$$V_{qs}^r = r_s I_{qs}^r + \omega_r L_d I_{ds}^r + \omega_r \lambda_m^{\prime r} \tag{6.4-1}$$

$$V_{ds}^r = r_s I_{ds}^r - \omega_r L_q I_{qs}^r \tag{6.4-2}$$

where uppercase letters denote steady-state (constant) quantities. It is clear that $\lambda_m^{\prime r}$ is always constant. The steady-state torque is expressed from (6.3-14) with uppercase letters as

$$T_e = \left(\frac{3}{2}\right)\left(\frac{P}{2}\right)[\lambda_m^{\prime r} I_{qs}^r + (L_d - L_q)I_{qs}^r I_{ds}^r] \tag{6.4-3}$$

It is possible to establish a phasor voltage equation from (6.4-1) and (6.4-2) similar to that for the synchronous machine. For this purpose let us write (6.3-22) as

$$\theta_{ev} = \theta_r + \phi_v \tag{6.4-4}$$

For steady-state operation ϕ_v is constant and represents the angular displacement between the peak value of the fundamental component of v_{as} and the q axis fixed in the rotor. If we reference the phasors to the q axis and let it be along the positive real axis of the "stationary" phasor diagram, then ϕ_v becomes the phase angle of $\tilde{V}_{as}$ and we can write

$$\tilde{V}_{as} = V_s e^{j\phi_v} = V_s \cos\phi_v + jV_s \sin\phi_v \tag{6.4-5}$$

Comparing (6.4-5) with (6.3-20) and (6.3-21), we see that

$$\sqrt{2}\tilde{V}_{as} = V_{qs}^r - jV_{ds}^r \tag{6.4-6}$$

If we go back and write equations for the 3-phase currents similar to (6.3-16)–(6.3-18) in terms of θ_{ei}, then (6.3-22) would be in terms of θ_{ei} and ϕ_i rather than θ_{ev} and ϕ_v; however, we would arrive at a similar relation for current as (6.4-6). In particular,

$$\sqrt{2}\tilde{I}_{as} = I_{qs}^r - jI_{ds}^r \tag{6.4-7}$$

and

$$j\sqrt{2}\tilde{I}_{as} = I_{ds} + jI_{qs}^r \tag{6.4-8}$$

Substituting (6.4-1) and (6.4-2) into (6.4-6) and using (6.4-7) and (6.4-8) yields

$$\tilde{V}_{as} = (r_s + j\omega_r L_q)\tilde{I}_{as} + \tilde{E}_a \tag{6.4-9}$$

where

$$\tilde{E}_a = \frac{1}{\sqrt{2}}\left[\omega_r(L_d - L_q)I_{ds}^r + \omega_r \lambda_m'^r\right]e^{j0} \tag{6.4-10}$$

In Chapter 5 the phasor diagram for the synchronous machine was referenced to the phasor of v_{as} ($\tilde{V}_{as}$), which was positioned along the real axis. Note that in the case of the synchronous machine the phase angle of $\tilde{E}_a$ was δ [see (5.9-18)], whereas $\tilde{E}_a$ for the brushless dc machine [see (6.4-10)] is along the reference (real) axis.

Common Operating Mode

From our earlier discussion we are aware that the values of V_{qs}^r and V_{ds}^r are determined by the firing of the driving inverter. However, the condition with $\phi_v = 0$, whereupon $V_{qs}^r = \sqrt{2}V_s$ and $V_{ds}^r = 0$, is used in most applications. In this case, (6.4-2) may be solved for I_{ds}^r in terms of I_{qs}^r.

$$I_{ds}^r = \frac{\omega_r L_q}{r_s} I_{qs}^r \qquad \text{for } \phi_v = 0 \tag{6.4-11}$$

Substituting (6.4-11) into (6.4-1) yields

$$V_{qs}^r = \left(\frac{r_s^2 + \omega_r^2 L_q L_d}{r_s} \right) I_{qs}^r + \omega_r \lambda_m^{\prime r} \qquad \text{for } \phi_v = 0 \qquad (6.4\text{-}12)$$

We now start to see a similarity between the voltage equation for the brushless dc machine operated in this mode ($\phi_v = 0$) and the dc machine discussed in Chapter 2. From (2.3-1) the steady-state armature voltage equation of a dc shunt machine is

$$V_a = r_a I_a + \omega_r L_{AF} I_f \qquad (6.4\text{-}13)$$

If we neglect $\omega_r^2 L_q L_d$ in (6.4-12) and if we assume the field current is constant in (6.4-13), then the two equations are identical in form. Let's note another similarity. If we neglect the inductances, or set $L_q = L_d$, then the expression for the torque given by (6.4-3) is identical in form to that of a dc shunt machine with a constant field current given by (2.3-5). We now see that the brushless dc motor is called a brushless dc motor, not because it has the same physical configuration as a dc machine but because its terminal characteristics may be made to resemble those of a dc machine. We must be careful, however, because in order for (6.4-12) and (6.4-13) to be identical in form, the term $\omega_r^2 L_q L_d$ must be significantly less than r_s. Let us see what effects this term has upon the torque versus speed characteristics. If we solve (6.4-12) for I_{qs}^r and we take that result along with (6.4-11) for I_{ds}^r and substitute these expressions into the expression for T_e (6.4-3) and if we assume that $L_q = L_d$, we obtain the following expression for torque:

$$T_e = \left(\frac{3}{2} \right) \left(\frac{P}{2} \right) \frac{r_s \lambda_m^{\prime r}}{r_s^2 + \omega_r^2 L_s^2} (V_{qs}^r - \omega_r \lambda_m^{\prime r}) \qquad \text{for } \phi_v = 0 \qquad (6.4\text{-}14)$$

We have used L_s for L_q and L_d because we have assumed that $L_q = L_d$.

The steady-state, torque–speed characteristics for a typical brushless dc motor are shown in Fig. 6.4-1. Therein $L_{mq} = L_{md}$ and $\phi_v = 0$; hence, Fig. 6.4-1 is a plot of (6.4-14). If $\omega_r^2 L_s^2$ is neglected, then (6.4-14) yields a straight line T_e versus ω_r characteristic for a constant V_{qs}^r. Thus, if $\omega_r^2 L_s^2$ could be neglected, the plot shown in Fig. 6.4-1 would be a straight line. Although the T_e versus ω_r is approximately linear over the region of motor operation where $T_e \geq 0$ and $\omega_r \geq 0$, it is not linear over the complete speed range. In fact, we see from Fig. 6.4-1 that there appears to be a maximum and minimum torque. Let us take the derivative of (6.4-14) with respect to ω_r and set the result to zero and solve for ω_r. Thus, zero slope of the torque versus speed characteristics for $L_q = L_d = L_s$, $V_{qs}^r = \sqrt{2} V_s$, and $V_{ds}^r = 0$ ($\phi_v = 0$) occurs at

$$\omega_{rMT} = \frac{V_{qs}^r}{\lambda_m^{\prime r}} \pm \sqrt{\left(\frac{V_{qs}^r}{\lambda_m^{\prime r}} \right)^2 + \left(\frac{r_s}{L_s} \right)^2} \qquad \text{for } \phi_v = 0 \qquad (6.4\text{-}15)$$

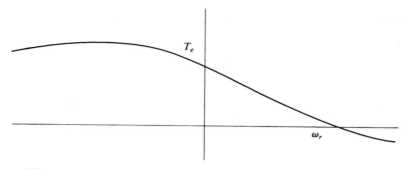

Figure 6.4-1 Torque–speed characteristics of a brushless dc motor with $L_{mq} = L_{md}$ ($L_q = L_d$), $\phi_v = 0$, $V_{qs}^r = \sqrt{2}V_s$, and $V_{ds}^r = 0$.

Due to the high reluctance of the magnetic material in the d axis, L_d is less than L_q for some machine designs. In most cases, whether $L_q > L_d$ or $L_q < L_d$, saliency has only secondary effects upon the torque–speed characteristics over the region of interest ($T_e \geq 0$ and $\omega_r \geq 0$). This is shown in Fig. 6.4-2 where $L_{mq} = 0.6L_{md}$ for the machine considered in Fig. 6.4-1.

Operating Modes Achievable by Phase Shifting the Applied Voltages

Let us return to (6.4-1)–(6.4-3), and this time we will not restrict ϕ_v to zero. Hence, from (6.4-2) we have

$$I_{ds}^r = \frac{V_{ds}^r + \omega_r L_q I_{qs}^r}{r_s} \tag{6.4-16}$$

Substituting into (6.4-1) yields

$$V_{qs}^r = \frac{r_s^2 + \omega_r^2 L_q L_d}{r_s} I_{qs}^r + \frac{\omega_r L_d}{r_s} V_{ds}^r + \omega_r \lambda_m'^r \tag{6.4-17}$$

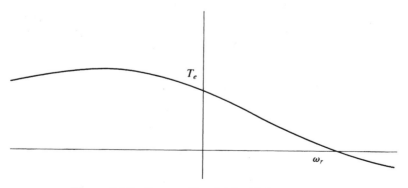

Figure 6.4-2   Same as Fig. 6.4-1 with $L_{mq} = 0.6L_{md}$.

Let us again set $L_q = L_d = L_s$, which simplifies our work. In this case, T_e and I_{qs}^r differ only by a constant multiplier. Solving (6.4-14) for I_{qs}^r yields

$$I_{qs}^r = \frac{r_s}{r_s^2 + \omega_r^2 L_s^2} \left(V_{qs}^r - \frac{L_s}{r_s} V_{ds}^r \omega_r - \lambda_m^{\prime r} \omega_r \right) \tag{6.4-18}$$

We see from (6.3-20) and (6.3-21) that if we consider only the fundamental component of the 3-phase applied stator voltages, then we have

$$V_{qs}^r = \sqrt{2} V_s \cos \phi_v \tag{6.4-19}$$

$$V_{ds}^r = -\sqrt{2} V_s \sin \phi_v \tag{6.4-20}$$

wherein ϕ_v is determined by the switching of the inverter. Substituting (6.4-19) and (6.4-20) into (6.4-18) yields

$$I_{qs}^r = \frac{\sqrt{2} r_s V_s}{r_s^2 + \omega_r^2 L_s^2} [\cos \phi_v + (\tau_s \sin \phi_v - \tau_v) \omega_r] \tag{6.4-21}$$

Here we are using time constants, or what appear to be time constants, in a steady-state equation. This is not normally done; however, it allows us to work more efficiently. The constant τ_s is L_s/r_s. It is the electrical or stator time constant which is analogous to the armature circuit time constant, τ_a, of a dc machine. The quantity τ_v is defined here as

$$\tau_v = \frac{\lambda_m^{\prime r}}{\sqrt{2} V_s} \tag{6.4-22}$$

It is introduced for our convenience. Be careful, however. Although τ_v has the unit of seconds, it is not a constant; it will change with V_s. In speed control systems the inverter supplying the brushless dc motor may provide a means to change the magnitude of V_s.

Equation (6.4-21) is an expression for I_{qs}^r, and thus T_e, in terms of the angle ϕ_v which is the phase shift of the fundamental component of the voltage v_{as} ahead of the q axis of the rotor. There are several pieces of information that can be gained from this equation. First let us find the maximum and minimum with respect to ϕ_v. This of course may be accomplished by taking the derivative of (6.4-21) with respect to ϕ_v and setting the result equal to zero. This yields

$$0 = -\sin \phi_v + \tau_s \omega_r \cos \phi_v \tag{6.4-23}$$

which says that at a given rotor speed a maximum or minimum value of $I_{qs}^r (T_e)$ occurs when

$$\phi_{vMT} = \tan^{-1} (\tau_s \omega_r) \tag{6.4-24}$$

Hence, the maximum or minimum steady-state electromagnetic torque, with ϕ_{vMT} given by (6.4-24), may be determined by substituting (6.4-24) into (6.4-21) and then substituting the results into (6.3-14) with $L_q = L_d = L_s$. Thus, the maximum or minimum steady electromagnetic torque for a given rotor speed may be expressed as

$$T_{eM} = \left(\frac{3}{2}\right)\left(\frac{P}{2}\right)\frac{\sqrt{2}V_s r_s \lambda_m^{'r}}{r_s^2 + \omega_r^2 L_s^2}\left[\frac{(r_s^2 + \omega_r^2 L_s^2)^{1/2}}{r_s} - \frac{\lambda_m^{'r}\omega_r}{\sqrt{2}V_s}\right] \qquad (6.4\text{-}25)$$

which may also be written as

$$T_{eM} = \left(\frac{3}{2}\right)\left(\frac{P}{2}\right)\frac{\sqrt{2}V_s^2\tau_v}{r_s(1 + \tau_s^2\omega_r^2)}[(1 + \tau_s^2\omega_r^2)^{1/2} - \tau_v\omega_r] \qquad (6.4\text{-}26)$$

Although (6.4-24)–(6.4-26) are valid for any speed, we are generally concerned with positive values of ω_r and T_e for steady-state operation. With this in mind, let us look back to (6.4-21) and assume that the speed is positive. With this assumption, T_e is positive if

$$\cos\phi_v + \tau_s\omega_r\sin\phi_v > \tau_v\omega_r \qquad (6.4\text{-}27)$$

Let us assume that ϕ_v is $\pi/2$, which means that $V_{qs}^r = 0$ and $V_{ds}^r = -\sqrt{2}V_s$. With $\phi_v = \pi/2$, (6.4-27) becomes

$$\tau_s > \tau_v \qquad (6.4\text{-}28)$$

Equation (6.4-28) tells us that if $\phi_v = \pi/2$ and if $\tau_s > \tau_v$, the torque will always be positive for $\omega_r > 0$. It follows that if $\phi_v = \pi/2$ and if $\tau_s = \tau_v$, the torque will be zero regardless of the speed of the rotor or if $\tau_s < \tau_v$ the torque will be negative for $\omega_r > 0$.

We will derive one more relationship from (6.4-21). For a fixed value of ϕ_v, the speeds at which the maximum and minimum torques occur may be determined by taking the derivative of (6.4-21) with respect to ω_r and setting the result to zero. This yields

$$\omega_{rMT} = \frac{1}{\tau_s\sin\phi_v - \tau_v}\left[-\cos\phi_v \pm \frac{1}{\tau_s}\sqrt{\tau_s^2 + \tau_v^2 - 2\tau_s\tau_v\sin\phi_v}\right] \qquad (6.4\text{-}29)$$

It is left to the reader to show that (6.4-29), with $\phi_v = 0$, is (6.4-15).

The steady-state, torque–speed characteristics for a brushless dc machine are shown in Fig. 6.4-3 for $L_q = L_d$ and for $L_{mq} = 0.6L_{md}$ in Fig. 6.4-4. In order to illustrate the limits of the torque–speed characteristics for the various possible values of ϕ_v, plots of torque are shown for $\phi_v = 0$, $\pm\pi/2$, π, and ϕ_{vMT}. The maximum torque for $\omega_r > 0$ is also plotted in Fig. 6.4-3. The machine parameters for the characteristics shown in Fig. 6.4-3 are $r_s = 3.4\,\Omega$, $L_{ls} = 1.1$ mH, and $L_{mq} = L_{md} = 11$ mH; thus $L_q = L_d = L_s = 12.1$ mH. The device is a 4-pole machine; and when it is driven at 1000 r/min, the open-circuit winding-to-winding voltage is sinusoidal

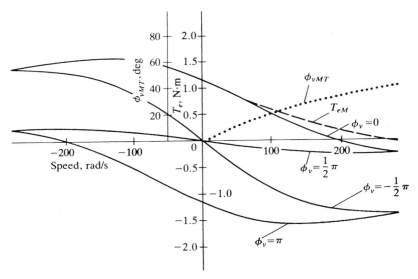

Figure 6.4-3 Torque–speed characteristics of a brushless dc machine with $L_{mq} = L_{md}(L_q = L_d)$.

with a peak-to-peak value of 60 V. From this, $\lambda_m'^r$ is calculated to be 0.0827 V·s. (The reader should verify this calculation.) The value of V_s is 11.25 V.

We see that this machine, which is commercially available, yields negative T_e for $\phi_v = \pi/2$ for $\omega_r > 0$. Therefore $\tau_s < \tau_v$. Let us see: $\tau_s = (12.1 \times 10^{-3})/3.4 = 3.56$ ms; $\tau_v = 0.0827/(\sqrt{2} \times 11.25) = 5.2$ ms. In references 3 and 4 it was

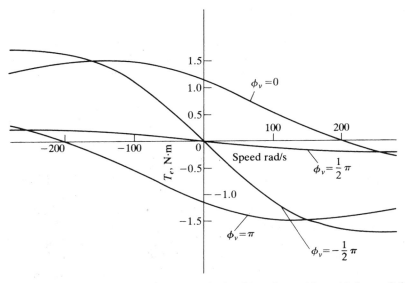

Figure 6.4-4 Torque–speed characteristics of a brushless dc machine with $L_{mq} = 0.6L_{md}$.

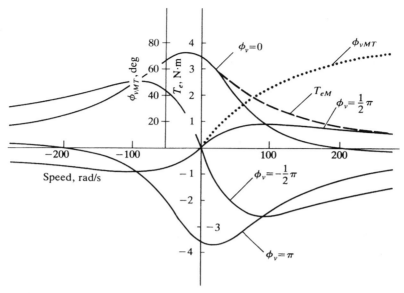

Figure 6.4-5   Torque–speed characteristics of a brushless dc motor with τ_s increased by a factor of three by decreasing r_s.

experimentally verified that advancing ϕ_v from zero, increased the torque at rotor speeds greater than zero. Therein, it is suggested that this might be a means of increasing the torque at high speeds. The parameters of the machine used in reference 4 are given and $\tau_s > \tau_v$. We now understand why. The influence of τ_s upon the steady-state torque–speed characteristics is illustrated by Figs. 6.4-5 and 6.4-6. In

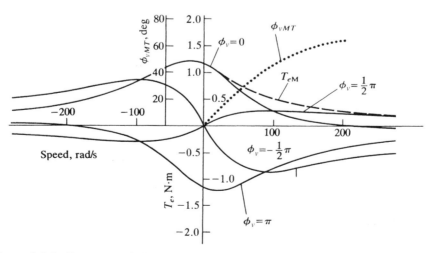

Figure 6.4-6   Torque–speed characteristics of a brushless dc motor with τ_s increased by a factor of three by increasing L_s.

Fig. 6.4-5, τ_s is increased by a factor of 3 ($\tau_s = 3 \times 3.56$ ms) by decreasing r_s. In Fig. 6.4-6, τ_s is increased by a factor of 3 by increasing L_s.

6.5 DYNAMIC PERFORMANCE

It is instructive to observe the machine variables during free acceleration and step changes in load torque with $\phi_v = 0$. The machine parameters are those given previously with $J = 1 \times 10^{-4}$ kg·m². In this section the dynamic performance is shown, by computer traces, for applied stator phase voltages that are sinusoidal

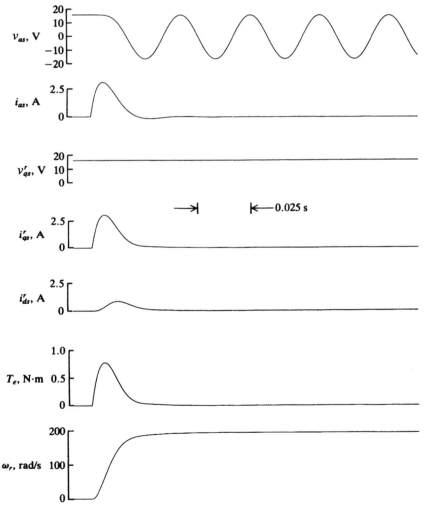

Figure 6.5-1   Free acceleration characteristics of a brushless dc motor ($\phi_v = 0$).

and that are stepped as would be the case if a typical six-step voltage source inverter were used to supply the machine [5–7]. Although it may seem inappropriate because we have yet to discuss the operation of the inverter, it does provide a first look at the machine variables in a drive application, and most important it provides a justification of the sinusoidal approximation of the actual six-step phase voltages. The brushless dc motor-inverter drives are considered in detail in Chapter 15.

Sinusoidal Phase Voltages

The free acceleration characteristics with sinusoidal phase voltages are shown in Fig. 6.5-1. The applied stator phase voltages are of the form given by (6.3-16)– (6.3-18) with $v_s = 11.25$ V. The phase voltage v_{as}, phase current i_{as}, q-axis voltage v_{qs}^r, q-axis current i_{qs}^r, d-axis current i_{ds}^r, electromagnetic torque T_e, and rotor speed ω_r in electrical rad/s are plotted in Fig. 6.5-1. It is clear that because $\phi_v = 0$, we have $v_{ds}^r = 0$. The device is a 4-pole machine; thus 200 electrical rad/s is 955 r/min. A plot of T_e versus ω_r is shown in Fig. 6.5-2 for the free acceleration depicted in Fig. 6.5-1. The steady-state, torque–speed characteristic is superimposed for purposes of comparison.

In Fig. 6.5-1, the rapid acceleration of the brushless dc motor is apparent. In fact the rotor reaches full speed in less than 0.05 s. The acceleration is so rapid that it is difficult to observe the change in frequency of v_{as} as the motor accelerates from stall. In order to illustrate the frequency change during the acceleration period, the free acceleration characteristics given in Figs. 6.5-3 and 6.5-4 are for an inertia of five times the inertia of the rotor (5×10^{-4} kg·m^2). It is important to note that the dynamic torque–speed characteristics shown in Figs 6.5-2 and 6.5-4 differ from the steady-state, torque–speed characteristics. One must be aware of

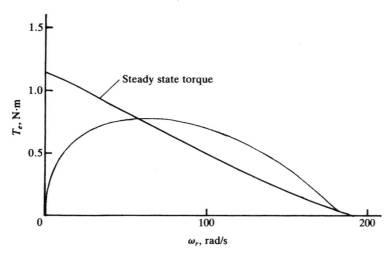

Figure 6.5-2 Torque–speed characteristics for free acceleration shown in Fig. 6.5-1.

this discrepancy, if one chooses to use the expression for the steady-state torque in a transfer function formulation describing the dynamic characteristics of a brushless dc motor.

The performance during step changes in load torque is illustrated in Fig. 6.5-5. Initially the machine is operating with $T_L = 0.1$ N·m. The load torque is suddenly stepped to 0.4 N·m. The machine slows down and once steady-state operation is

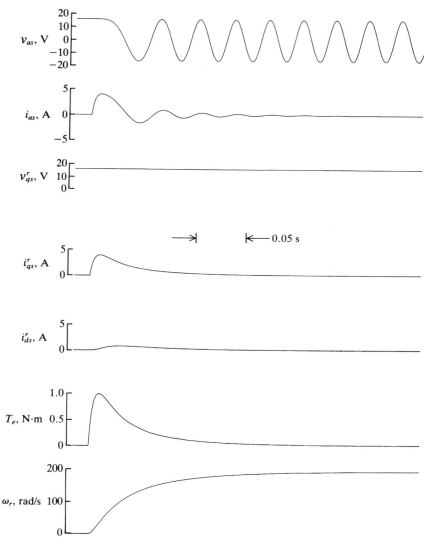

Figure 6.5-3   Free acceleration characteristics of a brushless dc motor ($\phi_v = 0$) with inertia equal to five times rotor inertia.

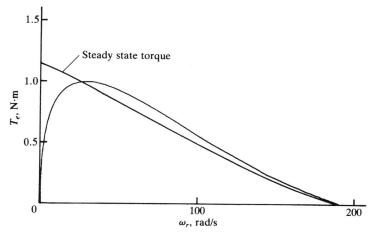

Figure 6.5-4 Torque–speed characteristics for free acceleration shown in Fig. 6.5-3.

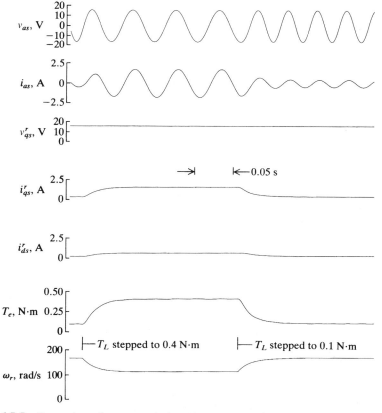

Figure 6.5-5 Dynamic performance of a brushless dc motor ($\phi_v = 0$) during step changes in load torque. Total inertia is twice rotor inertia.

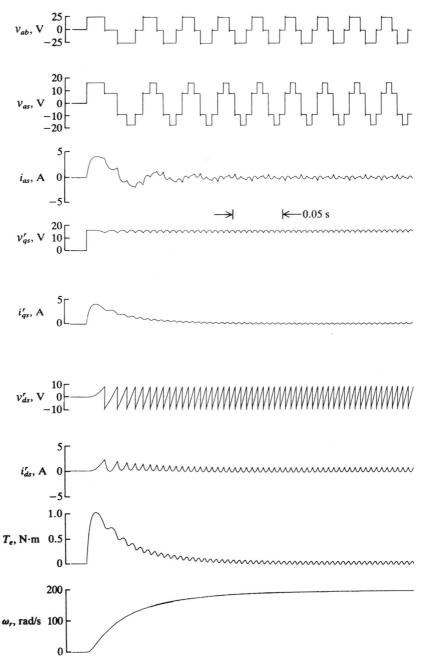

Figure 6.5-6 Free acceleration characteristics of a brushless dc motor supplied from a voltage source inverter ($\phi_v = 0$). Total inertia is five times rotor inertia. Compare with Fig. 6.5-3.

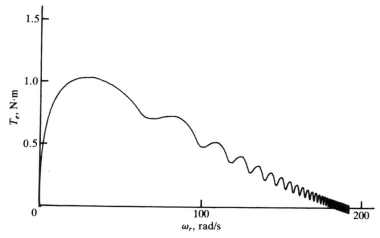

Figure 6.5-7 Torque–speed characteristics for free acceleration shown in Fig. 6.5-6.

established, the load torque is stepped back to 0.1 N·m. In these studies, the inertia is 2×10^{-4} kg·m^2, which is twice the inertia of the rotor.

Six-Step Phase Voltages

The free acceleration characteristics with the machine supplied from a typical voltage source inverter are shown in Fig. 6.5-6. The voltage v_{ds}^r is plotted in addition to the variables shown in Fig. 6.5-3. It is recalled that with the sinusoidal approximation, v_{ds}^r is zero for $\phi_v = 0$. We see in Fig. 6.5-6 that the average value of v_{ds}^r is zero. The torque versus speed characteristics for this free acceleration are shown in Fig. 6.5-7.

In order to compare with the sinusoidal approximation, the inverter voltage was selected so that the constant component of v_{qs}^r is equal to the value used in the case of ac applied stator voltages. It is interesting to note that during the initial acceleration period, in Fig. 6.5-6, before the first step (switching) occurs in v_{as}, the torque is larger than with ac voltages applied (Fig. 6.5-3). This is due to the fact that a constant voltage is applied to the phases until the first switching occurs, and this constant voltage is larger than the effective value of the ac voltage during the same interval.

The dynamic performance during load torque switching is shown in Fig. 6.5-8. As in Fig. 6.5-5 the inertia is twice the rotor inertia (2×10^{-4} kg · m^2) and the load torque is switched from 0.1 N · m to 0.4 N · m and then back to 0.1N · m. It is apparent that the sinusoidal approximation of the phase voltages is quite accurate in portraying dynamic and steady-state operation of a brushless dc motor supplied from a typical voltage source inverter. We will find that this fact is very useful for control design purposes because it justifies the neglecting of the harmonics due to the switching of the inverters.

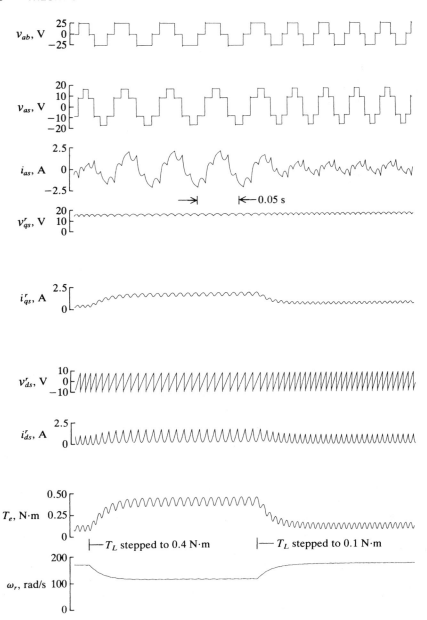

Figure 6.5-8 Dynamic performance during step changes in load torque of a brushless dc motor supplied from a voltage source inverter ($\phi_v = 0$). Total inertia is twice rotor inertia. Compare with Fig. 6.5-5.

REFERENCES

[1] P. L. Chapman and S. D. Sudhoff, Multiple Reference Frame Analysis of Brushless DC Machines with Non-Sinusoidal Back EMF Waveforms, *1998 SAE Transactions, Journal of Aerospace*, Section 1, pp. 145–153.

[2] P. L. Chapman, S. D. Sudhoff, and C. Whitcomb, Multiple Reference Frame Analysis of Non-Sinusoidal Brushless DC Drives, *IEEE Transactions on Energy Conversion*, Vol. 14, No. 3, September 1999, pp. 440–446.

[3] T. W. Nehl, F. A. Fouad, and N. A. Demerdash, Digital Simulation of Power Conditioner—Machine Interaction for Electronically Commutated DC Permanent Magnet Machines, *IEEE Transactions on Magnetics*, Vol. 17, November 1981, pp. 3284–3286.

[4] T. M. Jahns, Torque Production in Permanent-Magnet Sychronous Motor Drives with Rectangular Current Excitations, *IAS Conference Record*, October 1983, pp. 476–487.

[5] C. M. Ong, *Dynamic Simulation of Electric Machinery*, Prentice-Hall PTR, Upper Saddle River, NJ, 1998.

[6] J. S. Mayer and O. Wasynczuk, Analysis and Modeling of a Single-Phase Brushless DC Motor Drive System, *IEEE Transactions on Energy Conversion*, Vol. EC-4, September 1989, pp. 473–479.

[7] R. R. Nucera, S. D. Sudhoff, and P. C. Krause, Computation of Steady-State Performance of an Electronically Commutated Motor, *IEEE Transactions on Industry Applications*, Vol. 25, November/December 1989, pp. 1110–1117.

PROBLEMS

1 The device shown in Fig. 14.2-1 can be made to represent a 2-pole 2-phase brushless dc machine by eliminating the *cs* winding and moving the *bs* winding 30° clockwise. Following the procedure set forth in Section 6.2, write the voltage and torque equations in machine variables for this device.

2 Use the transformation given in Problem 1 of Chapter 3 to derive the voltage and torque equations in the rotor reference frame for the 2-phase, brushless dc machine described in Problem 1 above.

3 Derive the steady-state voltage equation for the 2-phase brushless dc machine.

4 Write the voltage equations given by (6.3-10)–(6.3-12) and the torque equation given by (6.3-13) in terms of flux linkages rather than currents.

5 Starting with the appropriate equations set forth in Section 6.3, develop a time-domain block diagram for the brushless dc machine. Consider v_{qs}^r, v_{ds}^r, and T_L as inputs.

6 Starting with the appropriate equations set forth in Section 6.3, develop the equations of state with v_{qs}^r, v_{ds}^r, and T_L as inputs.

7 A 4-pole, 2-phase brushless dc machine is driven by a mechanical source at $\omega_{rm} = 3600$ r/min. The open-circuit voltage across one phase is 50 V (rms). (*a*) Calculate λ_m'. The mechanical source is removed and the following voltages are applied: $V_{as} = \sqrt{2}\,25\cos\theta_r$; $V_{bs} = \sqrt{2}\,25\sin\theta_r$ where $\theta_r = \omega_r t$. (*b*) Let $B_m = 0$ and calculate the no-load rotor speed.

8 The parameters of a 4-pole, 3-phase brushless dc machine are $r_s = 3.4\,\Omega$, $L_{ls} = 1.1\,\text{mH}$ and $L_{mq} = L_{md} = 7.33\,\text{mH}$. When the device is driven by a mechanical source at $\omega_{rm} = 1000\,\text{r/min}$, the open-circuit winding-to-winding voltage is sinusoidal with a peak-to-peak value of 60 V. Determine λ'_m.

9 The parameters of a 2-pole, 2-phase brushless dc motor are as follows: $r_s = 2\,\Omega$, $\lambda'_m = 0.0707\,\text{V} \cdot \text{s/rad}$, $L_{ls} = 1\,\text{mH}$, $L_{mq} = L_{md} = 9\,\text{mH}$. The applied stator voltages are $V_{as} = \sqrt{2}\,20\cos{(\omega_r t + \phi_{vMT})}$ and $V_{bs} = \sqrt{2}\,20\sin{(\omega_r t + \phi_{vMT})}$. (a) Calculate $\tilde{V}_{as}$, $\tilde{I}_{as}$, and T_e when $\omega_r = 200\,\text{rad/s}$. (b) Draw the phasor diagram for this mode of operation; show all voltages.

10 The parameters of a 4-pole, 2-phase brushless dc motor are $r_s = 3.4\,\Omega$, $L_{ls} = 1.1\,\text{mH}$, $L_{mq} = L_{md} = 11\,\text{mH}$, and $\lambda'_m = 0.0826\,\text{V} \cdot \text{s/rad}$. The peak-to-peak value of the stator applied voltages in 45 V and $\phi_v = 0$. The load torque is 0.067 N·m and $B_m = 0$. Calculate and compare the steady-state value of ω_r with (a) the $\omega_r L_d I'_{ds}$ term neglected in the steady-state voltage equation and (b) this term included in the voltage equation.

11 Note in Fig. 6.4-3 that the steady-state torque versus speed plots for $\phi_v = 0$ and for $\phi_v = \pi/2$ intersect. Calculate the rotor speed at which this intersection occurs.

12 Show that the maximum torque plotted in Fig. 6.4-5 is three times the maximum torque plotted in Fig. 6.4-6.

13 Show that for ϕ_{vMT} set for maximum torque, the phase current leads the phase voltage when $\omega_r > \sqrt{2}V_s/2\cos\phi_v\lambda''^{r}_m$.

Chapter 7

MACHINE EQUATIONS IN OPERATIONAL IMPEDANCES AND TIME CONSTANTS

7.1 INTRODUCTION

In Chapter 5 we assumed that the electrical characteristics of the rotor of a synchronous machine could be portrayed by two windings in each axis. This type of a representation is sufficient for most applications; however, there are instances where a more refined model may be necessary. For example, when representing solid iron rotor machines it may be necessary to use three or more rotor windings in each axis.

R. H. Park [1] in his original paper did not specify the number of rotor circuits. Instead, he expressed the stator flux linkages in terms of operational impedances and a transfer function relating stator flux linkages to field voltage. In other words, Park recognized that, in general, the rotor of a synchronous machine appears as a distributed parameter system when viewed from the stator. The fact that an accurate, equivalent lumped parameter circuit representation of the rotor of a synchronous machine might require two, three, or four damper windings was more or less of academic interest until large digital computers became available. Prior to the 1970s the damper windings were seldom considered in stability studies; however, as the size of computers increased, it became desirable to represent the machine in more detail.

The standard short-circuit test, which involves monitoring the stator short-circuit currents, provides information from which the parameters of the field winding and one damper winding in the d axis can be determined. The parameters for the q-axis damper winding are calculated from design data. Due to the need for more accurate parameters, frequency-response data are now being used as means of measuring the operational impedances from which the parameters can be obtained for any number of rotor windings in both axes.

In this chapter the operational impedances as set forth by Park [1] are described. The standard and derived synchronous machine time constants are defined, and their relationship to the operational impedances is established. Finally, a method of approximating the measured operational impedances by lumped parameter rotor circuits is presented.

7.2   PARK'S EQUATIONS IN OPERATIONAL FORM

R. H. Park [1] published the original $qd0$-voltage equations in the form

$$v_{qs}^r = -r_s i_{qs}^r + \frac{\omega_r}{\omega_b}\psi_{ds}^r + \frac{p}{\omega_b}\psi_{qs}^r \tag{7.2-1}$$

$$v_{ds}^r = -r_s i_{ds}^r - \frac{\omega_r}{\omega_b}\psi_{qs}^r + \frac{p}{\omega_b}\psi_{ds}^r \tag{7.2-2}$$

$$v_{0s} = -r_s i_{0s} + \frac{p}{\omega_b}\psi_{0s} \tag{7.2-3}$$

where

$$\psi_{qs}^r = -X_q(p)i_{qs}^r \tag{7.2-4}$$

$$\psi_{ds}^r = -X_d(p)i_{ds}^r + G(p)v_{fd}^{\prime r} \tag{7.2-5}$$

$$\psi_{0s} = -X_{ls}i_{0s} \tag{7.2-6}$$

In these equations, the operator $X_q(p)$ is referred to as the q-axis operational impedance, $X_d(p)$ is the d-axis operational impedance, and $G(p)$ is a dimensionless transfer function relating stator flux linkages per second to field voltage.

With the equations written in this form the rotor of a synchronous machine can be considered as either a distributed or lumped parameter system. Over the years, the electrical characteristics of the rotor have often been approximated by three lumped parameter circuits, one field winding and two damper windings, one in each axis. Although this type of representation is generally adequate for salient-pole machines, it does not suffice for a solid iron rotor machine. It now appears that for dynamic and transient stability considerations, at least two and perhaps three damper windings should be used in the q axis for solid rotor machines with a field and two damper windings in the d axis [2].

7.3   OPERATIONAL IMPEDANCES AND $G(p)$ FOR A SYNCHRONOUS MACHINE WITH FOUR ROTOR WINDINGS

In Chapter 5 the synchronous machine was represented with a field winding and one damper winding in the d axis and with two damper windings in the q axis. It is helpful to determine $X_q(p)$, $X_d(p)$, and $G(p)$ for this type of rotor representation before

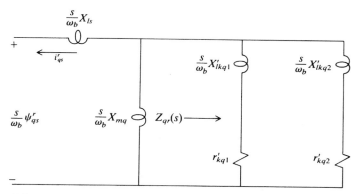

Figure 7.3-1   Equivalent circuit with two damper windings in the quadrature axis.

deriving the lumped parameter approximations from measured frequency-response data. For this purpose it is convenient to consider the network shown in Fig. 7.3-1. It is helpful in this and following derivations to express the input impedance of the rotor circuits in the form

$$Z_{qr}(s) = R_{eq} \frac{(1 + \tau_{qa}s)(1 + \tau_{qb}s)}{(1 + \tau_{Qa}s)} \tag{7.3-1}$$

Because it is customary to use the Laplace operator s rather than the operator p, Laplace notation will be employed hereafter. In (7.3-1)

$$R_{eq} = \frac{r'_{kq1} r'_{kq2}}{r'_{kq1} + r'_{kq2}} \tag{7.3-2}$$

$$\tau_{qa} = \frac{X'_{lkq1}}{\omega_b r'_{kq1}} \tag{7.3-3}$$

$$\tau_{qb} = \frac{X'_{lkq2}}{\omega_b r'_{kq2}} \tag{7.3-4}$$

$$\tau_{Qa} = \frac{X'_{lkq1} + X'_{lkq2}}{\omega_b(r'_{kq1} + r'_{kq2})} = R_{eq}\left(\frac{\tau_{qa}}{r'_{kq2}} + \frac{\tau_{qb}}{r'_{kq1}}\right) \tag{7.3-5}$$

From Fig. 7.3-1

$$\frac{sX_q(s)}{\omega_b} = \frac{sX_{ls}}{\omega_b} + \frac{(sX_{mq}/\omega_b)Z_{qr}(s)}{Z_{qr}(s) + (sX_{mq}/\omega_b)} \tag{7.3-6}$$

Solving the above equation for $X_q(s)$ yields the operational impedance for two damper windings in the q axis, which can be expressed as

$$X_q(s) = X_q \frac{1 + (\tau_{q4} + \tau_{q5})s + \tau_{q4}\tau_{q6}s^2}{1 + (\tau_{q1} + \tau_{q2})s + \tau_{q1}\tau_{q3}s^2} \tag{7.3-7}$$

where

$$\tau_{q1} = \frac{1}{\omega_b r'_{kq1}}(X'_{lkq1} + X_{mq}) \tag{7.3-8}$$

$$\tau_{q2} = \frac{1}{\omega_b r'_{kq2}}(X'_{lkq2} + X_{mq}) \tag{7.3-9}$$

$$\tau_{q3} = \frac{1}{\omega_b r'_{kq2}}\left(X'_{lkq2} + \frac{X_{mq}X'_{lkq1}}{X'_{lkq1} + X_{mq}}\right) \tag{7.3-10}$$

$$\tau_{q4} = \frac{1}{\omega_b r'_{kq1}}\left(X'_{lkq1} + \frac{X_{mq}X_{ls}}{X_{ls} + X_{mq}}\right) \tag{7.3-11}$$

$$\tau_{q5} = \frac{1}{\omega_b r'_{kq2}}\left(X'_{lkq2} + \frac{X_{mq}X_{ls}}{X_{ls} + X_{mq}}\right) \tag{7.3-12}$$

$$\tau_{q6} = \frac{1}{\omega_b r'_{kq2}}\left(X'_{lkq2} + \frac{X_{mq}X_{ls}X'_{lkq1}}{X_{mq}X_{ls} + X_{mq}X'_{lkq1} + X_{ls}X'_{lkq1}}\right) \tag{7.3-13}$$

The d-axis operational impedance $X_d(s)$ may be calculated for the machine with a field and a damper winding by the same procedure. In particular, from Fig. 7.3-2a

$$Z_{dr}(s) = R_{ed}\frac{(1 + \tau_{da}s)(1 + \tau_{db}s)}{(1 + \tau_{Da}s)} \tag{7.3-14}$$

where

$$R_{ed} = \frac{r'_{fd}r'_{kd}}{r'_{fd} + r'_{kd}} \tag{7.3-15}$$

$$\tau_{da} = \frac{X'_{lfd}}{\omega_b r'_{fd}} \tag{7.3-16}$$

$$\tau_{db} = \frac{X'_{lkd}}{\omega_b r'_{kd}} \tag{7.3-17}$$

$$\tau_{Da} = \frac{X'_{lfd} + X'_{lkd}}{\omega_b(r'_{fd} + r'_{kd})} = R_{ed}\left(\frac{\tau_{da}}{r'_{kd}} + \frac{\tau_{db}}{r'_{fd}}\right) \tag{7.3-18}$$

The operational impedance for a field and damper winding in the d axis can be obtained by setting v''^{r}_{fd} to zero and following the same procedure as in the case of the q axis. The final expression is

$$X_d(s) = X_d\frac{1 + (\tau_{d4} + \tau_{d5})s + \tau_{d4}\tau_{d6}s^2}{1 + (\tau_{d1} + \tau_{d2})s + \tau_{d1}\tau_{d3}s^2} \tag{7.3-19}$$

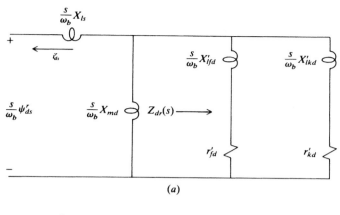

(a)

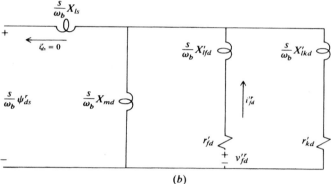

(b)

Figure 7.3-2 Calculation of $X_d(s)$ and $G(s)$ for two rotor windings in direct axis. (a) Calculation of $X_d(s)$; $v''_{fd} = 0$; (b) calculation of $G(s)$; $i^r_{ds} = 0$.

where

$$\tau_{d1} = \frac{1}{\omega_b r'_{fd}} (X'_{lfd} + X_{md}) \tag{7.3-20}$$

$$\tau_{d2} = \frac{1}{\omega_b r'_{kd}} (X'_{lkd} + X_{md}) \tag{7.3-21}$$

$$\tau_{d3} = \frac{1}{\omega_b r'_{kd}} \left(X'_{lkd} + \frac{X_{md} X'_{lfd}}{X'_{lfd} + X_{md}} \right) \tag{7.3-22}$$

$$\tau_{d4} = \frac{1}{\omega_b r'_{fd}} \left(X'_{lfd} + \frac{X_{md} X_{ls}}{X_{ls} + X_{md}} \right) \tag{7.3-23}$$

$$\tau_{d5} = \frac{1}{\omega_b r'_{kd}} \left(X'_{lkd} + \frac{X_{md} X_{ls}}{X_{ls} + X_{md}} \right) \tag{7.3-24}$$

$$\tau_{d6} = \frac{1}{\omega_b r'_{kd}} \left(X'_{lkd} + \frac{X_{md} X_{ls} X'_{lfd}}{X_{md} X_{ls} + X_{md} X'_{lfd} + X_{ls} X'_{lfd}} \right) \tag{7.3-25}$$

The transfer function $G(s)$ may be evaluated by expressing the relationship between stator flux linkages per second to field voltage, $v_{fd}^{\prime r}$, with i_{ds}^r equal to zero. Hence, from (7.2-5) we obtain

$$G(s) = \frac{\psi_{ds}^r}{v_{fd}^{\prime r}}\bigg|_{i_{ds}^r=0} \qquad (7.3\text{-}26)$$

From Fig. 7.3-2*b*, this yields

$$G(s) = \frac{X_{md}}{r_{fd}^{\prime}}\frac{1+\tau_{db}s}{1+(\tau_{d1}+\tau_{d2})s+\tau_{d1}\tau_{d3}s^2} \qquad (7.3\text{-}27)$$

where τ_{db} is defined by (7.3-17).

7.4 STANDARD SYNCHRONOUS MACHINE REACTANCES

It is instructive to set forth the commonly used reactances for the four-winding rotor synchronous machine and to relate these reactances to the operational impedances whenever appropriate. The *q*- and *d*-axis reactances are

$$X_q = X_{ls} + X_{mq} \qquad (7.4\text{-}1)$$

$$X_d = X_{ls} + X_{md} \qquad (7.4\text{-}2)$$

These reactances were defined in Section 5.5. They characterize the machine during balanced steady-state operation, whereupon variables in the rotor reference frame are constants. The zero frequency value of $X_q(s)$ or $X_d(s)$ is found by replacing the operator s with zero. Hence, the operational impedances for balanced steady-state operation are

$$X_q(0) = X_q \qquad (7.4\text{-}3)$$

$$X_d(0) = X_d \qquad (7.4\text{-}4)$$

Similarly, the steady-state value of the transfer function is

$$G(0) = \frac{X_{md}}{r_{fd}^{\prime}} \qquad (7.4\text{-}5)$$

The *q*- and *d*-axis transient reactances are defined as follows:

$$X_q' = X_{ls} + \frac{X_{mq}X_{lkq1}'}{X_{lkq1}' + X_{mq}} \qquad (7.4\text{-}6)$$

$$X_d' = X_{ls} + \frac{X_{md}X_{lfd}'}{X_{lfd}' + X_{md}} \qquad (7.4\text{-}7)$$

Although X'_q has not been defined previously, we did encounter the d-axis transient reactance in the derivation of the approximate transient torque–angle characteristic in Chapter 5.

The q- and d-axis subtransient reactances are defined as follows:

$$X''_q = X_{ls} + \frac{X_{mq}X'_{lkq1}X'_{lkq2}}{X_{mq}X'_{lkq1} + X_{mq}X'_{lkq2} + X'_{lkq1}X'_{lkq2}} \qquad (7.4\text{-}8)$$

$$X''_d = X_{ls} + \frac{X_{md}X'_{lfd}X'_{lkd}}{X_{md}X'_{lfd} + X_{md}X'_{lkd} + X'_{lfd}X'_{lkd}} \qquad (7.4\text{-}9)$$

These reactances are the high-frequency asymptotes of the operational impedances; that is,

$$X_q(\infty) = X''_q \qquad (7.4\text{-}10)$$

$$X_d(\infty) = X''_d \qquad (7.4\text{-}11)$$

The high-frequency response of the machine is characterized by these reactances. It is interesting that $G(\infty)$ is zero, which indicates that the stator flux linkages are essentially insensitive to high-frequency changes in field voltage. Primes are used to denote transient and subtransient quantities that can be confused with rotor quantities referred to the stator windings by a turns ratio. It is hoped that this confusion is minimized by the fact that X'_d and X'_q are the only single-primed parameters that are not referred impedances.

Although the steady-state and subtransient reactances can be related to the operational impedances, this is not the case with the transient reactances. It appears that the d-axis transient reactance evolved from Doherty and Nickle's [3] development of an approximate transient torque–angle characteristic where the effects of d-axis damper windings are neglected. The q-axis transient reactance has come into use only recently when it became desirable to portray more accurately the dynamic characteristics of the solid iron rotor machine in transient stability studies. In many of the early studies, only one damper winding was used to describe the electrical characteristics of the q axis, which is generally adequate in the case of salient-pole machines. In our earlier development, we implied a notational correspondence between the $kq1$ and the fd windings and between the $kq2$ and the kd windings. In this chapter we have associated the $kq1$ winding with the transient reactance (7.4-6) and associated the $kq2$ winding with the subtransient reactance (7.4-8). Therefore, it seems logical to use only the $kq2$ winding when one damper winding is deemed adequate to portray the electrical characteristics of the q axis. It is recalled that in Chapter 5 we chose to use the $kq2$ winding rather than the $kq1$ winding in the case of the salient-pole hydro turbine generator.

It is perhaps apparent that the subtransient reactances characterize the equivalent reactances of the machine during a very short period of time following an electrical disturbance. After a period, of perhaps a few milliseconds, the machine equivalent reactances approach thes of the transient reactances and even though they are not directly related to $X_q(s)$ and $X_d(s)$, their values lie between the subtransient and steady-states values. As more time elapses after a disturbance, the transient reactances give way to the steady-state reactances. In Chapter 5 we observed the impedance of the machine "changing" from transient to steady state following a system disturbance. Clearly, the use of the transient and subtransient quantities to portray the behavior of the machine over specific time intervals was a direct result of the need to simplify the machine equations so that precomputer computational techniques could be used.

7.5 STANDARD SYNCHRONOUS MACHINE TIME CONSTANTS

The standard time constants associated with a four-rotor winding synchronous machine are given in Table 7.5-1. These time constants are defined as follows:

τ'_{qo} and τ'_{do} are the q- and d-axis transient open-circuit time constants.

τ''_{qo} and τ''_{do} are the q- and d-axis subtransient open-circuit time constants.

Table 7.5-1   Standard Synchronous Machine Time Constants

Open-Circuit Time Constants

$$\tau'_{qo} = \frac{1}{\omega_b r'_{kq1}}\left(X'_{lkq1} + X_{mq}\right)$$

$$\tau'_{do} = \frac{1}{\omega_b r'_{fd}}\left(X'_{lfd} + X_{md}\right)$$

$$\tau''_{qo} = \frac{1}{\omega_b r'_{kq2}}\left(X'_{lkq2} + \frac{X_{mq}X'_{lkq1}}{X_{mq} + X'_{lkq1}}\right)$$

$$\tau''_{do} = \frac{1}{\omega_b r'_{kd}}\left(X'_{lkd} + \frac{X_{md}X'_{lfd}}{X_{md} + X'_{lfd}}\right)$$

Short-Circuit Time Constants

$$\tau'_q = \frac{1}{\omega_b r'_{kq1}}\left(X'_{lkq1} + \frac{X_{mq}X_{ls}}{X_{mq} + X_{ls}}\right)$$

$$\tau'_d = \frac{1}{\omega_b r'_{fd}}\left(X'_{lfd} + \frac{X_{md}X_{ls}}{X_{md} + X_{ls}}\right)$$

$$\tau''_q = \frac{1}{\omega_b r'_{kq2}}\left(X'_{lkq2} + \frac{X_{mq}X_{ls}X'_{lkq1}}{X_{mq}X_{ls} + X_{mq}X'_{lkq1} + X_{ls}X'_{lkq1}}\right)$$

$$\tau''_d = \frac{1}{\omega_b r'_{kd}}\left(X'_{lkd} + \frac{X_{md}X_{ls}X'_{lfd}}{X_{md}X_{ls} + X_{md}X'_{lfd} + X_{ls}X'_{lfd}}\right)$$

τ'_q and τ'_d are the q- and d-axis transient short-circuit time constants.
τ''_q and τ''_d are the q- and d-axis subtransient short-circuit time constants.

In the above definitions, open circuit and short circuit refer to the conditions of the stator circuits. All of these time constants are approximations of the actual time constants; and when used to determine the machine parameters, they can lead to substantial errors in predicting the dynamic behavior of a synchronous machine. More accurate expressions for the time constants are derived in the following section.

7.6 DERIVED SYNCHRONOUS MACHINE TIME CONSTANTS

The open-circuit time constants, which characterize the duration of transient changes of machine variables during open-circuit conditions, are the reciprocals of the roots of the characteristic equation associated with the operational impedances which, of course, are the poles of the operational impedances. The roots of the denominators of $X_q(s)$ and $X_d(s)$ can be found by setting these second-order polynomials equal to zero. From $X_q(s)$, given by (7.3-7), we have

$$s^2 + \frac{\tau_{q1} + \tau_{q2}}{\tau_{q1}\tau_{q3}} s + \frac{1}{\tau_{q1}\tau_{q3}} = 0 \tag{7.6-1}$$

From $X_d(s)$, given by (7.3-19), we obtain

$$s^2 + \frac{\tau_{d1} + \tau_{d2}}{\tau_{d1}\tau_{d3}} s + \frac{1}{\tau_{d1}\tau_{d3}} = 0 \tag{7.6-2}$$

The roots are of the form

$$s = -\frac{b}{2} \pm \frac{b}{2}\sqrt{1 - \frac{4c}{b^2}} \tag{7.6-3}$$

The exact solution of (7.6-3) is quite involved. It can be simplified, however, if the quantity $4c/b^2$ is much less than unity [4]. In the case of the q axis we have

$$\frac{4c}{b^2} = \frac{4\tau_{q1}\tau_{q3}}{\left(\tau_{q1} + \tau_{q2}\right)^2} \tag{7.6-4}$$

It can be shown that

$$\frac{4\tau_{q1}\tau_{q3}}{\left(\tau_{q1} + \tau_{q2}\right)^2} \approx \frac{4r'_{kq1}r'_{kq2}(X'_{lkq1} + X'_{lkq2})}{X_{mq}(r'_{kq1} + r'_{kq2})^2} \tag{7.6-5}$$

In the case of the d axis we have

$$\frac{4\tau_{d1}\tau_{d3}}{\left(\tau_{d1} + \tau_{d2}\right)^2} \approx \frac{4r'_{fd}r'_{kd}(X'_{lfd} + X'_{lkd})}{X_{md}(r'_{fd} + r'_{kd})^2} \tag{7.6-6}$$

In most cases the right-hand side of (7.6-5) and (7.6-6) is much less than unity. Hence, the solution of (7.6-3) with $4c/b^2 \ll 1$ and $c/b \ll b$ is obtained by employing the binomial expansion, from which

$$s_1 = -\frac{c}{b} \tag{7.6-7}$$

$$s_2 = -b \tag{7.6-8}$$

Now, the reciprocals of the roots are the time constants and if we define the transient open-circuit time constant as the largest time constant and the subtransient open-circuit time constant as the smallest, then

$$\tau'_{qo} = \frac{b}{c} = \tau_{q1} + \tau_{q2} \tag{7.6-9}$$

and

$$\tau''_{qo} = \frac{1}{b} = \frac{\tau_{q3}}{1 + \tau_{q2}/\tau_{q1}} \tag{7.6-10}$$

Similarly, the d-axis open-circuit time constants are

$$\tau'_{do} = \tau_{d1} + \tau_{d2} \tag{7.6-11}$$

$$\tau''_{do} = \frac{\tau_{d3}}{1 + \tau_{d2}/\tau_{d1}} \tag{7.6-12}$$

The above derived open-circuit time constants are expressed in terms of machine parameters in Table 7.6-1.

The short-circuit time constants are defined as the reciprocals of the roots of the numerator of the operational impedances. Although the stator resistance should be included in the calculation of the short-circuit time constants, its influence is generally small. From $X_q(s)$, given by (7.3-7), we have

$$s^2 + \frac{\tau_{q4} + \tau_{q5}}{\tau_{q4}\tau_{q6}} s + \frac{1}{\tau_{q4}\tau_{q6}} = 0 \tag{7.6-13}$$

From $X_d(s)$, given by (7.3-19), we obtain

$$s^2 + \frac{\tau_{d4} + \tau_{d5}}{\tau_{d4}\tau_{d6}} s + \frac{1}{\tau_{d4}\tau_{d6}} = 0 \tag{7.6-14}$$

The roots are of the form given by (7.6-3) and, as in the case of the open-circuit time constants, $4c/b^2 \ll 1$ and $c/b \ll b$. Hence

Table 7.6-1   Derived Synchronous Machine Time Constants

Open-Circuit Time Constants

$$\tau'_{qo} = \frac{1}{\omega_b r'_{kq1}}(X'_{lkq1} + X_{mq}) + \frac{1}{\omega_b r'_{kq2}}(X'_{lkq2} + X_{mq})$$

$$\tau'_{do} = \frac{1}{\omega_b r'_{fd}}(X'_{lfd} + X_{md}) + \frac{1}{\omega_b r'_{kd}}(X'_{lkd} + X_{md})$$

$$\tau''_{qo} = \frac{(1/\omega_b r'_{kq2})[X'_{lkq2} + (X_{mq}X'_{lkq1}/X'_{lkq1} + X_{mq})]}{1 + [(1/\omega_b r'_{kq2})(X'_{lkq2} + X_{mq})/(1/\omega_b r'_{kq1})(X'_{lkq1} + X_{mq})]}$$

$$\tau''_{do} = \frac{(1/\omega_b r'_{kd})[X'_{lkd} + (X_{md}X'_{lfd}/X'_{lfd} + X_{md})]}{1 + [(1/\omega_b r'_{kd})(X'_{lkd} + X_{md})/(1/\omega_b r'_{fd})(X'_{lfd} + X_{md})]}$$

Short-Circuit Time Constants

$$\tau'_q = \frac{1}{\omega_b r'_{kq1}}\left(X'_{lkq1} + \frac{X_{mq}X_{ls}}{X_{ls} + X_{mq}}\right) + \frac{1}{\omega_b r'_{kq2}}\left(X'_{lkq2} + \frac{X_{mq}X_{ls}}{X_{ls} + X_{mq}}\right)$$

$$\tau'_d = \frac{1}{\omega_b r'_{fd}}\left(X'_{lfd} + \frac{X_{md}X_{ls}}{X_{ls} + X_{md}}\right) + \frac{1}{\omega_b r'_{kd}}\left(X'_{lkd} + \frac{X_{md}X_{ls}}{X_{ls} + X_{md}}\right)$$

$$\tau''_q = \frac{(1/\omega_b r'_{lkq2})[X'_{lkq2} + (X_{mq}X_{ls}X'_{lkq1}/X_{mq}X_{ls} + X_{mq}X'_{lkq1} + X_{ls}X'_{lkq1})]}{1 + \{(1/\omega_b r'_{kq2})[X'_{lkq2} + (X_{mq}X_{ls}/X_{ls} + X_{mq})]/(1/\omega_b r'_{kq1})[X'_{lkq1} + (X_{mq}X_{ls}/X_{ls} + X_{mq})]\}}$$

$$\tau''_d = \frac{(1/\omega_b r'_{kd})[X'_{lkd} + (X_{md}X_{ls}X'_{lfd}/X_{md}X_{ls} + X_{md}X'_{lfd} + X_{ls}X'_{lfd})]}{1 + \{(1/\omega_b r'_{kd})[X'_{lkd} + (X_{md}X_{ls}/X_{ls} + X_{md})]/(1/\omega_b r'_{fd})[X'_{lfd} + (X_{md}X_{ls}/X_{ls} + X_{md})]\}}$$

$$\tau'_q = \tau_{q4} + \tau_{q5} \qquad (7.6\text{-}15)$$

$$\tau''_q = \frac{\tau_{q6}}{1 + \tau_{q5}/\tau_{q4}} \qquad (7.6\text{-}16)$$

$$\tau'_d = \tau_{d4} + \tau_{d5} \qquad (7.6\text{-}17)$$

$$\tau''_d = \frac{\tau_{d6}}{1 + \tau_{d5}/\tau_{d4}} \qquad (7.6\text{-}18)$$

The above derived synchronous machine time constants are given in Table 7.6-1 in terms of machine parameters. It is important to note that the standard machine time constants given in Table 7.5-1 are considerably different from the more accurate derived time constants. The standard time constants are acceptable approximations of the derived time constants if

$$r'_{kq2} \gg r'_{kq1} \qquad (7.6\text{-}19)$$

and

$$r'_{kd} \gg r'_{fd} \tag{7.6-20}$$

In the lumped parameter approximation of the rotor circuits, r'_{kd} is generally much larger than r'_{fd}, and therefore the standard d-axis time constants are often good approximations of the derived time constants. This is not the case for the q-axis lumped parameter approximation of the rotor circuits. That is, r'_{kq2} is seldom, if ever, larger than r'_{kq1}; hence the standard q-axis time constants are generally poor approximations of the derived time constants.

7.7 PARAMETERS FROM SHORT-CIRCUIT CHARACTERISTICS

It is common to determine the d-axis parameters from a short-circuit test performed on an unloaded synchronous machine [5]. The method of determining these parameters from test data is set forth in this section.

If the speed of the machine is constant, then (7.2-1)–(7.2-6) form a set of linear differential equations that can be solved using linear system theory. Prior to the short circuit of the stator terminals, the machine variables are in the steady state and the stator terminals are open-circuited. If the field voltage is held fixed at its prefault value, then the Laplace transform of the change in v''_{fd} is zero. Hence, if the terms involving r_s^2 are neglected, the Laplace transform of the fault currents, for the constant speed operation $(\omega_r = \omega_b)$, may be expressed as

$$i^r_{qs}(s) = -\frac{1/X_q(s)}{s^2 + 2\alpha s + \omega_b^2} \left[\frac{\omega_b^2 r_s v^r_{qs}(s)}{X_d(s)} + \omega_b s v^r_{qs}(s) - \omega_b^2 v^r_{ds}(s) \right] \tag{7.7-1}$$

$$i^r_{ds}(s) = -\frac{1/X_d(s)}{s^2 + 2\alpha s + \omega_b^2} \left[\frac{\omega_b^2 r_s v^r_{ds}(s)}{X_q(s)} + \omega_b s v^r_{ds}(s) + \omega_b^2 v^r_{qs}(s) \right] \tag{7.7-2}$$

where

$$\alpha = \frac{\omega_b r_s}{2} \left(\frac{1}{X_q(s)} + \frac{1}{X_d(s)} \right) \tag{7.7-3}$$

It is clear that the 0 quantities are zero for a 3-phase fault at the stator terminals. It is also clear that ω_r, ω_b, and ω_e are all equal in this example.

Initially the machine is operating open-circuited, hence

$$v^r_{qs} = \sqrt{2} V_s \tag{7.7-4}$$

$$v^r_{ds} = 0 \tag{7.7-5}$$

The 3-phase fault appears as a step decrease in v_{qs}^r to zero. Therefore, the Laplace transform of the change in the voltages from the prefault to fault values is

$$v_{qs}^r(s) = -\frac{\sqrt{2}V_s}{s} \qquad (7.7\text{-}6)$$

$$v_{ds}^r(s) = 0 \qquad (7.7\text{-}7)$$

If (7.7-6) and (7.7-7) are substituted into (7.7-1) and (7.7-2) and if the terms involving r_s are neglected except in α wherein the operational impedances are replaced by their high-frequency asymptotes, the Laplace transform of the short-circuit currents becomes

$$i_{qs}^r(s) = \frac{1/X_q(s)}{s^2 + 2\alpha s + \omega_b^2} \left(\omega_b \sqrt{2}V_s \right) \qquad (7.7\text{-}8)$$

$$i_{ds}^r(s) = \frac{1/X_d(s)}{s^2 + 2\alpha s + \omega_b^2} \left(\frac{\omega_b^2 \sqrt{2}V_s}{s} \right) \qquad (7.7\text{-}9)$$

where

$$\alpha = \frac{\omega_b r_s}{2} \left(\frac{1}{X_q(\infty)} + \frac{1}{X_d(\infty)} \right) \qquad (7.7\text{-}10)$$

Replacing the operational impedances with their high-frequency asymptotes in α is equivalent to neglecting the effects of the rotor resistances in α.

If we now assume that the electrical characteristics of the synchronous machine can be portrayed by two rotor windings in each axis, then we can express the operational impedances in terms of time constants. It is recalled that the open- and short-circuit time constants are, respectively, the reciprocals of the roots of the denominator and numerator of the operational impedances. Therefore, the reciprocals of the operational impedances may be expressed as

$$\frac{1}{X_q(s)} = \frac{1}{X_q} \frac{(1 + \tau_{qo}'s)(1 + \tau_{qo}''s)}{(1 + \tau_q's)(1 + \tau_q''s)} \qquad (7.7\text{-}11)$$

$$\frac{1}{X_d(s)} = \frac{1}{X_d} \frac{(1 + \tau_{do}'s)(1 + \tau_{do}''s)}{(1 + \tau_d's)(1 + \tau_d''s)} \qquad (7.7\text{-}12)$$

These expressions may be written as [6]

$$\frac{1}{X_q(s)} = \frac{1}{X_q} \left(1 + \frac{As}{1 + \tau_q's} + \frac{Bs}{1 + \tau_q''s} \right) \qquad (7.7\text{-}13)$$

$$\frac{1}{X_d(s)} = \frac{1}{X_d} \left(1 + \frac{Cs}{1 + \tau_d's} + \frac{Ds}{1 + \tau_d''s} \right) \qquad (7.7\text{-}14)$$

where

$$A = -\frac{\tau_q'(1 - \tau_{qo}'/\tau_q')(1 - \tau_{qo}''/\tau_q')}{1 - \tau_q''/\tau_q'} \qquad (7.7\text{-}15)$$

$$B = -\frac{\tau_q''(1 - \tau_{qo}'/\tau_q'')(1 - \tau_{qo}''/\tau_q'')}{1 - \tau_q'/\tau_q''} \qquad (7.7\text{-}16)$$

The constants C and D are identical to A and B, respectively, with the q subscript replaced by d in all time constants.

Because the subtransient time constants are considerably smaller than the transient time constants, (7.7-13) and (7.7-14) may be approximated by

$$\frac{1}{X_q(s)} = \frac{1}{X_q} + \left(\frac{\tau_{qo}'}{\tau_q'}\frac{1}{X_q} - \frac{1}{X_q}\right)\frac{\tau_q's}{1 + \tau_q's} + \left(\frac{1}{X_q''} - \frac{\tau_{qo}'}{\tau_q'}\frac{1}{X_q}\right)\frac{\tau_q''s}{1 + \tau_q''s} \qquad (7.7\text{-}17)$$

$$\frac{1}{X_d(s)} = \frac{1}{X_d} + \left(\frac{\tau_{do}'}{\tau_d'}\frac{1}{X_d} - \frac{1}{X_d}\right)\frac{\tau_d's}{1 + \tau_d's} + \left(\frac{1}{X_d''} - \frac{\tau_{do}'}{\tau_d'}\frac{1}{X_d}\right)\frac{\tau_d''s}{1 + \tau_d''s} \qquad (7.7\text{-}18)$$

Although the assumption that the subtransient time constants are much smaller than the transient time constants is appropriate in the case of the d-axis time constants, the difference is not as large in the case of the q-axis time constants. Hence, (7.7-17) is a less acceptable approximation than is (7.7-18). This inaccuracy will not influence our work in this section, however. Also, because we have not restricted the derivation as far as time constants are concerned, either the standard or derived time constants can be used in the equations given in this section. However, if the approximate standard time constants are used, $(\tau_{qo}'/\tau_q')(1/X_q)$ and $(\tau_{do}'/\tau_d')(1/X_d)$ can be replaced by $1/X_q'$ and $1/X_d'$, respectively.

If (7.7-17) and (7.7-18) are appropriately substituted into (7.7-8) and (7.7-9), the fault currents in terms of the Laplace operator become

$$i_{qs}^r(s) = \left(\frac{\sqrt{2}V_s}{s}\right)\left(\frac{\omega_b s}{s^2 + 2\alpha s + \omega_b^2}\right)\left[\frac{1}{X_q} + \left(\frac{\tau_{qo}'}{\tau_q'}\frac{1}{X_q} - \frac{1}{X_q}\right)\frac{\tau_q's}{1 + \tau_q's}\right.$$

$$\left. + \left(\frac{1}{X_q''} - \frac{\tau_{qo}'}{\tau_q'}\frac{1}{X_q}\right)\frac{\tau_q''s}{1 + \tau_q''s}\right] \qquad (7\text{-}7.19)$$

$$i_{ds}^r(s) = \left(\frac{\sqrt{2}V_s}{s}\right)\left(\frac{\omega_b^2}{s^2 + 2\alpha s + \omega_b^2}\right)\left[\frac{1}{X_d} + \left(\frac{\tau_{do}'}{\tau_d'}\frac{1}{X_d} - \frac{1}{X_d}\right)\frac{\tau_d's}{1 + \tau_d's}\right.$$

$$\left. + \left(\frac{1}{X_d''} - \frac{\tau_{do}'}{\tau_d'}\frac{1}{X_d}\right)\frac{\tau_d''s}{1 + \tau_d''s}\right] \qquad (7.7\text{-}20)$$

Equations (7.7-19) and (7.7-20) may be transformed to the time domain by the following inverse Laplace transforms. If a and α are much less than ω_b, then

$$L^{-1}\left[\frac{\omega_b s}{(s+a)(s^2+2\alpha s+\omega_b^2)}\right] = e^{-\alpha t}\sin\omega_b t \qquad (7.7\text{-}21)$$

$$L^{-1}\left[\frac{\omega_b^2}{(s+a)(s^2+2\alpha s+\omega_b^2)}\right] = e^{-at} - e^{-\alpha t}\cos\omega_b t \qquad (7.7\text{-}22)$$

If (7.7-21) is applied term by term to (7.7-19) with a set equal to zero for the term $1/X_q$ and then $1/\tau_q'$ and $1/\tau_q''$ for successive terms and if (7.7-22) is applied in a similar manner to (7.7-20), we obtain [6]

$$i_{qs}^r = \frac{\sqrt{2}V_s}{X_q''}e^{-\alpha t}\sin\omega_b t \qquad (7.7\text{-}23)$$

$$i_{ds}^r = \sqrt{2}V_s\left[\frac{1}{X_d} + \left(\frac{\tau_{do}'}{\tau_d'}\frac{1}{X_d} - \frac{1}{X_d}\right)e^{-t/\tau_d'}\right.$$
$$\left. + \left(\frac{1}{X_d''} - \frac{\tau_{do}'}{\tau_d'}\frac{1}{X_d}\right)e^{-t/\tau_d''}\right] - \frac{\sqrt{2}V_s}{X_d''}e^{-\alpha t}\cos\omega_b t \qquad (7.7\text{-}24)$$

It is clear that ω_b may be replaced by ω_e in the above equations.

Initially the machine is operating open-circuited with the time zero position of the q and d axis selected so that the a-phase voltage is maximum at the time the q axis coincides with the axis of the a phase. If we now select time zero at the instant of the short circuit and if the speed of the rotor is held fixed at synchronous speed, then

$$\theta_r = \omega_b t + \theta_r(0) \qquad (7.7\text{-}25)$$

where $\theta_r(0)$ is the position of the rotor relative to the magnetic axis of the as winding at the time of the fault. In other words, the point on the a-phase sinusoidal voltage relative to its maximum value. Substituting (7.7-25) into the transformation given by (3.3-6) yields the a-phase short-circuit current

$$i_{as} = \sqrt{2}V_s\left[\frac{1}{X_d} + \left(\frac{\tau_{do}'}{\tau_d'}\frac{1}{X_d} - \frac{1}{X_d}\right)e^{-t/\tau_d'} + \left(\frac{1}{X_d''} - \frac{\tau_{do}'}{\tau_d'}\frac{1}{X_d}\right)e^{-t/\tau_d''}\right]\sin[\omega_b t + \theta_r(0)]$$
$$- \frac{\sqrt{2}V_s}{2}\left(\frac{1}{X_d''} + \frac{1}{X_q''}\right)e^{-\alpha t}\sin\theta_r(0)$$
$$- \frac{\sqrt{2}V_s}{2}\left(\frac{1}{X_d''} - \frac{1}{X_q''}\right)e^{-\alpha t}\sin[2\omega_b t + \theta_r(0)] \qquad (7.7\text{-}26)$$

The short-circuit currents in phases b and c may be expressed by displacing each term of (7.7-26) by $-2\pi/3$ and $2\pi/3$ electrical degrees, respectively.

Let us take a moment to discuss the terms of (7.7-26) and their relationship to the terms of (7.7-23) and (7.7-24). Because the rotor speed is held fixed at synchronous, the rotor reference frame is the synchronously rotating reference frame. In Section 3.7, we showed that a balanced 3-phase set appears in the synchronously rotating reference frame as variables proportional to the amplitude of the 3-phase balanced set, (3.7-8) and (3.7-9), which may be time-varying. Therefore, we would expect that all terms on the right-hand side of (7.7-24), except the cosine term, would be the amplitude of the fundamental frequency balanced 3-phase set. We see from (7.7-26) that this is indeed the case. The amplitude of the balanced 3-phase set contains the information necessary to determine the d-axis parameters. Later we will return to describe the technique of extracting this information.

From the material presented in Section 3.10, we would expect the exponentially decaying offset occurring in the abc variables to appear as an exponentially decaying balanced 2-phase set in the synchronously rotating reference frame as illustrated by (3.10-10) and (3.10-11). In particular, if we consider only the exponentially decaying term of the abc variables, then

$$i_{as}^* = -\frac{\sqrt{2}V_s}{2}\left(\frac{1}{X_d''} + \frac{1}{X_q''}\right)e^{-\alpha t}\sin\theta_r(0) \tag{7.7-27}$$

$$i_{bs}^* = -\frac{\sqrt{2}V_s}{2}\left(\frac{1}{X_d''} + \frac{1}{X_q''}\right)e^{-\alpha t}\sin\left[\theta_r(0) - \frac{2\pi}{3}\right] \tag{7.7-28}$$

$$i_{cs}^* = -\frac{\sqrt{2}V_s}{2}\left(\frac{1}{X_d''} + \frac{1}{X_q''}\right)e^{-\alpha t}\sin\left[\theta_r(0) + \frac{2\pi}{3}\right] \tag{7.7-29}$$

where the asterisk is used to denote the exponentially decaying component of the short-circuit stator currents. If these currents are transformed to the rotor (synchronous) reference frame by (3.3-1), the following q- and d-axis currents are obtained:

$$i_{qs}^{r*} = \frac{\sqrt{2}V_s}{2}\left(\frac{1}{X_d''} + \frac{1}{X_q''}\right)e^{-\alpha t}\sin\omega_b t \tag{7.7-30}$$

$$i_{ds}^{r*} = -\frac{\sqrt{2}V_s}{2}\left(\frac{1}{X_d''} + \frac{1}{X_q''}\right)e^{-\alpha t}\cos\omega_b t \tag{7.7-31}$$

These expressions do not appear in this form in (7.7-23) and (7.7-24); however, before becoming too alarmed let us consider the double-frequency term occurring in the short-circuit stator currents. In particular, from (7.7-26) we have

$$i_{as}^{**} = -\frac{\sqrt{2}V_s}{2}\left(\frac{1}{X_d''} - \frac{1}{X_q''}\right)e^{-\alpha t}\sin[2\omega_b t + \theta_r(0)] \tag{7.7-32}$$

Therefore

$$i_{bs}^{**} = -\frac{\sqrt{2}V_s}{2}\left(\frac{1}{X_d''} - \frac{1}{X_q''}\right)e^{-\alpha t}\sin\left[2\omega_b t + \theta_r(0) - \frac{2\pi}{3}\right] \qquad (7.7\text{-}33)$$

$$i_{cs}^{**} = -\frac{\sqrt{2}V_s}{2}\left(\frac{1}{X_d''} - \frac{1}{X_q''}\right)e^{-\alpha t}\sin\left[2\omega_b t + \theta_r(0) + \frac{2\pi}{3}\right] \qquad (7.7\text{-}34)$$

where the superscript ** denotes the double-frequency components of the short-circuit stator currents. These terms form a double-frequency, balanced 3-phase set in the *abc* variables. We would expect this set to appear as a balanced 2-phase set of fundamental frequency in the synchronously rotating reference frame ($\omega = \omega_b$ or ω_e) and as decaying exponentials in a reference frame rotating at $2\omega_b$. Thus

$$i_{qs}^{r**} = -\frac{\sqrt{2}V_s}{2}\left(\frac{1}{X_d''} - \frac{1}{X_q''}\right)e^{-\alpha t}\sin\omega_b t \qquad (7.7\text{-}35)$$

$$i_{ds}^{r**} = -\frac{\sqrt{2}V_s}{2}\left(\frac{1}{X_d''} - \frac{1}{X_q''}\right)e^{-\alpha t}\cos\omega_b t \qquad (7.7\text{-}36)$$

We now see that if we add i_{qs}^r, (7.7-30), and i_{qs}^{r**}, (7.7-35), we obtain (7.7-23). Similarly, if we add i_{ds}^r, (7.7-31), and i_{ds}^{r**}, (7.7-36), we obtain the last term of (7.7-24). In other words, (7.7-23) and (7.7-24) can be written as

$$i_{qs}^r = \frac{\sqrt{2}V_s}{2}\left(\frac{1}{X_d''} + \frac{1}{X_q''}\right)e^{-\alpha t}\sin\omega_b t - \frac{\sqrt{2}V_s}{2}\left(\frac{1}{X_d''} - \frac{1}{X_q''}\right)e^{-\alpha t}\sin\omega_b t \qquad (7.7\text{-}37)$$

$$i_{ds}^r = \sqrt{2}V_s\left[\frac{1}{X_d} + \left(\frac{\tau_{do}'}{\tau_d'}\frac{1}{X_d} - \frac{1}{X_d}\right)e^{-t/\tau_d'} + \left(\frac{1}{X_d''} - \frac{\tau_{do}'}{\tau_d'}\frac{1}{X_d}\right)e^{-t/\tau_d''}\right]$$
$$- \frac{\sqrt{2}V_s}{2}\left(\frac{1}{X_d''} + \frac{1}{X_q''}\right)e^{-\alpha t}\cos\omega_b t - \frac{\sqrt{2}V_s}{2}\left(\frac{1}{X_d''} - \frac{1}{X_q''}\right)e^{-\alpha t}\cos\omega_b t \qquad (7.7\text{-}38)$$

Let us now return to the expression for the short-circuit current i_{as} given by (7.7-26). In most machines, X_d'' and X_q'' are comparable in magnitude; hence, the double frequency component of the short-circuit stator currents is small. Consequently, the short-circuit current is predominately the combination of a decaying fundamental frequency component and a decaying offset. We first observed the waveform of the short-circuit current in Figs. 5.11-1 and 5.11-3. Although the initial conditions were different in that the machine was loaded and the speed of the machine increased slightly during the 3-phase fault, the two predominate components of (7.7-26) are evident in these traces.

As mentioned previously, the amplitude or the envelope of the fundamental frequency component of each phase current contains the information necessary to determine the d-axis parameters. For purposes of explanation let

$$i_{sc} = \sqrt{2}V_s \left[\frac{1}{X_d} + \left(\frac{\tau'_{do}}{\tau'_d} \frac{1}{X_d} - \frac{1}{X_d} \right) e^{-t/\tau'_d} + \left(\frac{1}{X''_d} - \frac{\tau'_{do}}{\tau'_d} \frac{1}{X_d} \right) e^{-t/\tau''_d} \right] \qquad (7.7\text{-}39)$$

where i_{sc} is the envelope of the fundamental component of the short-circuit stator currents. This can be readily determined from a plot of any one of the instantaneous phase currents.

Now, at the instant of the fault we have

$$i_{sc}(t = 0^+) = \frac{\sqrt{2}V_s}{X''_d} \qquad (7.7\text{-}40)$$

At the final or steady-state value i_{sc} becomes

$$i_{sc}(t \to \infty) = \frac{\sqrt{2}V_s}{X_d} \qquad (7.7\text{-}41)$$

Hence, if we know the prefault voltage and if we can determine the initial and final values of the current envelope, X''_d and X_d can be calculated.

It is helpful to break up i_{sc} into three components:

$$i_{sc} = i_{ss} + i_t + i_{st} \qquad (7.7\text{-}42)$$

where i_{ss} is the steady-state component, i_t is the transient component that decays according to τ'_d, and i_{st} is the subtransient component with the time constant τ''_d. It is customary to subtract the steady-state component i_{ss} from the envelope and plot $i_t + i_{st}$ on semilog paper as illustrated in Fig. 7.7-1. Because $\tau'_d > \tau''_d$ the plot of $i_t + i_{st}$ is determined by i_t as time increases; and because the plot is on the semilog paper, this decay is a straight line. If the transient component is extended to the y axis as shown by the dashed line in Fig. 7.7-1, the initial value of the transient component is obtained:

$$i_t(t = 0^+) = \sqrt{2}V_s \left(\frac{\tau'_{do}}{\tau'_d} \frac{1}{X_d} - \frac{1}{X_d} \right) \qquad (7.7\text{-}43)$$

Because X_d is determined from (7.7-41), we can now determine $(\tau'_{do}/\tau'_d)(1/X_d)$ or if we choose to use the standard time constants, $(\tau'_{do}/\tau'_d)(1/X_d)$ is replaced by $1/X'_d$.

The time constant τ'_d can also be determined from the plot shown in Fig. 7.7-1. In particular, τ'_d is the time it takes for i_t to decrease to $1/e$ (0.368) of its original value.

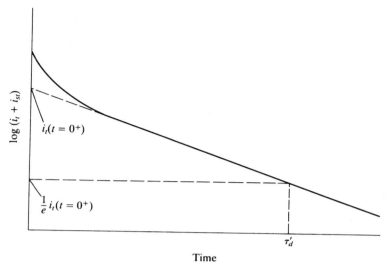

Figure 7.7-1   Plot of transient and subtransient components of the envelope of the short-circuit stator current.

Thus, we now know X_d'', X_d, and τ_d'. Also, X_d' is known if we wish to use the standard, approximate time constants; or τ_{do}' is known if we wish to use the derived time constants to calculate the d-axis parameters.

We can now extract the subtransient component from Fig. 7.7-1 by subtracting the dashed-line extension of the straight-line portion, which is i_t, from the plot of $i_t + i_{st}$. This difference will also yield a straight line when plotted on semilog paper from which the initial value of the subtransient component, $i_{st}(t = 0^+)$, and the time constant τ_d'' can be determined.

Thus we have determined X_d'', X_d, τ_d', τ_d'', and τ_{do}'. The stator leakage reactance X_{ls} can be calculated from the winding arrangement or from tests, or a reasonable value can be assumed. Hence, with a value of X_{ls} we can determine the d-axis parameters. If $r_{fd}' \ll r_{kd}'$, it is generally sufficient to use X_d' and the standard time constants, which, of course, markedly reduces the calculations involved.

7.8   PARAMETERS FROM FREQUENCY-RESPONSE CHARACTERISTICS

In the 1970s there was considerable interest in determining the machine parameters for dynamic and transient stability studies from measured frequency-response data [7–10]. These tests are generally performed by applying a low voltage across two terminals of the stator windings with the rotor at standstill and either the q or d axis aligned with the resultant magnetic axis established by the two stator windings. The frequency of the applied voltage is varied from a very low value of the order of

10^{-3} Hz up to approximately 100 Hz. From these data it is possible to extract $X_q(s)$, $X_d(s)$, and $G(s)$. Hence, we are able to gain information regarding both the q and d axes—unlike the short-circuit test, which provides information on the parameters of only the d axis. Moreover, the frequency-response test provides data from which the rotor can be represented by as many rotor windings in each axis as is required to obtain an acceptable match of the measured operational impedances and $G(s)$. This is a definite advantage over the short-circuit test, especially in the case of solid iron rotor machines where it is often necessary to represent the rotor with more than two rotor windings in each axis in order to portray accurately its electrical characteristics.

Plots of measured $X_q(s)$ and $X_d(s)$ versus frequency similar to those given in reference 10 are shown in Fig. 7.8-1 for a solid iron rotor machine. Figures 7.8-2 and 7.8-3 show, respectively, a two-rotor winding and a three-rotor winding approximation of $X_q(s)$. It is recalled from (7.7-11) that for two rotor windings

$$X_q(s) = X_q \frac{(1 + \tau_q's)(1 + \tau_q''s)}{(1 + \tau_{qo}'s)(1 + \tau_{qo}''s)} \qquad (7.8-1)$$

As illustrated in Figs. 7.8-2 and 7.8-3 the asymptotic approximation of $(1 + \tau s)$ is used to match the plot of the magnitude of $X_q(s)$ versus frequency. Although a computer program could be used to perform curve fitting, the asymptotic approximation is sufficient for our purposes. It is important, however, that regardless of the matching procedure employed, care must be taken to match the operational impedances as closely as possible over the frequency range from 0.05 to 5 Hz because it has been

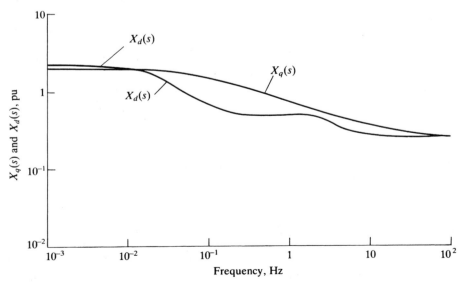

Figure 7.8-1 Plot of $X_q(s)$ and $X_d(s)$ versus frequency for a solid iron synchronous machine.

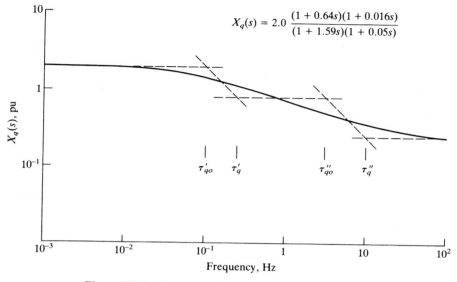

Figure 7.8-2 Two-rotor winding approximation of $X_q(s)$.

determined that matching over this range is critical in achieving accuracy in dynamic and transient stability studies [10].

The asymptotic approximation of $(1 + j\omega\tau)$, where s has been replaced by $j\omega$, is that for $\omega\tau < 1$, $(1 + j\omega\tau)$ is approximated by 1 and for $\omega\tau > 1$, $(1 + j\omega\tau)$ is approximated by $j\omega\tau$. The corner frequency or "breakpoint" is at $\omega\tau = 1$, from

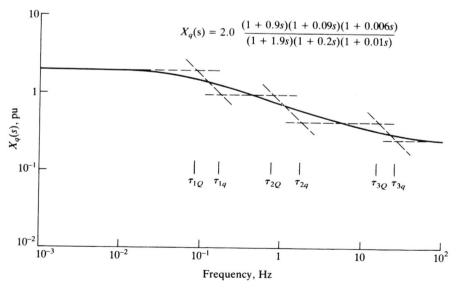

Figure 7.8-3 Three-rotor winding approximation of $X_q(s)$.

which the time constant may be determined. At the corner frequency the slope of the asymptotic approximation of $(1 + j\omega\tau)$ changes from zero to a positive value increasing by one decade in amplitude (a gain of 20 dB) for every decade increase in frequency. It follows that the asymptotic approximation of $(1 + j\omega\tau)^{-1}$ is a zero slope line to the corner frequency whereupon the slope becomes negative, decreasing in amplitude by one decade for every decade increase in frequency.

To obtain a lumped parameter approximation of $X_q(s)$ by using this procedure, we start at the low-frequency asymptote extending this zero slope line to a point where it appears that a breakpoint and thus a negative slope should occur in order to follow the measured value of $X_q(s)$. Because it is necessary that a negative slope occur after the breakpoint, a $(1 + \tau s)$ factor must be present in the denominator. Hence, this corner frequency determines the largest time constant in the denominator, which is τ'_{qo} in the case of the two-rotor winding approximation. We now continue on the negative slope asymptote until it is deemed necessary to again resume a zero slope asymptote in order to match the $X_q(s)$ plot. This swing back to a zero slope line gives rise to a $(1 + \tau s)$ factor in the numerator. This corner frequency determines the largest time constant in the numerator, τ'_q in the case of the two-rotor winding approximation. It follows that τ''_{qo} and τ''_q are determined by the same procedure.

The phase angle of $X_q(s)$ can also be measured at the same time that the magnitude of $X_q(s)$ is measured. However, the phase angle was not made use of in the curve-fitting process. Although the measured phase angle does provide a check on the asymptotic approximation of $X_q(s)$, it is not necessary in this "minimum phase" system where the magnitude of $X_q(s)$ as a function of frequency is sufficient to determine the phase $X_q(s)$ [9]. Hence, the asymptotic approximation provides an approximation of the magnitude and phase of $X_q(s)$.

The stator leakage reactance, X_{ls}, can be determined by tests or taken as the value recommended by the manufacturer, which is generally calculated or approximated from design data. The value of X_{ls} should not be larger than the subtransient reactances because this choice could result in negative rotor leakage reactances, which are not commonly used. For the machine under consideration, X_{ls} of 0.15 per unit is used. Once a value of X_{ls} is selected, the parameters may be determined from the information gained from the frequency-response tests. In particular, from Fig. 7.8-2 we have

$$X_q = 2 \text{ pu} \qquad X''_q = 0.25 \text{ pu}$$
$$\tau'_{qo} = 1.59 \text{ s} \qquad \tau''_{qo} = 0.05 \text{ s}$$
$$\tau'_q = 0.64 \text{ s} \qquad \tau''_q = 0.016 \text{ s}$$

With X_{ls} selected as 0.15 pu, X_{mq} becomes 1.85 pu. Four parameters remain to be determined: r'_{kq1}, X'_{lkq1}, r'_{kq2}, and X'_{lkq2}. These may be determined from the expressions of the derived q-axis time constants given in Table 7.6-1.

There is another approach by which the parameters of the lumped-circuit approximation of $X_q(s)$ may be determined, which is especially useful when it is necessary to represent the rotor with more than two windings in an axis. By a curve-fitting procedure such as illustrated in Figs. 7.8-2 and 7.8-3, it is possible to approximate $X_q(s)$ by

$$X_q(s) = X_q \frac{N_x(s)}{D_x(s)} \qquad (7.8\text{-}2)$$

where in general

$$N_x(s) = (1 + \tau_{1q}s)(1 + \tau_{2q}s)\cdots \qquad (7.8\text{-}3)$$
$$D_x(s) = (1 + \tau_{1Q}s)(1 + \tau_{2Q}s)\cdots \qquad (7.8\text{-}4)$$

The input impedance for a two-rotor winding circuit is expressed by (7.3-1). For any number of rotor circuits we have

$$Z_{qr}(s) = R_{eq}\frac{N_z(s)}{D_z(s)} \qquad (7.8\text{-}5)$$

where

$$\frac{1}{R_{eq}} = \frac{1}{R_{qa}} + \frac{1}{R_{qb}} + \cdots \qquad (7.8\text{-}6)$$
$$N_z(s) = (1 + \tau_{qa}s)(1 + \tau_{qb}s)\cdots \qquad (7.8\text{-}7)$$
$$D_z(s) = (1 + \tau_{Qa}s)(1 + \tau_{Qb}s)\cdots \qquad (7.8\text{-}8)$$

It is clear that (7.3-6) is valid regardless of the number of rotor windings. Thus if we substitute (7.8-2) into (7.3-6) and solve for $Z_{qr}(s)$, we obtain [7]

$$Z_{qr}(s) = \frac{sX_{mq}/\omega_b[N_x(s) - (X_{ls}/X_q)D_x(s)]}{D_x(s) - N_x(s)} \qquad (7.8\text{-}9)$$

Because the time constants of (7.8-3) and (7.8-4) can be obtained by a curve-fitting procedure and because X_q is readily obtained from $X_q(s)$, all elements of (7.8-9) are known once X_{ls} is selected. Hence, values can be substituted into (7.8-9); and after some algebraic manipulation it is possible to put (7.8-9) in the form of (7.8-5), whereupon R_{eq} and the time constants of (7.8-7) and (7.8-8) are known. The parameters of the lumped circuit approximation can then be determined. For example, in the case of the two-winding approximation we have

$$\begin{bmatrix} 1 & 1 \\ \tau_{qb} & \tau_{qa} \end{bmatrix} \begin{bmatrix} \frac{1}{r'_{kq1}} \\ \frac{1}{r'_{kq2}} \end{bmatrix} = \frac{1}{R_{eq}} \begin{bmatrix} 1 \\ \tau_{Qa} \end{bmatrix} \qquad (7.8\text{-}10)$$

where the second row of (7.8-10) is (7.3-5). Thus, r'_{kq1} and r'_{kq2} can be evaluated from (7.8-10), and then X'_{lkq1} and X'_{lkq2} can be evaluated from (7.3-3) and (7.3-4), respectively.

In the case of the three-rotor winding approximation in the q axis [7], (7.8-10) becomes

$$
\begin{bmatrix}
1 & 1 & 1 \\
\tau_{qb} + \tau_{qc} & \tau_{qa} + \tau_{qc} & \tau_{qa} + \tau_{qb} \\
\tau_{qb}\tau_{qc} & \tau_{qa}\tau_{qc} & \tau_{qa}\tau_{qb}
\end{bmatrix}
\begin{bmatrix}
\frac{1}{r'_{kq1}} \\
\frac{1}{r'_{kq2}} \\
\frac{1}{r'_{kq3}}
\end{bmatrix}
= \frac{1}{R_{eq}}
\begin{bmatrix}
1 \\
\tau_{Qa} + \tau_{Qb} \\
\tau_{Qa}\tau_{Qb}
\end{bmatrix}
\tag{7.8-11}
$$

It is left to the reader to express $Z_{qr}(s)$ for three-rotor windings.

In the development of the lumped parameter circuit approximation, there is generally no need to preserve the identity of a winding that might physically exist in the q axis of the rotor because the interest is to portray the electrical characteristics of this axis as viewed from the stator. However, in the d axis we view the characteristics of the rotor from the stator by the operation impedance $X_d(s)$ and the transfer function $G(s)$. If a lumped parameter circuit approximation is developed from only $X_d(s)$, the stator electrical characteristics may be accurately portrayed; however, the field-induced voltage during a disturbance could be quite different from that which occurs in the actual machine, especially if the $G(s)$ that is measured and the $G(s)$ that results when using only $X_d(s)$ do not correspond. A representation of this type, wherein only $X_d(s)$ is used to determine the lumped parameter approximation of the d axis and where the winding with the largest time constant is designated as the field winding, is quite adequate when the electrical characteristics of the field have only secondary influence upon the study being performed. Most dynamic and transient stability studies fall into this category.

When the induced field voltage is of interest, as in the rating and control of solid-state switching devices that might be used in fast response excitation systems, it may be necessary to represent more accurately the electrical characteristics of the field circuit. Several researchers have considered this problem [9,11,12]. I. M. Canay [11] suggested the use of an additional rotor leakage inductance whereupon the d-axis circuit for a two-rotor winding approximation would appear as shown in Fig. 7.8-4. The additional rotor leakage reactance or the "cross mutual" reactance provides a means to account for the fact that the mutual inductance between the rotor and the stator windings is not necessarily the same as that between the rotor field winding and equivalent damper windings [10]. I. M. Canay [11] showed that with additional rotor leakage reactance, both the stator and the field electrical variables could be accurately portrayed. However, in order to determine the parameters for this type of d-axis lumped parameter approximation, both $X_d(s)$ and $G(s)$ must be used [6,9].

There are several reasons for not considering the issue of the additional rotor leakage reactance further at this time. Instead, we will determine the lumped parameter circuit approximation for the d axis from only $X_d(s)$ using the same techniques as in the case of $X_q(s)$ and will designate the rotor winding with the largest time constant as the field winding. There are many cases where the measured $X_d(s)$ yields a winding arrangement that results in a $G(s)$ essentially the same as the measured $G(s)$; hence the additional rotor leakage reactance is small. Also, most studies do not

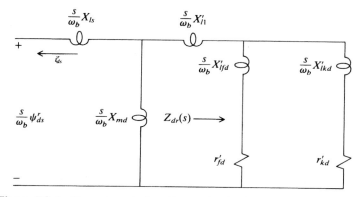

Figure 7.8-4   Two-rotor winding direct-axis circuit with unequal coupling.

require this degree of refinement in the machine representation; that is, the accuracy of the simulated field variables is of secondary or minor importance to the system performance of interest.

Recently finite-element techniques have been used to determine the machine parameters from the physical structure of the machine. Examples of this procedure are given in references 13 and 14.

REFERENCES

[1] R. H. Park, Two-Reaction Theory of Synchronous Machines—Generalized Method of Analysis, Part I, *AIEE Transactions*, Vol. 48, July 1929, pp. 716–727.

[2] R. P. Schulz, W. D. Jones, and D. W. Ewart, Dynamic Models of Turbine Generators Derived from Solid Rotor Equivalent Circuits, *IEEE Transactions on Power Apparatus and Systems*, Vol. 92, May/June 1973, pp. 926–933.

[3] R. E. Doherty and C. A. Nickle, Synchronous Machines—III, Torque–Angle Characteristics Under Transient Conditions, *AIEE Transactions*, Vol. 46, January 1927, pp. 1–8.

[4] G. Shackshaft, New Approach to the Determination of Synchronous Machine Parameters from Tests, *Proceedings of the IEE*, Vol. 121, November 1974, pp. 1385–1392.

[5] *IEEE Standard Dictionary of Electrical and Electronic Terms*, 2nd ed., John Wiley and Sons, New York, 1978.

[6] B. Adkins and R. G. Harley, *The General Theory of Alternating Current Machines*, Chapman and Hall, London, 1975.

[7] W. Watson and G. Manchur, Synchronous Machine Operational Impedance from Low Voltage Measurements at the Stator Terminals, *IEEE Transactions on Power Apparatus and Systems*, Vol. 93, May/June 1974, pp. 777–784.

[8] P. L. Dandeno and P. Kundur, Stability Performance of 555 MVA Turboalternators—Digital Comparisons with System Operating Tests, *IEEE Transactions on Power Apparatus and Systems*, Vol. 93, May/June 1974, pp. 767–776.

[9] S. D. Umans, J. A. Malleck, and G. L. Wilson, Modeling of Solid Rotor Turbogenerators, Parts I and II. *IEEE Transactions on Power Apparatus and Systems*, Vol. 97, January/February 1978, pp. 269–296.

[10] IEEE Committee Report, Supplementary Definitions and Associated Test Methods for Obtaining Parameters for Synchronous Machine Stability and Study Simulations, *IEEE Transactions on Power Apparatus and Systems*, Vol. 99, July/August 1980, pp. 1625–1633.

[11] I. M. Canay, Causes of Discrepancies on Calculation of Rotor Quantities and Exact Equivalent Diagrams of the Synchronous Machine, *IEEE Transactions on Power Apparatus and Systems*, Vol. 88, July 1969, pp. 1114–1120.

[12] Y. Tabeda and B. Adkins, Determination of Synchronous Machine Parameters Allowing for Unequal Mutual Inductances, *Proceedings of the IEE*, Vol. 121, December 1974, pp. 1501–1504.

[13] M. P. Krefta and O. Wasynczuk, A Finite Element Based State Model of Solid Rotor Synchronous Machines, *IEEE Transactions on Energy Conversion*, Vol. EC-2, No. 1, March 1987, pp. 21–30.

[14] K. Hameyer and R. Belmans, *Numerical Modelling and Design of Electrical Machines and Devices*, WIT Press, Southampton, UK, 1999.

PROBLEMS

1 Derive expressions for the short-circuit time constants with the stator resistance included.

2 Calculate and compare the standard and derived time constants for the hydro turbine generator given in Chapter 5.

3 Repeat Problem 2 for the steam turbine generator given in Chapter 5.

4 Derive an expression for the instantaneous electromagnetic torque during a 3-phase short circuit at the terminals. Assume that the stator terminals of the machine are initially open-circuited and that the speed does not change during the fault.

5 Derive an expression for the instantaneous field current for a 3-phase short circuit at the terminals. As in Problem 4, assume that the machine is initially operating with the stator open-circuited and that the speed remains constant during the fault.

6 Consider the short-circuit stator currents shown in Fig. 7.P-1. The machine is originally operating open-circuited at rated voltage. The speed is fixed during the fault. Assume $r_s = 0.0037$ pu, $X_d = 1.7$ pu, and $X_{ls} = 0.19$ pu. Determine the remaining d-axis circuit parameters using (*a*) the derived time constants and (*b*) the standard time constants.

7 For two-rotor windings in the d axis, show that $i_{fd}^{\prime r} = pG(p)i_{ds}^r$ for $v_{fd}^{\prime r} = 0$.

8 Determine r_{kq1}', X_{lkq1}', r_{kq2}', and X_{lkq2}' for the two-rotor winding approximation of $X_q(s)$ given in Fig. 7.8-2 by using (*a*) the derived time constants and (*b*) (7.8-10), (7.3-3), and (7.3-4).

9 Determine the parameters of a two-rotor winding approximation of $X_d(s)$ given in Fig. 7.8-1.

10 Express $Z_{qr}(s)$ for a three-rotor winding approximation. Compare the terms in the denominator to the last two rows of (7.8-11).

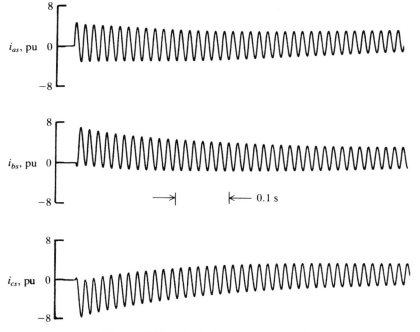

Figure 7.P-1   Short-circuit stator currents.

11 Determine the time constants of $X_d(s)$ given in Fig. 7.8-1 for a three-rotor winding approximation.

12 Determine the parameters of a three-rotor winding approximation of $X_q(s)$ shown in Fig. 7.8-2.

13 Repeat Problem 12 for $X_d(s)$.

14 Write Park's equations for a synchronous machine represented by three damper windings in the q axis and two damper windings and a field winding in the d axis.

15 Derive $X_d(s)$ and $G(s)$ for the d-axis circuit with the additional rotor leakage reactance shown in Fig. 7.8-4. Show that both have the same denominator.

Chapter 8

LINEARIZED MACHINE EQUATIONS

8.1 INTRODUCTION

The equations that describe the behavior of the induction and synchronous machines are nonlinear and can be solved only with the aid of a computer. However, considerable insight can be gained regarding the small-excursion behavior from the linearized version of these equations. Although standard computer algorithms may be used to automatically linearize these equations, it is important to be aware of the steps necessary to perform linearization. This procedure is set forth by applying Taylor's expansion about an operating point. The resulting set of linear differential equations describes the dynamic behavior of small displacements or small excursions about this operating point. The machines can then be treated as linear systems with regard to small disturbances, whereupon basic linear system theory can be used to calculate eigenvalues and to establish transfer functions for use in the design of controls associated with the machines.

In this chapter the nonlinear equations of the induction and synchronous machines are linearized. Although these equations are valid for operation with stator voltages of any frequency, as in variable-frequency drive systems, only rated frequency operation is considered in detail. Eigenvalues are determined for both induction and synchronous machines, and transfer functions are formulated for the purpose of establishing the basis from which control analysis and design may begin. With present-day computer languages, linearization, calculation of eigenvalues, and formulation of transfer functions can be done automatically from the nonlinear equations that describe these machines [1–3]. The material presented in this chapter serves to inform the reader of the steps necessary in order to "automatically" characterize small-excursion operation of induction and synchronous machines.

8.2 MACHINE EQUATIONS TO BE LINEARIZED

The linearized machine equations are conveniently derived from voltage equations expressed in terms of constant parameters with constant driving forces, independent of time. During steady-state balanced conditions these requirements are satisfied, in the case of the induction machine, by the voltage equations expressed in the synchronously rotating reference frame, and in the case of the synchronous machine they are satisfied by the voltage equations in the rotor reference frame. Because the currents and flux linkages are not independent variables, the machine equations can be written with either currents or flux linkages, or flux linkages per second, as state variables. The choice is generally determined by the application. Currents are selected here. Formulating the small-displacement equations in terms of flux linkages per second is left as an exercise for the reader.

Induction Machine

The voltage equations for the induction machine with currents as state variables may be written in the synchronously rotating reference frame from (4.5-34) by setting $\omega = \omega_e$:

$$
\begin{bmatrix} v_{qs}^e \\ v_{ds}^e \\ v_{qr}^{\prime e} \\ v_{dr}^{\prime e} \end{bmatrix} = \begin{bmatrix} r_s + \frac{p}{\omega_b}X_{ss} & \frac{\omega_e}{\omega_b}X_{ss} & \frac{p}{\omega_b}X_M & \frac{\omega_e}{\omega_b}X_M \\ -\frac{\omega_e}{\omega_b}X_{ss} & r_s + \frac{p}{\omega_b}X_{ss} & -\frac{\omega_e}{\omega_b}X_M & \frac{p}{\omega_b}X_M \\ \frac{p}{\omega_b}X_M & s\frac{\omega_e}{\omega_b}X_M & r_r' + \frac{p}{\omega_b}X_{rr}' & s\frac{\omega_e}{\omega_b}X_{rr}' \\ -s\frac{\omega_e}{\omega_b}X_M & \frac{p}{\omega_b}X_M & -s\frac{\omega_e}{\omega_b}X_{rr}' & r_r' + \frac{p}{\omega_b}X_{rr}' \end{bmatrix} \begin{bmatrix} i_{qs}^e \\ i_{ds}^e \\ i_{qr}^{\prime e} \\ i_{dr}^{\prime e} \end{bmatrix}
$$

$$(8.2-1)$$

where s is the slip defined by (4.9-13) and the 0 quantities have been omitted because only balanced conditions are considered. The reactances X_{ss} and X_{rr}' are defined by (4.5-35) and (4.5-36), respectively.

Because we have selected currents as state variables, the electromagnetic torque is most conveniently expressed as

$$T_e = X_M(i_{qs}^e i_{dr}^{\prime e} - i_{ds}^e i_{qr}^{\prime e})   \qquad (8.2-2)$$

Here, the per unit version of (4.6-2) is selected for compactness. The per unit relationship between torque and speed is (4.8-10), which is written here for convenience:

$$T_e = 2Hp\frac{\omega_r}{\omega_b} + T_L   \qquad (8.2-3)$$

Synchronous Machine

The voltage equations for the synchronous machine in the rotor reference frame may be written from (5.5-38) for balanced conditions as

$$
\begin{bmatrix} v_{qs}^r \\ v_{ds}^r \\ v_{kq1}'^r \\ v_{kq2}'^r \\ e_{xfd}'^r \\ v_{kd}'^r \end{bmatrix} =
\begin{bmatrix}
-r_s - \frac{p}{\omega_b}X_q & -\frac{\omega_r}{\omega_b}X_d & \frac{p}{\omega_b}X_{mq} & \frac{p}{\omega_b}X_{mq} & \frac{\omega_r}{\omega_b}X_{md} & \frac{\omega_r}{\omega_b}X_{md} \\
\frac{\omega_r}{\omega_b}X_q & -r_s - \frac{p}{\omega_b}X_d & -\frac{\omega_r}{\omega_b}X_{mq} & -\frac{\omega_r}{\omega_b}X_{mq} & \frac{p}{\omega_b}X_{md} & \frac{p}{\omega_b}X_{md} \\
-\frac{p}{\omega_b}X_{mq} & 0 & r_{kq1}' + \frac{p}{\omega_b}X_{kq1}' & \frac{p}{\omega_b}X_{mq} & 0 & 0 \\
-\frac{p}{\omega_b}X_{mq} & 0 & \frac{p}{\omega_b}X_{mq} & r_{kq2}' + \frac{p}{\omega_b}X_{kq2}' & 0 & 0 \\
0 & -\frac{X_{md}}{r_{fd}'}\left(\frac{p}{\omega_b}X_{md}\right) & 0 & 0 & \frac{X_{md}}{r_{fd}'}\left(r_{fd}' + \frac{p}{\omega_b}X_{fd}'\right) & \frac{X_{md}}{r_{fd}'}\left(\frac{p}{\omega_b}X_{md}\right) \\
0 & -\frac{p}{\omega_b}X_{md} & 0 & 0 & \frac{p}{\omega_b}X_{md} & r_{kd}' + \frac{p}{\omega_b}X_{kd}'
\end{bmatrix}
\begin{bmatrix} i_{qs}^r \\ i_{ds}^r \\ i_{kq1}'^r \\ i_{kq2}'^r \\ i_{fd}'^r \\ i_{kd}'^r \end{bmatrix}
$$

$$(8.2\text{-}4)$$

where the reactances are defined by (5.5-39)–(5.5-44).

With the currents as state variables the per unit electromagnetic torque is expressed from (5.6-2) as

$$
T_e = X_{md}(-i_{ds}^r + i_{fd}'^r + i_{kd}')i_{qs}^r - X_{mq}(-i_{qs}^r + i_{kq1}'^r + i_{kq2}'^r)i_{ds}^r \tag{8.2-5}
$$

The per unit relationship between torque and rotor speed is given by (5.8-3), which is

$$
T_e = -2Hp\frac{\omega_r}{\omega_b} + T_I \tag{8.2-6}
$$

The rotor angle is expressed from (5.7-1) as

$$
\delta = \frac{\omega_b}{p}\left(\frac{\omega_r - \omega_e}{\omega_b}\right) \tag{8.2-7}
$$

It is necessary, in the following analysis, to relate variables in the synchronously rotating reference frame to variables in the rotor reference frame. This is accomplished by using (5.7-2) with the 0 quantities omitted. Thus

$$
\begin{bmatrix} f_{qs}^r \\ f_{ds}^r \end{bmatrix} =
\begin{bmatrix} \cos\delta & -\sin\delta \\ \sin\delta & \cos\delta \end{bmatrix}
\begin{bmatrix} f_{qs}^e \\ f_{ds}^e \end{bmatrix} \tag{8.2-8}
$$

8.3 LINEARIZATION OF MACHINE EQUATIONS

There are two procedures that can be followed to obtain the linearized machine equations. One is to employ Taylor's expansion about a fixed value or operating point. That is, any machine variable f_i can be written in terms of a Taylor expansion about its fixed value, f_{io}, as

$$g(f_i) = g(f_{io}) + g'(f_{io})\Delta f_i + \frac{g''(f_{io})}{2!}\Delta f_i^2 + \cdots \qquad (8.3\text{-}1)$$

where

$$f_i = f_{io} + \Delta f_i \qquad (8.3\text{-}2)$$

If only a small excursion from the fixed point is experienced, all terms higher than the first order may be neglected and $g(f_i)$ may be approximated by

$$g(f_i) \approx g(f_{io}) + g'(f_{io})\Delta f_i \qquad (8.3\text{-}3)$$

Hence, the small-displacement characteristics of the system are given by the first-order terms of Taylor's series; that is,

$$\Delta g(f_i) = g'(f_{io})\Delta f_i \qquad (8.3\text{-}4)$$

For functions of two variables the same argument applies:

$$g(f_1,f_2) \approx g(f_{1o},\, f_{2o}) + \frac{\partial}{\partial f_1}g(f_{1o},\, f_{2o})\Delta f_1 + \frac{\partial}{\partial f_2}g(f_{1o},\, f_{2o})\Delta f_2 \qquad (8.3\text{-}5)$$

where $\Delta g(f_1,\, f_2)$ is the last two terms of (8.3-5).

If, for example, we apply this method to the expression for induction machine torque, (8.2-2), then

$$T_e(i_{qs}^e, i_{ds}^e, i_{qr}^{\prime e}, i_{dr}^{\prime e}) \approx T_e(i_{qso}^e, i_{dso}^e, i_{qro}^{\prime e}, i_{dro}^{\prime e}) + \frac{\partial T_e(i_{qso}^e, i_{dso}^e, i_{qro}^{\prime e}, i_{dro}^{\prime e})}{\partial i_{qs}^e}\Delta i_{qs}^e + \text{etc.}$$

$$(8.3\text{-}6)$$

whereupon the small-displacement expression for torque becomes

$$\Delta T_e = X_M(i_{qso}^e\Delta i_{dr}^{\prime e} + i_{dro}^{\prime e}\Delta i_{qs}^e - i_{dso}^e\Delta i_{qr}^{\prime e} - i_{qro}^{\prime e}\Delta i_{ds}^e) \qquad (8.3\text{-}7)$$

where the added subscript o denotes steady-state quantities.

An equivalent method of linearizing nonlinear equations is to write all variables in the form given by (8.3-2). If all multiplications are then performed and the steady-state expressions are canceled from both sides of the equations and if all products of small-displacement terms ($\Delta f_1 \Delta f_2$, for example) are neglected, the small-displacement equations are obtained. It is left to the reader to obtain (8.3-7) by this technique.

Induction Machine

If either of the above-described methods of linearization is employed for (8.2-1)–(8.2-3), the linear differential equations of an induction machine become

$$
\begin{bmatrix}
\Delta v_{qs}^e \\
\Delta v_{ds}^e \\
\Delta v_{qr}'^e \\
\Delta v_{dr}'^e \\
\Delta T_L
\end{bmatrix}
=
\begin{bmatrix}
r_s + \frac{p}{\omega_b}X_{ss} & \frac{\omega_e}{\omega_b}X_{ss} & \frac{p}{\omega_b}X_M & \frac{\omega_e}{\omega_b}X_M & 0 \\
-\frac{\omega_e}{\omega_b}X_{ss} & r_s + \frac{p}{\omega_b}X_{ss} & -\frac{\omega_e}{\omega_b}X_M & \frac{p}{\omega_b}X_M & 0 \\
\frac{p}{\omega_b}X_M & s_o\frac{\omega_e}{\omega_b}X_M & r_r' + \frac{p}{\omega_b}X_{rr}' & s_o\frac{\omega_e}{\omega_b}X_{rr}' & -X_M i_{dso}^e - X_{rr}' i_{dro}'^e \\
-s_o\frac{\omega_e}{\omega_b}X_M & \frac{p}{\omega_b}X_M & -s_o\frac{\omega_e}{\omega_b}X_{rr}' & r_r' + \frac{p}{\omega_b}X_{rr}' & X_M i_{qso}^e + X_{rr}' i_{qro}'^e \\
X_M i_{dro}'^e & -X_M i_{qro}'^e & -X_M i_{dso}^e & X_M i_{qso}^e & -2HP
\end{bmatrix}
\begin{bmatrix}
\Delta i_{qs}^e \\
\Delta i_{ds}^e \\
\Delta i_{qr}'^e \\
\Delta i_{dr}'^e \\
\frac{\Delta \omega_r}{\omega_b}
\end{bmatrix}
$$

$$(8.3\text{-}8)$$

where

$$s_o = \frac{\omega_e - \omega_{ro}}{\omega_e} \qquad (8.3\text{-}9)$$

It is clear that with applied voltages of rated frequency the ratio of ω_e to ω_b is unity. However, (8.3-8) and (8.3-9) are written with ω_e included explicitly so as to accommodate applied voltages of a constant frequency other than rated as would occur in variable-speed drive systems. The frequency of the applied stator voltages in variable-speed drive systems is varied by controlling the firing of the converter. Therefore, in some applications the frequency of the stator voltages is used as a controlled variable. It is recalled from Chapter 3 that variable-frequency operation may be investigated in the synchronously rotating reference frame by simply changing the speed of the reference frame corresponding to the change in frequency. Therefore, if frequency is a system input variable, then a small displacement in frequency may be taken into account by allowing the reference-frame speed to change by replacing ω_e with $\omega_{eo} + \Delta\omega_e$. This feature could be incorporated in (8.3-8); however, it is of importance only in variable-speed drive systems.

It is convenient to separate out the derivative terms and write (8.3-8) in the form

$$\mathbf{E}p\mathbf{x} = \mathbf{F}\mathbf{x} + \mathbf{u} \qquad (8.3\text{-}10)$$

where

$$(\mathbf{x})^T = \left[\Delta i_{qs}^e \ \Delta i_{ds}^e \ \Delta i_{qr}'^e \ \Delta i_{dr}'^e \ \frac{\Delta\omega_r}{\omega_b}\right] \qquad (8.3\text{-}11)$$

$$(\mathbf{u})^T = \left[\Delta v_{qs}^e \ \Delta v_{ds}^e \ \Delta v_{qr}'^e \ \Delta v_{dr}'^e \ \Delta T_L\right] \qquad (8.3\text{-}12)$$

$$\mathbf{E} = \frac{1}{\omega_b}
\begin{bmatrix}
X_{ss} & 0 & X_M & 0 & 0 \\
0 & X_{ss} & 0 & X_M & 0 \\
X_M & 0 & X_{rr}' & 0 & 0 \\
0 & X_M & 0 & X_{rr}' & 0 \\
0 & 0 & 0 & 0 & -2H\omega_b
\end{bmatrix} \qquad (8.3\text{-}13)$$

$$
\mathbf{F} = -\begin{bmatrix}
r_s & \frac{\omega_e}{\omega_b}X_{ss} & 0 & \frac{\omega_e}{\omega_b}X_M & 0 \\[2ex]
-\frac{\omega_e}{\omega_b}X_{ss} & r_s & -\frac{\omega_e}{\omega_b}X_M & 0 & 0 \\[2ex]
0 & s_o\frac{\omega_e}{\omega_b}X_M & r_r' & s_o\frac{\omega_e}{\omega_b}X_{rr}' & -X_M i_{dso}^e - X_{rr}' i_{dro}^{\prime e} \\[2ex]
-s_o\frac{\omega_e}{\omega_b}X_M & 0 & -s_o\frac{\omega_e}{\omega_b}X_{rr}' & r_r' & X_M i_{qso}^e + X_{rr}' i_{qro}^{\prime e} \\[2ex]
X_M i_{dro}^{\prime e} & -X_M i_{qro}^{\prime e} & -X_M i_{dso}^e & X_M i_{qso}^e & 0
\end{bmatrix}
$$

$$(8.3\text{-}14)$$

In the analysis of linear systems it is convenient to express the linear differential equations in the form

$$p\mathbf{x} = \mathbf{A}\mathbf{x} + \mathbf{B}\mathbf{u} \qquad (8.3\text{-}15)$$

Equation (8.3-15) is the fundamental form of the linear differential equations. It is commonly referred to as the state equation.

Equation (8.3-10) may be written as

$$p\mathbf{x} = (\mathbf{E})^{-1}\mathbf{F}\mathbf{x} + (\mathbf{E})^{-1}\mathbf{u} \qquad (8.3\text{-}16)$$

which is in the form of (8.3-15) with

$$\mathbf{A} = (\mathbf{E})^{-1}\mathbf{F} \qquad (8.3\text{-}17)$$

$$\mathbf{B} = (\mathbf{E})^{-1} \qquad (8.3\text{-}18)$$

Synchronous Machines

Linearizing (8.2-4)–(8.2-8) yields (8.3-19) (see next page). Because the steady-state damper winding currents ($i_{kdo}^{\prime r}$, $i_{kq1o}^{\prime r}$, and $i_{kq2o}^{\prime r}$) are zero, they are not included in (8.3-19). Because the synchronous machine is generally connected to an electric system such as a power system and because it is advantageous to linearize the system voltage equations in the synchronously rotating reference frame, it is convenient to include the relationship between $\Delta\omega_r$ and $\Delta\delta$ in (8.3-19). As in the case of linearized equations for the induction machine, ω_e is included explicitly in (8.3-19) so that the equations are in a form convenient for voltages of any constant frequency. Small controlled changes in the frequency of the applied stator voltages, as is possible in variable-speed drive systems, may be taken into account analytically by replacing ω_e with $\omega_{eo} + \Delta\omega_e$ in the expression for δ given by (8.2-7).

$$
\begin{bmatrix}
\Delta v_{qs}^{r} \\[2pt]
\Delta v_{ds}^{r} \\[2pt]
\Delta v_{kq1}^{\prime r} \\[2pt]
\Delta v_{kq2}^{\prime r} \\[2pt]
\Delta e_{xfd}^{\prime r} \\[2pt]
\Delta v_{kd}^{\prime r} \\[2pt]
\Delta T_l \\[2pt]
0
\end{bmatrix}
=
\begin{bmatrix}
-r_s-\dfrac{p}{\omega_b}X_q & -\dfrac{\omega_e}{\omega_b}X_d & \dfrac{p}{\omega_b}X_{mq} & \dfrac{p}{\omega_b}X_{mq} & \dfrac{\omega_e}{\omega_b}X_{md} & \dfrac{\omega_e}{\omega_b}X_{md} & -X_d i_{dso}^{r}+X_{md}i_{fdo}^{\prime r} & 0 \\[10pt]
\dfrac{\omega_e}{\omega_b}X_q & -r_s-\dfrac{p}{\omega_b}X_d & -\dfrac{\omega_e}{\omega_b}X_{mq} & -\dfrac{\omega_e}{\omega_b}X_{mq} & \dfrac{p}{\omega_b}X_{md} & \dfrac{p}{\omega_b}X_{md} & X_q i_{qso}^{r} & 0 \\[10pt]
-\dfrac{p}{\omega_b}X_{mq} & 0 & r_{kq1}'+\dfrac{p}{\omega_b}X_{kq1}' & \dfrac{p}{\omega_b}X_{mq} & 0 & 0 & 0 & 0 \\[10pt]
-\dfrac{p}{\omega_b}X_{mq} & 0 & \dfrac{p}{\omega_b}X_{mq} & r_{kq2}'+\dfrac{p}{\omega_b}X_{kq2}' & 0 & 0 & 0 & 0 \\[10pt]
0 & -\dfrac{X_{md}}{r_{fd}'}\left(\dfrac{p}{\omega_b}X_{md}\right) & 0 & 0 & \dfrac{X_{md}}{r_{fd}'}\left(r_{fd}'+\dfrac{p}{\omega_b}X_{fd}'\right) & \dfrac{X_{md}}{r_{fd}'}\left(\dfrac{p}{\omega_b}X_{md}\right) & 0 & 0 \\[10pt]
0 & -\dfrac{p}{\omega_b}X_{md} & 0 & 0 & \dfrac{p}{\omega_b}X_{md} & r_{kd}'+\dfrac{p}{\omega_b}X_{kd}' & 0 & 0 \\[10pt]
X_{mq}i_{dso}^{r}-X_{md}(i_{dso}^{r}-i_{fdo}^{\prime r}) & -X_{md}i_{qso}^{r}+X_{mq}i_{qso}^{r} & -X_{mq}i_{dso}^{r} & -X_{mq}i_{dso}^{r} & X_{md}i_{qso}^{r} & X_{md}i_{qso}^{r} & 2Hp & 0 \\[10pt]
0 & 0 & 0 & 0 & 0 & 0 & -\omega_b & p
\end{bmatrix}
\begin{bmatrix}
\Delta i_{qs}^{r} \\[2pt]
\Delta i_{ds}^{r} \\[2pt]
\Delta i_{kq1}^{\prime r} \\[2pt]
\Delta i_{kq2}^{\prime r} \\[2pt]
\Delta i_{fd}^{\prime r} \\[2pt]
\Delta i_{kd}^{\prime r} \\[2pt]
\dfrac{\Delta\omega_r}{\omega_b} \\[2pt]
\Delta\delta
\end{bmatrix}
$$

(8.3-19)

Equation (8.3-19) can also be written in the form

$$\mathbf{E}p\mathbf{x} = \mathbf{F}\mathbf{x} + \mathbf{u} \tag{8.3-20}$$

where

$$(\mathbf{x})^T = \left[\Delta i_{qs}^r \;\; \Delta i_{ds}^r \;\; \Delta i_{kq1}^{\prime r} \;\; \Delta i_{kq2}^{\prime r} \;\; \Delta i_{fd}^{\prime r} \;\; \Delta i_{kd}^{\prime r} \;\; \frac{\Delta \omega_r}{\omega_b} \;\; \Delta \delta \right] \tag{8.3-21}$$

$$(\mathbf{u})^T = \left[\Delta v_{qs}^r \;\; \Delta v_{ds}^r \;\; \Delta v_{kq1}^{\prime r} \;\; \Delta v_{kq2}^{\prime r} \;\; \Delta e_{xfd}^{\prime} \;\; \Delta v_{kd}^{\prime r} \;\; \Delta T_I \;\; 0 \right] \tag{8.3-22}$$

and

$$\mathbf{E} = \frac{1}{\omega_b}
\begin{bmatrix}
-X_q & 0 & -X_{mq} & X_{mq} & 0 & 0 & 0 & 0 \\
0 & -X_d & 0 & 0 & X_{md} & X_{md} & 0 & 0 \\
-X_{mq} & 0 & X_{kq1}' & X_{mq} & 0 & 0 & 0 & 0 \\
-X_{mq} & 0 & X_{mq} & X_{kq2}' & 0 & 0 & 0 & 0 \\
0 & -\frac{X_{md}^2}{r_{fd}'} & 0 & 0 & \frac{X_{md}X_{fd}'}{r_{fd}'} & \frac{X_{md}^2}{r_{fd}'} & 0 & 0 \\
0 & -X_{md} & 0 & 0 & X_{md} & X_{kd}' & 0 & 0 \\
0 & 0 & 0 & 0 & 0 & 0 & 2H\omega_b & 0 \\
0 & 0 & 0 & 0 & 0 & 0 & 0 & \omega_b
\end{bmatrix} \tag{8.3-23}$$

$$\mathbf{F} = -
\begin{bmatrix}
-r_s & -\frac{\omega_e}{\omega_b}X_d & 0 & 0 & \frac{\omega_e}{\omega_b}X_{md} & \frac{\omega_e}{\omega_b}X_{md} & -X_d i_{dso}^r + X_{md} i_{fdo}^{\prime r} & 0 \\
\frac{\omega_e}{\omega_b}X_q & -r_s & -\frac{\omega_e}{\omega_b}X_{mq} & -\frac{\omega_e}{\omega_b}X_{mq} & 0 & 0 & X_q i_{qso}^r & 0 \\
0 & 0 & r_{kq1}' & 0 & 0 & 0 & 0 & 0 \\
0 & 0 & 0 & r_{kq2}' & 0 & 0 & 0 & 0 \\
0 & 0 & 0 & 0 & X_{md} & 0 & 0 & 0 \\
0 & 0 & 0 & 0 & 0 & r_{kd}' & 0 & 0 \\
\begin{array}{l} X_{mq} i_{dso}^r - \\ X_{md}(i_{dso}^r - i_{fdo}^{\prime r}) \end{array} & \begin{array}{l} -X_{md} i_{qso}^r + \\ X_{mq} i_{qso}^r \end{array} & -X_{mq} i_{dso}^r & -X_{mq} i_{dso}^r & X_{md} i_{qso}^r & X_{md} i_{qso}^r & 0 & 0 \\
0 & 0 & 0 & 0 & 0 & 0 & -\omega_b & 0
\end{bmatrix} \tag{8.3-24}$$

Equations (8.3-21)–(8.3-24) are sufficient to describe the small-displacement dynamic behavior during isolated operation with the synchronous machine connected only to a passive load. In most cases, however, the synchronous machine is connected to a power system whereupon the voltage v_{qs}^r and v_{ds}^r, which are functions of the state variable δ, will vary as the rotor angle varies during a disturbance. It is of course necessary to account for the dependence of the driving forces upon the state variables before expressing the linear differential equations in fundamental form.

In power system analysis it is often assumed that some place in the system there is a balanced source that can be considered a constant-amplitude, constant-frequency, zero-impedance source (infinite bus). This would be a balanced independent driving force that would be represented as constant voltages in the synchronously rotating reference frame. Hence, it is necessary to relate the synchronously rotating reference-frame variables, where the independent driving force exists, to the variables in the rotor reference frame. The transformation given by (8.2-8) is nonlinear. In order to incorporate it into a linear set of differential equations, it must be linearized. By employing the approximations that $\cos \Delta \delta = 1$ and $\sin \Delta \delta = \Delta \delta$, the linearized version of (8.2-8) is

$$\begin{bmatrix} \Delta f_{qs}^r \\ \Delta f_{ds}^r \end{bmatrix} = \begin{bmatrix} \cos \delta_o & -\sin \delta_o \\ \sin \delta_o & \cos \delta_o \end{bmatrix} \begin{bmatrix} \Delta f_{qs}^e \\ \Delta f_{ds}^e \end{bmatrix} + \begin{bmatrix} -f_{dso}^r \\ f_{qso}^r \end{bmatrix} \Delta \delta \qquad (8.3\text{-}25)$$

Linearizing the inverse transformation yields

$$\begin{bmatrix} \Delta f_{qs}^e \\ \Delta f_{ds}^e \end{bmatrix} = \begin{bmatrix} \cos \delta_o & \sin \delta_o \\ -\sin \delta_o & \cos \delta_o \end{bmatrix} \begin{bmatrix} \Delta f_{qs}^r \\ \Delta f_{ds}^r \end{bmatrix} + \begin{bmatrix} f_{dso}^e \\ -f_{qso}^e \end{bmatrix} \Delta \delta \qquad (8.3\text{-}26)$$

It is convenient to write the above equations in the form

$$\Delta \mathbf{f}_{qds}^r = \mathbf{T} \Delta \mathbf{f}_{qds}^e + \mathbf{F}^r \Delta \delta \qquad (8.3\text{-}27)$$

$$\Delta \mathbf{f}_{qds}^e = (\mathbf{T})^{-1} \Delta \mathbf{f}_{qds}^r + \mathbf{F}^e \Delta \delta \qquad (8.3\text{-}28)$$

Before proceeding with the substitution and arranging of the above equations into the fundamental form, it is instructive to view the interconnections of the above relationships as shown in Fig. 8.3-1. With the equations as shown in Fig. 8.3-1, a change in $\Delta \mathbf{v}_{qds}^e$ is reflected through the transformation to the voltage equations in the rotor reference frame and finally back to the synchronously rotating reference-frame currents $\Delta \mathbf{i}_{qds}^e$. The detail shown in Fig. 8.3-1 is more than is generally necessary. If, for example, the objective is to study the small-displacement dynamics of a synchronous machine with its terminals connected to an infinite bus, then $\Delta \mathbf{v}_{qds}^e$ is zero and $\Delta \mathbf{v}_{qds}^r$ changes due only to $\Delta \delta$. Also, in this case it is unnecessary to transform the rotor reference-frame currents to the synchronously rotating reference frame because the source (infinite bus) has zero impedance.

If the machine is connected through a transmission line to a large system (infinite bus), the small-displacement dynamics of the transmission system must be taken into account. If only the machine is connected to the transmission line and if it is not equipped with a voltage regulator, then it is convenient to transform the equations of the transmission line to the rotor reference frame, whereupon the machine and transmission line can be considered in much the same way as a machine connected to an infinite bus. If, however, the machine is equipped with an automatic voltage regulator or more than one machine is connected to the same transmission line, it is generally preferable to express the dynamics of the transmission system in the synchronously rotating reference frame and transform to and from the rotor reference frame of each machine as depicted in Fig 8.3-1.

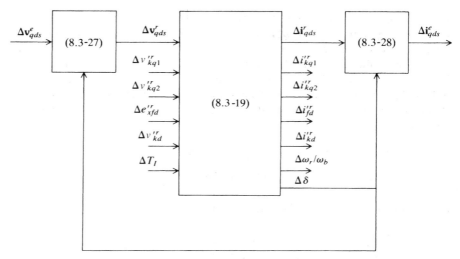

Figure 8.3-1 Interconnection of small-displacement equations of a synchronous machine: Park's equations.

If the machine is equipped with a voltage regulator, the dynamic behavior of the regulator will affect the dynamic characteristics of the machine. Therefore, the small-displacement dynamics of the regulator must be taken into account. When regulators are employed, the change in field voltage $\Delta e''^r_{xfd}$ is dynamically related to the change in terminal voltage, which is a function of $\Delta \mathbf{v}^e_{qds}$ (or $\Delta \mathbf{v}^r_{qds}$), the change in field current $\Delta \mathbf{i}''_{fd}$, and perhaps the change in rotor speed $\Delta \omega_r / \omega_b$ if the excitation system is equipped with a control to help damp rotor oscillations by means of field voltage control. This type of damping control is referred to as a power system stabilizer (PSS).

In some investigations it is necessary to incorporate the small-displacement dynamics of the prime mover system. The change of input torque ΔT_I is a function of the change in rotor speed $\Delta \omega_r / \omega_b$, which in turn is a function of the dynamics of the masses, shafts, and damping associated with the mechanical system and, if long-term transients are of interest, the steam dynamics or hydrodynamics and associated controls.

Although a more detailed discussion of the dynamics of the excitation and prime mover systems would be helpful, it is clear, from the above discussion, that the equations that describe the operation and control of a synchronous machine equipped with a voltage regulator and a prime mover system are very involved. This becomes readily apparent when it is necessary to arrange the small-displacement equations of the complete system into the fundamental form. However, one would not perform this task by hand because computer languages are readily available which can be used to formulate small-displacement equations along with the interconnecting transformations of the complete system into the fundamental form [1–3].

Although our attention has been diverted somewhat, let us now return to (8.3-19). The first step toward a set of linear differential equations arranged in fundamental form (state equation) is to transform the stator voltages and currents to the synchronously rotating reference frame. Therefore, if (8.3-25) is substituted for both Δv_{qds}^r and Δi_{qds}^r, (8.3-19) becomes (8.3-29).

$$
\begin{bmatrix}
\cos\delta_o\Delta v_{qs}^e - \sin\delta_o\Delta v_{ds}^e \\
\sin\delta_o\Delta v_{qs}^e + \cos\delta_o\Delta v_{ds}^e \\
\Delta v_{kq1}'^r \\
\Delta v_{kq2}'^r \\
\Delta e_{xfd}'^r \\
\Delta v_{kd}'^r \\
\Delta T_I \\
0
\end{bmatrix}
=
\begin{bmatrix}
\begin{array}{c}
\left(r_s + \frac{P}{\omega_b}X_q\right)i_{dso}^r + \frac{\omega_e}{\omega_b}X_d i_{qso}^r + v_{dso}^r \\[4pt]
\left(-r_s - \frac{P}{\omega_b}X_d\right)i_{qso}^r + \frac{\omega_e}{\omega_b}X_q i_{dso}^r - v_{qso}^r \\[4pt]
\text{First} \qquad \frac{P}{\omega_b}X_{mq}i_{dso}^r \\[4pt]
\text{seven} \qquad \frac{P}{\omega_b}X_{mq}i_{dso}^r \\[4pt]
\text{columns} \qquad -\frac{X_{md}}{r_{fd}'}\left(\frac{P}{\omega_b}X_{md}\right)i_{qso}^r \\[4pt]
\text{same as} \qquad -\frac{P}{\omega_b}X_{md}i_{qso}^r \\[4pt]
(8.3.19) \qquad -i_{dso}^r[X_{md}i_{dso}^r - X_{md}(i_{dso}^r - i_{fdo}'^r)] \\[4pt]
-i_{qso}^r(X_{md}i_{qso}^r - X_{mq}i_{qso}^r) \\[4pt]
p
\end{array}
\end{bmatrix}
\begin{bmatrix}
\cos\delta_o\Delta i_{qs}^e - \sin\delta_o\Delta i_{ds}^e \\
\sin\delta_o\Delta i_{qs}^e + \cos\delta_o\Delta i_{ds}^e \\
\Delta i_{kq1}'^r \\
\Delta i_{kq2}'^r \\
\Delta i_{fd}'^r \\
\Delta i_{kd}'^r \\
\frac{\Delta\omega_r}{\omega_b} \\
\Delta\delta
\end{bmatrix}
$$

$$(8.3\text{-}29)$$

Equation (8.3-29) may be partitioned and written as

$$
\begin{bmatrix}
\mathbf{T}\Delta\mathbf{v}_{qds}^e \\
\Delta\mathbf{v}_{rr}
\end{bmatrix}
=
\begin{bmatrix}
\mathbf{W} & \mathbf{Y} \\
\mathbf{Q} & \mathbf{S}
\end{bmatrix}
\begin{bmatrix}
\mathbf{T}\Delta\mathbf{i}_{qds}^e \\
\Delta\mathbf{i}_{rr}
\end{bmatrix}
\qquad (8.3\text{-}30)
$$

where all matrices and vectors are readily ascertained from (8.3-29). The above equations may be arranged as

$$
\begin{bmatrix}
\Delta\mathbf{v}_{qds}^e \\
\Delta\mathbf{v}_{rr}
\end{bmatrix}
=
\begin{bmatrix}
(\mathbf{T})^{-1}\mathbf{W}\mathbf{T} & (\mathbf{T})^{-1}\mathbf{Y} \\
\mathbf{Q}\mathbf{T} & \mathbf{S}
\end{bmatrix}
\begin{bmatrix}
\Delta\mathbf{i}_{qds}^e \\
\Delta\mathbf{i}_{rr}
\end{bmatrix}
\qquad (8.3\text{-}31)
$$

Equation (8.3-31) can now be written in the form

$$\mathbf{E}p\mathbf{x} = \mathbf{F}\mathbf{x} + \mathbf{u} \qquad (8.3\text{-}32)$$

where

$$
(\mathbf{x})^T = \left[(\Delta\mathbf{i}_{qds}^e)^T \quad (\Delta\mathbf{i}_{rr})^T \right]
$$

$$
= \left[\Delta i_{qs}^e \quad \Delta i_{ds}^e \quad \Delta i_{kq1}'^r \quad \Delta i_{kq2}'^r \quad \Delta i_{fd}'^r \quad \Delta i_{kd}'^r \quad \frac{\Delta\omega_r}{\omega_b} \quad \Delta\delta \right] \qquad (8.3\text{-}33)
$$

$$
(\mathbf{u})^T = \left[(\Delta\mathbf{v}_{qds}^e)^T \quad (\Delta\mathbf{v}_{rr})^T \right]
$$

$$
= \left[\Delta v_{qs}^e \quad \Delta v_{ds}^e \quad \Delta v_{kq1}'^r \quad \Delta v_{kq2}'^r \quad \Delta e_{xfd}'^r \quad \Delta v_{kd}'^r \quad \Delta T_I \quad 0 \right] \qquad (8.3\text{-}34)
$$

It is important to note that $\Delta\mathbf{i}_{rr}(\Delta\mathbf{v}_{rr})$ contains variables other than rotor currents (voltages). Although it may be preferable to reserve $\Delta\mathbf{i}_{rr}$ and $\Delta\mathbf{v}_{rr}$ only for rotor currents and voltages, this would require separating out $\Delta\omega_r/\omega_b$, $\Delta\delta$, and ΔT_L, thereby increasing the terms in the partitioned matrix. The notation used here is solely for the purpose of compactness.

The $\mathbf{E}$ and $\mathbf{F}$ matrices are

$$\mathbf{E} = \begin{bmatrix} (\mathbf{T})^{-1}\mathbf{W}_p\mathbf{T} & (\mathbf{T})^{-1}\mathbf{Y}_p \\ \mathbf{Q}_p\mathbf{T} & \mathbf{S}_p \end{bmatrix} \tag{8.3-35}$$

$$\mathbf{F} = -\begin{bmatrix} (\mathbf{T})^{-1}\mathbf{W}_k\mathbf{T} & (\mathbf{T})^{-1}\mathbf{Y}_k \\ \mathbf{Q}_k\mathbf{T} & \mathbf{S}_k \end{bmatrix} \tag{8.3-36}$$

In the above equations the matrices with the subscript p contain only the terms associated with the derivatives while the k subscript denotes the matrices containing the remaining terms. In particular,

$$\mathbf{W}_p = \frac{1}{\omega_b}\begin{bmatrix} -X_q & 0 \\ 0 & -X_d \end{bmatrix} \tag{8.3-37}$$

$$\mathbf{Y}_p = \frac{1}{\omega_b}\begin{bmatrix} X_{mq} & X_{mq} & 0 & 0 & 0 & X_q i^r_{dso} \\ 0 & 0 & X_{md} & X_{md} & 0 & -X_d i^r_{qso} \end{bmatrix} \tag{8.3-38}$$

$$(\mathbf{Q}_p)^T = \frac{1}{\omega_b}\begin{bmatrix} -X_{mq} & -X_{mq} & 0 & 0 & 0 & 0 \\ 0 & 0 & -\frac{X^2_{md}}{r'_{fd}} & -X_{md} & 0 & 0 \end{bmatrix} \tag{8.3-39}$$

$$\mathbf{S}_p = \frac{1}{\omega_b}\begin{bmatrix} X'_{kq1} & X_{mq} & 0 & 0 & 0 & X_{mq} i^r_{dso} \\ X_{mq} & X'_{kq2} & 0 & 0 & 0 & X_{mq} i^r_{dso} \\ 0 & 0 & \frac{X_{md}X'_{fd}}{r'_{fd}} & \frac{X^2_{md}}{r'_{fd}} & 0 & -\frac{X^2_{md}}{r'_{fd}} i^r_{qso} \\ 0 & 0 & X_{md} & X'_{kd} & 0 & -X_{md} i^r_{qso} \\ 0 & 0 & 0 & 0 & 2H\omega_b & 0 \\ 0 & 0 & 0 & 0 & 0 & 1 \end{bmatrix} \tag{8.3-40}$$

$$\mathbf{W}_k = \begin{bmatrix} -r_s & -\frac{\omega_e}{\omega_b}X_d \\ \frac{\omega_e}{\omega_b}X_q & -r_s \end{bmatrix} \tag{8.3-41}$$

$$\mathbf{Y}_k = \begin{bmatrix} 0 & 0 & \frac{\omega_e}{\omega_b}X_{md} & \frac{\omega_e}{\omega_b}X_{md} & -X_d i^r_{dso} + X_{md} i^r_{fdo} & r_s i^r_{dso} - \frac{\omega_e}{\omega_b}X_d i^r_{qso} + v^r_{dso} \\ -\frac{\omega_e}{\omega_b}X_{mq} & -\frac{\omega_e}{\omega_b}X_{mq} & 0 & 0 & X_q i^r_{qso} & -r_s i^r_{qso} - \frac{\omega_e}{\omega_b}X_q i^r_{dso} - v^r_{qso} \end{bmatrix} \tag{8.3-42}$$

$$(\mathbf{Q}_k)^T = \begin{bmatrix} 0 & 0 & 0 & 0 & X_{mq} i^r_{dso} - X_{md}(i^r_{dso} - i^{\prime r}_{fdo}) & 0 \\ 0 & 0 & 0 & 0 & -X_{md} i^r_{qso} + X_{mq} i^r_{qso} & 0 \end{bmatrix} \tag{8.3-43}$$

$$
S_k = \begin{bmatrix}
r'_{kq1} & 0 & 0 & 0 & 0 & 0 \\
0 & r'_{kq2} & 0 & 0 & 0 & 0 \\
0 & 0 & X_{md} & 0 & 0 & 0 \\
0 & 0 & 0 & r'_{kd} & 0 & 0 \\
-X_{mq}i^r_{dso} & -X_{mq}i^r_{dso} & X_{md}i^r_{qso} & X_{md}i^r_{qso} & 0 & \begin{matrix} -i^r_{dso}[X_{mq}i^r_{dso} \\ -X_{md}(i^r_{dso}-i^r_{fdo})] \\ -i^r_{qso}(X_{md}i^r_{qso} \\ -X_{mq}i^{\prime r}_{qso}) \end{matrix} \\
0 & 0 & 0 & 0 & -\omega_b & 0
\end{bmatrix}
$$

$$(8.3\text{-}44)$$

Equation (8.3-32) may be written in the fundamental form

$$p\mathbf{x} = \mathbf{A}\mathbf{x} + \mathbf{B}\mathbf{u} \qquad (8.3\text{-}45)$$

with

$$\mathbf{A} = (\mathbf{E})^{-1}\mathbf{F} \qquad (8.3\text{-}46)$$

$$\mathbf{B} = (\mathbf{E})^{-1} \qquad (8.3\text{-}47)$$

where $\mathbf{E}$ is given by (8.3-35) and $\mathbf{F}$ by (8.3-36).

8.4 SMALL-DISPLACEMENT STABILITY: EIGENVALUES

With the linear differential equations written in state variable form, the $\mathbf{u}$ vector represents the forcing functions. If $\mathbf{u}$ is set equal to zero, the general solution of the homogeneous or force-free linear differential equations becomes

$$\mathbf{x} = e^{\mathbf{A}t}\mathbf{K} \qquad (8.4\text{-}1)$$

where $\mathbf{K}$ is a vector formed by an arbitrary set of initial conditions. The exponential $e^{\mathbf{A}t}$ represents the unforced response of the system. It is called the state transition matrix. Small-displacement stability is assured if all elements of the transition matrix approach zero asymptotically as time approaches infinity. Asymptotic behavior of all elements of the matrix occurs whenever all of the roots of the characteristic equation of $\mathbf{A}$ have negative real parts where the characteristic equation of $\mathbf{A}$ is defined as

$$\det(\mathbf{A} - \lambda\mathbf{I}) = 0 \qquad (8.4\text{-}2)$$

In (8.4-2), $\mathbf{I}$ is the identity matrix and λ are the roots of the characteristic equation of $\mathbf{A}$ referred to as characteristic roots, latent roots, or eigenvalues. Herein, we will use the latter designation. One should not confuse the λ used here to denote eigenvalues with the same notation used earlier to denote flux linkages.

The eigenvalues provide a simple means of predicting the behavior of an induction or synchronous machine at any balanced operating condition. Eigenvalues may be either real or complex; when complex, they occur as conjugate pairs signifying a mode of oscillation of the state variables. Negative real parts correspond to state variables or oscillations of state variables which decrease exponentially with time. Positive real parts indicate an exponential increase with time, an unstable condition.

8.5 EIGENVALUES OF TYPICAL INDUCTION MACHINES

The eigenvalues of an induction machine can be obtained by using a standard eigenvalue computer routine to calculate the roots of **A** given by (8.3-17) [1–3]. The eigenvalues given in Table 8.5-1 are for the machines listed in Table 4.10-1. The induction machine, as we have perceived it, is described by five state variables and hence five eigenvalues. Sets of eigenvalues for each machine at stall, rated, and no-load speeds are given in Table 8.5-1 for rated frequency operation. Plots of the eigenvalues (real part and only the positive imaginary part) for rotor speeds from stall to synchronous are given in Figs. 8.5-1 and 8.5-2 for the 3- and 2250-hp induction motors, respectively.

At stall, the two complex conjugate pairs of eigenvalues both have a frequency (imaginary part) corresponding to ω_b. The frequency of one complex conjugate pair decreases as the speed increases from stall, whereas the frequency of the other complex conjugate pair remains at approximately ω_b— in fact, nearly equal to ω_b for the larger-horsepower machines. The eigenvalues are dependent upon the parameters of the machine, and it is difficult to relate analytically a change in an eigenvalue with a change in a specific machine parameter. It is possible, however, to identify an

Table 8.5-1 Induction Machine Eigenvalues

Rating (hp)	Stall	Rated Speed	No Load
3	$-4.57 \pm j377$	$-85.6 \pm j313$	$-89.2 \pm j316$
	$-313 \pm j377$	$-223 \pm j83.9$	$-218 \pm j60.3$
	1.46	-16.8	-19.5
50	$-2.02 \pm j377$	$-49.4 \pm j356$	$-50.1 \pm j357$
	$-198 \pm j377$	$-142 \pm j42.5$	$-140 \pm j18.2$
	1.18	-14.4	-17.0
500	$-0.872 \pm j377$	$-41.8 \pm j374$	$-41.8 \pm j374$
	$-70.3 \pm j377$	$-15.4 \pm j41.5$	$-14.3 \pm j42.8$
	0.397	-27.5	-29.6
2250	$-0.428 \pm j377$	$-24.5 \pm j376$	$-24.6 \pm j376$
	$-42.6 \pm j377$	$-9.36 \pm j41.7$	$-9.05 \pm j42.5$
	0.241	-17.9	-18.5

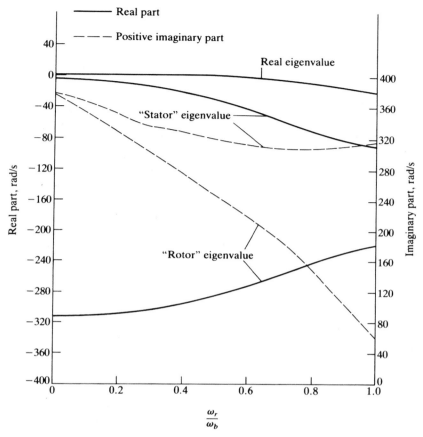

Figure 8.5-1   Plot of eigenvalues for a 3-hp motor.

association between eigenvalues and the machine variables. For example, the complex conjugate pair that remains at a frequency close to ω_b is primarily associated with the transient offset currents in the stator windings; this reflects into the synchronously rotating reference frame as a decaying 60-Hz variation. We will find that this complex conjugate pair, which is denoted as the "stator" eigenvalues in Figs. 8.5-1 and 8.5-2, is not present when the electric transients are neglected in the stator voltage equations. It follows that the transient response of the machine is influenced by this complex conjugate eigenvalue pair whenever a disturbance causes a transient offset in the stator currents. It is recalled that, in Chapter 4, we noted a transient pulsation in the instantaneous torque of 60 Hz during free acceleration and following a 3-phase fault at the terminals with the machine initially operating at near rated conditions. We also noted that the pulsations were more damped in the case of the smaller-horsepower machines than for the larger-horsepower machines. It is noted in Table 8.5-1 that the magnitudes of the real part of the complex eigenvalues

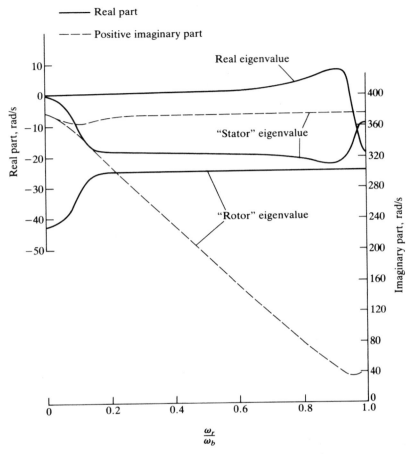

Figure 8.5-2   Plot of eigenvalues for a 2250-hp induction motor.

with a frequency corresponding to ω_b are larger, signifying more damping, for the smaller-horsepower machines than for the larger machines.

The complex conjugate pair that changes in frequency as the rotor speed varies is associated primarily with the electric transients in the rotor circuits and denoted in Figs. 8.5-1 and 8.5-2 as the "rotor" eigenvalue. This complex conjugate pair is not present when the rotor electric transients are neglected. The damping associated with this complex conjugate pair is less for the larger-horsepower machines than for the smaller machines. It is recalled that, during free acceleration, the 3- and 50-hp machines approached synchronous speed in a well-damped manner whereas the 500- and the 2250-hp machines demonstrated damped oscillations about synchronous speed. Similar behavior was noted as the machines approached their final operating point following a load torque change or a 3-phase terminal fault. This behavior corresponds to that predicted by this eigenvalue. It is interesting to note that this eigenvalue is reflected noticeably into the rotor speed, whereas the higher-frequency "stator" eigenvalue is not. This, of course, is due to the fact that for a given inertia and torque amplitude, a low-frequency torque component will cause a larger amplitude variation in rotor speed than will a high-frequency component.

The real eigenvalue signifies an exponential response. It would characterize the behavior of the induction machine equations if all electric transients are mathematically neglected or if, in the actual machine, the electric transients are highly damped as in the case of the smaller-horsepower machines. Perhaps the most interesting feature of this eigenvalue, which is denoted as the real eigenvalue in Figs. 8.5-1 and 8.5-2, is that it can be related to the steady-state torque–speed curve. If we think for a moment about the torque–speed characteristics, we realize that an induction machine can operate stably only in the negative-slope portion of the torque–speed curve. If we were to assume an operating point on the positive-slope portion of the torque–speed curve, we would find that a small disturbance would cause the machine to move away from this operating point, either accelerating to the negative-slope region or decelerating to stall and perhaps reversing direction of rotation depending upon the nature of the load torque. A positive eigenvalue signifies a system that would move away from an assumed operating point. Note that this eigenvalue is positive over the positive-slope region of the torque–speed curve, becoming negative after maximum steady-state torque.

8.6 EIGENVALUES OF TYPICAL SYNCHRONOUS MACHINES

The eigenvalues of a synchronous machine can be obtained by using a standard eigenvalue computer routine to calculate the roots of $\mathbf{A}$ given by (8.3-46) [1–3]. The eigenvalues of the two synchronous machines studied in Section 5.10 are given in Table 8.6-1 for rated operation.

The complex conjugate pair with the frequency (imaginary part) approximately equal to ω_b is associated with the transient offset currents in the stator windings that cause the 60-Hz pulsation in electromagnetic torque. This 60-Hz pulsation in torque is evident in the computer traces of a 3-phase fault at the machine terminals shown in Figs. 5.11-1 and 5.11-3. Although operation in Figs. 5.11-1 and 5.11-3 is initially at

Table 8.6-1 Synchronous Machine Eigenvalues for Rated Conditions

Hydro Turbine Generator	Steam Turbine Generator
$-3.58 \pm j77$	$-4.45 \pm j377$
$-1.33 \pm j8.68$	$-1.70 \pm j10.5$
-24.4	-32.2
-22.9	-11.1
-0.453	-0.349
	-0.855

rated conditions, the 3-phase fault and subsequent switching causes the operating condition to change significantly from rated conditions. Nevertheless, we note that the 60-Hz pulsation is damped slightly more in the case of the steam turbine generator than in the case of the hydro turbine generator. Correspondingly, the relative values of the real parts of the "stator" eigenvalues given in Table 8.6-1 indicate that the stator electric transients of the steam unit are damped more than the stator transients of the hydro unit.

The remaining complex conjugate pair is similar to the "rotor" eigenvalue in the case of the induction machine. However, in the case of the synchronous machine this mode of oscillation is commonly referred to as the hunting or swing mode, which is the principal mode of oscillation of the rotor of the machine relative to the electrical angular velocity of the electrical system (the infinite bus in the case of studies made in Chapter 5). This mode of oscillation is apparent in the machine variables, especially the rotor speed, in Figs. 5.11-1 and 5.11-3 during the "settling out" period

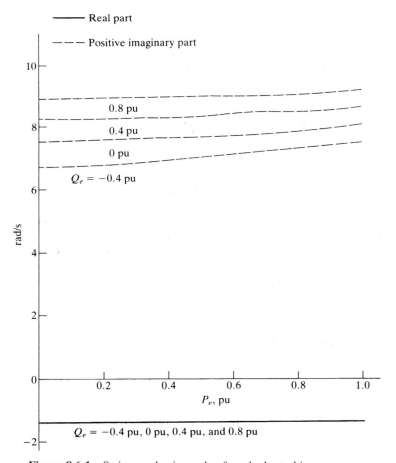

Figure 8.6-1 Swing mode eigenvalue for a hydro turbine generator.

following reclosing. As indicated by this complex conjugate eigenvalue, the "settling out" rotor oscillation of the steam unit (Fig. 5.11-3) is more damped and of higher frequency than the corresponding rotor oscillation of the hydro unit.

The real eigenvalues are associated with the decay of the offset currents in the rotor circuits and therefore associated with the inverse of the effective time constant of these circuits. It follows that because the field winding has the largest time constant, it gives rise to the smallest of the real eigenvalues.

In the case of the two synchronous machines being considered, the "stator" eigenvalue and the real eigenvalues do not change significantly in value as the real and reactive power loading conditions change. It is interesting, however, to observe the change in the "rotor" eigenvalue associated with the swing mode for various loading conditions. Plots of this eigenvalue are shown in Fig. 8.6-1 for the hydro turbine generator and are shown in Fig. 8.6-2 for the steam turbine generator. The real power is varied from 0 to 1.0 pu with the reactive power requirements varied from -0.4 pu (consuming reactive power) to 0.8 pu. The changes are not drastic, especially in the case of the damping. The frequency (imaginary part) of the swing mode increases slightly for a given reactive power as the real power output increases.

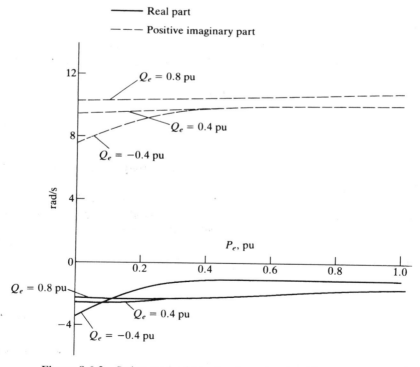

Figure 8.6-2 Swing mode eigenvalue for a steam turbine generator.

8.7 TRANSFER FUNCTION FORMULATION

The linearized equations of induction and synchronous machines are often used in the analysis and design of controls associated with the machine. For example, the design of speed controls for induction machines in variable-speed drive systems, as well as the design of voltage regulators for synchronous machines, is facilitated by the linearized equations written in the form of a transfer function relating the output variable being controlled to the controlling input variable. To a lesser extent, transfer functions of the linearized equations are also used to study the small-displacement behavior about an operating point rather than using the detailed non-linear equations. It is not our intent to become involved in control design or to determine the accuracy of the small-displacement equations when subjected to large excursions in input variables. It is, however, important to be aware of the steps necessary to formulate various transfer functions in terms of state variables, which is readily performed by standard computer languages [1–3].

It is convenient to employ basic linear control theory to establish the transfer functions. The linear system dynamical equations are

$$p\mathbf{x} = \mathbf{A}\mathbf{x} + \mathbf{B}\mathbf{u} \tag{8.7-1}$$

$$\mathbf{y} = \mathbf{C}\mathbf{x} + \mathbf{D}\mathbf{u} \tag{8.7-2}$$

Equation (8.7-1) is the state equation or the fundamental form of the linearized system equations. Equation (8.7-2) is the output equation where $\mathbf{y}$ is the output vector which is expressed as a linear combination of the state vector $\mathbf{x}$ and the input vector $\mathbf{u}$.

If (8.7-1) is solved for $\mathbf{x}$ and the result is substituted into (8.7-2), we have

$$\mathbf{y} = \mathbf{C}(s\mathbf{I} - \mathbf{A})^{-1}\mathbf{B}\mathbf{u} + \mathbf{D}\mathbf{u} \tag{8.7-3}$$

where the operator p has been replaced by the Laplace operator s commonly used in transfer function formulation.

In the derivation of a transfer function we are generally interested in the relationship between the input that can be a single input variable or a linear combination of several input variables and the system output $\mathbf{y}$ that can be a single output state variable or a linear combination of several output variables together with an input variable or a linear combination of input variables. In this formulation it is often convenient to express $\mathbf{u}$ as

$$\mathbf{u} = \mathbf{G}\,\Delta f_i \tag{8.7-4}$$

where $\mathbf{G}$ is a column matrix and Δf_i is an input variable such as load torque, for example, or a linear combination of several input variables.

Substituting (8.7-4) into (8.7-3) yields

$$\mathbf{y} = \mathbf{C}(s\mathbf{I} - \mathbf{A})^{-1}\mathbf{B}\mathbf{G}\,\Delta f_i + \mathbf{D}\mathbf{G}\,\Delta f_i \tag{8.7-5}$$

The formulation of the transfer function is completed once $\mathbf{C}$ and $\mathbf{D}$ are selected to give the desired output variable and once $\mathbf{G}$ is selected to give the desired input variable.

Example 8A   As an example let us formulate the transfer function between a change in the magnitude of terminal voltages Δv_s of an induction machine and a change in electromagnetic torque ΔT_e. Before proceeding, it is helpful to recall some relationships that we will need. In particular, (3.7-8) and (3.7-9) are

$$f_{qs}^e = \sqrt{2}f_s\cos[\theta_{ef}(0) - \theta_e(0)] \tag{8A-1}$$

$$f_{ds}^e = -\sqrt{2}f_s\sin[\theta_{ef}(0) - \theta_e(0)] \tag{8A-2}$$

Our objective is to formulate the transfer function $\Delta T_e/\Delta v_s$. From (8.7-4) we obtain

$$\mathbf{u} = \mathbf{G}\,\Delta f_i = \mathbf{G}\,\Delta v_s \tag{8A-3}$$

In terms of per unit voltages, (8A-1) and (8A-2) become

$$v_{qs}^e = v_s\cos\alpha \tag{8A-4}$$

$$v_{ds}^e = -v_s\sin\alpha \tag{8A-5}$$

where for compactness we have

$$\alpha = \theta_{ev}(0) - \theta_e(0) \tag{8A-6}$$

It is assumed that the voltage changes in magnitude and not in phase. Therefore

$$\Delta v_{qs}^e = \Delta v_s\cos\alpha \tag{8A-7}$$

$$\Delta v_{ds}^e = -\Delta v_s\sin\alpha \tag{8A-8}$$

It follows that

$$(\mathbf{G})^T = [\cos\alpha \quad -\sin a \quad 0 \quad 0 \quad 0] \tag{8A-9}$$

From (8.3-7) we can write ΔT_e as

$$\Delta T_e = X_M(i_{dro}^{'e}\Delta i_{qs}^e - i_{qro}^{'e}\Delta i_{ds}^e - i_{dso}^e\Delta i_{qr}^{'e} + i_{qso}^e\Delta i_{dr}^{'e}) \tag{8A-10}$$

If we put (8A-10) in the form of (8.7-2), we obtain

$$\Delta T_e = \mathbf{C}\mathbf{x}$$

where $\mathbf{D}$ in (8.7-2) is zero and

$$\mathbf{C} = X_M[i_{dro}^{\prime e} \quad -i_{qro}^{\prime e} \quad -i_{dso}^{e} \quad i_{qso}^{e} \quad 0] \tag{8A-11}$$

Hence, from (8.7-5)

$$\frac{\Delta T_e}{\Delta v_s} = \mathbf{C}(s\mathbf{I} - \mathbf{A})^{-1}\mathbf{B}\mathbf{G} \tag{8A-12}$$

where $\mathbf{A}$ and $\mathbf{B}$ are defined respectively by (8.3-17) and (8.3-18) and $\mathbf{C}$ and $\mathbf{G}$ are defined by (8A-11) and (8A-9), respectively.

The form of this transfer function, with the representation of the induction machine as we have perceived it, is

$$\frac{\Delta T_e}{\Delta v_s} = \frac{A(s - a_1)(s - a_2)\cdots(s - a_4)}{(s - b_1)(s - b_2)\cdots(s - b_s)} \tag{8A-13}$$

For the 3-hp induction machine (data given in Table 4.10-1) operating at rated conditions the gain A and the zeros $(a_1, \ldots, a_4)$ and poles $(b_1, \ldots, b_5)$ are

$$A = 5.02 \times 10^3$$

$$a_1, a_2 = -7.33 \pm j95.8 \qquad a_3 = -329 \qquad a_4 = 0$$

$$b_1, b_2 = -223 \pm j83.9 \qquad b_3 = -16.8 \qquad b_4, b_5 = -85.6 \pm j313$$

It is clear that the poles are the eigenvalues that were given previously in Table 8.5-1 for rated operation.

For the 2250-hp induction machine (data given in Table 4.10-1) operating at rated conditions

$$A = 2.48 \times 10^3$$

$$a_1, a_2 = -10.4 \pm j20.5 \qquad a_3 = 9.44 \times 10^{-5} \qquad a_4 = -80.0$$

$$b_1, b_2 = -9.37 \pm j41.7 \qquad b_3 = -17.9 \qquad b_4, b_5 = -24.6 \pm j376$$

Example 8B   As a second example let us repeat Example 8A for a synchronous machine. Because we are interested only in the transfer function between a change in the magnitude of the terminal voltages Δv_s and a change in the electromagnetic torque ΔT_e, it is unnecessary to transform the rotor reference frame currents $\Delta\mathbf{i}_{qds}^r$ to the synchronously rotating reference frame. Hence for this example, (8.3-29) may be written as

$$
\begin{bmatrix}
\cos\delta_o\Delta v_{qs}^e - \sin\delta_o\Delta v_{ds}^e \\
\sin\delta_o\Delta v_{qs}^e + \cos\delta_o\Delta v_{ds}^e \\
\Delta v_{kq1}^{\prime r} \\
\Delta v_{kq2}^{\prime r} \\
\Delta e_{xfd}^{\prime r} \\
\Delta v_{kd}^{\prime r} \\
\Delta T_l \\
0
\end{bmatrix}
=
\begin{bmatrix}
& & & & & & & v_{dso}^{\prime r} \\
& & & & & & & -v_{qso}^r \\
& & \text{First} & & & & & 0 \\
& & \text{seven} & & & & & 0 \\
& & \text{columns} & & & & & 0 \\
& & \text{same as} & & & & & 0 \\
& & (8.3\text{-}19) & & & & & 0 \\
& & & & & & & p
\end{bmatrix}
\begin{bmatrix}
\Delta i_{qs}^r \\
\Delta i_{ds}^r \\
\Delta i_{kq1}^{\prime r} \\
\Delta i_{kq2}^{\prime r} \\
\Delta i_{fd}^{\prime r} \\
\Delta i_{kd}^{\prime r} \\
\frac{\Delta\omega_r}{\omega_b} \\
\Delta\delta
\end{bmatrix}
\tag{8B-1}
$$

The above equation may be written as

$$
\begin{bmatrix}
\Delta\mathbf{v}_{qds}^e \\
\Delta\mathbf{v}_{rr}
\end{bmatrix}
=
\begin{bmatrix}
(\mathbf{T})^{-1}\mathbf{W} & (\mathbf{T})^{-1}\mathbf{Y} \\
\mathbf{Q} & \mathbf{S}
\end{bmatrix}
\begin{bmatrix}
\Delta\mathbf{i}_{qds}^r \\
\Delta\mathbf{i}_{rr}
\end{bmatrix}
\tag{8B-2}
$$

where $\mathbf{T}$ is defined by (8.3-27). Equation (8B-2) may be written in the form

$$
\mathbf{E}p\mathbf{x} = \mathbf{F}\mathbf{x} + \mathbf{u}
\tag{8B-3}
$$

where $\mathbf{x}$ and $\mathbf{u}$ are defined by (8B-2). The $\mathbf{E}$ and $\mathbf{F}$ matrices are

$$
\mathbf{E} =
\begin{bmatrix}
(\mathbf{T})^{-1}\mathbf{W}_p & (\mathbf{T})^{-1}\mathbf{Y}_p \\
\mathbf{Q}_p & \mathbf{S}_p
\end{bmatrix}
\tag{8B-4}
$$

$$
\mathbf{F} =
\begin{bmatrix}
(\mathbf{T})^{-1}\mathbf{W}_k & (\mathbf{T})^{-1}\mathbf{Y}_k \\
\mathbf{Q}_k & \mathbf{S}_k
\end{bmatrix}
\tag{8B-5}
$$

where $\mathbf{W}_p$, $\mathbf{Q}_p$, $\mathbf{W}_k$, and $\mathbf{Q}_k$ are defined by (8.3-37), (8.3-39), (8.3-41), and (8.4-43), respectively. The matrix $\mathbf{Y}_p$ in (8B-5) is defined by (8.3-38) with the last column replaced by zeros. Likewise, $\mathbf{S}_p$ is (8.4-40) with the last column replaced by zeros except for the last term, which remains as 1. Also, $\mathbf{Y}_k$ is (8.3-42) with $r_s i_{dso}^r - (\omega_e/\omega_b)X_d i_{qso}^r$ eliminated from the first term of the last column and $-r_s i_{qso}^r + (\omega_e/\omega_b)X_q i_{dso}^r$ eliminated from the second. $\mathbf{S}_k$ in (8B-5) is (8.3-44) with only zeros in the last column. We can now write (8B-3) in the fundamental form given by (8.3-45) where $\mathbf{A}$ and $\mathbf{B}$ are defined by (8.3-46) and (8.3-48), respectively.

We can make use of some of the material in Example 8A. In particular, the evaluation of $\mathbf{G}$ for the synchronous machine is the same as for the induction machine because we have expressed the terminal voltages of the synchronous machine in terms of synchronously rotating reference-frame variables $\Delta \mathbf{v}_{qds}^e$. Hence, from (8A-9) we obtain

$$(\mathbf{G})^T = [\cos \alpha \quad -\sin \alpha \quad 0 \quad 0 \quad 0 \quad 0 \quad 0 \quad 0] \tag{8B-6}$$

Because

$$\Delta T_I = \Delta T_e + 2Hp \frac{\Delta \omega_r}{\omega_b} \tag{8B-7}$$

we can extract ΔT_e from (8.3-19) and write in the form

$$\Delta T_e = \mathbf{Cx} \tag{8B-8}$$

where

$$\mathbf{C} = [X_{mq} i_{dso}^r - X_{md}(i_{dso}^r - i_{fdo}^{'r}) \quad -X_{md} i_{qso}^r + X_{mq} i_{qso}^r \quad -X_{mq} i_{dso}^r$$

$$-X_{mq} i_{dso}^r \quad X_{md} i_{qso}^r \quad X_{md} i_{qso}^r \quad 0 \quad 0] \tag{8B-9}$$

Hence, from (8.7-5) with $\mathbf{D}$ equal to zero we obtain

$$\frac{\Delta T_e}{\Delta v_s} = \mathbf{C}(s\mathbf{I} - \mathbf{A})^{-1}\mathbf{BG} \tag{8B-10}$$

where $\mathbf{C}$ and $\mathbf{G}$ are defined by (8B-9) and (8B-6), respectively.

The transfer function $\Delta T_e/\Delta v_s$ for a synchronous machine with $kq1$, $kq2$, fd, and kd windings will be of the form

$$\frac{\Delta T_e}{\Delta v_s} = \frac{A(s - a_1)(s - a_2) \cdots (s - a_7)}{(s - b_1)(s - b_2) \cdots (s - b_8)} \tag{8B-11}$$

The hydro turbine generator (data given in Section 5.10) does not have a $kq1$ winding; hence, a_7 and b_8 are not present. For operation at rated conditions we have

$$A = -2.08 \times 10^3$$

$a_1 = -20.5$	$a_2 = -14.5$	$a_3 = -1.46$
$a_4 = -1.82 \times 10^{-3}$	$a_5 = 1.82 \times 10^{-3}$	$a_6 = 53.8$
$b_1, b_2 = -1.33 \pm j8.68$	$b_3 = -24.4$	$b_4 = -22.9$
$b_5 = -0.453$	$b_6, b_7 = -3.58 \pm j377$	

For the steam turbine generator (data given in Section 5.10) operating at rated conditions we have

$$A = -1.66 \times 10^3$$

$a_1, a_2 = -1.27 \pm j4.22$	$a_3 = 52.6$	$a_4 = -0.559$
$a_5 = -41.3$	$a_6 = 6.25 \times 10^{-2}$	$a_7 = -7.10 \times 10^{-2}$
$b_1, b_2 = -1.70 \pm j10.5$	$b_3 = -32.2$	$b_4 = -11.1$
$b_5 = -0.349$	$b_6 = -0.855$	$b_7, b_8 = -4.45 \pm j377$

In Examples 8A and 8B it is assumed that the machine is connected to an infinite bus. In other words the terminal voltages cannot be changed by any machine variable. Hence, voltage regulation is irrelevant; however, when the synchronous machine is connected to an actual system, voltage regulation may be required, in which case it is often desirable to formulate a transfer function between a change in field voltage $\Delta e''_{xfd}$ and a change in terminal voltage Δv_s. We note, however, that $\Delta e''_{xfd}$ is an input variable and Δv_s is a combination of input variables. Therefore it is necessary to express the change in terminal voltage as a function of state variables. This requires considerable matrix manipulation. The problem can be simplified, however, by assuming that large resistors or small capacitors exist across the terminals of the machine. This assumption markedly reduces the matrix manipulations and calculations involved, and in most cases it introduces negligible error.

REFERENCES

[1] C. M. Ong, *Dynamic Simulation of Electric Machinery*, Prentice-Hall PTR, Upper Saddle River, NJ, 1998.
[2] Advanced Continuous Simulation Language (ACSL), Reference Manual, Version 11, MGA Software, 1995.
[3] Simulink: Dynamic System Simulation for Matlab, Using Simulink Version 4, The MathWorks Inc., 2000.

PROBLEMS

1 Derive the small-displacement equations of an induction machine with flux linkages per second as state variables. Express the equations in fundamental form.

2 Repeat Problem 1 for a synchronous machine.

3 Write the small-displacement equations for an induction machine with currents as state variables and with a small displacement in ω_e, where $\Delta\omega_e$ is an input variable. It is clear that the resulting equations are valid for small changes in the frequency of the applied stator voltages.

4 Repeat Problem 3 for a synchronous machine.

5 A synchronous machine is being driven at one-half rated speed with the stator terminals open circuited. A step voltage is applied to the field winding, and the speed is held fixed. Express the magnitude of the terminal voltage in terms of the state variables.

6 Calculate the eigenvalues for the 3-hp induction motor for stator voltages of frequencies of $\frac{1}{4}$, $\frac{1}{2}$, and $\frac{3}{4}$ rated. Assume the amplitude of the applied voltages decreases linearly with frequency.

7 Formulate (a) $(\Delta\omega_r/\omega_b)/\Delta T_L$ and (b) $\Delta P_e/\Delta v_s$ for an induction machine.

8 Formulate (a) $(\Delta\omega_r/\omega_b)/\Delta T_I$, (b) $(\Delta P_e/\Delta v_s)$, and (c) $\Delta i_s/\Delta e''^r_{xfd}$ for a synchronous machine.

9 Repeat Example 8A using flux linkages per second as state variables.

10 Repeat Example 8B using flux linkages per second as state variables.

11 Using the results of Problem 3, formulate the transfer function $\Delta\omega_r/\Delta\omega_e$.

12 Repeat Problem 11 for a synchronous machine using the results from Problem 4.

13 Formulate the transfer function between the field voltage and the amplitude of the terminal voltages of a synchronous machine connected to an infinite bus through a transmission line having the resistance r_l and inductive reactance X_l, per phase.

14 Repeat Problem 13 with a balanced set of resistors connected, in wye, across the machine terminals.

15 Repeat Problem 14 with the resistors replaced by a balanced set of capacitors.

16 In this chapter we have dealt with the 3-phase induction machine. Explain the changes that must be made in the equations in order for the relationships to apply to a 2-phase induction machine.

17 Repeat Problem 16 for a 2-phase synchronous machine.

Chapter 9

REDUCED-ORDER MACHINE EQUATIONS

9.1 INTRODUCTION

Over the years there has been considerable attention given to the development of simplified models primarily for the purpose of predicting the dynamic behavior of electric machines during large excursions in some or all of the machine variables. Before the 1960s the dynamic behavior of induction machines was generally predicted with the steady-state voltage equations and the dynamic relationship between rotor speed and torque. Similarly, the large-excursion behavior of synchronous machines was predicted by a set of steady-state voltage equations with modifications to account for transient conditions, as presented in Chapter 5, along with the dynamic relationship between rotor angle and torque.

With the advent of computers, these models have given way to more accurate representations. In some cases the machine equations are programmed in detail; however, in the vast majority of cases a reduced-order model is used in computer simulations of power systems. In particular, it is standard to neglect the electric transients in the stator voltage equations of all machines and in the voltage equations of all power system components connected to the stator (transformers, transmission lines, etc.). By using a static representation for the 60-Hz part of the system, the required number of integrations is drastically reduced. However, neglecting stator electric transients is generally not appropriate when considering power electronic-based electric drive systems wherein electronic switching occurs in the stator phases.

Because "neglecting stator electric transients" is an important aspect of machine analysis primarily for the power system engineer, it seems appropriate to devote some time to this subject. In the beginning of this chapter the theory of neglecting electric transients is established following the development given in reference 1. The voltage equations for an induction and synchronous machine are given in the arbitrary reference frame with the stator electric transients neglected. The large-excursion behavior of these machines as predicted by these reduced-order models is compared with the behavior predicted by the complete equations given in Chapters 4 and 5. Next, the eigenvalues are calculated using the reduced-order models and compared with those given in Chapter 8 for the detailed equations. From these comparisons, we not only become aware of the inaccuracies involved when using the reduced-order models but we also are able to observe the influence that the electric transients have on the dynamic behavior of induction and synchronous machines. Although the reduced-order equations are valid for operation with stator voltages of any frequency, attention is focused on rated operation. The main advantage of these models is the order reduction it provides when simulating a large power system.

The nonlinear differential equations that describe the behavior of an electric machine prohibit a rigorous analytical explanation of the large-excursion inaccuracies of the model with the stator transients neglected; however, an analytical explanation of the inaccuracies of this reduced-order model in predicting the eigenvalues of an induction machine is given in reference 2. Because this approach is quite lengthy and of most interest to the specialist, it is omitted and only the results of this analysis are discussed.

9.2 REDUCED-ORDER EQUATIONS

In the case of the induction machine there are two reduced-order models commonly employed to calculate the electromagnetic torque during large transient excursions. The most elementary of these is the one wherein the electric transients are neglected in both the stator and rotor circuits. We are familiar with this steady-state model from the information presented in Chapter 4. The reduced-order model of present interest is the one wherein the electric transients are neglected only in the stator voltage equations.

In the case of the synchronous machine, there are a number of reduced-order models used to predict its large-excursion dynamic behavior. Perhaps the best known is the voltage-behind-transient reactance reduced-order model, which was discussed in Chapter 5. The reduced-order model that is now being widely used is the one wherein the electric transients of the stator voltage equations are neglected. This type of reduced-order model is considered in this chapter.

The theory of neglecting electric transients is set forth in reference 1. To establish this theory let us return for a moment to the work in Section 3.4 where the variables associated with stationary resistive, inductive, and capacitive elements were transformed to the arbitrary reference frame. It is obvious that the instantaneous voltage equations for the 3-phase resistive circuit are the same form for either transient or steady-state conditions. However, it is not obvious that the equations describing the behavior of linear symmetrical inductive and capacitive elements with the electric transients neglected (steady-state behavior) may be arranged so that the instantaneous voltages and currents are related algebraically without the operator d/dt.

Because the derivation to establish these equations is analogous for inductive and capacitive elements, it will be carried out for an inductive circuit.

First let us express the voltage equations of the 3-phase inductive circuit in the synchronously rotating reference frame. From (3.4-11) and (3.4-12) with $\omega = \omega_e$ and for balanced conditions we have

$$v_{qs}^e = \omega_e \lambda_{ds}^e + p\lambda_{qs}^e \qquad (9.2\text{-}1)$$

$$v_{ds}^e = -\omega_e \lambda_{qs}^e + p\lambda_{ds}^e \qquad (9.2\text{-}2)$$

For balanced steady-state conditions, the variables in the synchronously rotating reference frame are constants. Hence, we can neglect the electric transients by neglecting $p\lambda_{qs}^e$ and $p\lambda_{ds}^e$. Our purpose is to obtain algebraically related instantaneous voltage equations in the arbitrary reference frame which may be used to portray the behavior with the electric transients neglected (steady-state behavior). To this end, it is helpful to determine the arbitrary reference-frame equivalent of neglecting $p\lambda_{qs}^e$ and $p\lambda_{ds}^e$. This may be accomplished by noting from (3.6-1) that the synchronously rotating and arbitrary reference-frame variables are related by

$$\mathbf{f}_{qd0s} = {}^e\mathbf{K}\mathbf{f}_{qd0s}^e \qquad (9.2\text{-}3)$$

From (3.6-1) we obtain

$$
{}^e\mathbf{K} = \begin{bmatrix} \cos(\theta - \theta_e) & -\sin(\theta - \theta_e) & 0 \\ \sin(\theta - \theta_e) & \cos(\theta - \theta_e) & 0 \\ 0 & 0 & 1 \end{bmatrix} \qquad (9.2\text{-}4)
$$

It is recalled that the arbitrary reference-frame variables do not carry a raised index. If (9.2-1) and (9.2-2) are appropriately substituted in (9.2-3), the arbitrary reference-frame voltage equations may be written as

$$v_{qs} = -(\omega_e - \omega)\lambda_{ds} + \omega_e\lambda_{ds} + p\lambda_{qs} \qquad (9.2\text{-}5)$$

$$v_{ds} = (\omega_e - \omega)\lambda_{qs} - \omega_e\lambda_{qs} + p\lambda_{ds} \qquad (9.2\text{-}6)$$

These equations are identical to (3.4-11) and (3.4-12) but written in a form that preserves the identity of $p\lambda_{qs}^e$ and $p\lambda_{ds}^e$. In particular, the first and third terms on the right-hand side of (9.2-5) and (9.2-6) result from transforming $p\lambda_{qs}^e$ and $p\lambda_{ds}^e$ to the arbitrary reference frame. Thus, for balanced conditions, neglecting the electric transients in the arbitrary reference frame is achieved by neglecting these terms. The resulting equations are

$$v_{qs} = \omega_e\lambda_{ds} \qquad (9.2\text{-}7)$$

$$v_{ds} = -\omega_e\lambda_{qs} \qquad (9.2\text{-}8)$$

These equations, taken as a set, describe the behavior or a linear symmetrical inductive circuit in any reference frame with the electric transients neglected. They could not be deduced from the equations written in the form of (3.4-11) and (3.4-12), and at first glance one might question their validity. Although one recognizes that these equations are valid for neglecting the electric transients in the synchronously rotating reference frame, it is more difficult to accept the fact that these equations are also valid in an asynchronously rotating reference frame where the balanced steady-state variables are sinusoidal. However, these steady-state variables form orthogonal balanced sinusoidal sets for a symmetrical system. Therefore, the $\lambda_{ds}(\lambda_{qs})$ appearing in the $v_{qs}(v_{ds})$ equation provides the reactance voltage drop. It is left to the reader to show that the linear algebraic equations in the arbitrary reference frame for a linear symmetrical capacitive circuit are

$$i_{qs} = \omega_e q_{ds} \tag{9.2-9}$$

$$i_{ds} = -\omega_e q_{qs} \tag{9.2-10}$$

Let us consider what we have done; the arbitrary reference-frame voltage equations have been established for inductive circuits with the electric transients neglected, by neglecting the change of flux linkages in the synchronously rotating reference frame. This is the same as neglecting the offsets that occur in the actual currents as a result of a system disturbance. However, during unbalanced conditions, such as unbalanced voltages applied to the stator circuits, the voltages in the synchronously rotating reference frame will vary with time. For example, 60-Hz unbalanced stator voltages give rise to a constant and a double-frequency voltage in the synchronously rotating reference frame. Therefore, the flux linkages in the synchronously rotating reference frame will also contain a double-frequency component. It follows that, during unbalanced conditions, neglecting the change in the synchronously rotating reference frame flux linkages results in neglecting something more than just the electric transients. Therefore, the voltage equations that have been derived by neglecting the change in the flux linkages in the synchronously rotating reference frame apply for balanced or symmetrical conditions such as simultaneous application of balanced voltages, a change in either load or input torque, and a 3-phase fault. Consequently, the 0 quantities are not included in the machine equations given in this chapter.

Induction Machine

The voltage equations written in the arbitrary reference frame for an induction machine with the electric transients of the stator voltage equations neglected may be written from (4.5-22)–(4.5-33) with the 0 quantities eliminated and (9.2-7) and (9.2-8) appropriately taken into account.

$$v_{qs} = r_s i_{qs} + \frac{\omega_e}{\omega_b}\psi_{ds} \tag{9.2-11}$$

$$v_{ds} = r_s i_{ds} + \frac{\omega_e}{\omega_b}\psi_{qs} \tag{9.2-12}$$

$$v'_{qr} = r'_r i'_{qr} + \left(\frac{\omega - \omega_r}{\omega_b}\right)\psi'_{dr} + \frac{p}{\omega_b}\psi'_{qr} \tag{9.2-13}$$

$$v'_{dr} = r'_r i'_{dr} - \left(\frac{\omega - \omega_r}{\omega_b}\right)\psi'_{qr} + \frac{p}{\omega_b}\psi'_{dr} \tag{9.2-14}$$

where

$$\psi_{qs} = X_{ls}i_{qs} + X_M(i_{qs} + i'_{qr}) \tag{9.2-15}$$

$$\psi_{ds} = X_{ls}i_{ds} + X_M(i_{ds} + i'_{dr}) \tag{9.2-16}$$

$$\psi'_{qr} = X'_{lr}i'_{qr} + X_M(i_{qs} + i'_{qr}) \tag{9.2-17}$$

$$\psi'_{dr} = X'_{lr}i'_{qr} + X_M(i_{ds} + i'_{dr}) \tag{9.2-18}$$

Although the reference-frame speed appears in the speed voltages in the rotor voltage equations, it does not appear in the stator voltage equations.

The voltage equations may be expressed in terms of currents by appropriately replacing the flux linkages per second in (9.2-11)–(9.2-14) with (9.2-15)–(9.2-18) or directly from (4.5-34) with the $0s$ and $0r$ quantities and all derivatives in the v_{qs} and v_{ds} voltage equations eliminated and with ω set equal to ω_e in the v_{qs} and v_{ds} voltage equations. Hence

$$
\begin{bmatrix}
v_{qs} \\
v_{ds} \\
v'_{qr} \\
v'_{dr}
\end{bmatrix}
=
\begin{bmatrix}
r_s & \frac{\omega_e}{\omega_b}X_{ss} & 0 & \frac{\omega_e}{\omega_b}X_M \\
-\frac{\omega_e}{\omega_b}X_{ss} & r_s & -\frac{\omega_e}{\omega_b}X_M & 0 \\
\frac{p}{\omega_b}X_M & \left(\frac{\omega-\omega_r}{\omega_b}\right)X_M & r'_r + \frac{p}{\omega_b}X'_{rr} & \left(\frac{\omega-\omega_r}{\omega_b}\right)X'_{rr} \\
-\left(\frac{\omega-\omega_r}{\omega_b}\right)X_M & \frac{p}{\omega_b}X_M & -\left(\frac{\omega-\omega_r}{\omega_b}\right)X'_{rr} & r'_r + \frac{p}{\omega_b}X'_{rr}
\end{bmatrix}
\begin{bmatrix}
i_{qs} \\
i_{ds} \\
i'_{qr} \\
i'_{dr}
\end{bmatrix}
$$

$$\tag{9.2-19}$$

where X_{ss} and X'_{rr} are defined by (4.5-35) and (4.5-36), respectively. It is important to note that a derivative of $i_{qs}(i_{ds})$ appears in $v'_{qr}(v'_{dr})$; however, i_{qs} and i_{ds} are algebraically related to i'_{qr} and i'_{dr} by the equations for v_{qs} and v_{ds}. Hence, one might conclude that i'_{qr} and i'_{dr} may be selected as independent or state variables. This is not the case. In the following section we will see that all currents are both algebraically and dynamically related to the stator voltages. Thus, if i'_{qr} and i'_{dr} are selected as state variables, the state equation must be written in a nonstandard form with a derivative of the stator voltages on the right-hand side of (8.7-1). This is not the most convenient form for computer simulation or for transfer function formulation.

The voltage equations may be written in terms of flux linkages per second by solving (9.2-15)–(9.2-18) for the currents [which would yield (4.5-38) with the $0s$ and $0r$ quantities omitted] and substituting the results into (9.2-11)–(9.2-14). Clearly, the same voltage equations may be obtained directly from (4.5-40) by eliminating the $0s$ and $0r$ quantities and the derivatives in v_{qs} and v_{ds} and replacing ω with ω_e in v_{qs} and v_{ds}. Thus

$$
\begin{bmatrix}
v_{qs} \\
v_{ds} \\
v'_{qr} \\
v'_{dr}
\end{bmatrix}
=
\begin{bmatrix}
\frac{r_sX'_{rr}}{D} & \frac{\omega_e}{\omega_b} & -\frac{r_sX_M}{D} & 0 \\
-\frac{\omega_e}{\omega_b} & \frac{r_sX'_{rr}}{D} & 0 & -\frac{r_sX_M}{D} \\
-\frac{r'_rX_M}{D} & 0 & \frac{r'_rX_{ss}}{D} + \frac{p}{\omega_b} & \frac{\omega-\omega_r}{\omega_b} \\
0 & -\frac{r'_rX_M}{D} & -\frac{\omega-\omega_r}{\omega_b} & \frac{r'_rX_{ss}}{D} + \frac{p}{\omega_b}
\end{bmatrix}
\begin{bmatrix}
\psi_{qs} \\
\psi_{ds} \\
\psi'_{qr} \\
\psi'_{dr}
\end{bmatrix}
\tag{9.2-20}
$$

Equation (4.5-39) is repeated here:

$$D = X_{ss}X'_{rr} - X_M^2 \tag{9.2-21}$$

It is important to note that derivatives of ψ_{qs} and ψ_{ds} do not appear in (9.2-20); only the derivatives of ψ'_{qr} and ψ'_{dr} appear. We will see that ψ'_{qr} and ψ'_{dr} are indeed independent or state variables. This set of equations is the most convenient for computer simulation and transfer function formulation.

Synchronous Machine

The stator voltage equations of the synchronous machine written in the arbitrary reference frame are given by (5.4-1). As illustrated by (9.2-7) and (9.2-8), the electric transients are neglected in the stator voltage equations in the arbitrary reference frame by neglecting the derivative of flux linkages and setting $\omega = \omega_e$. Thus, in terms of flux linkages per second and with the electric transients neglected, the stator voltage equations of the synchronous machine expressed in the arbitrary reference frame are of the same form as (9.2-11) and (9.2-12) with a negative ir drop because positive current is assumed out of the terminals of the synchronous machine. It follows that the voltage equations for the synchronous machine in the rotor reference frame with the stator electric transients neglected are obtained by neglecting the derivative of the flux linkages in Park's equations and setting $\omega_r = \omega_e$. Thus, with the 0s quantities omitted, we obtain

$$v_{qs}^r = -r_s i_{qs}^r + \frac{\omega_e}{\omega_b}\psi_{ds}^r \tag{9.2-22}$$

$$v_{ds}^r = -r_s i_{ds}^r - \frac{\omega_e}{\omega_b}\psi_{qs}^r \tag{9.2-23}$$

$$v_{kq1}^{\prime r} = r'_{kq1} i_{kq1}^{\prime r} + \frac{p}{\omega_b}\psi_{kq1}^{\prime r} \tag{9.2-24}$$

$$v_{kq2}^{\prime r} = r'_{kq2} i_{kq2}^{\prime r} + \frac{p}{\omega_b}\psi_{kq2}^{\prime r} \tag{9.2-25}$$

$$v_{fd}^{\prime r} = r'_{fd} i_{fd}^{\prime r} + \frac{p}{\omega_b}\psi_{fd}^{\prime r} \tag{9.2-26}$$

$$v_{kd}^{\prime r} = r'_{kd} i_{kd}^{\prime r} + \frac{p}{\omega_b}\psi_{kd}^{\prime r} \tag{9.2-27}$$

where

$$\psi_{qs}^r = -X_{ls} i_{qs}^r + X_{mq}(-i_{qs}^r + i_{kq1}^{\prime r} + i_{kq2}^{\prime r}) \tag{9.2-28}$$

$$\psi_{ds}^r = -X_{ls} i_{ds}^r + X_{md}(-i_{ds}^r + i_{fd}^{\prime r} + i_{kd}^{\prime r}) \tag{9.2-29}$$

$$\psi_{kq1}^{\prime r} = X'_{lkq1} i_{kq1}^{\prime r} + X_{mq}(-i_{qs}^r + i_{kq1}^{\prime r} + i_{kq2}^{\prime r}) \tag{9.2-30}$$

$$\psi_{kq2}^{\prime r} = X'_{lkq2} i_{kq2}^{\prime r} + X_{mq}(-i_{qs}^r + i_{kq1}^{\prime r} + i_{kq2}^{\prime r}) \tag{9.2-31}$$

$$\psi_{fd}^{\prime r} = X'_{lfd} i_{fd}^{\prime r} + X_{md}(-i_{ds}^r + i_{fd}^{\prime r} + i_{kd}^{\prime r}) \tag{9.2-32}$$

$$\psi_{kd}^{\prime r} = X'_{lkd} i_{kd}^{\prime r} + X_{md}(-i_{ds}^r + i_{fd}^{\prime r} + i_{kd}^{\prime r}) \tag{9.2-33}$$

As in the case of the induction machine the voltage equations for the synchronous machine may be written in terms of the currents. Hence, (9.2-34) results from the substitution of (9.2-28)–(9.2-33) into (9.2-22)–(9.2-27) or directly from (5.5-38) with the 0s quantities and derivatives in the v_{qs}^r and v_{ds}^r voltage equations eliminated and with ω_r set equal to ω_e in the v_{qs}^r and v_{ds}^r voltage equations.

$$
\begin{bmatrix} v_{qs}^r \\ v_{ds}^r \\ v_{kq1}^{\prime r} \\ v_{kq2}^{\prime r} \\ e_{xfd}^{\prime r} \\ v_{kd}^{\prime r} \end{bmatrix} =
\begin{bmatrix}
-r_s & -\frac{\omega_e}{\omega_b}X_d & 0 & 0 & \frac{\omega_e}{\omega_b}X_{md} & \frac{\omega_e}{\omega_b}X_{md} \\
\frac{\omega_e}{\omega_b}X_q & -r_s & -\frac{\omega_e}{\omega_b}X_{mq} & -\frac{\omega_e}{\omega_b}X_{mq} & 0 & 0 \\
-\frac{P}{\omega_b}X_{mq} & 0 & r_{kq1}'+\frac{P}{\omega_b}X_{kq1}' & \frac{P}{\omega_b}X_{mq} & 0 & 0 \\
-\frac{P}{\omega_b}X_{mq} & 0 & \frac{P}{\omega_b}X_{mq} & r_{kq2}'+\frac{P}{\omega_b}X_{kq2}' & 0 & 0 \\
0 & -\frac{X_{md}}{r_{fd}'}\left(\frac{P}{\omega_b}X_{md}\right) & 0 & 0 & \frac{X_{md}}{r_{fd}'}\left(r_{fd}'+\frac{P}{\omega_b}X_{fd}'\right) & \frac{X_{md}}{r_{fd}'}\left(\frac{P}{\omega_b}X_{md}\right) \\
0 & -\frac{P}{\omega_b}X_{md} & 0 & 0 & \frac{P}{\omega_b}X_{md} & r_{kd}'+\frac{P}{\omega_b}X_{kd}'
\end{bmatrix}
\begin{bmatrix} i_{qs}^r \\ i_{ds}^r \\ i_{kq1}^{\prime r} \\ i_{kq2}^{\prime r} \\ i_{fd}^{\prime r} \\ i_{kd}^{\prime r} \end{bmatrix}
$$

$$(9.2\text{-}34)$$

The reactances given in (9.2-34) are defined by (5.5-39)–(5.5-44). As in the case of the induction machine, the formulation of the voltage equations in terms of currents, with the stator electric transients neglected, has limited usefulness because all machine currents are both algebraically and dynamically related to the stator voltages. This gives rise to an inconvenient (nonstandard) form of the state equation.

The voltage equations may be written in terms of flux linkages per second most conveniently by modifying (5.5-54). Hence, (9.2-35) results if, in (5.5-54), ω_r is set equal to ω_e and the derivatives are eliminated in the voltage equations of v_{qs}^r and v_{ds}^r and the 0s quantities are eliminated.

$$
\begin{bmatrix} v_{qs}^r \\ v_{ds}^r \\ v_{kq1}^{\prime r} \\ v_{kq2}^{\prime r} \\ e_{xfd}^{\prime r} \\ v_{kd}^{\prime r} \end{bmatrix} =
\begin{bmatrix}
-r_s a_{11} & \frac{\omega_e}{\omega_b} & -r_s a_{12} & -r_s a_{13} & 0 & 0 \\
-\frac{\omega_e}{\omega_b} & r_s b_{11} & 0 & 0 & -r_s b_{12} & -r_s b_{13} \\
r_{kq1}' a_{21} & 0 & r_{kq1}' a_{22}+\frac{P}{\omega_b} & r_{kq1}' a_{23} & 0 & 0 \\
r_{kq2}' a_{31} & 0 & r_{kq2}' a_{32} & r_{kq2}' a_{33}+\frac{P}{\omega_b} & 0 & 0 \\
0 & X_{md}b_{21} & 0 & 0 & X_{md}b_{22}+\frac{X_{md}}{r_{fd}'}\frac{P}{\omega_b} & X_{md}b_{23} \\
0 & r_{kd}' b_{31} & 0 & 0 & r_{kd}' b_{32} & r_{kd}' b_{33}+\frac{P}{\omega_b}
\end{bmatrix}
\begin{bmatrix} \psi_{qs}^r \\ \psi_{ds}^r \\ \psi_{kq1}^{\prime r} \\ \psi_{kq2}^{\prime r} \\ \psi_{fd}^{\prime r} \\ \psi_{kd}^{\prime r} \end{bmatrix}
$$

$$(9.2\text{-}35)$$

In (9.2-35) a_{ij} and b_{ij} are elements of (5.5-49) and (5.5-50), respectively. We will see that $\psi_{kq1}^{\prime r}$, $\psi_{kq2}^{\prime r}$, $\psi_{fd}^{\prime r}$, and $\psi_{kd}^{\prime r}$ are independent or state variables.

9.3 INDUCTION MACHINE LARGE-EXCURSION BEHAVIOR PREDICTED BY REDUCED-ORDER EQUATIONS

It is instructive to compare the large-excursion induction machine behavior predicted by the reduced-order equations with that predicted by the detailed equations given in Chapter 4. As mentioned earlier, our main interest is the behavior predicted with the electric transients of the stator voltage equations neglected. The material presented in Chapter 4 has already made us aware of the inaccuracies involved

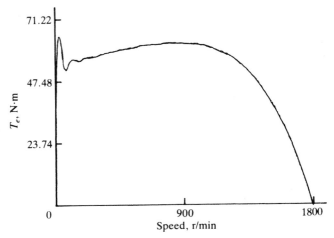

Figure 9.3-1 Torque–speed characteristics during free acceleration predicted with stator electric transients neglected: 3-hp induction motor.

when both the stator and rotor transients are neglected. Additional information regarding the large-excursion accuracy of the reduced-order models with only stator electric transients neglected and with both stator and rotor electric transients neglected is given in reference 2.

Free Acceleration Characteristics

The free acceleration characteristics predicted for the 3- and 2250-hp induction motors with the electric transients neglected in the stator voltage equations are given in Figs. 9.3-1 through 9.3-4. The parameters and operating conditions are identical

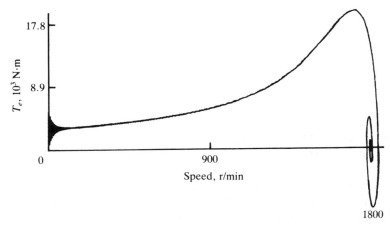

Figure 9.3-2 Torque–speed characteristics during free acceleration predicted with stator electric transients neglected: 2250-hp induction motor.

to those used in Chapter 4. A comparison of the torque versus speed characteristics shown in Figs. 9.3-1 and 9.3-2 with those shown in Figs. 4.10-1 and 4.10-4 reveals that the only significant difference is in the initial starting transient. Although a transient occurs in the instantaneous starting torque, it is much less pronounced when the stator electric transients are neglected. Our first reaction is to assume that the transient that remains is due to the rotor circuits. Although this is essentially the case, we

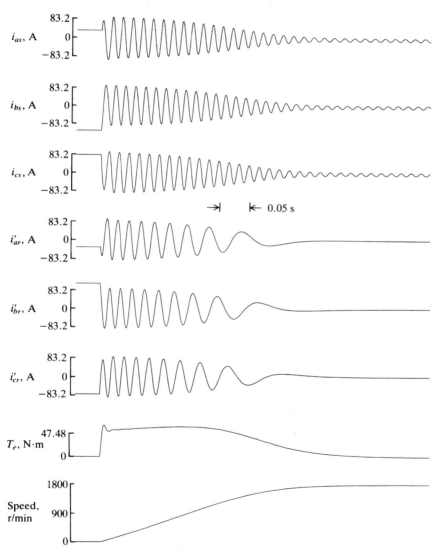

Figure 9.3-3 Machine variables during free acceleration of a 3-hp induction machine predicted with stator electric transients neglected.

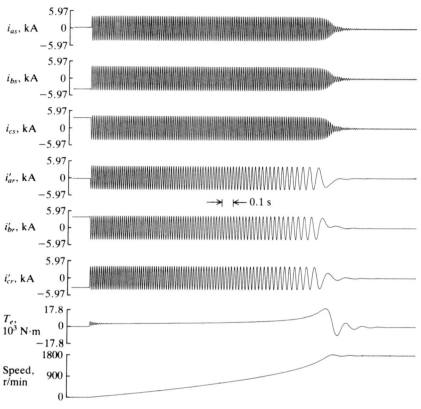

Figure 9.3-4 Machine variables during free acceleration of a 2250-hp induction machine predicted with stator electric transients neglected.

must be careful with such an interpretation because we are imposing a condition upon the voltage equations which could not be realized in practice. We are aware from our earlier analysis that the stator electric transient gives rise to a 60-Hz pulsating torque and a complex conjugate eigenvalue pair with a frequency (imaginary part) of approximately ω_e. Because we are neglecting the electric transients in the stator voltage equations, we would expect discrepancies to occur whenever this transient is excited and whenever it influences the behavior of the machine. In effect, this is what we observe when comparing Figs. 9.3-1 and 9.3-2 with Figs. 4.10-1 and 4.10-4. Once the stator electric transient subsides, the torque–speed characteristics are identical for all practical purposes. For the machine studied, the speed is not significantly influenced by the 60-Hz transient torque during free acceleration. If the inertia were relatively small or if the stator voltages were of frequency considerably less than rated, as occurs in variable-speed drive systems, the pulsating electromagnetic torque could have a significant influence upon the behavior of the machine.

The oscillation about synchronous speed, which is determined primarily by the rotor circuits, is still present as is clearly illustrated in the case of the 2250-hp machine (Figs. 9.3-2 and 9.3-4). This oscillation does not occur when the electric transients of the rotor circuits are neglected.

Another interesting feature regarding the transient characteristics of the induction machine is apparent in Figs. 9.3-3 and 9.3-4. The varying envelope of the machine currents during free acceleration does not occur when the stator electric transients are neglected. Therefore, we must conclude that the varying current envelope depicted in Figs. 4.10-5 and 4.10-6 occurs due to the interaction of stator and rotor electric transients.

A word of explanation regarding Figs. 9.3-3 and 9.3-4 is necessary. In the computer simulation the terminal voltages are applied in the initial condition mode. Although this has no influence upon the solution which follows, we can see the ambiguity that occurs when imposing impossible restrictions upon the behavior of electric circuits. Here we see that the stator voltages are algebraically related to all machine currents because the stator and rotor currents change instantaneously when the stator voltages are applied in the initial condition mode. It is interesting that this situation, which is impossible practically, does not give rise to an initial torque.

Changes in Load Torque

As noted in Chapter 4, the steady-state voltage equations, along with the dynamic relationship between torque and speed, can generally be used to predict the dynamic response to changes in load torque for small-horsepower induction machines. However, this reduced model, wherein both the stator and rotor electric transients are neglected, cannot adequately predict the dynamic response of large horsepower induction machines to load torque disturbances. On the other hand, with only the stator electric transients neglected, the predicted dynamic response to load torque changes in the vicinity of rated torque is, for all practical purposes, identical to that predicted by the detailed model. We would expect this because, due to the inertia of the mechanical system, a change in load torque would normally excite a negligibly small transient offset in the stator currents.

3-Phase Fault at Machine Terminals

The dynamic behavior of the 3- and 2250-hp induction machines during and following a 3-phase fault at the terminals, predicted with the stator electric transients neglected, is given in Figs. 9.3-5 and 9.3-6, respectively. An indication of the accuracy of this reduced-order model can be ascertained by comparing these plots with those given in Figs. 4.13-1 and 4.13-2. The same parameters and operating conditions are used in each case. Initially each machine is operating at base torque. A 3-phase fault at the terminals is simulated by setting v_{as}, v_{bs}, and v_{cs} to zero at the instant v_{as} passes through zero going positive. After six cycles, the source voltages are reapplied. Note the step change in all machine currents and torque at the instant the fault is applied and again at the instant the fault is removed and the stator voltages reapplied. The algebraic relationship between the machine currents and stator voltages is clear.

With the electric transients neglected in the stator voltage equations, transient offset currents will appear only in the rotor circuits. At the beginning of the 3-phase fault and following the reapplication of the terminal voltages, the rotor offset

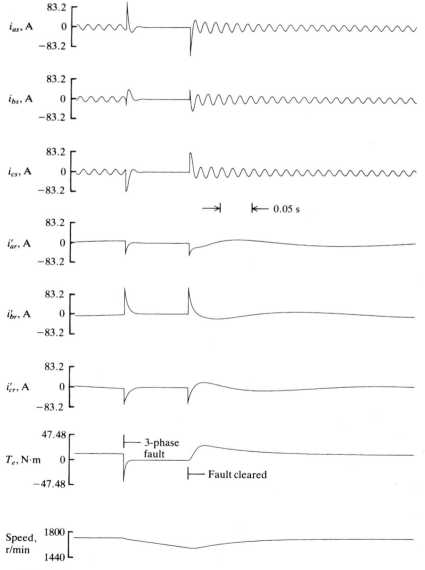

Figure 9.3-5 Dynamic performance of a 3-hp induction motor during a 3-phase fault at the terminals predicted with stator electric transients neglected.

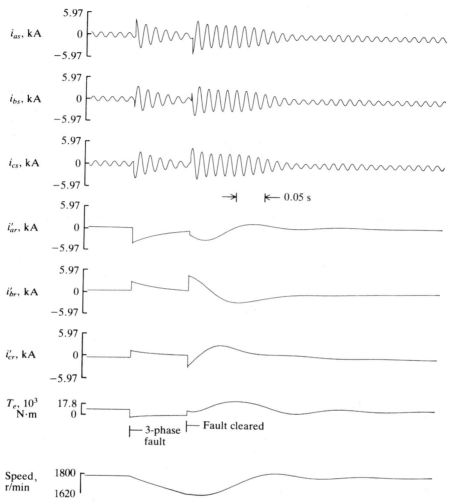

Figure 9.3-6   Dynamic performance of a 2250-hp induction motor during a 3-phase fault at the terminals predicted with stator electric transients neglected.

currents will reflect to the stator as balanced decaying sinusoidal currents with a frequency corresponding to the rotor speed. However, because transient offsets are neglected in the stator circuits, the rotor currents will not contain a sinusoidal component due to the reflection of the stator offset currents into the rotor circuits. This is particularly evident in the case of the 2250-hp machine.

With the stator electric transients absent, a decaying 60-Hz pulsating electromagnetic torque does not occur. A comparison of the response predicted with the stator electric transients neglected and that predicted by the detailed model reveals the error that can occur in rotor speed when the 60-Hz torque is neglected. In most power system applications this error would not be sufficient to warrant the use of the detailed model; however, for investigations of speed control of variable-speed drive systems this simplified model would probably not suffice. Also, if instantaneous shaft torques are of interest, the detailed model must be used.

If the reduced-order model with both stator and rotor electric transients neglected is used, the machine currents, and thus the torque, would instantaneously become zero at the time of the fault. The calculated speed would decrease, obeying the dynamic relationship between the load torque and rotor speed. The currents and torque would instantaneously assume their steady-state values, for the specific rotor speed, at the time the stator voltages are reapplied.

9.4   SYNCHRONOUS MACHINE LARGE-EXCURSION BEHAVIOR PREDICTED BY REDUCED-ORDER EQUATIONS

The reduced-order model of the synchronous machine obtained by neglecting the stator electric transients is used widely in the power industry as an analysis tool. Therefore, it is important to compare the performance of the synchronous machine predicted by the reduced-order equations with that predicted by the detailed model (Chapter 5), especially for disturbances common in transient stability studies. Additional information regarding the large-excursion accuracy of the reduced-order model is found in reference 1, including the accuracy in predicting the free acceleration characteristics.

Changes in Input Torque

A change in input torque would not normally excite an appreciable transient offset in the stator currents. Therefore, neglecting the electric transients in the stator voltage equations has negligible effect upon the accuracy in predicting the dynamic behavior of the typical synchronous machine during normal input torque disturbances.

3-Phase Fault at Machine Terminals

The dynamic behavior of the hydro turbine generator during and following a 3-phase fault at the terminals, predicted with the electric transients neglected in the stator voltage equations, is shown in Figs. 9.4-1 and 9.4-2. The behavior predicted for the steam turbine generator is shown in Figs. 9.4-3 and 9.4-4. An indication of the accuracy of this reduced-order model can be obtained by comparing the behavior depicted in these figures to that shown in Figs. 5.11-1 through 5.11-4. The machines and operating conditions are identical in both cases. Initially, each machine is connected to an infinite bus delivering rated MVA at rated power factor. (Machine data are given in Section 5.10.) In the case of the hydro turbine generator the input torque is held constant at $(0.85)\, 27.6 \times 10^6$ N·m with E'_{xfd} fixed at $(1.6)\, \sqrt{\tfrac{2}{3}}20$ kV; for the

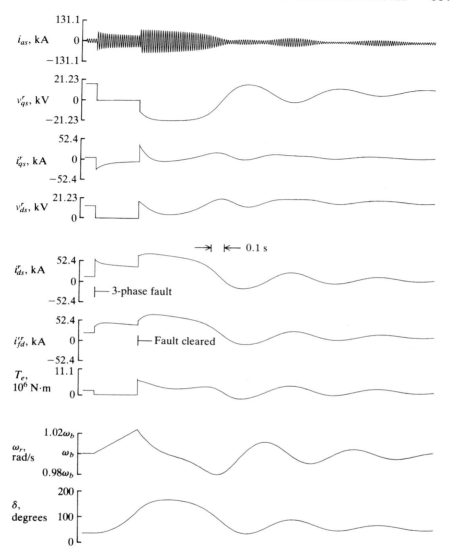

Figure 9.4-1 Dynamic performance of a hydro turbine generator during a 3-phase fault at the terminals predicted with stator electric transients neglected.

steam turbine generator $T_I = (0.85)2.22 \times 10^6$ N·m and $E'_{xfd} = (2.48)\sqrt{\frac{2}{3}}26$ kV. With the machines operating in this steady-state condition, a 3-phase fault at the terminals is simulated by setting v_{as}, v_{bs}, and v_{cs} to zero at the instant v_{as} passes through zero going positive. The instantaneous changes in the machine currents and torque at the initiation and removal of the fault demonstrate the algebraic relationship between stator voltages and machine currents.

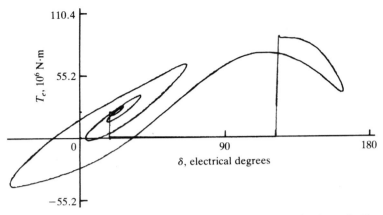

Figure 9.4-2   Torque versus rotor angle characteristics for the study shown in Fig. 9.4-1.

With the stator electric transients neglected, the offset transients do not appear and consequently the 60-Hz pulsating electromagnetic torque is not present during and following the 3-phase fault. The absence of the 60-Hz transient torque is especially apparent in the torque versus rotor angle characteristics given in Figs. 9.4-2 and 9.4-4.

From all outward appearances it would seem that other than the pulsating electromagnetic torque there is little difference between the behavior predicted using the reduced or detailed models. Therefore, one would expect the reduced-order model to be sufficiently accurate in predicting this performance during and following a 3-phase fault. There is, however, a difference that occurs when determining the critical clearing time. The situation portrayed in Figs. 5.11-1 through 5.11-4 and likewise in Figs. 9.4-1 through 9.4-4 is one where only a slight increase in the fault time would cause the machines to become unstable. That is, if the 3-phase fault were allowed to persist slightly longer, the rotor speed would not return to synchronous after the fault is cleared. Using the detailed model in Chapter 5, this critical clearing time was determined to be 0.466 s for the hydro unit and 0.362 s for the steam unit. For the reduced model, the critical clearing time for the hydro unit is 0.424 s and 0.334 s for the steam unit. The longer critical clearing times predicted by the detailed models is due primarily to the pulsating 60-Hz torque that occurs immediately following the occurrence of the fault. The initial torque pulsation causes the rotor to slow down very slightly, which has the effect of delaying the increase in the "average" rotor speed. This effect can be observed in Figs. 5.11-1 and 5.11-3. With the stator electric transients neglected, the 60-Hz pulsating torque is absent and hence the initial "backswing" does not occur [3]. There are two points that warrant mentioning in defense of this apparent inaccuracy of the reduced-order model. First, it will always yield conservative results. Second, a 3-phase fault seldom, if ever, occurs instantaneously. That is, a fault generally starts as a single line-to-ground fault or as a phase-to-phase fault, and then it may progress rapidly to a 3-phase fault. Hence, the instantaneous pulsation in electromagnetic torque is generally not sufficient in the practical case to cause the slowing down of the rotor as depicted in Figs. 5.11-1 and 5.11-3. In this regard, the effects of the 60-Hz pulsating torque resulting from an instantaneous 3-phase fault is perhaps more of academic than of practical interest.

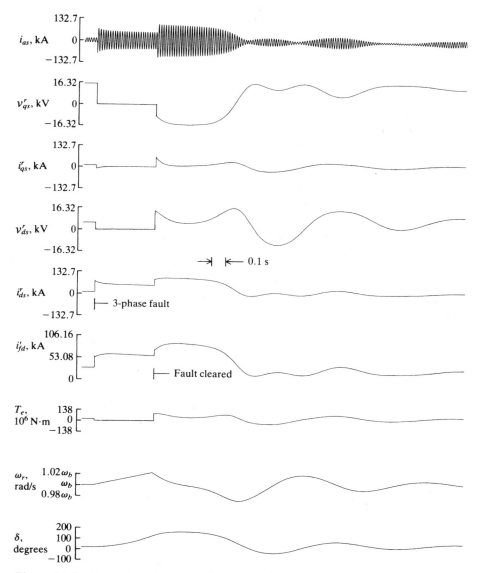

Figure 9.4-3 Dynamic performance of a steam turbine generator during a 3-phase fault at the terminals predicted with stator electric transients neglected.

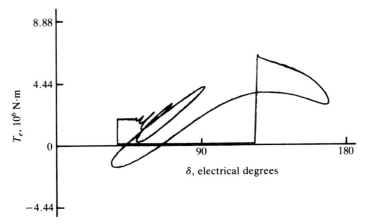

Figure 9.4-4 Torque versus rotor angle characteristics for the study shown in Fig. 9.4-3.

9.5 LINEARIZED REDUCED-ORDER EQUATIONS

In Chapter 8 the detailed differential equations that describe the induction and synchronous machines were linearized. It was pointed out that, with present-day computer languages, this linearization as well as the calculation of the eigenvalues and formulation of the transfer function for control design purpose can all be performed using standard computer routines. Therefore, the linearization techniques presented in Chapter 8 were primarily for information purposes.

Although the reduced-order equations may be linearized following the procedure used in Chapter 8, this formulation will not be set forth. The interested reader is directed to reference 4 where this development is given in detail. It is, however, of interest to mention that because there is both an algebraic and dynamic relationship between stator voltages and machine currents in the reduced-order machine equations, it is advantageous to use flux linkages per second as state variables rather than currents as was done in the case of the detailed equations (Chapter 8). Nevertheless, linearization, eigenvalue calculations, and transfer function formulation may all be achieved for the reduced order-equations using standard computer languages [5–7].

9.6 EIGENVALUES PREDICTED BY LINEARIZED
REDUCED-ORDER EQUATIONS

Even though the linearized reduced-order equations are not set forth, it is instructive to give the eigenvalues calculated using the reduced-order equations. The eigenvalues for the four example induction machines calculated with the stator electric transients neglected are given in Table 9.6-1. Comparison with the eigenvalues

Table 9.6-1 Induction Machine Eigenvalues Calculated with Stator Electric Transients Neglected

Rating (hp)	Stall	Speed Rated	No Load
3	$-192 \pm j431$	$-182 \pm j76.0$	$-181 \pm j52.0$
	1.46	-16.7	-19.6
50	$-142 \pm j397$	$-134 \pm j41.0$	$-133 \pm j17.9$
	1.18	-14.4	-17.0
500	$-29.4 \pm j380$	$-15.6 \pm j41.3$	$-14.5 \pm j42.5$
	0.397	-27.2	-29.4
2250	$-18.5 \pm j378$	$-9.49 \pm j41.6$	$-9.18 \pm j42.5$
	0.240	-17.8	-18.5

calculated using the five-state variable model given in Table 8.5-1 reveals that the real eigenvalue is nearly identical in both cases, and the "rotor" eigenvalue is nearly the same in both cases in the vicinity of normal operation (between rated and no load). It is reported in reference 2 that for rated frequency operation the agreement between the detailed and reduced models improves as the horsepower rating increases, which in effect says that the accuracy improves as the leakage reactance to resistance ratio increases. Additional information regarding the influence of the machine parameters upon the accuracy of the eigenvalues calculated using the reduced-order model is given in reference 2 for rated frequency operation.

The eigenvalues for the hydro and steam turbine generators calculated with the stator electric transients neglected are given in Table 9.6-2. Comparison of the real eigenvalues and the eigenvalue of the swing mode given in Table 9.6-2 to those given in Table 8.6-1 for the detailed model reveals that they are nearly identical. In fact, accuracy is not sacrificed by neglecting the stator electric transients for operation at rated frequency and for normal loading conditions.

9.7 SIMULATION OF REDUCED-ORDER MODELS

The reduced-order model, wherein the stator electric transients are neglected, is considered the most sophisticated of these reduced-order models for representing the

Table 9.6-2 Synchronous Machine Eigenvalues for Rated Conditions Calculated with Stator Electric Transients Neglected

Hydro Turbine Generator	Steam Turbine Generator
$-1.33 \pm j8.68$	$-1.70 \pm j10.5$
-24.4	-32.2
-22.9	-11.1
-0.453	-0.855
	-0.350

behavior of induction and synchronous machines in power system studies. In computer programs used for transient stability studies of power systems, it is standard to neglect the electric transients in the stator voltage equations of all machines and in the voltage equations of all power system components connected to the machines [1]. It is important to realize that once it is decided to neglect the electric transients in the representation of any part of the network, the electric transients should be neglected in the complete network; otherwise, erroneous electric transients will appear in the solution.

Induction Machine

The qs and ds voltage equations in the arbitrary reference frame with the stator electric transients neglected are given by (9.2-11) and (9.2-12) for an induction machine. If the voltages are the input variables to the simulation, then these equations may be solved for ψ_{ds} and ψ_{qs}, respectively. Thus

$$\psi_{ds} = \frac{\omega_e}{\omega_b}\left[v_{qs} + \frac{r_s}{X_{ls}}(\psi_{mq} - \psi_{qs})\right] \tag{9.7-1}$$

$$\psi_{qs} = \frac{\omega_e}{\omega_b}\left[v_{ds} - \frac{r_s}{X_{ls}}(\psi_{md} - \psi_{ds})\right] \tag{9.7-2}$$

where (4.14-1) and (4.14-2) have been substituted for i_{qs} and i_{ds}, respectively. The above equations may be used to replace (4.14-9) and (4.14-10) in the computer representation. As pointed out, only balanced or symmetrical conditions are considered when the electric transients are neglected; therefore zero quantities do not exist. In power-system applications, ω_e is constant and generally equal to ω_b. In some cases, it may be convenient to write the above equations in the following form:

$$\psi_{qs} = \frac{X_{ls}}{r_s}\left(v_{qs} - \frac{\omega_e}{\omega_b}\psi_{ds}\right) + \psi_{mq} \tag{9.7-3}$$

$$\psi_{ds} = \frac{X_{ls}}{r_s}\left(v_{ds} + \frac{\omega_e}{\omega_b}\psi_{qs}\right) + \psi_{md} \tag{9.7-4}$$

Either (9.7-1) and (9.7-2) or (9.9-3) and (9.7-4) may be readily incorporated into the simulation shown in Fig. 4.14-1.

It is also possible to have currents as the input variables. In this case, the voltages are solved from (9.2-11) and (9.2-12) with ψ_{qs} and ψ_{ds} calculated from the familiar relations

$$\psi_{qs} = X_{ls}i_{qs} + \psi_{mq} \tag{9.7-5}$$

$$\psi_{ds} = X_{ls}i_{ds} + \psi_{md} \tag{9.7-6}$$

Equations (9.2-11), (9.2-12), (9.2-5), and (9.2-6) may be readily incorporated into the computer simulation shown in Fig. 4.14-1 to represent the induction machine with the stator electric transients neglected and currents as the input variables. It is clear that to make the block diagram in Fig. 4.14-1 valid for currents as the input variables, the roles of the stator voltages and currents must be reversed. That is, $\mathbf{i}_{abcs}$ is transformed to $\mathbf{i}_{qd0s}$, which is an input to the computational block; $\mathbf{v}_{qd0s}$ is then an output of the computational block, and these voltages are transformed to $\mathbf{v}_{abcs}$. Although we have used the 0s quantities in the above notations, we realize that these quantities are zero.

Synchronous Machine

The qs- and ds-voltage equations of a synchronous machine written in the rotor reference frame with the stator transients neglected are given by (9.2-22) and (9.2-23). The flux linkages may be computed from these equations with voltages as the input variables. The resulting equations are in the same form as (9.7-1) and (9.7-2) or (9.7-3) and (9.7-4), which should be used to replace (5.16-1) and (5.16-2), respectively, to represent the synchronous machine in the rotor reference frame with the electric transients neglected.

If currents are the input variables, (9.2-22) and (9.2-23) may be used to calculate the voltages along with

$$\psi^r_{qs} = -X_{ls}i^r_{qs} + \psi^r_{mq} \qquad (9.7\text{-}7)$$
$$\psi^r_{ds} = -X_{ls}i^r_{ds} + \psi^r_{md} \qquad (9.7\text{-}8)$$

These equations may be incorporated into either Fig. 5.16-1a or 5.16-1b to simulate the synchronous machine with the stator transients neglected and with currents as the input variables. It is clear that the roles of the voltages and currents are reversed in Fig. 5.16-1 with currents as the input variables.

The stator voltage equations written in the arbitrary reference frame with the electric transients neglected are

$$v_{qs} = -r_s i_{qs} + \frac{\omega_e}{\omega_b}\psi_{ds} \qquad (9.7\text{-}9)$$

$$v_{ds} = -r_s i_{ds} - \frac{\omega_e}{\omega_b}\psi_{qs} \qquad (9.7\text{-}10)$$

With voltages as input variables, these equations may be solved for the flux linkages and used to replace (5.16-20) and (5.16-21). The block diagram shown in Fig. 5.16-2 may then be used to simulate the synchronous machine with the electric transients neglected and the stator voltage equations written in the arbitrary reference frame.

When currents are the input variables, (9.7-9) and (9.7-10) are used to calculate the voltages. In this case, these equations are incorporated into the computational block shown in Fig. 5.16-2. The roles of $\mathbf{v}_{abcs}$ and $\mathbf{i}_{abcs}$ as well as $\mathbf{v}_{qd0s}$ and $\mathbf{i}_{qd0s}$ are reversed. Also the roles of ψ_{qds} and $\mathbf{i}_{qds}$ as well as ψ^r_{qds} and $\mathbf{i}^r_{qds}$ are interchanged. The flux linkages, ψ^r_{qs} and ψ^r_{ds}, are computed in the main computational block using

$$\psi^r_{qs} = -X_{ls}i^r_{qs} + \psi^r_{mq} \qquad (9.7\text{-}11)$$
$$\psi^r_{ds} = -X_{ls}i^r_{ds} + \psi^r_{md} \qquad (9.7\text{-}12)$$

9.8 CLOSING COMMENTS AND GUIDELINES

It must be kept in mind that one facility of neglecting stator electric transients is to enhance a convenient computer simulation of large-scale power systems for studying transient stability. In this application, only the electromechanical dynamics of large machines (induction and synchronous) are of interest. If the reduced-order equations are simulated with flux linkages as state variables, the accuracy of this reduced-order model is quite adequate for normal, small, or large excursions of a large-scale power system about an operating point. The reduction in the number of integrations brought about by neglecting the stator and network electric transients far outweighs any inaccuracy that may occur. Moreover, one must be aware that the system database would likely be far more inaccurate than the inaccuracies brought about due to neglecting stator and network electric transients.

Although the reduced-order model obtained by neglecting stator electric transients is generally quite adequate for transient stability studies of large-scale power systems subjected to symmetrical disturbances, we are imposing impossible electrical conditions upon the voltage equations. When we attempt to "fool Mother nature" in this way, we should proceed with caution. In particular, using the reduced-order models for induction and synchronous machines in low-power, small-system, or electric drive applications may not be advisable. If the machines are small in horsepower or if the machines are operated over a relatively wide frequency range, one should not use the reduced-order model without first comparing its predications of system response to that of the detailed model. This is perhaps the most reasonable guideline to offer because it would be nearly impossible and certainly inappropriate to establish the accuracy of the reduced-order model obtained by neglecting the stator electric transients for all conceivable applications. The reader may wish to refer to references 8 and 9 where a more extensive treatment of reduced-order modeling is presented including a reduced-order model obtained by dynamic response separation techniques.

REFERENCES

[1] P. C. Krause, F. Nozari, T. L. Skvarenina, and D. W. Olive, The Theory of Neglecting Stator Transients, *IEEE Transactions on Power Apparatus and Systems*, Vol. 98, January/February 1979, pp. 141–148.

[2] T. L. Skvarenina and P. C. Krause, Accuracy of a Reduced Order Model of Induction Machines in Dynamic Stability Studies, *IEEE Transactions on Power Apparatus and Systems*, Vol. 98, July/August 1979, pp. 1192–1197.

[3] R. G. Harley and B. Adkins, Calculations of the Angular Back Swing Following a Short-Circuit of a Loaded Alternator, *Proceedings of the IEE*, Vol. 117, No. 2, February 1970, pp. 377–386.

[4] P. C. Krause, O. Wasynczuk, and S. D. Sudhoff, *Analysis of Electric Machinery*, IEEE Press, Piscataway, NJ, 1995.

[5] C. M. Ong, *Dynamic Simulation of Electric Machinery*, Prentice-Hall PTR, Upper Saddle River, NJ, 1998.

[6] Advanced Continuous Simulation Language (ACSL), Reference Manual, Version 11, MGA Software, 1995.

[7] Simulink: Dynamic System Simulation for Matlab, Using Simulink Version 4, The MathWorks Inc., 2000.

[8] O. Wasynczuk, Y. M. Diao, and P. C. Krause, Theory and Comparison of Reduced-Order Models of Induction Machines, *IEEE Transactions on Power Apparatus and Systems*, Vol. 104, March 1985, pp. 598–606.

[9] F. D. Rodriguez and O. Wasynczuk, A Refined Method of Deriving Reduced Order Models of Induction Machines, *IEEE Transactions on Energy Conversion*, Vol. EC-2, No. 1, March 1987, pp. 31–37.

PROBLEMS

1 Express the rotor voltage equations of an induction machine in the arbitrary reference frame with the electric transients neglected. Appropriately combine these equations with (9.2-11) and (9.2-12) to obtain the standard steady-state equivalent circuit given in Fig. 4.9-1.

2 Neglect the electric transients in the stator and rotor of a synchronous machine. From these voltage equations derive the familiar steady-state voltage equation given by (5.9-19).

3 Determine the initial condition currents for Figs. 9.3-3 and 9.3-4. Show that the initial condition torque is zero.

4 Although flux linkages per second are generally used as state variables when linearizing the reduced-order equations, it is instructive to neglect the time rate of change of stator currents in (8.3-8) and express the result in the form

$$\begin{bmatrix} \Delta \mathbf{v}_{qds}^e \\ \Delta \mathbf{v}_{rr} \end{bmatrix} = \begin{bmatrix} \mathbf{W} & \mathbf{Y} \\ \mathbf{Q} & \mathbf{S} \end{bmatrix} \begin{bmatrix} \Delta \mathbf{i}_{qds}^e \\ \Delta \mathbf{i}_{rr} \end{bmatrix}$$

where

$$\Delta \mathbf{v}_{qds}^e = \begin{bmatrix} \Delta v_{qs}^e & \Delta v_{ds}^e \end{bmatrix}^T$$

Express all other matrices.

5 Using the results of Problem 4 above, show that we can arrive at two equations for $p\Delta \mathbf{i}_{qds}^e$ which are inconsistent. In particular,

$$p\Delta \mathbf{i}_{qds}^e = -(\mathbf{W}_p)^{-1}\mathbf{Y}_p p\Delta \mathbf{i}_{rr}$$

$$p\Delta \mathbf{i}_{qds}^e = (\mathbf{W}_k)^{-1} p\Delta \mathbf{v}_{qds}^e - (\mathbf{W}_k)^{-1}\mathbf{Y}_k p\Delta \mathbf{i}_{rr}$$

where

$$\mathbf{W} = \mathbf{W}_p + \mathbf{W}_k$$

$$\mathbf{Y} = \mathbf{Y}_p + \mathbf{Y}_k$$

In the above equations the matrices with the subscript p contain only terms associated with the derivatives while the k subscript denotes the matrices containing the remaining terms.

6 In this chapter we have considered only 3-phase machines. Explain the changes that must be made so that this material will apply to (*a*) 2-phase induction machines and (*b*) 2-phase synchronous machines.

7 It is appropriate to neglect stator transients when analyzing an induction machine with unbalanced stator applied voltages? Why?

Chapter 10

SYMMETRICAL AND UNSYMMETRICAL 2-PHASE INDUCTION MACHINES

10.1 INTRODUCTION

The analysis of a symmetrical 3-phase induction machine is set forth for balanced conditions in Chapter 4. It has been mentioned that the analysis of a symmetrical 2-phase induction machine is essentially the same as that of its 3-phase counterpart. The first part of this chapter is devoted to a brief discussion of this similarity. The reader who has followed the material given in Chapter 4 will find this presentation to be more or less a review or logical extension of concepts already covered. However, the analysis of unbalanced stator conditions is presented in detail sufficient for the reader to analyze unbalanced operation of both symmetrical and unsymmetrical 2-phase induction machines.

The symmetrical and unsymmetrical 2-phase induction machines are generally used in single-phase applications such as refrigerators, washing machines, clothes dryers, furnace fans, garbage disposals, and air conditioners. In the case of the symmetrical 2-phase induction machine a capacitor is connected in series with one of the stator windings to shift the phase of one stator current relative to the other, thereby producing an average starting torque. In the case of the unsymmetrical 2-phase induction machine a starting torque is produced by designing the stator windings to be electrically different and, in some cases, placing a capacitor in series with one of the windings.

In this chapter considerable time is devoted to the analysis of unsymmetrical 2-phase induction machines which is often accomplished by employing revolving-field theory [1–3]. This approach is not used here; instead, reference-frame theory is used to establish the voltage equations in the stationary reference frame which are valid for transient and steady-state conditions. The phasor equations valid for predicting steady-state operation are established from these equations for stator voltages of any periodic waveform. The standard steady-state equivalent circuits are then derived from these equations using symmetrical component theory. Computer traces are given to demonstrate the free acceleration characteristics of two types of single-phase induction motors; the capacitor-start and the capacitor-start, capacitor-run.

10.2 ANALYSIS OF SYMMETRICAL 2-PHASE INDUCTION MACHINES

The analysis of 2-phase symmetrical induction machines is similar in many respects to that of 3-phase symmetrical induction machines. This fact is mentioned from time to time in Chapter 4 where several problems are given for the purpose of emphasizing this similarity. Therefore most of the material given in this section is a gathering and/or discussion of information that has been established previously. The development of the machine equations is not repeated; however, the analysis of unbalanced stator conditions is set forth in detail. This allows the reader to analyze this mode of operation for both symmetrical and unsymmetrical 2-phase induction machines without becoming involved in the analysis of unbalanced 3-phase operation of an induction machine [4].

Voltage and Torque Equations in Machine Variables

A 2-pole, 2-phase symmetrical induction machine is shown in Fig. 4.15-1. The voltage and torque equations expressed in machine variables may be obtained by following the procedure used for the 3-phase machine in Section 4.2.

Voltage and Torque Equations in Arbitrary Reference-Frame Variables

A change of variables which formulates a transformation of the 2-phase variables of stationary circuit elements to the arbitrary reference is given in the first problem of Chapter 3. This transformation is repeated here for convenience. In particular,

$$\mathbf{f}_{qds} = \mathbf{K}_{2s}\mathbf{f}_{abs} \tag{10.2-1}$$

where

$$(\mathbf{f}_{qds})^T = [f_{qs} \quad f_{ds}] \tag{10.2-2}$$

$$(\mathbf{f}_{abs})^T = [f_{as} \quad f_{bs}] \tag{10.2-3}$$

$$\mathbf{K}_{2s} = \begin{bmatrix} \cos\theta & \sin\theta \\ \sin\theta & -\cos\theta \end{bmatrix} \tag{10.2-4}$$

$$\omega = \frac{d\theta}{dt} \tag{10.2-5}$$

The transformation matrix and its inverse are equal; that is,

$$\mathbf{K}_{2s} = (\mathbf{K}_{2s})^{-1} \tag{10.2-6}$$

A change of variables which formulates a transformation of the 2-phase variables of the rotor circuits to the arbitrary reference frame is

$$\mathbf{f}_{qdr} = \mathbf{K}_{2r}\mathbf{f}_{abr} \tag{10.2-7}$$

where

$$(\mathbf{f}_{qdr})^T = [\,f_{qr}\quad f_{dr}\,] \tag{10.2-8}$$

$$(\mathbf{f}_{abr})^T = [\,f_{ar}\quad f_{br}\,] \tag{10.2-9}$$

$$\mathbf{K}_{2r} = (\mathbf{K}_{2r})^{-1} = \begin{bmatrix} \cos\beta & \sin\beta \\ \sin\beta & -\cos\beta \end{bmatrix} \tag{10.2-10}$$

$$\beta = \theta - \theta_r \tag{10.2-11}$$

The angular displacement θ is defined by (10.2-5); θ_r is

$$\omega_r = \frac{d\theta_r}{dt} \tag{10.2-12}$$

The voltage equations expressed in arbitrary reference-frame variables are given in Section 4.5 for the 3-phase induction machine. These equations, which are given in terms of currents by (4.5-34) and in terms of flux linkages by (4.5-40), may also be used for the 2-phase symmetrical induction machine with the following changes. First only the qs, qr, ds, and dr variables exist in the case of a 2-phase machine; all $0s$ and $0r$ variables should be ignored. In the case of the 3-phase machine, $L_M = \frac{3}{2}L_{ms}(X_M = \frac{3}{2}X_{ms})$; for the 2-phase machine, $L_M(X_M)$ should be replaced by $L_{ms}(X_{ms})$ in all equations where $L_M(X_M)$ appears. Similarly, the electromagnetic torque equations given in Section 4.6 may be used for the 2-phase machine if we replace $L_M(X_M)$ with $L_{ms}(X_{ms})$ and eliminate the $\frac{3}{2}$ factor. For example, (4.6-2) written for a 2-phase machine becomes

$$T_e = \left(\frac{P}{2}\right) L_{ms}\left(i_{qs}i'_{dr} - i_{ds}i'_{qr}\right) \tag{10.2-13}$$

Balanced Steady-State Operation

The equivalent circuit for balanced steady-state operation of a 3-phase symmetrical induction machine shown in Fig. 4.9-1 may be used for a 2-phase symmetrical machine if X_M is replaced with X_{ms}. The expressions for steady-state electromagnetic torque given in Section 4.9 for the 3-phase machine may be used for the

2-phase machine if X_M is replaced by X_{ms} and the factor 3 replaced by 2 whenever phasors are used. In particular, for a 2-phase symmetrical induction machine, (4.9-14) becomes

$$T_e = 2\left(\frac{P}{2}\right)\left(\frac{X_{ms}}{\omega_b}\right)\mathrm{Re}[j\tilde{I}_{as}^* \tilde{I}_{ar}'] \tag{10.2-14}$$

As pointed out in the discussion of Fig. 4.9-3, when the 2-phase symmetrical induction machine is used as a servomotor in a positioning control system, it is designed with a relatively large rotor resistance in order to achieve the maximum possible starting torque.

Linearized Equations and Transfer Function Formulation

The linearized or small-displacement equations for a 3-phase symmetrical induction machine are set forth in Chapter 8. Therein, the linearized voltage and torque equations are written in the synchronously rotating reference frame [see (8.3-8)]. If X_M is replaced with X_{ms}, these equations are valid for a 2-phase symmetrical induction machine. The difference occurs, of course, when one relates the synchronously rotating reference-frame variables to the phase variables. In one case a 2-phase transformation is used; in the other the 3-phase transformation is used.

It follows that the state equations given by (8.3-16) are also valid for a 2-phase induction machine with the appropriate change in X_M. Similarly, the transfer function formulation given in Section 8.7 for a 3-phase induction machine applies also to the 2-phase machine. In fact, the formulation given in Example 8A applies directly to a 2-phase symmetrical induction machine.

The linearized equations and transfer function formulation are given for a 3-phase symmetrical induction machine with the stator electric transients neglected in reference 4. Clearly these equations, as well as the large-excursion equations, are equally valid for the 2-phase counterpart; however, the accuracy of this reduced-order model may not be acceptable for small-horsepower 2-phase machines.

Unbalanced Steady-State Operation

The arbitrary reference frame can be used to establish the method of symmetrical components for the purpose of analyzing unbalanced conditions of a 3-phase induction machine [4]. A parallel development is possible for a 2-phase machine. For this purpose, let us assume that the 2-phase stator variables may be expressed as

$$f_{as} = \sum_{k=0}^{\infty} (f_{ask\alpha}\cos\omega_{esk}t + f_{ask\beta}\sin\omega_{esk}t) \tag{10.2-15}$$

$$f_{bs} = \sum_{k=0}^{\infty} (f_{bsk\alpha}\cos\omega_{esk}t + f_{bsk\beta}\sin\omega_{esk}t) \tag{10.2-16}$$

In the above series expansions, all ω_{esk} are constant. The α and β subscripts denote, respectively, the coefficients of the cosine and sine terms. Nonsinusoidal components are the coefficients of the cosine terms with $k = 0$ because, by definition, $\omega_{es0} = 0$. Similarly, $k = 1$ is generally used to denote the coefficients associated with the fundamental frequency. In other words, ω_{es1} is generally ω_e, and in most applications this is the only frequency present in the stator voltages for balanced rotor conditions except in variable-frequency electric drive systems where the stator is generally connected to a converter. Also, the variables are written as instantaneous quantities, and the coefficients may be constant or time-varying. Actually we will only consider steady-state conditions whereupon the coefficients are constants and (10.2-15) and (10.2-16) would be expressed by uppercase letters; however, we will continue with the lowercase notation for a moment.

Transforming (10.2-15) and (10.2-16) to the arbitrary reference frame by (10.2-1) yields f_{qs} and f_{ds}. In particular,

$$f_{qs} = \sum_{k=0}^{\infty} [f_{qskA} \cos(\omega_{esk}t - \theta) + f_{qskB} \sin(\omega_{esk}t - \theta)]$$

$$+ \sum_{k=0}^{\infty} [f_{qskC} \cos(\omega_{esk}t + \theta) + f_{qskD} \sin(\omega_{esk}t + \theta)] \qquad (10.2\text{-}17)$$

$$f_{ds} = \sum_{k=0}^{\infty} [f_{dskA} \cos(\omega_{esk}t - \theta) + f_{dskB} \sin(\omega_{esk}t - \theta)]$$

$$+ \sum_{k=0}^{\infty} [f_{dskC} \cos(\omega_{esk}t + \theta) + f_{dskD} \sin(\omega_{esk}t + \theta)] \qquad (10.2\text{-}18)$$

where θ is defined by (10.2-5). For the 2-phase machine the *ABCD* quantities are defined as follows:

$$f_{qskA} = \frac{1}{2}(f_{ask\alpha} + f_{bsk\beta}) = -f_{dskB} \qquad (10.2\text{-}19)$$

$$f_{qskB} = \frac{1}{2}(f_{ask\beta} - f_{bsk\alpha}) = f_{dskA} \qquad (10.2\text{-}20)$$

$$f_{qskC} = \frac{1}{2}(f_{ask\alpha} - f_{bsk\beta}) = f_{dskD} \qquad (10.2\text{-}21)$$

$$f_{qskD} = \frac{1}{2}(f_{ask\beta} + f_{bsk\alpha}) = -f_{dskC} \qquad (10.2\text{-}22)$$

It is important to note that regardless of the waveforms of the 2-phase stator variables, (10.2-17) and (10.2-18) reveal that the *qs* and *ds* variables form a series of 2-phase balanced sets in the arbitrary reference frame. The only restriction is that the 2-phase stator variables must be of the general form given by (10.2-15) and (10.2-16). For $k > 0$ the balanced sets with the argument $(\omega_{esk}t - \theta)$ may be considered as positive sequence or positively rotating sets because they produce counterclockwise rotating air-gap MMFs relative to the stator windings. It follows that the balanced sets with the argument $(\omega_{esk}t + \theta)$ can be thought of as negative-sequence or negatively rotating sets for $k > 0$.

If we use uppercase letters to denote steady-state variables, we can write the qs and ds variables in phasor form, for $k > 0$, as

$$\tilde{F}_{qsk}^s = \tilde{F}_{qs+k}^s + \tilde{F}_{qs-k}^s \tag{10.2-23}$$

$$\tilde{F}_{dsk}^s = \tilde{F}_{ds+k}^s + \tilde{F}_{ds-k}^s \tag{10.2-24}$$

where (10.2-23) comes from (10.2-17) and (10.2-24) from (10.2-18). The phasors with the $+k$ subscripts come from the first summation on the right-hand side of (10.2-17) and (10.2-18); the phasors with the $-k$ subscript come from the second summation. It should be pointed out, however, that the above equations are not valid for constant (dc) quantities where $k = 0$. This of course is not a serious restriction because a dc voltage is generally applied to the stator only for the purpose of dynamic braking. The analysis for this type of operation may be handled as a special case.

From (10.2-17)–(10.2-22) we can write

$$\sqrt{2}\tilde{F}_{qs+k}^s = F_{qskA} - jF_{qskB} \tag{10.2-25}$$

$$\sqrt{2}\tilde{F}_{ds+k}^s = F_{dskA} - jF_{dskB} = j\sqrt{2}\tilde{F}_{qs+k}^s \tag{10.2-26}$$

$$\sqrt{2}\tilde{F}_{qs-k}^s = F_{qskC} - jF_{qskD} \tag{10.2-27}$$

$$\sqrt{2}\tilde{F}_{ds-k}^s = F_{dskC} - jF_{dskD} = -j\sqrt{2}\tilde{F}_{qs-k}^s \tag{10.2-28}$$

If we substitute (10.2-26) and (10.2-28) into (10.2-24), we can write (10.2-23) and (10.2-24) as

$$\begin{bmatrix} \tilde{F}_{qsk}^s \\ \tilde{F}_{dsk}^s \end{bmatrix} = \begin{bmatrix} 1 & 1 \\ j1 & -j1 \end{bmatrix} \begin{bmatrix} \tilde{F}_{qs+k}^s \\ \tilde{F}_{qs-k}^s \end{bmatrix} \tag{10.2-29}$$

The equation establishes a complex transformation for a 2-phase system. Actually, (10.2-29) is the symmetrical component transformation; and in order to comply with the convention that we have established, we should write (10.2-29) as

$$\tilde{\mathbf{F}}_{qs\pm k}^s = \mathbf{S}_{2qd}\tilde{\mathbf{F}}_{qdsk}^s \tag{10.2-30}$$

where

$$\left(\tilde{\mathbf{F}}_{qs\pm k}^s\right)^T = \begin{bmatrix} \tilde{F}_{qs+k}^s & \tilde{F}_{qs-k}^s \end{bmatrix} \tag{10.2-31}$$

$$\left(\tilde{\mathbf{F}}_{qdsk}^s\right)^T = \begin{bmatrix} \tilde{F}_{qsk}^s & \tilde{F}_{dsk}^s \end{bmatrix} \tag{10.2-32}$$

$$\mathbf{S}_{2qd} = \begin{bmatrix} 1 & 1 \\ j1 & -j1 \end{bmatrix}^{-1} = \frac{1}{2}\begin{bmatrix} 1 & -j1 \\ 1 & j1 \end{bmatrix} \tag{10.2-33}$$

In this analysis of unbalanced stator conditions it is assumed that the external rotor circuits are symmetrical. Moreover, if the machine is doubly fed, the voltages applied to the rotor phases are balanced and contain only one frequency. With these constraints the stator and rotor variables will be in the same form when expressed in the arbitrary reference frame. Hence, (10.2-17)–(10.2-22) may also be used to express the rotor variables in the arbitrary reference frame by simply replacing the s subscript with r, except in the case of ω_{esk}, and adding a prime to denote variables referred by a turns ratio. Therefore we can write

$$\tilde{\mathbf{F}}'^{s}_{qr\pm k} = \mathbf{S}_{2qd}\tilde{\mathbf{F}}'^{s}_{qdrk} \tag{10.2-34}$$

where the vectors are defined by (10.2-31) and (10.2-32) with the appropriate change in subscripts.

The steady-state voltage equations of a 2-phase symmetrical induction machine expressed in the stationary reference frame may be obtained from (4.5-34) by setting $\omega = 0$ with the $0s$ and $0r$ quantities omitted and X_M replaced by X_{ms} and by setting $p = j\omega_{esk}$. If we then substitute the inverse of (10.2-30) and the inverse of (10.2-34) into the resulting equations, we obtain

$$
\begin{bmatrix}
\tilde{V}^s_{qs+k} \\
\dfrac{\tilde{V}'^s_{qr+k}}{s_k} \\
\tilde{V}^s_{qs-k} \\
\dfrac{\tilde{V}'^s_{qr-k}}{2-s_k}
\end{bmatrix}
=
\begin{bmatrix}
r_s + j\dfrac{\omega_{esk}}{\omega_b}X_{ss} & j\dfrac{\omega_{esk}}{\omega_b}X_{ms} & 0 & 0 \\[2mm]
j\dfrac{\omega_{esk}}{\omega_b}X_{ms} & \dfrac{r'_r}{s_k} + j\dfrac{\omega_{esk}}{\omega_b}X'_{rr} & 0 & 0 \\[2mm]
0 & 0 & r_s + j\dfrac{\omega_{esk}}{\omega_b}X_{ss} & j\dfrac{\omega_{esk}}{\omega_b}X_{ms} \\[2mm]
0 & 0 & j\dfrac{\omega_{esk}}{\omega_b}X_{ms} & \dfrac{r'_r}{2-s_k} + j\dfrac{\omega_{esk}}{\omega_b}X'_{rr}
\end{bmatrix}
\begin{bmatrix}
\tilde{I}^s_{qs+k} \\
\tilde{I}'^s_{qr+k} \\
\tilde{I}^s_{qs-k} \\
\tilde{I}'^s_{qr-k}
\end{bmatrix}
$$
$$\tag{10.2-35}$$

where

$$s_k = \frac{\omega_{esk} - \omega_r}{\omega_{esk}} \tag{10.2-36}$$

We must now relate $\tilde{\mathbf{F}}_{qs\pm k}$ to $\tilde{\mathbf{F}}_{absk}$. Because $\mathbf{K}^s_{2s}$ is algebraic, it may be used to transform phasors; thus from (10.2-1) we obtain

$$\tilde{\mathbf{F}}^s_{qdsk} = \mathbf{K}^s_{2s}\tilde{\mathbf{F}}_{absk} \tag{10.2-37}$$

With $\theta(0) = 0$ we have

$$\mathbf{K}^s_{2s} = \begin{bmatrix} 1 & 0 \\ 0 & -1 \end{bmatrix} \tag{10.2-38}$$

Substituting (10.2-37) into (10.2-30) yields

$$\tilde{\mathbf{F}}^s_{qs\pm k} = \mathbf{S}_{2qd}\mathbf{K}^s_{2s}\tilde{\mathbf{F}}_{absk}$$
$$= \mathbf{S}_2\tilde{\mathbf{F}}_{absk} \tag{10.2-39}$$

where

$$S_2 = \frac{1}{2}\begin{bmatrix} 1 & j1 \\ 1 & -j1 \end{bmatrix}$$ (10.2-40)

Clearly, (10.2-39) is the desired relationship between steady-state phase variables and the component variables in the stationary reference frame. It is important to mention that the reason for the difference between S_{2qd} and S_2 is the fact that we have selected the transformation K_{2s} so that, when $\omega = 0$ and $\theta(0) = 0$, the magnetic axis of the ds winding in the stationary reference frame is opposite that of the bs winding. In other words, $f_{ds}^s = -f_{bs}$ for $\omega = 0$ and $\theta(0) = 0$. Traditionally, the symmetrical components transformation, S_2, relates phase and component variables. For this reason, S_{2qd} was used to denote the "symmetrical component transformation" between qd variables and component variables, thereby distinguishing it from S_2.

From (4.6-2) we can express the electromagnetic torque for a 2-phase induction machine as

$$T_e = \left(\frac{P}{2}\right)\left(\frac{X_{ms}}{\omega_b}\right)(i_{qs}i_{dr}' - i_{ds}i_{qr}')$$ (10.2-41)

Now, i_{qs} and i_{ds} may be expressed by (10.2-17) and (10.2-18); and as we just mentioned, i_{qr}' and i_{dr}' may be expressed in the same form with s replaced by r except in ω_{esk}. If these expressions for the arbitrary reference frame currents are substituted into (10.2-41), we obtain

$$T_e = \left(\frac{P}{2}\right)\left(\frac{X_{ms}}{\omega_b}\right)\sum_{k=0}^{\infty}\sum_{K=0}^{\infty}[(i_{qskA}i_{qrKB}' - i_{qskB}i_{qrKA}'$$

$$- i_{qskC}i_{qrKD}' + i_{qskD}i_{qrKC}')\cos(\omega_{esk} - \omega_{esK})t$$

$$+ (-i_{qskA}i_{qrKD}' + i_{qskC}i_{qrKB}' - i_{qskB}i_{qrKC}' + i_{qskD}i_{qrKA}')\cos(\omega_{esk} + \omega_{esK})t$$

$$+ (i_{qskA}i_{qrKA}' + i_{qskB}i_{qrKB}' - i_{qskC}i_{qrKC}' - i_{qskD}i_{qrKD}')\sin(\omega_{esk} - \omega_{esK})t$$

$$+ (i_{qskA}i_{qrKC}' - i_{qskB}i_{qrKD}' - i_{qskC}i_{qrKA}' + i_{qskD}i_{qrKB}')\sin(\omega_{esk} + \omega_{esK})t]$$

(10.2-42)

In (10.2-42) it is necessary to use the uppercase index K in i_{qr}' and i_{dr}' because a double summation is required.

Equations (10.2-25)–(10.2-28) provide relationships between $\tilde{F}_{qs+k}^s$, $\tilde{F}_{qs-k}^s$, $\tilde{F}_{ds+k}^s$, and $\tilde{F}_{ds-k}^s$ and the associated ABCD quantities that appear in (10.2-42). Identical expressions can be written between $\tilde{F}_{qr+k}^s$, $\tilde{F}_{qr-k}^s$, $\tilde{F}_{dr+k}^s$, and $\tilde{F}_{dr-k}^s$ and the associated ABCD quantities. With these relationships substituted into (10.2-42) and with considerable work, we can write

$$
T_e = 2\left(\frac{P}{2}\right)\left(\frac{X_{ms}}{\omega_b}\right)\sum_{k=1}^{\infty}\sum_{K=1}^{\infty}\{\mathrm{Re}[j(\tilde{I}_{qs+k}^{s*}\tilde{I}_{qr+K}^{\prime s} - \tilde{I}_{qs-k}^{s*}\tilde{I}_{qr-K}^{\prime s})]\cos(\omega_{esk} - \omega_{esK})t
$$

$$
+ \mathrm{Re}[j(-\tilde{I}_{qs+k}^{s}\tilde{I}_{qr-K}^{\prime s} + \tilde{I}_{qs-k}^{s}\tilde{I}_{qr+K}^{\prime s})]\cos(\omega_{esk} + \omega_{esK})t
$$

$$
+ \mathrm{Re}[\tilde{I}_{qs+k}^{s*}\tilde{I}_{qr+K}^{\prime s} - \tilde{I}_{qs-k}^{s*}\tilde{I}_{qr-K}^{\prime s}]\sin(\omega_{esk} - \omega_{esK})t
$$

$$
+ \mathrm{Re}[\tilde{I}_{qs+k}^{s}\tilde{I}_{qr-K}^{\prime s} - \tilde{I}_{qs-k}^{s}\tilde{I}_{qr+K}^{\prime s}]\sin(\omega_{esk} + \omega_{esK})t\} \tag{10.2-43}
$$

where the asterisk denotes the conjugate. The expression for torque may also be written in terms of phasors representing stator phase variables by (10.2-39) and the equivalent for the phasors representing rotor phase variables which follows from (10.2-34).

The analysis of unbalanced stator applied voltages wherein v_{as} and v_{bs} are unbalanced due to unbalanced source voltages is quite straightforward [3]. Here, we need only to substitute the actual phasor values of $\tilde{V}_{ask}$ and $\tilde{V}_{bsk}$ into (10.2-39) to determine $\tilde{V}_{qs+k}^{s}$ and $\tilde{V}_{qs-k}^{s}$, which in turn are substituted into (10.2-35) and the component currents are calculated. The torque may then be calculated by substituting the component currents into (10.2-43).

The analysis of an unbalanced stator impedance requires some modification of (10.2-35) [3]. In this case let

$$
e_{ga} = i_{as}z(p) + v_{as} \tag{10.2-44}
$$

$$
e_{gb} = v_{bs} \tag{10.2-45}
$$

where e_{ga} and e_{gb} are the source voltages (balanced or unbalanced) and $z(p)$ is an impedance in series with phase a of the machine. The phasor equivalents are

$$
\tilde{V}_{as} = \tilde{E}_{ga} - \tilde{I}_{as}Z \tag{10.2-46}
$$

$$
\tilde{V}_{bs} = \tilde{E}_{gb} \tag{10.2-47}
$$

Substituting into (10.2-39) yields

$$
\tilde{V}_{qs+}^{s} = \frac{1}{2}[\tilde{E}_{ga} + j\tilde{E}_{gb} - \tilde{I}_{as}Z] \tag{10.2-48}
$$

$$
\tilde{V}_{qs-}^{s} = \frac{1}{2}[\tilde{E}_{ga} - j\tilde{E}_{gb} - \tilde{I}_{as}Z] \tag{10.2-49}
$$

For simplicity, we have assumed that the source voltages contain only one frequency; thus the subscript 1 is omitted from the above equations. It is clear from the inverse of (10.2-39) that

$$
\tilde{I}_{as} = \tilde{I}_{qs+}^{s} + \tilde{I}_{qs-}^{s} \tag{10.2-50}
$$

Substitution of (10.2-48)–(10.2-50) into (10.2-35) yields

$$
\begin{bmatrix} \tilde{E} \\ 0 \\ \tilde{E}^* \\ 0 \end{bmatrix} = \begin{bmatrix} \frac{1}{2}Z + r_s + j\frac{\omega_e}{\omega_b}X_{ss} & j\frac{\omega_e}{\omega_b}X_{ms} & \frac{1}{2}Z & 0 \\ j\frac{\omega_e}{\omega_b}X_{ms} & \frac{r'_r}{s} + j\frac{\omega_e}{\omega_b}X'_{rr} & 0 & 0 \\ \frac{1}{2}Z & 0 & \frac{1}{2}Z + r_s + j\frac{\omega_e}{\omega_b}X_{ss} & j\frac{\omega_e}{\omega_b}X_{ms} \\ 0 & 0 & j\frac{\omega_e}{\omega_b}X_{ms} & \frac{r'_r}{2-s} + j\frac{\omega_e}{\omega_b}X'_{rr} \end{bmatrix} \begin{bmatrix} \tilde{I}^s_{qs+} \\ \tilde{I}'^s_{qr+} \\ \tilde{I}^s_{qs-} \\ \tilde{I}'^s_{qr-} \end{bmatrix}
$$

$$(10.2\text{-}51)$$

where $\tilde{E}^*$ is the conjugate of $\tilde{E}$ and

$$\tilde{E} = \frac{1}{2}(\tilde{E}_{ga} + j\tilde{E}_{gb}) \tag{10.2-52}$$

It is clear that we are assuming that the machine is not doubly fed because $\tilde{V}'^s_{qr+}$ and $\tilde{V}'^s_{qr-}$ have been made zero.

For the analysis of an open-circuited stator phase let us assume that $i_{as}(i^s_{qs})$ is zero [5]. Hence

$$v_{as} = v^s_{qs} = \frac{p}{\omega_b}\psi^s_{qs} \tag{10.2-53}$$

Because i^s_{qs} is zero we obtain

$$\psi^s_{qs} = X_{ms}i'^s_{qr} \tag{10.2-54}$$

Hence

$$v_{as} = \frac{p}{\omega_b}X_{ms}i'^s_{qr} \tag{10.2-55}$$

$$v_{bs} = e_{gb} \tag{10.2-56}$$

If we substitute the phasor equivalent of (10.2-55) and (10.2-56) into (10.2-39), we obtain

$$\tilde{V}^s_{qs+} = \frac{1}{2}j\frac{\omega_e}{\omega_b}X_{ms}\tilde{I}'^s_{qr} + \frac{1}{2}j\tilde{E}_{gb} \tag{10.2-57}$$

$$\tilde{V}^s_{qs-} = \frac{1}{2}j\frac{\omega_e}{\omega_b}X_{ms}\tilde{I}'^s_{qr} - \frac{1}{2}j\tilde{E}_{gb} \tag{10.2-58}$$

The subscript 1 has been dropped because it is assumed that the source has only one frequency. From the inverse of (10.2-34) we obtain

$$\tilde{I}'^s_{qr} = \tilde{I}'^s_{qr+} + \tilde{I}'^s_{qr-} \tag{10.2-59}$$

and because $\tilde{I}^s_{qs} = 0$ we have

$$\tilde{I}^s_{qs-} = -\tilde{I}^s_{qs+} \tag{10.2-60}$$

Substituting (10.2-59) into (10.2-57) and (10.2-58) and then substituting the results into (10.2-35) and including (10.2-60), we can write

$$\begin{bmatrix} \frac{1}{2}j\tilde{E}_{gb} \\ 0 \\ 0 \end{bmatrix} = \begin{bmatrix} r_s + j\frac{\omega_e}{\omega_b}X_{ss} & j\frac{1}{2}\frac{\omega_e}{\omega_b}X_{ms} & -j\frac{1}{2}\frac{\omega_e}{\omega_b}X_{ms} \\ j\frac{\omega_e}{\omega_b}X_{ms} & \frac{r'_r}{s} + j\frac{\omega_e}{\omega_b}X'_{rr} & 0 \\ -j\frac{\omega_e}{\omega_b}X_{ms} & 0 & \frac{r'_r}{2-s} + j\frac{\omega_e}{\omega_b}X'_{rr} \end{bmatrix} \begin{bmatrix} \tilde{I}^s_{qs+} \\ \tilde{I}'^s_{qr+} \\ \tilde{I}'^s_{qr-} \end{bmatrix} \tag{10.2-61}$$

Here again we are assuming the machine is not doubly fed.

10.3   VOLTAGE AND TORQUE EQUATIONS IN MACHINE VARIABLES FOR UNSYMMETRICAL 2-PHASE INDUCTION MACHINES

A 2-pole, 2-phase unsymmetrical induction machine is shown in Fig. 10.3-1. The stator windings are nonidentical sinusoidally distributed windings arranged in space quadrature. The as winding is assumed to have N_s equivalent turns with resistance r_s.

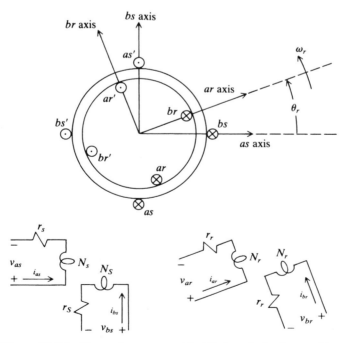

Figure 10.3-1   Two-pole, unsymmetrical 2-phase induction machine.

The *bs* winding has N_S equivalent turns with resistance r_S. The rotor windings may be considered as two identical sinusoidally distributed windings arranged in space quadrature. Each rotor winding has N_r equivalent turns with resistance r_r. The voltage equations expressed in machine variables may be written as

$$\mathbf{v}_{abs} = \mathbf{r}_s \mathbf{i}_{abs} + p\boldsymbol{\lambda}_{abs} \tag{10.3-1}$$

$$\mathbf{v}_{abr} = \mathbf{r}_r \mathbf{i}_{abr} + p\boldsymbol{\lambda}_{abr} \tag{10.3-2}$$

where

$$(\mathbf{f}_{abs})^T = [f_{as} \quad f_{bs}] \tag{10.3-3}$$

$$(\mathbf{f}_{abr})^T = [f_{ar} \quad f_{br}] \tag{10.3-4}$$

$$\mathbf{r}_s = \operatorname{diag}[r_s \quad r_S] \tag{10.3-5}$$

The resistance matrix $\mathbf{r}_r$ is diagonal with equal nonzero elements. The flux linkages may be expressed as

$$\begin{bmatrix} \boldsymbol{\lambda}_{abs} \\ \boldsymbol{\lambda}_{abr} \end{bmatrix} = \begin{bmatrix} \mathbf{L}_s & \mathbf{L}_{sr} \\ (\mathbf{L}_{sr})^T & \mathbf{L}_r \end{bmatrix} \begin{bmatrix} \mathbf{i}_{abs} \\ \mathbf{i}_{abr} \end{bmatrix} \tag{10.3-6}$$

where

$$\mathbf{L}_s = \begin{bmatrix} L_{ls} + L_{ms} & 0 \\ 0 & L_{lS} + L_{mS} \end{bmatrix} \tag{10.3-7}$$

$$\mathbf{L}_r = \begin{bmatrix} L_{lr} + L_{mr} & 0 \\ 0 & L_{lr} + L_{mr} \end{bmatrix} \tag{10.3-8}$$

$$\mathbf{L}_{sr} = \begin{bmatrix} L_{sr}\cos\theta_r & -L_{sr}\sin\theta_r \\ L_{Sr}\sin\theta_r & L_{Sr}\cos\theta_r \end{bmatrix} \tag{10.3-9}$$

In (10.3-7), L_{ls} and L_{ms} (L_{lS} and L_{mS}) are, respectively, the leakage and magnetizing inductances of the *as* winding (*bs* winding). In the case of the identical rotor windings, L_{lr} and L_{mr} are the leakage and magnetizing inductances, respectively. In (10.3-9), L_{sr} (L_{Sr}) is the amplitude of the mutual inductance between the *as* winding (*bs* winding) and the rotor windings. The rotor angular displacement is defined in Fig. 10.3-1.

The expression for electromagnetic torque may be obtained from (4.3-6). Thus

$$T_e = \left(\frac{P}{2}\right)(\mathbf{i}_{abs})^T \frac{\partial}{\partial\theta_r}[(\mathbf{L}_{sr})^T]\mathbf{i}_{abr} \tag{10.3-10}$$

In expanded form, (10.3-10) which is positive for motor action becomes

$$T_e = \left(\frac{P}{2}\right)[L_{sr}i_{as}(-i_{ar}\sin\theta_r - i_{br}\cos\theta_r) + L_{Sr}i_{bs}(i_{ar}\cos\theta_r - i_{br}\sin\theta_r)] \tag{10.3-11}$$

The torque and rotor speed are related by

$$T_e = J\left(\frac{2}{P}\right)p\omega_r + T_L \tag{10.3-12}$$

where J is the inertia of the rotor and the connected load. The load torque is T_L; it is positive for motor action.

10.4 VOLTAGE AND TORQUE EQUATIONS IN STATIONARY REFERENCE-FRAME VARIABLES FOR UNSYMMETRICAL 2-PHASE INDUCTION MACHINES

Because the stator windings of the unsymmetrical 2-phase induction machine are not identical, it is necessary to transform all machine variables to the stationary reference frame in order to obtain voltage equations with constant coefficients. Recalling the properties of the transformation set forth in Chapter 3, the voltage equations in terms of stationary reference-frame variables ($\omega = 0$) become

$$\mathbf{v}_{qds}^s = \mathbf{r}_s\mathbf{i}_{qds}^s + p\boldsymbol{\lambda}_{qds}^s \tag{10.4-1}$$

$$\mathbf{v}_{qdr}^s = \mathbf{r}_r\mathbf{i}_{qdr}^s - \omega_r\boldsymbol{\lambda}_{dqr}^s + p\boldsymbol{\lambda}_{qdr}^s \tag{10.4-2}$$

where

$$(\boldsymbol{\lambda}_{dqr}^s)^T = [\,\lambda_{dr}^s \quad -\lambda_{qr}^s\,] \tag{10.4-3}$$

Substituting the equations of transformation given by (10.2-1) into the flux linkage equations yields

$$\begin{bmatrix} \boldsymbol{\lambda}_{qds}^s \\ \boldsymbol{\lambda}_{qdr}^s \end{bmatrix} = \begin{bmatrix} \mathbf{K}_{2s}^s\mathbf{L}_s(\mathbf{K}_{2s}^s)^{-1} & \mathbf{K}_{2s}^s\mathbf{L}_{sr}(\mathbf{K}_{2r}^s)^{-1} \\ \mathbf{K}_{2r}^s(\mathbf{L}_{sr})^T(\mathbf{K}_{2s}^s)^{-1} & \mathbf{K}_{2r}^s\mathbf{L}_r(\mathbf{K}_{2r}^s)^{-1} \end{bmatrix}\begin{bmatrix} \mathbf{i}_{qds}^s \\ \mathbf{i}_{qdr}^s \end{bmatrix} \tag{10.4-4}$$

where $\mathbf{L}_s$, $\mathbf{L}_r$, and $\mathbf{L}_{sr}$ are defined, respectively, by (10.3-7)–(10.3-9). With $\theta(0)$ set equal to zero we obtain

$$\mathbf{K}_{2s}^s = \begin{bmatrix} 1 & 0 \\ 0 & -1 \end{bmatrix} \tag{10.4-5}$$

Therefore

$$\mathbf{K}_{2s}^s\mathbf{L}_s(\mathbf{K}_{2s}^s)^{-1} = \mathbf{L}_s \tag{10.4-6}$$

$$\mathbf{K}_{2r}^s\mathbf{L}_r(\mathbf{K}_{2r}^s)^{-1} = \mathbf{L}_r \tag{10.4-7}$$

$$\mathbf{K}_{2s}^s\mathbf{L}_{sr}(\mathbf{K}_{2r}^s)^{-1} = \mathbf{K}_{2r}^s(\mathbf{L}_{sr})^T(\mathbf{K}_{2s}^s)^{-1}$$

$$= \begin{bmatrix} L_{sr} & 0 \\ 0 & L_{Sr} \end{bmatrix} \tag{10.4-8}$$

In expanded form we have

$$v_{qs}^s = r_s i_{qs}^s + p\lambda_{qs}^s \tag{10.4-9}$$

$$v_{ds}^s = r_s i_{ds}^s + p\lambda_{ds}^s \tag{10.4-10}$$

$$v_{qr}^s = r_r i_{qr}^s - \omega_r \lambda_{dr}^s + p\lambda_{qr}^s \tag{10.4-11}$$

$$v_{dr}^s = r_r i_{dr}^s + \omega_r \lambda_{qr}^s + p\lambda_{dr}^s \tag{10.4-12}$$

where

$$\lambda_{qs}^s = L_{ss} i_{qs}^s + L_{sr} i_{qr}^s \tag{10.4-13}$$

$$\lambda_{ds}^s = L_{SS} i_{ds}^s + L_{Sr} i_{dr}^s \tag{10.4-14}$$

$$\lambda_{qr}^s = L_{rr} i_{qr}^s + L_{sr} i_{qs}^s \tag{10.4-15}$$

$$\lambda_{dr}^s = L_{rr} i_{dr}^s + L_{Sr} i_{ds}^s \tag{10.4-16}$$

where $L_{ss} = L_{ls} + L_{ms}$, $L_{SS} = L_{lS} + L_{mS}$, and $L_{rr} = L_{lr} + L_{mr}$. The inductances L_{sr} and L_{Sr} are defined by (10.3-9).

In developing equivalent circuits of electric machines it is customary to refer the rotor variables to the stator windings by a turns ratio. The stator windings of an unsymmetrical 2-phase induction machine do not have the same number of effective turns. Later we will find it desirable to refer all rotor and stator variables to the *as* winding; however, at this time it is convenient to refer all q variables to the *as* winding with N_s effective turns and all d variables to the *bs* winding with N_S effective turns. It is clear that this type of referencing is best performed now, after implementing a change of variables to the stationary reference frame with $\theta(0)$ set equal to zero. If all q variables are referred to the winding with N_s effective turns (*as* winding) and all d variables are referred to the winding with N_S effective turns (*bs* winding) the voltage equations become

$$v_{qs}^s = r_s i_{qs}^s + p\lambda_{qs}^s \tag{10.4-17}$$

$$v_{ds}^s = r_s i_{ds}^s + p\lambda_{ds}^s \tag{10.4-18}$$

$$v_{qr}^{\prime s} = r_r' i_{qr}^{\prime s} - \frac{N_s}{N_S} \omega_r \lambda_{dr}^{\prime s} + p\lambda_{qr}^{\prime s} \tag{10.4-19}$$

$$v_{dr}^{\prime s} = r_R' i_{dr}^{\prime s} + \frac{N_S}{N_s} \omega_r \lambda_{qr}^{\prime s} + p\lambda_{dr}^{\prime s} \tag{10.4-20}$$

where

$$\lambda_{qs}^s = L_{ls} i_{qs}^s + L_{ms}(i_{qs}^s + i_{qr}^{\prime s}) \tag{10.4-21}$$

$$\lambda_{ds}^s = L_{lS} i_{ds}^s + L_{mS}(i_{ds}^s + i_{dr}^{\prime s}) \tag{10.4-22}$$

$$\lambda_{qr}^{\prime s} = L_{lr}' i_{qr}^{\prime s} + L_{ms}(i_{qs}^s + i_{qr}^{\prime s}) \tag{10.4-23}$$

$$\lambda_{dr}^{\prime s} = L_{lR}' i_{dr}^{\prime s} + L_{mS}(i_{ds}^s + i_{dr}^{\prime s}) \tag{10.4-24}$$

in which

$$v_{qr}^{\prime s} = \frac{N_s}{N_r} v_{qr}^{s} \tag{10.4-25}$$

$$i_{qr}^{\prime s} = \frac{N_r}{N_s} i_{qr}^{s} \tag{10.4-26}$$

$$v_{dr}^{\prime s} = \frac{N_s}{N_r} v_{dr}^{s} \tag{10.4-27}$$

$$i_{dr}^{\prime s} = \frac{N_r}{N_s} i_{dr}^{s} \tag{10.4-28}$$

$$r_r^{\prime} = \left(\frac{N_s}{N_r}\right)^2 r_r \tag{10.4-29}$$

$$L_{lr}^{\prime} = \left(\frac{N_s}{N_r}\right)^2 L_{lr} \tag{10.4-30}$$

$$r_R^{\prime} = \left(\frac{N_s}{N_r}\right)^2 r_r \tag{10.4-31}$$

$$L_{lR}^{\prime} = \left(\frac{N_s}{N_r}\right)^2 L_{lr} \tag{10.4-32}$$

$$L_{ms} = \frac{N_s}{N_r} L_{sr} \tag{10.4-33}$$

$$L_{mS} = \frac{N_s}{N_r} L_{Sr} \tag{10.4-34}$$

The above equations suggest the equivalent circuits shown in Fig. 10.4-1.

It is often convenient to express the voltage and flux linkages equations in terms of reactances rather than inductances. Hence, (10.4-17)–(10.4-24) are often written as

$$v_{qs}^{s} = r_s i_{qs}^{s} + \frac{p}{\omega_b} \psi_{qs}^{s} \tag{10.4-35}$$

$$v_{ds}^{s} = r_s i_{ds}^{s} + \frac{p}{\omega_b} \psi_{ds}^{s} \tag{10.4-36}$$

$$v_{qr}^{\prime s} = r_r^{\prime} i_{qr}^{\prime s} - \frac{N_s \omega_r}{N_s \omega_b} \psi_{dr}^{\prime s} + \frac{p}{\omega_b} \psi_{qr}^{\prime s} \tag{10.4-37}$$

$$v_{dr}^{\prime s} = r_R^{\prime} i_{dr}^{\prime s} + \frac{N_s \omega_r}{N_s \omega_b} \psi_{qr}^{\prime s} + \frac{p}{\omega_b} \psi_{dr}^{\prime s} \tag{10.4-38}$$

where ω_b is the base electrical angular velocity used to calculate the inductive reactances. The flux linkage equations are now written in terms of flux linkages

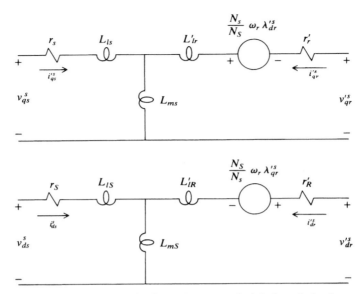

Figure 10.4-1 Equivalent circuits for an unsymmetrical 2-phase induction machine.

per second. Thus

$$\psi^s_{qs} = X_{ls}i^s_{qs} + X_{ms}(i^s_{qs} + i'^s_{qr}) \tag{10.4-39}$$

$$\psi^s_{ds} = X_{lS}i^s_{ds} + X_{mS}(i^s_{ds} + i'^s_{dr}) \tag{10.4-40}$$

$$\psi'^s_{qr} = X'_{lr}i'^s_{qr} + X_{ms}(i^s_{qs} + i'^s_{qr}) \tag{10.4-41}$$

$$\psi'^s_{dr} = X'_{lR}i'^s_{dr} + X_{mS}(i^s_{ds} + i'^s_{dr}) \tag{10.4-42}$$

The equations may also be written in matrix form as

$$\begin{bmatrix} v^s_{qs} \\ v^s_{ds} \\ v'^s_{qr} \\ v'^s_{dr} \end{bmatrix} = \begin{bmatrix} r_s + \frac{p}{\omega_b}X_{SS} & 0 & \frac{p}{\omega_b}X_{ms} & 0 \\ 0 & r_s + \frac{p}{\omega_b}X_{SS} & 0 & \frac{p}{\omega_b}X_{mS} \\ \frac{p}{\omega_b}X_{ms} & -\frac{1}{n}\frac{\omega_r}{\omega_b}X_{ms} & r'_r + \frac{p}{\omega_b}X'_{rr} & -\frac{1}{n}\frac{\omega_r}{\omega_b}X'_{RR} \\ n\frac{\omega_r}{\omega_b}X_{ms} & \frac{p}{\omega_b}X_{mS} & n\frac{\omega_r}{\omega_b}X'_{rr} & r'_R + \frac{p}{\omega_b}X'_{RR} \end{bmatrix} \begin{bmatrix} i^s_{qs} \\ i^s_{ds} \\ i'^s_{qr} \\ i'^s_{dr} \end{bmatrix} \tag{10.4-43}$$

For compactness, the turns ratio N_S/N_s has been replaced by n and

$$X_{ss} = X_{ls} + X_{ms} \tag{10.4-44}$$

$$X_{SS} = X_{lS} + X_{mS} \tag{10.4-45}$$

$$X'_{rr} = X'_{lr} + X_{ms} \tag{10.4-46}$$

$$X'_{RR} = X'_{lR} + X_{mS} \tag{10.4-47}$$

The instantaneous electromagnetic torque may be expressed in terms of the substitute variables as

$$T_e = \frac{P}{2}(L_{sr}i_{qs}^s i_{dr}^s - L_{Sr}i_{ds}^s i_{qr}^s)$$ (10.4-48)

The inductances can be expressed in terms of turns and reluctance. Therefore, it can be shown that

$$L_{sr} = \frac{N_r}{N_s}L_{ms}$$ (10.4-49)

$$L_{Sr} = \frac{N_S N_r}{N_s^2}L_{ms}$$ (10.4-50)

It follows that (10.4-48) may also be written as

$$T_e = \left(\frac{P}{2}\right)\left(\frac{N_S}{N_s}\right)\left(\frac{X_{ms}}{\omega_b}\right)(i_{qs}^s i_{dr}^{\prime s} - i_{ds}^s i_{qr}^{\prime s})$$ (10.4-51)

Because the stator windings are unsymmetrical, time-varying coefficients will appear in the synchronously rotating reference frame. Therefore, linearization, eigenvalue calculation, and transfer function formulation are not possible following the procedures set forth in Chapter 8. However, the concept of multiple reference frames is used in reference 5 to achieve these small-displacement characteristics. This information is invaluable to the engineer interested in the control of single-phase applications of induction machine.

10.5 ANALYSIS OF STEADY-STATE OPERATION OF UNSYMMETRICAL 2-PHASE INDUCTION MACHINES

The unsymmetrical 2-phase induction machine is designed to operate from a single-phase source. Hence, it is always subjected to unbalanced applied stator voltages. In other words, the voltages applied to the *as* and *bs* windings do not normally form a balanced 2-phase set. It would first appear desirable to employ the method of symmetrical components along with the concept of the arbitrary reference frame to express the variables as positively and negatively rotating components. Although this approach will be used, it does not offer the same convenience of analysis as in the case of the symmetrical induction machine. In particular, we will find that in the case of the unsymmetrical machine the voltage equations are not decoupled by the symmetrical component transformation. Before employing the concept of symmetrical components to obtain the voltage equations expressed in their traditional form we will show that, in the case of the unsymmetrical machine, the voltage equations in stationary reference-frame variables are equally convenient for analyzing steady-state operation.

Voltage Equations in qs and ds Variables

The phasor voltage equations may be written from (10.4-43) with p set equal to $j\omega_{esk}$. Thus

$$
\begin{bmatrix} \tilde{V}^s_{qsk} \\ \tilde{V}^s_{dsk} \\ 0 \\ 0 \end{bmatrix} = \begin{bmatrix} r_s + j\frac{\omega_{esk}}{\omega_b}X_{ss} & 0 & j\frac{\omega_{esk}}{\omega_b}X_{ms} & 0 \\ 0 & r_S + j\frac{\omega_{esk}}{\omega_b}X_{SS} & 0 & j\frac{\omega_{esk}}{\omega_b}X_{ms} \\ j\frac{\omega_{esk}}{\omega_b}X_{ms} & -\frac{1}{n}\frac{\omega_r}{\omega_b}X_{ms} & r'_r + j\frac{\omega_{esk}}{\omega_b}X'_{rr} & -\frac{1}{n}\frac{\omega_r}{\omega_b}X'_{RR} \\ n\frac{\omega_r}{\omega_b}X_{ms} & j\frac{\omega_{esk}}{\omega_b}X_{mS} & n\frac{\omega_r}{\omega_b}X'_{rr} & r'_R + j\frac{\omega_{esk}}{\omega_b}X'_{RR} \end{bmatrix} \begin{bmatrix} \tilde{I}^s_{qsk} \\ \tilde{I}^s_{dsk} \\ \tilde{I}'^s_{qrk} \\ \tilde{I}'^s_{drk} \end{bmatrix}
$$

$$(10.5\text{-}1)$$

The rotor voltages are set equal to zero because unsymmetrical 2-phase induction machines are only equipped with squirrel-cage rotors. It is clear that the qs and ds variables may be related to the phase variables by

$$\tilde{\mathbf{F}}_{qdsk} = \mathbf{K}^s_{2s}\tilde{\mathbf{F}}_{absk} \qquad (10.5\text{-}2)$$

where $\mathbf{K}^s_{2s}$ is defined by (10.4-5) for $\theta(0) = 0$. Because computers are readily available, it may be preferable to use (10.5-1) directly. On the other hand, there are advantages, with regard to hand calculations and interpretation, to reduce (10.5-1) to a 2×2 matrix. For this purpose it is convenient to use the following substitutions primarily for compactness:

$$Z_s = r_s + j\frac{\omega_{esk}}{\omega_b}X_{ss} \qquad (10.5\text{-}3)$$

$$Z_S = r_S + j\frac{\omega_{esk}}{\omega_b}X_{SS} \qquad (10.5\text{-}4)$$

$$Z'_r = r'_r + j\frac{\omega_{esk}}{\omega_b}X'_{rr} \qquad (10.5\text{-}5)$$

$$f_e = \frac{\omega_{esk}}{\omega_b} \qquad (10.5\text{-}6)$$

$$f_r = \frac{\omega_r}{\omega_b} \qquad (10.5\text{-}7)$$

Also, it is clear that the following equalities exist:

$$Z'_R = n^2 Z'_r \qquad (10.5\text{-}8)$$

$$X_{mS} = n^2 X_{ms} \qquad (10.5\text{-}9)$$

If the above are substituted into (10.5-1), the voltage equations become

$$
\begin{bmatrix} \tilde{V}^s_{qsk} \\ \tilde{V}^s_{dsk} \\ 0 \\ 0 \end{bmatrix} = \begin{bmatrix} Z_s & 0 & jf_e X_{ms} & 0 \\ 0 & Z_S & 0 & jn^2 f_e X_{ms} \\ jf_e X_{ms} & -nf_r X_{ms} & Z'_r & -nf_r X'_{rr} \\ nf_r X_{ms} & jn^2 f_e X_{ms} & nf_r X'_{rr} & n^2 Z'_r \end{bmatrix} \begin{bmatrix} \tilde{I}^s_{qsk} \\ \tilde{I}^s_{dsk} \\ \tilde{I}'^s_{qrk} \\ \tilde{I}'^s_{drk} \end{bmatrix} \qquad (10.5\text{-}10)
$$

This equation can be partitioned into 2×2 matrices and written as

$$\begin{bmatrix} \tilde{\mathbf{V}}^s_{qdsk} \\ 0 \end{bmatrix} = \begin{bmatrix} \mathbf{A} & \mathbf{B} \\ \mathbf{C} & \mathbf{D} \end{bmatrix} \begin{bmatrix} \tilde{\mathbf{I}}^s_{qdsk} \\ \tilde{\mathbf{I}}'^s_{qdrk} \end{bmatrix} \tag{10.5-11}$$

It follows that

$$\begin{aligned} \tilde{\mathbf{I}}'^s_{qdrk} &= -(\mathbf{D})^{-1}\mathbf{C}\tilde{\mathbf{I}}^s_{qdsk} \\ &= \begin{bmatrix} -Z_A & nZ_B \\ -\dfrac{Z_B}{n} & -Z_A \end{bmatrix} \begin{bmatrix} \tilde{I}^s_{qsk} \\ \tilde{I}^s_{dsk} \end{bmatrix} \end{aligned} \tag{10.5-12}$$

where

$$Z_A = \frac{X_{ms}}{Z_r'^2 + f_r^2 X_{rr}'^2}[(f_r^2 - f_e^2)X_{rr}' + jr_r'f_e] \tag{10.5-13}$$

$$Z_B = \frac{X_{ms}f_r r_r'}{Z_r'^2 + f_r^2 X_{rr}'^2} \tag{10.5-14}$$

If (10.5-12) is substituted into the first row of (10.5-11) we can write

$$\tilde{\mathbf{V}}^s_{qdsk} = [\mathbf{A} - \mathbf{B}(\mathbf{D})^{-1}\mathbf{C}]\tilde{\mathbf{I}}^s_{qdsk} \tag{10.5-15}$$

which can be written as

$$\begin{bmatrix} \tilde{V}^s_{qsk} \\ \tilde{V}^s_{dsk} \end{bmatrix} = \begin{bmatrix} Z_s - jf_eX_{ms}Z_A & jnf_eX_{ms}Z_B \\ -jnf_eX_{ms}Z_B & Z_S - jn^2 f_eX_{ms}Z_A \end{bmatrix} \begin{bmatrix} \tilde{I}^s_{qsk} \\ \tilde{I}^s_{dsk} \end{bmatrix} \tag{10.5-16}$$

Voltage Equations in Component Form

The voltage equations given in (10.5-16) are in a form convenient for hand calculations, and one need not look to other forms of these equations for computational purposes. However, over the years, revolving-field theory [1–3] has been used to establish an equivalent circuit that has become a standard for analyzing the steady-state operation of unsymmetrical 2-phase induction machines. This equivalent circuit may be derived from (10.5-16) by using 2-phase symmetrical components. However, in order to express the equations in the form normally used, it is necessary to refer the ds variables to the qs winding (as winding). It is recalled that, after all variables were transformed to the stationary reference frame with $\theta(0)$ set equal to zero, the qr variables were referred to the qs winding (N_s turns) and the dr variables were referred to the ds windings (N_S turns). The ds variables are referred to the qs winding by the following relations:

$$\tilde{I}'^s_{ds} = n\tilde{I}^s_{ds} \tag{10.5-17}$$

$$\tilde{V}'^s_{ds} = \frac{1}{n}\tilde{V}^s_{ds} \tag{10.5-18}$$

Substituting (10.5-17) and (10.5-18) into (10.5-16) yields

$$
\begin{bmatrix} \tilde{V}^s_{qsk} \\ \tilde{V}'^s_{dsk} \end{bmatrix} = \begin{bmatrix} Z_s - jf_e X_{ms} Z_A & jf_e X_{ms} Z_B \\ -jf_e X_{ms} Z_B & \frac{Z_s}{n^2} - jf_e X_{ms} Z_A \end{bmatrix} \begin{bmatrix} \tilde{I}^s_{qsk} \\ \tilde{I}'^s_{dsk} \end{bmatrix}
\tag{10.5-19}
$$

The above equation may be written in component form by substituting (10.2-30) for the voltages and currents. Actually we are substituting for $\tilde{F}'^s_{ds}$ rather than $\tilde{F}^s_{ds}$ as indicated by the notation used in (10.2-30); however, it is understood that the transformation is equally valid for $\tilde{F}'^s_{ds}$. Therefore, if the voltages and currents are expressed in component form by (10.2-30) and if the equations are arranged so that the operator j does not appear in the voltage and current vectors, we can write

$$
\begin{bmatrix} \tilde{V}^s_{qs+k} + \tilde{V}^s_{qs-k} \\ \tilde{V}^s_{qs+k} - \tilde{V}^s_{qs-k} \end{bmatrix} = \begin{bmatrix} Z_s - jf_e X_{ms} Z_A & -f_e X_{ms} Z_B \\ -f_e X_{ms} Z_B & \frac{Z_s}{n^2} - jf_e X_{ms} Z_A \end{bmatrix} \begin{bmatrix} \tilde{I}^s_{qs+k} + \tilde{I}^s_{qs-k} \\ \tilde{I}^s_{qs+k} - \tilde{I}^s_{qs-k} \end{bmatrix}
\tag{10.5-20}
$$

In revolving-field theory the $+$ and $-$ quantities are often called, respectively, the forward and backward quantities associated with the forward and backward revolving fields. The coefficients of the currents may now be gathered, and (10.5-20) may be written as

$$
\begin{bmatrix} \tilde{V}^s_{qs+k} + \tilde{V}^s_{qs-k} \\ \tilde{V}^s_{qs+k} - \tilde{V}^s_{qs-k} \end{bmatrix} = \begin{bmatrix} Z_s - f_e X_{ms} Z_B - jf_e X_{ms} Z_A & Z_s + f_e X_{ms} Z_B - jf_e X_{ms} Z_A \\ \frac{Z_s}{n^2} - f_e X_{ms} Z_B - jf_e X_{ms} Z_A & -\frac{Z_s}{n^2} - f_e X_{ms} Z_B + jf_e X_{ms} Z_A \end{bmatrix} \begin{bmatrix} \tilde{I}^s_{qs+k} \\ \tilde{I}^s_{qs-k} \end{bmatrix}
\tag{10.5-21}
$$

Adding and subtracting the above voltage equations yields

$$
\begin{bmatrix} \tilde{V}^s_{qs+k} \\ \tilde{V}^s_{qs-k} \end{bmatrix} = \frac{1}{2} \begin{bmatrix} Z_s + \frac{Z_s}{n^2} - 2f_e X_{ms} Z_B - j2f_e X_{ms} Z_A & Z_s - \frac{Z_s}{n^2} \\ Z_s - \frac{Z_s}{n^2} & Z_s + \frac{Z_s}{n^2} + 2f_e X_{ms} Z_B - j2f_e X_{ms} Z_A \end{bmatrix} \begin{bmatrix} \tilde{I}^s_{qs+k} \\ \tilde{I}^s_{qs-k} \end{bmatrix}
\tag{10.5-22}
$$

It is left to the reader to show that (10.5-22) may be represented by the equivalent circuit shown in Fig. 10.5-1. Therein

$$
s_k = \frac{f_e - f_r}{f_e}
\tag{10.5-23}
$$

where f_e and f_r are defined, respectively, by (10.5-6) and (10.5-7). In Fig. 10.5-1, the parallel combination of $jf_e X_{ms}$ and $(r_r/s_k + jf_e X'_{lr})$ is commonly referred to as the forward-field impedance, and the parallel combination of $jf_e X_{ms}$ and $[r_r/(2 - s_k) + jf_e X'_{lr}]$ is called the backward-field impedance. We, of course, recognize these impedances of the symmetrical rotor circuits which are associated with the positive and negative sequence voltage equations.

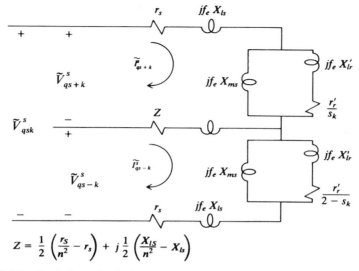

$$Z = \frac{1}{2}\left(\frac{r_s}{n^2} - r_s\right) + j\frac{1}{2}\left(\frac{X_{ls}}{n^2} - X_{ls}\right)$$

Figure 10.5-1 Equivalent circuit for steady-state operation of an unsymmetrical 2-phase induction machine.

The qs and ds currents may be expressed in terms of forward and backward quantities by employing the following relationships. From the inverse of (10.2-30) we obtain

$$\tilde{I}^s_{qsk} = \tilde{I}^s_{qs+k} + \tilde{I}^s_{qs-k} \tag{10.5-24}$$

$$\tilde{I}^s_{dsk} = j(\tilde{I}^s_{qs+k} - \tilde{I}^s_{qs-k}) \tag{10.5-25}$$

It is clear that

$$\tilde{I}^s_{dsk} = \frac{1}{n}\tilde{I}'^s_{dsk} = j\frac{1}{n}(\tilde{I}^s_{qs+k} - \tilde{I}^s_{qs-k}) \tag{10.5-26}$$

We can also write

$$\tilde{I}'^s_{qrk} = \tilde{I}'^s_{qr+k} + \tilde{I}'^s_{qr-k} \tag{10.5-27}$$

$$\tilde{I}'^s_{drk} = j\frac{1}{n}(\tilde{I}'^s_{qr+k} - \tilde{I}'^s_{qr-k}) \tag{10.5-28}$$

It is evident from Fig. 10.5-1 that

$$\tilde{I}'^s_{qr+k} = -\frac{jf_e X_{ms}}{r'_r/s_k + jf_e X'_{rr}}\tilde{I}^s_{qs+k} \tag{10.5-29}$$

$$\tilde{I}'^s_{qr-k} = -\frac{jf_e X_{ms}}{r'_r/(2 - s_k) + jf_e X'_{rr}}\tilde{I}^s_{qs-k} \tag{10.5-30}$$

The stator phase currents may be written from (10.5-24) and (10.5-26) because $\tilde{F}_{qs}^s = \tilde{F}_{as}$ and $\tilde{F}_{ds}^s = -\tilde{F}_{bs}$.

Torque Equation

By following a procedure similar to that used to derive (10.2-42), the steady-state electromagnetic torque may be expressed in terms of phase currents convenient for determining the torque if (10.5-16) is used to calculate the phase currents in phasor form. In particular,

$$
\begin{aligned}
T_e = {}& \left(\frac{P}{2}\right)\left(\frac{N_S}{N_s}\right)\left(\frac{X_{ms}}{\omega_b}\right) \sum_{k=0}^{\infty} \sum_{K=0}^{\infty} [\mathrm{Re}(\tilde{I}_{qsk}^{s*}\tilde{I}_{drK}^{\prime s} - \tilde{I}_{dsk}^{s*}\tilde{I}_{qrK}^{\prime s})\cos(\omega_{esk} - \omega_{esK})t \\
& + \mathrm{Re}(\tilde{I}_{qsk}^{s}\tilde{I}_{drK}^{\prime s} - \tilde{I}_{dsk}^{s}\tilde{I}_{qrK}^{\prime s})\cos(\omega_{esk} + \omega_{esK})t \\
& + \mathrm{Re}(-j\tilde{I}_{qsk}^{s*}\tilde{I}_{drK}^{\prime s} + j\tilde{I}_{dsk}^{s*}\tilde{I}_{qrK}^{\prime s})\sin(\omega_{esk} - \omega_{esK})t \\
& + \mathrm{Re}(-j\tilde{I}_{qsk}^{s}\tilde{I}_{drK}^{\prime s} + j\tilde{I}_{dsk}^{s}\tilde{I}_{qrK}^{\prime s})\sin(\omega_{esk} + \omega_{esK})t]
\end{aligned}
\tag{10.5-31}
$$

The uppercase index K is used for the rotor currents because a double summation is necessary and the asterisk denotes the conjugate. Note that (10.5-31) is in terms of $\tilde{I}_{dsk}^s$ rather than $\tilde{I}_{dsk}^{\prime s}$. If the component currents are calculated from (10.5-22), then (10.2-43) may be used to determine the torque if (10.2-43) is multiplied by (N_S/N_s).

Conditions for Balanced Operation

The purpose of the auxiliary winding (ds winding) and/or placing of capacitors in series with this winding is to shift the current so that it leads the current flowing in the qs winding. In this way a leading or "in-quadrature" current is established in the ds winding relative to the qs winding, whereupon an average torque is produced. It is interesting to look at the possibility of "balanced" operation a little closer. If we assume that

$$
\tilde{I}_{dsk}^{\prime s} = j\tilde{I}_{qsk}^s
\tag{10.5-32}
$$

which may be written as

$$
\tilde{I}_{dsk}^{s} = j\frac{1}{n}\tilde{I}_{qsk}^s
\tag{10.5-33}
$$

then the currents flowing in the stator windings produce only a positively rotating magnetic field (forward) for each frequency k. We will refer to this condition as "balanced" even though the actual currents are not equal in magnitude. Let us see the conditions that will give rise to (10.5-32) or (10.5-33). If (10.5-33) is substituted into the voltage equations given by (10.5-16) and if therein $\tilde{V}_{dsk}^s$ is set equal to

$\tilde{V}_{qsk}^s$, which is the case when both windings are supplied from the same single-phase source, we obtain

$$\tilde{V}_{qsk}^s = [Z_s - f_e X_{ms}(Z_B + jZ_A)]\tilde{I}_{qsk}^s \qquad (10.5\text{-}34)$$

$$\tilde{V}_{qsk}^s = \left[j\frac{Z_S}{n} - jnf_e X_{ms}(Z_B + jZ_A)\right]\tilde{I}_{qsk}^s \qquad (10.5\text{-}35)$$

Hence, in order for only the forward magnetic field to exist, the above two equations must be equal. Therefore

$$Z_s - f_e X_{ms}(Z_B + jZ_A) = j\frac{Z_S}{n} - jnf_e X_{ms}(Z_B + jZ_A) \qquad (10.5\text{-}36)$$

This equation can be expressed in terms of slip, (10.5-23), as

$$Z_s - \frac{jf_e^2 X_{ms}^2}{f_e X_{rr}' - j(r_r'/s_k)} = j\frac{Z_S}{n} + \frac{nf_e^2 X_{ms}^2}{f_e X_{rr}' - j(r_r'/s_k)} \qquad (10.5\text{-}37)$$

Equation (10.5-37) must be satisfied in order for the machine to operate in a "balanced" condition. It is clear that for each frequency component k, this can occur at only one rotor speed for a given set of machine parameters. Recall that

$$Z_s = r_s + jf_e(X_{ls} + X_{ms}) \qquad (10.5\text{-}38)$$

$$Z_S = r_S + jf_e(X_{lS} + n^2 X_{ms}) \qquad (10.5\text{-}39)$$

Therefore, we can also account for external impedance in series with either winding by simply replacing r_s with $r_s + r_A$ (r_S with $r_S + r_B$) and X_{ls} with $X_{ls} + X_A$ (X_{lS} with $X_{lS} + X_B$), where the A and B subscripts denote elements external to the as and bs windings, respectively.

10.6 SINGLE-PHASE INDUCTION MACHINES

The unsymmetrical 2-phase induction machine is designed for single-phase applications. There are several types of single-phase induction machines; the split-phase, the capacitor-start, and the capacitor-start, capacitor-run are perhaps the most common. It will become evident that the symmetrical 2-phase induction machine can also be used as capacitor-start and capacitor-start, capacitor-run machines. The equations given in this chapter may be simply modified to describe the behavior of all of these machines [6].

The unsymmetrical 2-phase induction machine used to demonstrate the performance of the different types of single-phase machines is a 4-pole, $\frac{1}{4}$-hp, 110-V,

60-Hz machine with the following parameters expressed in ohms [1] and [7].

$$
\begin{array}{lll}
r_s = 2.02 & X_{ms} = 66.8 & r'_r = 4.12 \\
X_{ls} = 2.79 & & X'_{lr} = 2.12 \\
r_S = 7.14 & X_{mS} = 92.9 & r'_R = 5.74 \\
X_{lS} = 3.22 & & X'_{lR} = 2.95
\end{array}
$$

The total inertia is $J = 1.46 \times 10^{-2}$ kg·m^2 and $N_S/N_s = 1.18$. Although we will consider other modes of operation of this machine, it is designed to operate as a capacitor-start, capacitor-run machine. The impedance of the start capacitor at 60 Hz is $3 - j14.5$ ohms; the run capacitor is $9 - j172$ ohms.

Single-Phase Stator Winding

The induction machine with only one stator winding energized will not develop an average torque at zero rotor speed. This fact is quite easy to establish. With a sinusoidal voltage applied to only one stator winding, the air-gap magnetomotive force (MMF) pulsates at the frequency of the source. This pulsating MMF may be thought of as two equal MMFs rotating in opposite directions. When the rotor is at stall the torques due to the interaction of the induced rotor currents with these two MMFs are equal and opposite. Although this device can develop an average torque, once it has started, to rotate in either direction, it cannot be started from stall without some auxiliary means.

A starting torque is developed if both stator windings are appropriately connected to the same single-phase source and if the stator windings, which are in space quadrature, are not symmetrical. With the stator windings unsymmetrical the current in one stator winding is shifted in phase with respect to the other stator winding current giving rise to stator current components which are in time quadrature. This results in an average torque. Consequently, both stator windings are used during starting and then, in the case of the split-phase or capacitor-start machines, one of the windings is disconnected from the source when the machine reaches 60% to 80% of synchronous speed. Thus, the normal mode of operation of many single-phase induction machines involves only one stator winding. The stator winding that remains connected to the source is generally referred to as the main or run winding; the winding that is used only for starting purposes in the case of the split-phase or capacitor-start machines is referred to as the auxiliary or start winding. We will let the *as* or *qs* winding be the main winding and let the *bs* or *ds* winding be the auxiliary winding.

Except for the starting period, the split-phase and capacitor-start single-phase induction machines operate with only one stator winding excited. Therefore it is important to establish the equations that can be used to predict this mode of operation. The voltage equations for only the *qs* winding may be written from the voltage equations given in the previous sections with the *ds* variables eliminated. Thus, the steady-state voltage equation from (10.5-16) is

$$
\tilde{V}^s_{qsk} = (Z_s - jf_e X_{ms} Z_A)\tilde{I}^s_{qsk} \tag{10.6-1}
$$

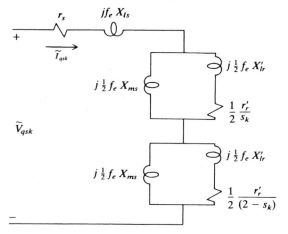

Figure 10.6-1 Equivalent circuit for single-phase stator winding.

It is left to the reader to show that the above equation is portrayed by the equivalent circuit shown in Fig. 10.6-1, where s_k is defined by (10.5-23).

The electromagnetic torque for this device may be calculated using the expressions for torque given previously. It is only necessary to set the *ds* current to zero and N_S equal to N_s ($n = 1$). The steady-state torque–speed characteristics of the example single-phase machine with rated voltage applied to the *qs* winding (*as* winding) are shown in Fig. 10.6-2. For this calculation v_{as} is set equal to $\sqrt{2}110\cos 377t$. Thus, $k = 1$ and $\omega_{esk} = \omega_{es1} = 377$ rad/s. The average torque $T_{e(avg)}$, along with the zero to peak amplitude of the double-frequency pulsating torque component $|T_{e(pul)}|$, is plotted.

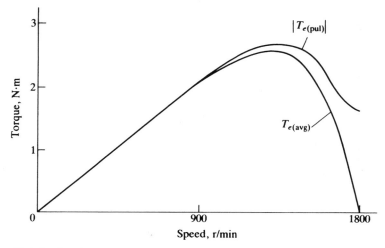

Figure 10.6-2 Steady-state torque–speed characteristics for single-phase stator winding.

Split-Phase Induction Motor

The split-phase induction machine is equipped with unsymmetrical stator windings. This machine is started with both stator windings energized from a common single-phase source. Once the rotor has reached 60% to 80% of synchronous speed, the auxiliary or start winding (*bs* or *ds* winding) is disconnected from the source by a centrifugal type disconnect switch. The voltage and torque relationship set forth in the previous section may be applied directly to analyze this machine during the time both windings are energized, and the relationships given for the single-phase stator winding may be employed once the auxiliary winding is disconnected. The steady-state torque–speed characteristics for the example machine with both stator windings connected are shown in Fig. 10.6-3. The parameters of the start winding are

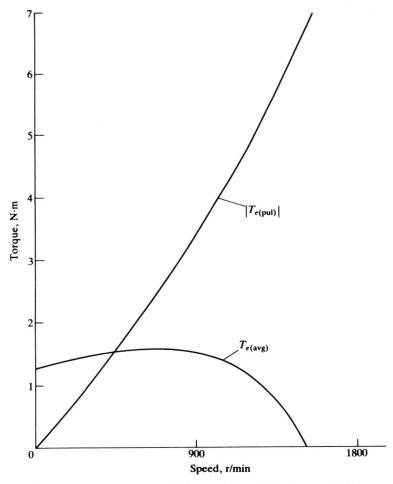

Figure 10.6-3   Steady-state torque–speed characteristics with both stator windings energized.

selected so that $\tilde{I}_{ds}^s$ leads $\tilde{I}_{qs}^s$. Plotted in Fig. 10.6-3 are the average torque $T_{e(\text{avg})}$ and the zero to peak amplitude of the double-frequency pulsating torque component $|T_{e(\text{pul})}|$. For this calculation, v_{as} is set equal to $\sqrt{2}110\cos 377t$ ($k = 1$ and $\omega_{es1} = 377\text{rad/s}$) and, in order to achieve counterclockwise rotation of the rotor, $v_{bs} = -v_{as}$; thus, $v_{ds}^s = v_{qs}^s$.

The example machine is a capacitor-start, capacitor-run induction machine and, although it can be operated as a split-phase machine (without the capacitors), it is clear from a comparison of Figs. 10.6-2 and 10.6-3 that it is not designed for this purpose. Nevertheless, the machine, when operated as a split-phase machine, will develop an average starting torque larger than rated torque. However, at approximately 40% of synchronous speed the average torque decreases to less than the average torque that the machine could produce with only the *as* winding energized (Fig. 10.6-2). In other words, $\tilde{I}_{ds}^s$ lags $\tilde{I}_{qs}^s$ after the machine has reached approximately 40% of synchronous speed. A machine designed for operation as a split-phase machine would generally produce a steady-state staring torque of 2 to 3 times rated, and it would reach 60% to 80% of synchronous speed before its average torque would become less than that which could be produced with only one stator winding.

Capacitor-Start Induction Motor

Placing a capacitor in series with the start winding (*bs* or *ds* winding) has the effect of increasing the magnitude of the leading "in-quadrature" component of $\tilde{I}_{ds}^s$, which in turn increases the average electromagnetic torque during the starting period. It is clear that this same technique may be used to develop a starting torque with a symmetrical 2-phase machine supplied from a single-phase source. With the reference frame stationary, the behavior of the unsymmetrical or symmetrical 2-phase induction machine with a capacitor connected in series with the start winding may be described by simply replacing r_S with $r_S + r_C$, where r_C is the resistance of the capacitor, and $(p/\omega_b)X_{SS}$ with $(p/\omega_b)X_{SS} + (\omega_b/p)X_C$ in the voltage equations. For the steady-state voltage equations, $j(\omega_{esk}/\omega_b)X_{SS}$ should be replaced by $j[(\omega_{esk}/\omega_b)X_{SS} - (\omega_b/\omega_{esk})X_C]$.

The steady-state torque–speed characteristics for the example machine with the start capacitor connected in series with the *bs* or *ds* winding are shown in Fig. 10.6-4. In this case the value of r_C and X_C are 3 and 14.5 ohms, respectively. These values correspond to the values used during the starting period of the example machine. The average torque $T_{e(\text{avg})}$ and $|T_{e(\text{pul})}|$ are plotted in Fig. 10.6-4. The applied voltages are the same as in the case of the split-phase induction motor. The free-acceleration characteristics from stall are shown, by computer traces, in Figs. 10.6-5 and 10.6-6 for a capacitor-start, single-phase induction motor [7, 8]. The variables v_{as}, i_{as}, v_{bs}, i_{bs}, v_c, T_e, and speed are plotted. The voltage v_c is the total instantaneous voltage across the series-connected resistor r_C and capacitor X_C. The complete free acceleration period is shown in Fig. 10.6-5. The machine variables are shown with an expanded scale in Fig. 10.6-6 to illustrate the switching of the start winding which is disconnected from the source at a normal current zero once the rotor reached 75% of synchronous speed. It is clear that the steady-state, torque–speed characteristics

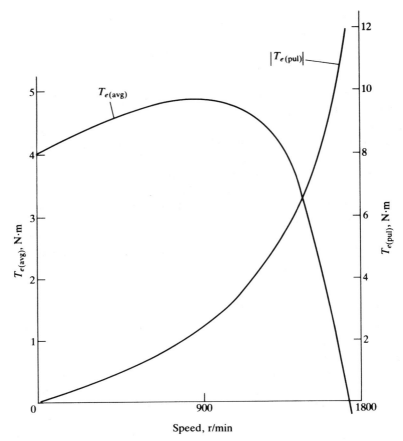

Figure 10.6-4   Steady-state torque–speed characteristics with the series capacitor equal to the start capacitor.

for the capacitor-start machine may be obtained by an appropriate combination of the steady-state torque characteristics given in Fig. 10.6-4 for the starting period and Fig. 10.6-2 after the start winding is disconnected from the source. In Fig. 10.6-6 the voltage across the capacitor is shown to remain constant at its value when the start winding is disconnected from the source. In practice, the voltage would slowly decay due to leakage currents within the capacitor which are not considered in this analysis. Also, it must again be recalled that the machine is designed as a capacitor-start, capacitor-run machine. The average torque developed when operated with the start capacitor in series with the start winding (Fig. 10.6-4) is larger than the average torque produced with only the run winding energized (Fig. 10.6-2) until approximately 90% of synchronous speed. Therefore, disconnecting the start winding at 75% of synchronous speed results in a drop in the average torque and thus a decrease in the acceleration which can be seen in Fig. 10.6-5 or 10.6-6.

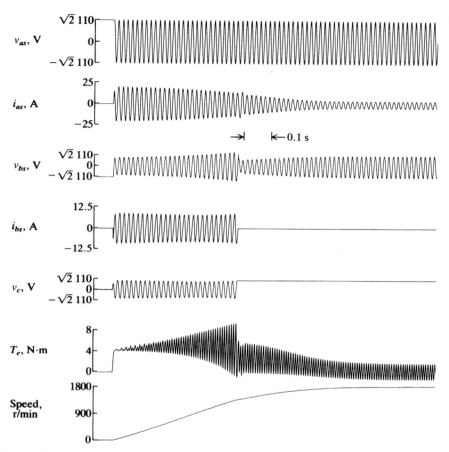

Figure 10.6-5   Machine variables during free acceleration of a capacitor-start, single-phase induction motor.

Capacitor-Start, Capacitor-Run Induction Machine

In the case of capacitor-start, capacitor-run, single-phase induction motor the auxiliary or start winding is not disconnected. Instead, the value of the series capacitor is altered. In other words, there are two values of series capacitances: one during the start period and one during the period of normal operation. These two values of capacitances are obtained by using two capacitors connected in parallel with provisions to open-circuit one of the parallel paths once the rotor speed has reached 60% to 80% of synchronous speed. Clearly, the purpose of the run capacitor is to establish a leading "in-quadrature" component of $\tilde{I}_{ds}^s$ during normal loads, thereby achieving improved operation over that which is possible with only one stator winding. Although we will consider only an unsymmetrical induction machine, it is clear that the capacitor-start, capacitor-run scheme can be used equally well with a symmetrical 2-phase induction machine.

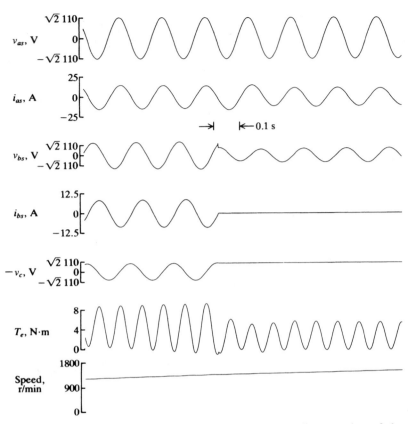

Figure 10.6-6 Expanded plot of Fig. 10.6-5 illustrating the disconnecting of the start winding.

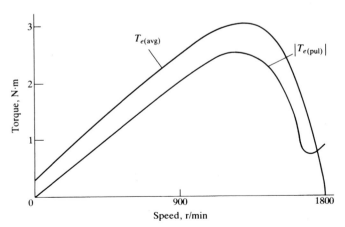

Figure 10.6-7 Steady-state torque–speed characteristics with the series capacitor equal to the run capacitor.

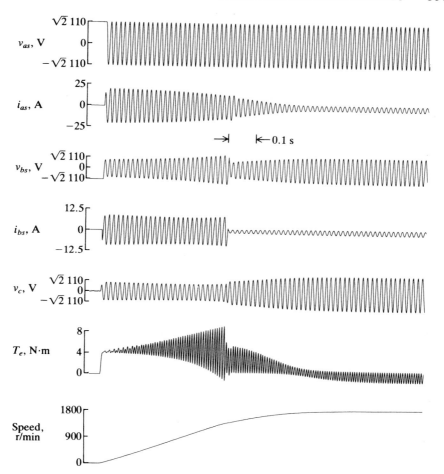

Figure 10.6-8 Machine variables during free acceleration of capacitor-start, capacitor-run, single-phase induction motor.

The steady-state torque–speed characteristics of the example machine with the capacitor connected in series with the start winding equal to the run capacitor are shown in Fig. 10.6-7. In this case the values of r_C and X_C associated with the run capacitor equal 9 and 172 ohms, respectively. Here again, $T_{e(\text{avg})}$ and $|T_{e(\text{pul})}|$ are plotted with the applied voltages the same as in the case of the split-phase motor. Figures 10.6-8 and 10.6-9 portray the free acceleration characteristics of the capacitor-start, capacitor-run, single-phase induction machine motor. The value of the capacitances was changed by switching out one of the parallel capacitors at the first normal current zero after the rotor reached 75% of synchronous speed. The steady-state torque–speed characteristics corresponding to this mode of operation may be obtained from Fig. 10.6-4 for the starting period and from Fig. 10.6-7 for speeds

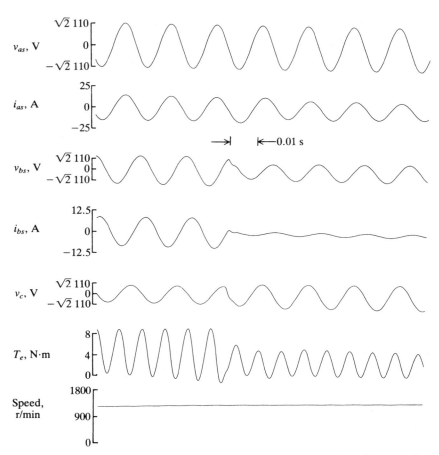

Figure 10.6-9   Expanded plot of Fig. 10.6-8 illustrating the switching of the capacitors.

above 75% of synchronous speed. Actually we see that perhaps a speed closer to 80% or 85% of synchronous could have been selected as the speed at which the capacitors were switched because the average torque with the start capacitor was considerably larger at 75% synchronous speed than the average torque with the run capacitor.

The variables v_{as}, i_{as}, v_{bs}, i_{bs}, v_c, T_e, and ω_r/ω_b are plotted in Figs. 10.6-8 and 10.6-9. It is important to note that the value of run capacitance has been selected so as to reduce the pulsating torque in the vicinity of rated torque (approximately 1.0 N·m). It is seen by a comparison of Figs. 10.6-2 and 10.6-7 that the run capacitor in series with the auxiliary winding increases the average torque somewhat in the vicinity of normal operation; however, the additional capacitor and winding combination markedly reduces the pulsating torque, making it a smooth and quiet operating machine. Selecting the value of capacitances to maintain permissible capacitor voltages while improving starting torque and minimizing the pulsating torque during normal operation are objectives involved in the design of this type of induction motor. It is clear that the reduction in pulsating torque occurs as a result of minimizing the unbalancing of the stator currents. This of course has the effect of reducing the losses in the rotor circuits due to the negatively rotating magnetic field established by unbalanced stator currents.

REFERENCES

[1] A. E. Fitzgerald, C. Kingsley, Jr., S. D. Umans, *Electric Machinery*, 5th ed., McGraw-Hill, New York, 1990.

[2] S. J. Chapman, *Electric Machinery Fundamentals*, 3rd ed., McGraw-Hill, New York, 1999.

[3] G. McPherson and R. D. Laramore, *An Introduction to Electrical Machines and Transformers*, 2nd ed., John Wiley and Sons, New York, 1990.

[4] P. C. Krause, O. Wasynczuk, and S. D. Sudhoff, *Analysis of Electric Machinery*, IEEE Press, Piscataway, NJ, 1995.

[5] S. D. Sudhoff, Multiple Reference Frame Analysis of an Unsymmetrical Induction Machine, *IEEE Transactions on Energy Conversion*, Vol. 8, No. 3, September 1993, pp. 425–432.

[6] P. C. Krause and O. Wasynczuk, *Electromechanical Motion Devices*, Purdue Press, Purdue University, *http://www.adpc.purdue.edu/PrintServ/krauseordfrm.htm*; or call 765-494-0314.

[7] P. C. Krause, Simulation of Unsymmetrical 2-phase Induction Machines, *IEEE Transactions on Power Apparatus and Systems*, Vol. 84, November 1965, pp. 1025–1037.

[8] C. M. Ong, *Dynamic Simulation of Electric Machinery*, Prentice-Hall PTR, Upper Saddle River, NJ, 1998.

PROBLEMS

1 Calculate $\tilde{F}^s_{qs+}$ and $\tilde{F}^s_{qs-}$ for the following sets.
 (a) $\tilde{F}_{as} = 10\underline{/30^\circ}$, $\tilde{F}_{bs} = 30\underline{/-60^\circ}$.
 (b) $\tilde{F}_{as} = 10\underline{/0^\circ}$, $\tilde{F}_{bs} = 0$.
 (c) $F_{as} = \cos(\omega_e t + 45^\circ)$, $F_{bs} = \cos(\omega_e t - 45^\circ)$.

2 Calculate the coefficients, $f_{as\alpha}, f_{as\beta}, f_{bs\alpha}$, and $f_{bs\beta}$ given in (10.2-15) and (10.2-16) for each set given in Problem 1.

3 Derive (10.2-35).

4 Unbalanced voltages of fundamental frequency are applied to a 2-phase symmetrical induction machine. (a) Express I_{as} and I_{bs} in the form given by (10.2-15) and (10.2-16). (b) Express T_e in a form reduced from (10.2-42). (c) Repeat (b) from (10.2-43).

5 An impedance Z_a is connected in series with the *as* winding of a symmetrical 2-phase machine, and an impedance Z_b is connected in series with the *bs* winding. Derive the phasor voltage equations.

6 Express (10.2-61) with $\tilde{I}^s_{qr+}$ and $\tilde{I}^s_{qr-}$ eliminated.

7 Derive the voltage equations for an open-circuited b phase of a symmetrical 2-phase induction machine.

8 When evaluating (10.4-4) why is it necessary to set $\theta(0) = 0$? Why isn't it necessary to set $\theta_r(0) = 0$?

9 Write the voltage equations of a 2-phase unsymmetrical induction machine in terms of flux linkages per second.

10 Show that in the case of the unsymmetrical induction machine the electromagnetic torque may be expressed as follows only if $L_{lS} = n^2 L_{ls}$:

$$T_e = \left(\frac{P}{2}\right)\left(\frac{1}{n}\lambda^s_{ds}i^s_{qs} - n\lambda^s_{qs}i^s_{ds}\right)$$

11 Show that the steady-state pulsating torque of an unsymmetrical induction machine is zero if the rotor speed is zero.

12 Derive (10.5-31).

13 Repeat Problem 5 using a modification of (10.5-22).

14 Assume there is no external impedance in series with the as winding of the example machine given in Section 10.6. For 60-Hz operation, calculate the values of r_B and X_B for "balanced" conditions to occur with $\tilde{V}_{ds} = \tilde{V}_{qs}$ at (a) $s = 1$, (b) $s = 0.05$, and (c) $s = 0$.

15 Equations (10.2-50)–(10.2-58) set forth a method of analyzing an open-circuit stator phase of a symmetrical 2-phase induction machine. Repeat this analysis for an unsymmetrical 2-phase induction machine with the auxiliary winding (bs winding) open-circuited.

16 A 4-pole, $\frac{1}{4}$-hp, 110-V, 60-Hz, split-phase induction motor has the following parameters expressed in ohms at 60 Hz:

$$r_s = 2.50\,\Omega \qquad X_{ms} = 60.1\,\Omega \qquad r'_r = 2.4\,\Omega$$
$$X_{ls} = 2.90\,\Omega \qquad\qquad\qquad X'_{lr} = 1.7\,\Omega$$
$$r_S = 10.7\,\Omega \qquad n = 0.9 \qquad X_{lS} = 2.5\,\Omega$$

Calculate the steady-state starting torque and compare this value to rated horsepower.

Chapter 11

SEMICONTROLLED BRIDGE CONVERTERS

11.1 INTRODUCTION

A brief analysis of single- and 3-phase semicontrolled bridge converters is presented in this chapter. In many respects, this analysis for constant output current is the same for each converter. Although this results in some repetition, the analysis is carried out for each converter because it allows their independent study. The objective is to provide a basic background in converter operation without becoming overly involved. For this reason, only the constant-current mode is analyzed while other modes of operation are demonstrated by computer traces without lengthy discussion. A more detailed analysis of these and other converters can be found in references 1–4 at the end of this chapter. Finally, to set the stage for the analysis of dc and ac drive systems in later chapters, an average value model of the three-phase semicontrolled bridge converter is derived. This model can be used to predict the average value performance during steady-state and dynamical operating conditions.

11.2 SINGLE-PHASE LOAD COMMUTATED CONVERTER

A single-phase line-commutated full converter is shown in Fig. 11.2-1. The ac source voltage and current are denoted e_{ga} and i_{ga}, respectively. The series inductance (commutating inductance) is denoted l_c. The SCRs or thyristors are numbered $T1$ through $T4$, and the associated gating or firing signals are denoted e_{f1} through e_{f4}. The converter output voltage and current are denoted v_d and i_d, respectively. The following simplifying assumptions are made in this analysis: (1) The ac source contains only one frequency, (2) the output current i_d is constant, (3) the thyristor is

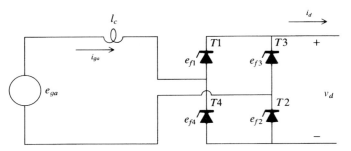

Figure 11.2-1   Single-phase full converter.

an infinite impedance device when in the reverse bias mode (cathode positive) or when the gating signal blocks current flow, and (4) in the conducting mode the voltage drop across the thyristor is negligibly small.

Operation without Commutating Inductance and Phase Delay

It is convenient to analyze converter operation in steps starting with the simplest case where the commutating inductance is not present and there is no phase delay. In this case, it can be assumed that the gating signals are always present, whereupon the thyristors will conduct whenever they become forward biased (anode positive) just as if they were diodes. Converter operation for constant i_d with $l_c = 0$ and without phase delay is depicted in Fig. 11.2-2. The thyristors in the upper part of the converter ($T1$ and $T3$) conduct whenever the anode of one of this group becomes more positive than the other. Similarly, the thyristor that conducts in the lower part of the converter ($T2$ or $T4$) is the one whose cathode is the most negative. In this case, the converter operates as a full-wave rectifier. If we assume that

$$e_{ga} = \sqrt{2}E \cos \omega_e t \tag{11.2-1}$$

then, for the interval $-\pi/2 \le \omega_e t \le \pi/2$, we have

$$v_d = e_{ga} \tag{11.2-2}$$

The output voltage is made up of two identical π intervals per cycle of the source voltage. Hence, the average output voltage may be determined by finding the average of (11.2-1) over the interval of $-\pi/2 \le \omega_e t \le \pi/2$. Thus

$$V_{d0} = \frac{1}{\pi} \int_{-\pi/2}^{\pi/2} \sqrt{2}E \cos \omega_e t \, d(\omega_e t) = \frac{2}{\pi}(\sqrt{2}E) \tag{11.2-3}$$

where E is the root mean square (rms) value of the source voltage. We will use V_{d0} to denote the average output voltage without commutation and without phase delay.

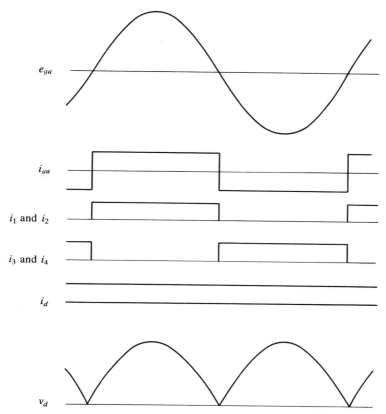

Figure 11.2-2 Single-phase, full-converter operation for constant output current without l_c and phase delay.

Operation with Commutating Inductance and without Phase Delay

When l_c is zero, the process of "current switching" from one thyristor to the other in either the upper or lower part of the converter ($T1$ to $T3$ to $T1$ to ..., and $T2$ to $T4$ to $T2$ to ...) takes place instantaneously. In practice, instantaneous commutation cannot occur because there is always some inductance between the source and the converter. Converter operation with commutating inductance and without phase delay is shown in Fig. 11.2-3. During commutation, the source is short-circuited simultaneously through $T1$ and $T3$ and through $T2$ and $T4$. Hence, if we consider the commutation from $T1$ to $T3$ and $T2$ to $T4$ and if we assume that the short-circuit current during commutation is positive through $T3$, then

$$e_{ga} = -l_c \frac{di_{sc}}{dt} \qquad (11.2\text{-}4)$$

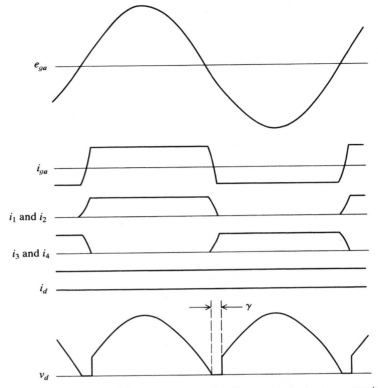

Figure 11.2-3   Single-phase, full-converter operation for constant output current with l_c and without phase delay.

where i_{sc} is the short-circuit current. Substituting (11.2-1) into (11.2-4) and solving for i_{sc} yields

$$i_{sc} = -\frac{1}{l_c} \int \sqrt{2}E \cos \omega_e t \, dt = -\frac{\sqrt{2}E}{\omega_e l_c} \sin \omega_e t + C \qquad (11.2\text{-}5)$$

At $\omega_e t = \pi/2$, $i_{sc} = 0$, therefore

$$C = \frac{\sqrt{2}E}{\omega_e l_c} \qquad (11.2\text{-}6)$$

whereupon

$$i_{sc} = \frac{\sqrt{2}E}{\omega_e l_c}(1 - \sin \omega_e t) \qquad (11.2\text{-}7)$$

At the end of commutation $\omega_e t = \pi/2 + \gamma$ and $i_{sc} = I_d$, therefore

$$I_d = \frac{\sqrt{2}E}{\omega_e l_c}(1 - \cos\gamma) \tag{11.2-8}$$

where γ is the overlap angle (Fig. 11.2-3). An uppercase letter (I_d) is used to denote constant or average-value quantities. During commutation, the converter output voltage v_d is zero. Once commutation is completed, the short-circuit paths are broken and the output voltage jumps to the value of the source voltage because a constant i_d is assumed and hence i_{ga} is constant after commutation and there is zero voltage dropped across the inductance l_c. V_{d0} given by (11.2-3) is the average converter output voltage when l_c is zero. When l_c is considered, the output voltage is zero during commutation. Hence, the average output voltage is less than without commutation. The average converter output voltage may be determined by

$$\begin{aligned} V_d &= \frac{1}{\pi} \int_{-\pi/2+\gamma}^{\pi/2} \sqrt{2}E \cos\omega_e t \, d(\omega_e t) \\ &= \frac{V_{d0}}{2}(1 + \cos\gamma) \end{aligned} \tag{11.2-9}$$

If (11.2-8) is solved for $\cos\gamma$ and the result is substituted into (11.2-9), the average converter output voltage with commutating inductance but without phase delay becomes

$$V_d = V_{d0} - \frac{\omega_e l_c}{\pi} I_d \tag{11.2-10}$$

In the previous equation, commutation appears as a voltage drop as if the converter had an internal resistance of $\omega_e l_c/\pi$ ohms. However, this is not a resistance in the sense that it dissipates energy. If the thyristors are assumed ideal, the converter does not dissipate any energy.

Operation without Commutating Inductance and with Phase Delay

Thus far, we have considered the thyristor as a diode and hence only rectifier operation of the converter. However, we realize that the thyristor will conduct only if the anode voltage is positive and if it has received a gating signal. Hence, the conduction of a thyristor may be delayed after the anode has become positive by delaying the gating signal (firing pulse). Converter operation with phase delay but without commutating inductance is shown in Fig. 11.2-4.

We can determine the average output by

$$V_d = \frac{1}{\pi} \int_{-\pi/2+\alpha}^{\pi/2+\alpha} \sqrt{2}E \cos\omega_e t \, d(\omega_e t) = \frac{2}{\pi}(\sqrt{2}E)\cos\alpha = V_{d0}\cos\alpha \tag{11.2-11}$$

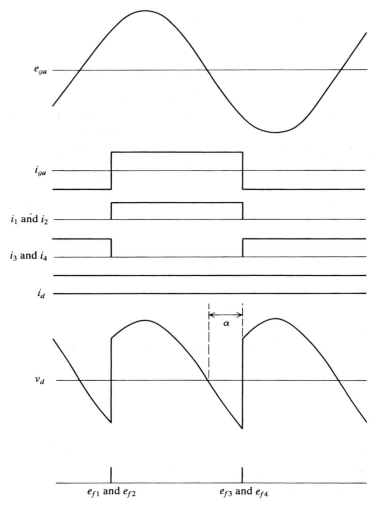

Figure 11.2-4   Single-phase, full-converter operation for constant output current without l_c and with phase delay.

where α is the delay angle (Fig. 11.2-4). If the current is maintained constant, the average output voltage will become negative for α greater than $\pi/2$. This is referred to as inverter operation where power is being transferred from the dc system to the ac system.

Operation with Commutating Inductance and Phase Delay

Converter operation with both commutating inductance and phase delay is shown in Fig. 11.2-5. The calculation of I_d and V_d are identical to that given by (11.2-4)–(11.2-10), only the intervals of evaluation are different. In particular, (11.2-5)

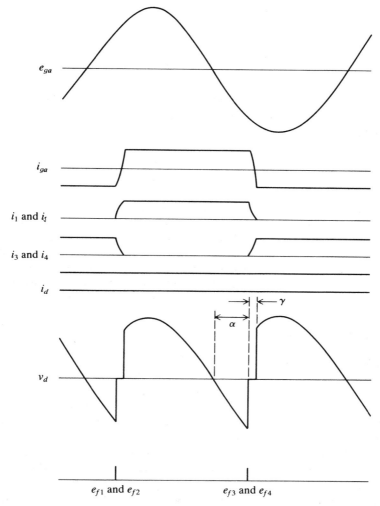

Figure 11.2-5   Single-phase, full-converter operation for constant output current with l_c and phase delay.

applies, but at $\omega_e t = \pi/2 + \alpha$ we have $i_{sc} = 0$; thus

$$C = \frac{\sqrt{2}E}{\omega_e l_c} \cos\alpha \qquad (11.2\text{-}12)$$

Commutation ends at $\omega_e t = \pi/2 + \alpha + \gamma$, whereupon $i_{sc} = I_d$; thus

$$I_d = \frac{\sqrt{2}E}{\omega_e l_c} [\cos\alpha - \cos(\alpha + \gamma)] \qquad (11.2\text{-}13)$$

From (11.2-9) we obtain

$$V_d = \frac{1}{\pi}\int_{-\pi/2+\alpha+\gamma}^{\pi/2+\alpha} \sqrt{2}E\cos\omega_e t \, d(\omega_e t) = \frac{V_{d0}}{2}[\cos\alpha + \cos(\alpha+\gamma)] \quad (11.2\text{-}14)$$

Solving (11.2-13) for $\cos(\alpha+\gamma)$ and substituting the results into (11.2-14) yields the following expression for the average output voltage with commutating inductance and phase delay:

$$V_d = V_{d0}\cos\alpha - \frac{\omega_e l_c}{\pi}I_d \quad (11.2\text{-}15)$$

The equivalent circuit suggested by (11.2-15) is shown in Fig. 11.2-6.

The average-value relations and corresponding equivalent circuit depicted in Fig. 11.2-6 were developed based upon the assumptions that (1) the rms amplitude of the alternating-current (ac) source voltage E is constant, and (2) the direct-current (dc) load current i_d is constant. This equivalent circuit provides a reasonable approximation of the average dc voltage even if E and i_d vary with respect to time, provided that the variations are small from one conduction interval to the next.

Modes of Operation

In the preceding analysis, it was assumed that the dc (load) current is constant over the averaging period. If the converter is connected to a dc load in which the current is not constant, the analytical determination of the response and corresponding average-value relationships is, in general, much more difficult and, in many cases, impractical. In these cases, the response can be established using computer simulation. In this regard, there are numerous tools and techniques that can be used to simulate the steady-state or dynamic characteristics of power electronic systems using ideal or nonideal models of the switching elements. Several are listed in references 6–12 provided at the end of this chapter. Although each of these techniques has specific advantages and disadvantages, if properly used, they will all yield identical results for a given system.

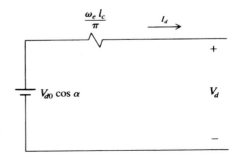

Figure 11.2-6 Average value equivalent circuit for a single-phase full converter.

Various modes of operation of a single-phase, full converter are illustrated by computer traces in Figs. 11.2-7 through 11.2-9. The source voltage is 280 V (rms) and the commutating inductance in 1.4 mH. In each case e_{ga}, i_{ga}, i_1, i_3, v_d, and i_d are plotted where i_1 and i_3 are the currents through thyristors $T1$ and $T3$, respectively. In Fig. 11.2-7, the converter is operating with a series RL load connected across the output terminals where $R = 3\ \Omega$ and $L = 40$ mH. In Fig. 11.2-7a, the converter is operating without phase delay. In Fig. 11.2-7b, the phase delay angle is 45°. In Fig. 11.2-7c, the phase delay is slightly less than 90°; the current i_d is discontinuous. The output current is nearly constant when the converter is operating without phase delay due to the large-load inductance. When i_1 becomes zero, the output voltage "jumps" to zero. The magnitude of the jump in voltages is $L(di_d/dt)$. Thereafter, both v_d and i_d remain zero until the firing signals cause $T2$ and $T4$ to start conducting.

In the case shown in Fig. 11.2-8, the combination of a series RL ($R = 3\ \Omega$, $L = 40$ mH) connected in series with a constant 200-V source is connected across the output terminals of the converter. The dc source is connected so that it opposes a positive v_d. In Fig. 11.2-8a the converter is operating without phase delay, whereas in Fig. 11.2-8b the phase delay angle is 60°. Therein, when i_1 becomes zero the

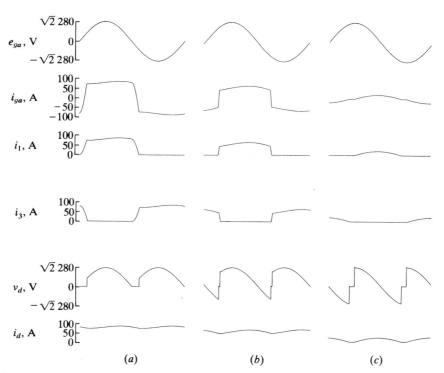

Figure 11.2-7   Single-phase, full-converter operation with RL load. (a) $\alpha = 0°$; (b) $\alpha = 45°$; (c) $\alpha \approx 90°$, discontinuous operation.

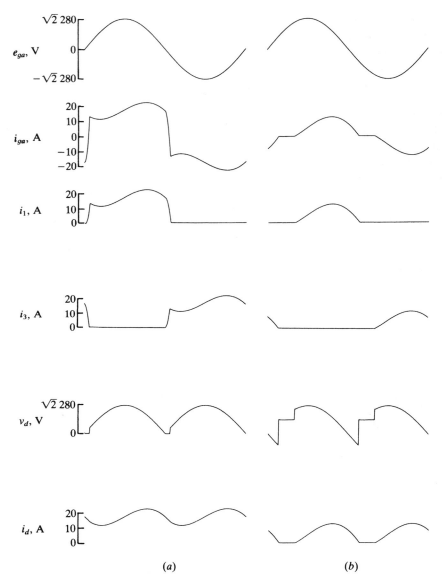

Figure 11.2-8 Single-phase, full-converter operation with RL and an opposing dc source connected in series across the converter terminals. (*a*) $\alpha = 0°$; (*b*) $\alpha = 60°$.

jump in v_d is again $L(di_d/dt)$ as in Fig. 11.2-7*c*. However, during the zero current mode of operation, v_d is equal to 200 V, the magnitude of the series connected dc source.

Inverter operation is depicted in Fig. 11.2-9. In this case the combination of the *RL* load and dc source is still connected across the output terminals of the converter

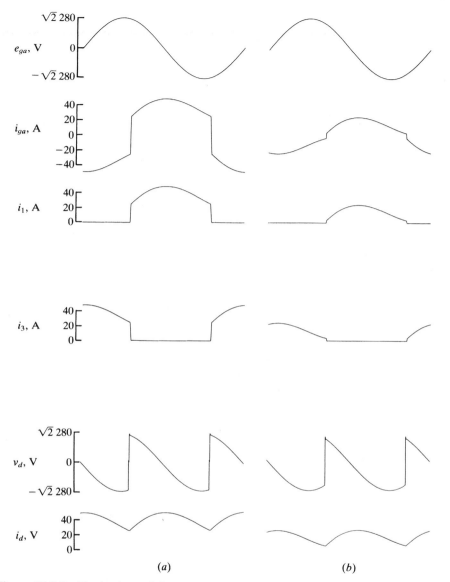

Figure 11.2-9 Single-phase, full-converter operation with RL and an aiding dc source connected in series across the converter terminals. (*a*) $\alpha = 180°$; (*b*) $\alpha = 116°$.

but the polarity of the dc source is reversed. In Fig. 11.2-9*a* the phase delay angle is 108°. In Fig. 11.2-9*b* the phase delay angle is 126°.

Although (11.2-15) was derived for a constant output current, it is quite accurate for determining the average values of converter voltage and current, especially if the current is not discontinuous. The reader should take the time to compare the calculated converter output voltage and current using (11.2-15) with the average values shown in Figs. 11.2-7 through 11.2-9 and to justify qualitatively any differences that may occur.

11.3   3-PHASE LOAD COMMUTATED CONVERTER

A 3-phase, line-commutated, full converter is shown in Fig. 11.3-1. The voltages of the 3-phase ac source are denoted e_{ga}, e_{gb}, and e_{gc}; the phase currents are denoted i_{ga}, i_{gb}, and i_{gc}. The series inductance (commutating inductance) is denoted l_c. The thyristors are numbered $T1$ through $T6$, and the gating or firing signals for the thyristors are e_{f1} through e_{f6}. The converter output voltage and current are denoted v_d and i_d, respectively. The following simplifying assumptions are made in this analysis: (1) The 3-phase source is balanced, (2) the current i_d is constant, (3) the thyristor is an infinite impedance device when in the reverse bias mode (cathode positive) or when the gating signal blocks current flow, and (4) in the conducting mode the voltage drop across the thyristor is negligibly small.

Operation without Commutating Inductance and Phase Delay

As before, it is convenient to analyze converter operation in steps starting with the simplest case where the commutating inductance is not present and there is no phase delay [1]. In this case, it can be assumed that the gating signals are always present, whereupon the thyristors will conduct whenever they are forward biased (anode positive) just as if they were diodes. Converter operation for constant i_d and with $l_c = 0$ and without phase delay is depicted in Fig. 11.3-2. The thyristors in the upper part of the converter ($T1$, $T3$, and $T5$) conduct whenever the anode of one of this group of three becomes more positive than the other two. Similarly, the thyristor that conducts in the lower part of the converter ($T2$, $T4$, or $T6$) is the one whose cathode is the most negative.

The average of the output voltage v_d can be found by noting that the voltage appearing at the output is always a line-to-line voltage. For example, during the

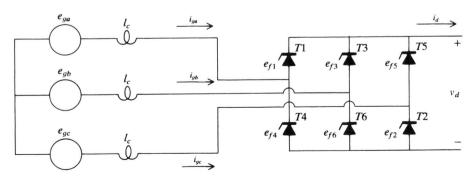

Figure 11.3-1   A 3-phase full converter.

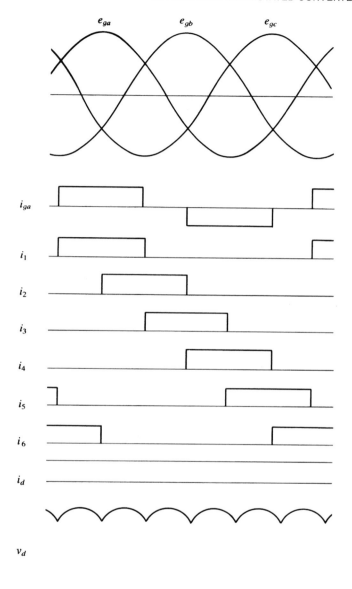

Figure 11.3-2   A 3-phase, full-converter operation for constant output current without l_c and phase delay.

interval when $T1$ and $T2$ are conducting, the output voltage is the voltage between phases a and c. For an abc sequence with

$$e_{ga} = \sqrt{2}E\cos\omega_e t \tag{11.3-1}$$

then, for the interval of $0 \le \omega_e t \le \pi/3$ we have

$$v_d = v_{ac} = e_{ga} - e_{gc} = \sqrt{6}E\cos\left(\omega_e t - \frac{\pi}{6}\right) \tag{11.3-2}$$

The output voltage is made up of six identical $\pi/3$ intervals per cycle of the source voltage. Hence, the average output voltage may be determined by finding the average of (11.3-2) over the interval of $\pi/3$ radians. Thus

$$V_{d0} = \frac{3}{\pi}\int_0^{\pi/3} \sqrt{6}E\cos\left(\omega_e t - \frac{\pi}{6}\right) d(\omega_e t) = \frac{3\sqrt{3}}{\pi}(\sqrt{2}E) \tag{11.3-3}$$

where E is the rms value of the source voltage. We will use V_{d0} to denote the average output voltage without commutation and without phase delay.

Operation with Commutating Inductance and without Phase Delay

When l_c is zero, the process of "current switching" from one thyristor to the other in either the upper or lower part of the converter ($T1$ to $T3$ to $T5$ to $T1$ to $\ldots$, and $T2$ to $T4$ to $T6$ to $T2$ to $\ldots$) takes place instantaneously. It is clear that instantaneous commutation cannot occur in practice because there is always some inductance between the source and converter. The operation of the converter with commutating inductance and without phase delay is shown in Fig. 11.3-3. Therein, let us consider the commutation of the current from $T1$ to $T3$. With zero phase delay, commutation begins at point a where the anode of $T3$ becomes more positive than the anode of $T1$. During commutation, shown by path a to b, the voltages e_{ga} and e_{gb} are short-circuited through the two inductances. The voltage equation around this path is

$$e_{gb} - e_{ga} = 2l_c \frac{di_{sc}}{dt} \tag{11.3-4}$$

where i_{sc} is the short-circuit current considered positive through $T3$. Assuming e_{ga} is given by (11.3-1), then

$$e_{gb} - e_{ga} = \sqrt{6}E\cos\left(\omega_e t - \frac{5}{6}\pi\right) \tag{11.3-5}$$

Solving (11.3-4) for i_{sc} yields

$$i_{sc} = \frac{1}{2l_c}\int \sqrt{6}E\cos\left(\omega_e t - \frac{5}{6}\pi\right) dt = \frac{\sqrt{6}E}{2\omega_e l_c}\sin\left(\omega_e t - \frac{5}{6}\pi\right) + C \tag{11.3-6}$$

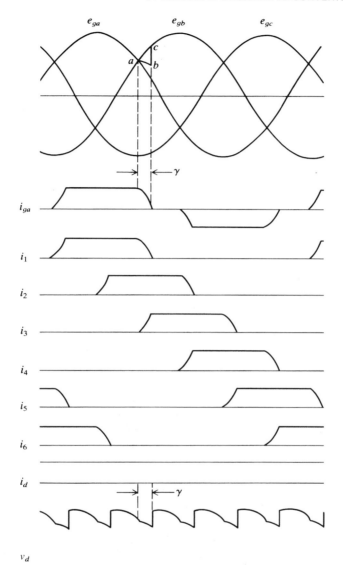

Figure 11.3-3 A 3-phase, full-converter operation for constant output current with l_c and without phase delay.

At $\omega_e t = \pi/3$, $i_{sc} = 0$. Thus

$$C = \frac{\sqrt{6}E}{2\omega_e l_c} \tag{11.3-7}$$

whereupon

$$i_{sc} = \frac{\sqrt{6}E}{2\omega_e l_c} \left[1 + \sin\left(\omega_e t - \frac{5\pi}{6} \right) \right] \tag{11.3-8}$$

At the end of commutation, $\omega_e t = \pi/3 + \gamma$ and $i_{sc} = I_d$; therefore

$$I_d = \frac{\sqrt{6}E}{2\omega_e l_c} (1 - \cos\gamma) \tag{11.3-9}$$

where γ is the overlap angle (Fig. 11.3-3). An uppercase letter (I_d) is used to denote constant or average-value quantities. In normal operation, γ is less than $\pi/3$. If the overlap angle exceeds $\pi/3$, then two commutations will take place simultaneously. This mode of operation is not considered here. An analysis of the operation in this and other modes may be found in reference 2.

Once i_1 becomes zero, commutation is completed; $T1$ stops conducting. The short-circuit path is broken and the voltage jumps from b to c (Fig. 11.3-3). Commutation occurs six times per cycle of the source voltages. Recall that the average output voltage V_{d0} without commutation was obtained by finding the average of the output voltage over one of the six identical $\pi/3$ intervals (11.3-3). With commutation, the area abc (Fig. 11.3-3) must be subtracted from this averaging process. Thus, the average output voltage may be obtained by subtracting from V_{d0} the area abc averaged over a $\pi/3$ interval. The area abc is $\frac{1}{2}(e_{gb} - e_{ga})$ over the interval from $\pi/3$ to $\pi/3 + \gamma$ which may be averaged over a $\pi/3$ interval by dividing by $\pi/3$. Therefore, with $(e_{gb} - e_{ga})$ expressed by (11.3-5), the average output voltage with commutation but without phase delay becomes

$$\begin{aligned} V_d &= \frac{3\sqrt{6}E}{\pi} - \frac{3}{2\pi} \int_{\pi/3}^{\pi/3+\gamma} \sqrt{6}E\cos\left(\omega_e t - \frac{5}{6}\pi \right) d(\omega_e t) \\ &= \frac{1}{2} V_{d0}(1 + \cos\gamma) \end{aligned} \tag{11.3-10}$$

If (11.3-9) is solved for $\cos\gamma$ and the result is substituted into (11.3-10), we obtain an expression for the average output voltage with commutating inductance but without phase delay.

$$V_d = V_{d0} - \frac{3\omega_e l_c}{\pi} I_d \tag{11.3-11}$$

As before, commutation appears as a voltage drop as if the converter had an internal resistance of $3\omega_e l_c / \pi$ ohms. Again, this is not a resistance in the sense that it dissipates energy.

Operation without Commutating Inductance and with Phase Delay

Thus far we have considered the thyristor as a diode and hence only rectifier operation of the converter. However, we realize that the thyristor will conduct only if the anode voltage is positive and if it has received a gating signal. Hence, the conduction of a thyristor may be delayed after the anode has become positive by delaying the gating signal (firing pulse). Converter operation with phase delay but without commutating inductance is shown in Fig. 11.3-4.

The average output voltage may be calculated as in (11.3-3) except the limits of integration are different, in particular,

$$V_d = \frac{3}{\pi} \int_{\alpha}^{\pi/3+\alpha} \sqrt{6}E \cos\left(\omega_e t - \frac{\pi}{6}\right) d(\omega_e t) = \frac{3\sqrt{6}E}{\pi} \cos\alpha = V_{d0} \cos\alpha \quad (11.3\text{-}12)$$

where α is the delay angle (Fig. 11.3-4). If the current is maintained constant, the average output voltage will become negative for α greater than $\pi/2$. This is referred to as inverter operation where power is being transferred from the dc system to the ac system.

Operation with Commutating Inductance and Phase Delay

Converter operation with both commutating inductance and phase delay is shown in Fig. 11.3-5. The calculation of I_d and V_d are identical to that given by (11.3-4)–(11.3-11); only the intervals of evaluation are different. In particular, (11.3-6) applies, but at $\omega_e t = \pi/3 + \alpha$ we have $i_{sc} = 0$; thus

$$C = \frac{\sqrt{6}E}{2\omega_e l_c} \cos\alpha \quad (11.3\text{-}13)$$

Commutation ends at $\omega_e t = \pi/3 + \alpha + \gamma$, whereupon $i_{sc} = I_d$; thus

$$I_d = \frac{\sqrt{3}E}{2\omega_e l_c} [\cos\alpha - \cos(\alpha + \gamma)] \quad (11.3\text{-}14)$$

From (11.3-10)

$$V_d = V_{d0} \cos\alpha - \frac{3}{2\pi} \int_{\pi/3+\alpha}^{\pi/3+\alpha+\gamma} \sqrt{6}E \cos\left(\omega_e t - \frac{5}{6}\pi\right) d(\omega_e t)$$

$$= \frac{1}{2} V_{d0} [\cos\alpha + \cos(\alpha + \gamma)] \quad (11.3\text{-}15)$$

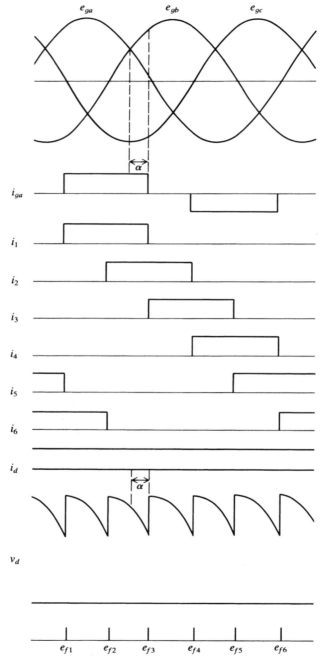

Figure 11.3-4 A 3-phase, full-converter operation for constant output current without l_c and with phase delay.

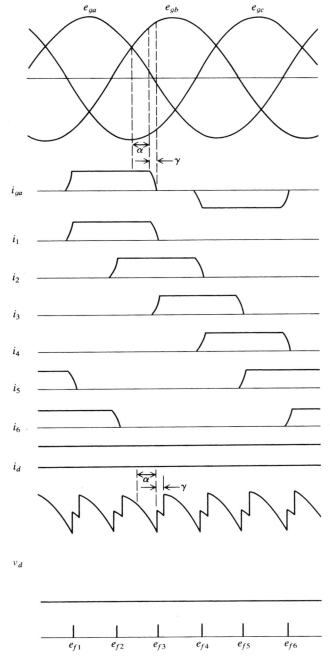

Figure 11.3-5 A 3-phase, full-converter operation for constant output current with I_c and phase delay.

Solving (11.3-14) for $\cos(\alpha + \gamma)$ and substituting the results into (11.3-15) yields the following expression for the average output voltage with commutating inductance and phase delay

$$V_d = V_{d0}\cos\alpha - \frac{3\omega_e l_c}{\pi}I_d \tag{11.3-16}$$

The equivalent circuit, suggested by (11.3-16), is of the same form as given in Fig. 11.2-6.

Modes of Operation

If the converter is connected to a dc load in which the current is not constant, the previous relationships are only approximate and an exact analytical determination of the response and corresponding average-value relationships is, in general, much more difficult and, in many cases, impractical. In these cases, the response is more readily established by computer simulation using any of the tools or techniques listed in the references.

Several modes of operation of a 3-phase, full converter are illustrated in Figs. 11.3-6 through 11.3-8. The line-to-line ac source voltage is 208 V (rms) and the commutating inductance is 0.045 mH. In each case, e_{ga}, i_{ga}, i_1, i_3, v_d, and i_d are

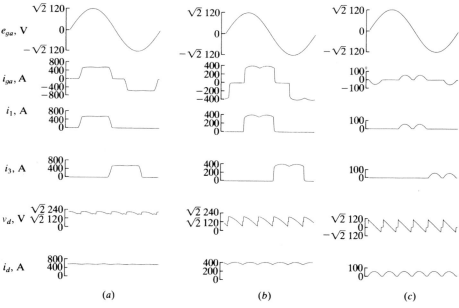

Figure 11.3-6   A 3-phase, full-converter operation with RL load. (a) $\alpha = 0°$; (b) $\alpha = 45°$; (c) $\alpha = 90°$.

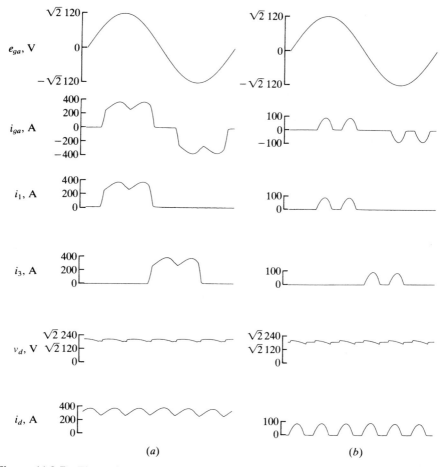

Figure 11.3-7 Three-phase, full-converter operation with RL and an opposing dc source connected in series across the converter terminals. (a) $\alpha = 0°$; (b) $\alpha = 35°$.

plotted where the currents i_1 and i_3 are the currents through thyristors $T1$ and $T3$, respectively. In Fig. 11.3-6, the converter is operating with a series RL load connected across the output terminals where $R = 0.5\,\Omega$ and $L = 1.33$ mH. In Fig. 11.3-6a, the converter is operating without phase delay. It is interesting to note that the output current is nearly constant during this mode of operation. The phase delay angle is $45°$ in Fig. 11.3-6b and $90°$ in Fig. 11.3-6c where the output current i_d is discontinuous. Note that when i_d is zero, v_d is also zero.

In the case depicted in Fig. 11.3-7, the series combination of an RL branch and a 260-V dc source is connected across the output terminals of the converter. The dc source is connected so that it opposes a positive v_d. The resistance is $0.05\,\Omega$ and the inductance is 0.133 mH. In Fig. 11.3-7a, the converter is operating without

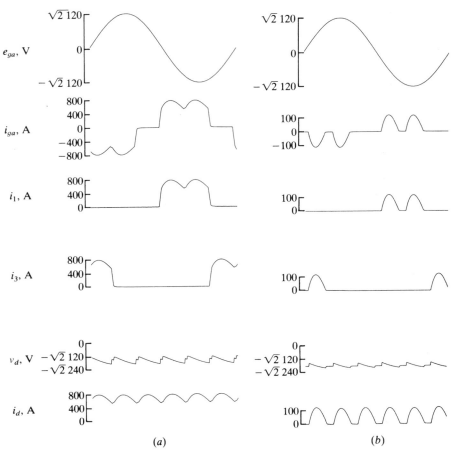

Figure 11.3-8 A 3-phase, full-converter operation with RL and an aiding dc source connected in series across the converter terminals. (*a*) $\alpha = 140°$; (*b*) $\alpha = 160°$.

phase delay. In Fig. 11.3-7*b*, the phase delay angle is 35° and the output current is discontinuous. Note that when i_d is zero, v_d is 260 V.

Inverter operation is illustrated in Fig. 11.3-8. In this case, the combination of the RL-dc source is still connected across the output terminals of the converter, but the polarity of the dc source is reversed. In Fig. 11.3-8*a*, the delay angle is 140°. The delay angle in Fig. 11.3-8*b* is 160° where discontinuous output current occurs. Clearly, when i_d is zero, v_d is −260 V.

Although (11.3-16) was derived for a constant output current, it is quite accurate for determining the average values of converter voltage. The reader should take the time to compare the calculated converter output voltage and current using (11.3-16) with the average values shown in Figs. 11.3-6 through 11.3-8 and to justify qualitatively any differences that may occur.

Dynamic Average-Value Model

The average-value relations and corresponding equivalent circuit depicted in Fig. 11.2-6 were developed based upon the assumptions that (1) the rms amplitude of the ac source voltages E is constant, (2) the dc load current i_d is constant, and (3) only one commutation takes place at a time. This equivalent circuit provides an approximation of the average voltage even if E and i_d vary with respect to time, provided that the variations are small from one $60°$ conduction interval to the next. The equivalent circuit, however, does not predict the amplitude or the phase angle of the ac source currents. Moreover, if the dc load current i_d varies with time, the calculated dc voltage in (11.3-16) is only an approximation of the actual averaged dc voltage. In this section, a dynamic average value model of the load-commutated inverter is derived that more accurately predicts the average dc voltage as well as the average q- and d-axis components of the ac source currents. These q- and d-axis components may be used, in turn, to calculate the real, apparent, and/or reactive power supplied by the ac source using formulas developed in Chapter 3.

A circuit diagram of the load-commutated converter analyzed herein is shown in Fig. 11.3-9. The ac source voltages are denoted v_{ag}, v_{bg}, and v_{cg}, and l_c is the source (commutating) inductance. This circuit also includes a dc inductor (resistor) L_{dc} (r_{dc}) that may represent, for example, the armature inductance (resistance) of a dc machine or the inductance (resistance) of a filtering circuit. Likewise, the voltage e_d may represent, for example, the back emf of a dc machine or the capacitor voltage in a dc filter. The ac source voltages may be expressed as

$$v_{ag} = \sqrt{2}E\cos\theta_g \tag{11.3-17}$$

$$v_{bg} = \sqrt{2}E\cos\left(\theta_g - \frac{2\pi}{3}\right) \tag{11.3-18}$$

$$v_{cg} = \sqrt{2}E\cos\left(\theta_g + \frac{2\pi}{3}\right) \tag{11.3-19}$$

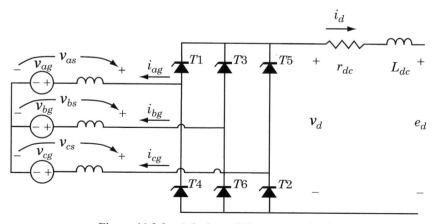

Figure 11.3-9 A 3-phase, full converter circuit.

where E is the rms magnitude and θ_g is the angular displacement of the source voltages. The source frequency is $\omega_g = p\theta_g$. Applying the reference frame transformation (3.3-1) with $\theta = \theta_g$ yields

$$v_{qg}^g = \sqrt{2}E \tag{11.3-20}$$

$$v_{dg}^g = 0 \tag{11.3-21}$$

The superscript g is used to denote the reference frame in which the d-axis component of voltage is identically zero. If the q- and d-axis components of the source voltages are given in the arbitrary reference frame, the transformation into the reference frame wherein $v_{dg}^g = 0$ may be deduced from the diagram depicted in Fig. 11.3-10. In particular,

$$\begin{bmatrix} f_q^g \\ f_d^g \end{bmatrix} = \begin{bmatrix} \cos\phi_g & \sin\phi_g \\ -\sin\phi_g & \cos\phi_g \end{bmatrix} \begin{bmatrix} f_q^a \\ f_d^a \end{bmatrix} \tag{11.3-22}$$

where f can be v or i. In Fig. 11.3-10, the raised superscript a is used to emphasize those variables that are expressed in the arbitrary reference frame. Specifically, f_q^a and f_d^a are the q- and d-axis components of the selected variables in the arbitrary reference frame and f_q^g and f_d^g are the q- and d-axis components in the reference frame wherein $v_{dg}^g = 0$. The angle between the f_q^a–f_d^a and f_q^g–f_d^g axes may be deduced from Fig. 11.3-10. In particular,

$$\phi_g = \tan^{-1}(v_{dg}^a / v_{qg}^a) \tag{11.3-23}$$

where v_{qg}^a and v_{dg}^a are the q- and d-axis components of the source voltages (11.3-17)–(11.3-19) expressed in the arbitrary reference frame. In the event that $v_{qg}^a = 0$, (11.3-23) cannot be used directly to establish ϕ_g. More generally, ϕ_g can be

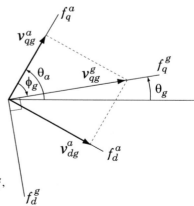

Figure 11.3-10   Relationship between f_q^a, f_d^a, f_q^g, and f_d^g.

established as the angle of a complex vector whose real component is v_{qg}^a and whose imaginary component is v_{dg}^a with positive angles measured, as usual, counterclockwise relative to the real axis. Specifically,

$$\phi_g = \text{ang}(v_{qg}^a + jv_{dg}^a) \tag{11.3-24}$$

Symbolically, (11.3-22) can be expressed as $\mathbf{f}_{qd}^g = {}^a\mathbf{K}^g\mathbf{f}_{qd}^a$, where

$$
{}^a\mathbf{K}^g = \begin{bmatrix} \cos\phi_g & \sin\phi_g \\ -\sin\phi_g & \cos\phi_g \end{bmatrix} \tag{11.3-25}
$$

The inverse relationship may be expressed as $\mathbf{f}_{qd}^a = {}^g\mathbf{K}^a\mathbf{f}_{qd}^g$, where

$$
{}^g\mathbf{K}^a = [{}^a\mathbf{K}^g]^{-1} = \begin{bmatrix} \cos\phi_g & -\sin\phi_g \\ \sin\phi_g & \cos\phi_g \end{bmatrix} \tag{11.3-26}
$$

The dc voltage and the q- and d-axis components of the ac currents are periodic in $\pi/3$ radians of θ_g. Thus, the average values may be established for any $\pi/3$ interval of θ_g. It is convenient to consider the $\pi/3$ interval that begins when $T3$ begins to conduct and ends when $T4$ begins to conduct. The average dc voltage over this interval may be expressed as

$$\bar{v}_d = \frac{3}{\pi}\int_{\frac{\pi}{3}+\alpha}^{\frac{2\pi}{3}+\alpha}(v_{bs} - v_{cs})\,d\theta_g \tag{11.3-27}$$

In (11.3-27), the overbar is used to denote the average value during dynamical conditions wherein the dc current i_d and/or the amplitude of the ac voltages E are allowed to vary, provided that the variation from one switching interval to the next is relatively small. In other words, the averaging interval in (11.3-27) is assumed to be small relative to the longer-term dynamics associated with the variations in E and/or i_d. Thus, (11.3-27) may be interpreted as the short-term or fast average of v_d. Likewise, the fast average of i_d (average of i_d over a $\pi/3$ interval) will be denoted as $\bar{i}_d$. The firing angle α in (11.3-27) is defined such that $T3$ fires when

$$\theta_g = \frac{\pi}{3} + \alpha \tag{11.3-28}$$

The average dc voltage indicated in (11.3-27) may be evaluated by noting from Fig. 11.3-9 that

$$v_{as} = v_{ag} + l_c\frac{di_{ag}}{dt} \tag{11.3-29}$$

$$v_{bs} = v_{bg} + l_c\frac{di_{bg}}{dt} \tag{11.3-30}$$

$$v_{cs} = v_{cg} + l_c\frac{di_{cg}}{dt} \tag{11.3-31}$$

Substituting (11.3-30)–(11.3-31) into (11.3-27) yields

$$\bar{v}_d = \frac{3}{\pi} \int_{\frac{\pi}{3}+\alpha}^{\frac{2\pi}{3}+\alpha} (v_{bg} - v_{cg}) \, d\theta_g + \frac{3}{\pi} l_c \omega_g (i_{bg} - i_{cg}) \Big|_{\frac{\pi}{3}+\alpha}^{\frac{2\pi}{3}+\alpha} \tag{11.3-32}$$

Substituting (11.3-18)–(11.3-19) into (11.3-32) and simplifying yields

$$\bar{v}_d = \frac{3\sqrt{3}}{\pi} \sqrt{2} E \cos \alpha + \frac{3}{\pi} l_c \omega_g (i_{bg} - i_{cg}) \Big|_{\frac{\pi}{3}+\alpha}^{\frac{2\pi}{3}+\alpha} \tag{11.3-33}$$

Further simplification can be obtained by observing that, prior to the instant when $T3$ begins to conduct, only $T1$ and $T2$ are conducting. Therefore,

$$\mathbf{i}_{abcg}\big|_{\theta_g=\frac{\pi}{3}+\alpha} = [-\bar{i}_d \quad 0 \quad \bar{i}_d]^T \tag{11.3-34}$$

Similarly, immediately prior to the instant when $T4$ begins to conduct, only $T2$ and $T3$ are on; therefore

$$\mathbf{i}_{abcg}\big|_{\theta_g=\frac{2\pi}{3}+\alpha} = [0 \quad -\bar{i}_d - \Delta \bar{i}_d \quad \bar{i}_d + \Delta \bar{i}_d]^T \tag{11.3-35}$$

In (11.3-35), $\Delta \bar{i}_d$ represents the change in average dc current over the given conduction interval due to long-term dynamics. It follows from this definition that the derivative of the fast-average rectifier current may be approximated as

$$\frac{d\bar{i}_d}{dt} = \frac{\Delta \bar{i}_d}{\pi/3} \omega_g \tag{11.3-36}$$

Substituting (11.3-34)–(11.3-36) into (11.3-33) and simplifying, we obtain

$$\bar{v}_d = \frac{3\sqrt{3}}{\pi} \sqrt{2} E \cos \alpha - \frac{3}{\pi} l_c \omega_g \bar{i}_d - 2 l_c \frac{d\bar{i}_d}{dt} \tag{11.3-37}$$

This equation is similar to (11.3-11) with the exception of the last term in (11.3-37). Under the assumption that the dc current is constant, (11.3-37) is equivalent to (11.3-11). However, during transients, the average dc current will, in general, change with respect to time and (11.3-37) will provide a more accurate expression of the average dc voltage.

From Fig. 11.3-9, $\bar{v}_d$ can be related to $\bar{i}_d$ and e_{dc} using

$$\bar{v}_d = r_{dc} \bar{i}_d + L_{dc} \frac{d\bar{i}_d}{dt} + e_d \tag{11.3-38}$$

Combining (11.3-37) and (11.3-38) yields

$$\frac{d\bar{i}_d}{dt} = \frac{\frac{3\sqrt{3}}{\pi}\sqrt{2}E\cos\alpha - (r_{dc} + \frac{3}{\pi}l_c\omega_g)\bar{i}_d - e_d}{L_{dc} + 2l_c} \tag{11.3-39}$$

To establish the average q- and d-axis components of the ac currents, the dc current is assumed constant throughout the interval and equal to $\bar{i}_d$. For convenience, the averaging interval is divided into two subintervals: the commutation interval during which the current is transferred from $T1$ to $T3$, and the conduction interval during which only $T2$ and $T3$ are conducting. During the commutation interval, $T1$, $T2$, and $T3$ are conducting. Therefore, the current into the ac source must be of the form

$$\mathbf{i}_{abcg} = \begin{bmatrix} i_{ag} & -\bar{i}_d - i_{ag} & \bar{i}_d \end{bmatrix}^T \tag{11.3-40}$$

and

$$v_{as} = v_{bs} = 0 \tag{11.3-41}$$

Substituting (11.3-29), (11.3-30), and (11.3-40) into (11.3-41) and solving for the a-phase current yields

$$i_{ag}(\theta_g) = -\bar{i}_d + \frac{1}{l_c\omega_e}\frac{\sqrt{3}}{2}\sqrt{2}E\left[\cos(\alpha) - \cos\left(\theta_g - \frac{\pi}{3}\right)\right] \tag{11.3-42}$$

The commutation subinterval ends when the current in $T1$, which is the a-phase current, becomes zero. The angle from the time $T3$ is turned on and $T1$ is turned off is known as the commutation angle u. It can be found by setting (11.3-42) equal to zero. In particular,

$$u = -\alpha + \arccos\left[\cos\alpha - \frac{l_c\omega_g\bar{i}_d}{E}\right] \tag{11.3-43}$$

After commutation, the a-phase current remains at zero; therefore

$$\mathbf{i}_{abcg} = \begin{bmatrix} 0 & -\bar{i}_d & \bar{i}_d \end{bmatrix}^T \tag{11.3-44}$$

Equations (11.3-40), (11.3-42), and (11.3-44) specify the ac currents throughout the $\pi/3$ interval that begins when $T3$ is fired. The average q- and d-axis components can be established using

$$\bar{i}_{qg}^g = \frac{3}{\pi}\int_{\frac{\pi}{3}+\alpha}^{\frac{2\pi}{3}+\alpha} i_{qg}^g(\theta_g)\, d\theta_g \tag{11.3-45}$$

$$\bar{i}_{dg}^g = \frac{3}{\pi}\int_{\frac{\pi}{3}+\alpha}^{\frac{2\pi}{3}+\alpha} i_{dg}^g(\theta_g)\, d\theta_g \tag{11.3-46}$$

Because the expressions for the ac currents are different during the conduction interval than in the commutation interval, it is convenient to break up (11.3-45) and (11.3-46) into components corresponding to these two intervals. In particular,

$$\bar{i}_{qg}^g = \bar{i}_{qg,\text{com}}^g + \bar{i}_{qg,\text{cond}}^g \tag{11.3-47}$$

$$\bar{i}_{dg}^g = \bar{i}_{dg,\text{com}}^g + \bar{i}_{dg,\text{cond}}^g \tag{11.3-48}$$

where

$$\bar{i}_{qg,\text{com}}^g = \frac{3}{\pi} \int_{\frac{\pi}{3}+\alpha}^{\frac{\pi}{3}+\alpha+u} i_{qg,\text{com}}^g(\theta_g)\, d\theta_g \tag{11.3-49}$$

$$\bar{i}_{qg,\text{cond}}^g = \frac{3}{\pi} \int_{\frac{\pi}{3}+\alpha+u}^{\frac{2\pi}{3}+\alpha} i_{qg,\text{cond}}^g(\theta_g)\, d\theta_g \tag{11.3-50}$$

$$\bar{i}_{dg,\text{com}}^g = \frac{3}{\pi} \int_{\frac{\pi}{3}+\alpha}^{\frac{\pi}{3}+\alpha+u} i_{dg,\text{com}}^g(\theta_g)\, d\theta_g \tag{11.3-51}$$

$$\bar{i}_{dg,\text{cond}}^g = \frac{3}{\pi} \int_{\frac{\pi}{3}+\alpha+u}^{\frac{2\pi}{3}+\alpha} i_{dg,\text{cond}}^g(\theta_g) d\theta_g \tag{11.3-52}$$

The commutation component of the current may be found by substituting (11.3-42) into (11.3-40), applying the reference-frame transformation (3.3-1) with $\theta = \theta_g$, and integrating in accordance with (11.3-49) and (11.3-51). After extensive effort we obtain

$$\bar{i}_{qg,\text{com}}^g = \frac{2\sqrt{3}}{\pi}\bar{i}_d \left[\sin\left(u+\alpha-\frac{5\pi}{6}\right) - \sin\left(\alpha-\frac{5\pi}{6}\right)\right] \frac{3}{\pi}\frac{\sqrt{3}E}{l_c\omega_g}\cos\alpha[\cos(u+\alpha)$$

$$ - \cos\alpha] + \frac{1}{4}\frac{3}{\pi}\frac{\sqrt{2}E}{l_c\omega_g}[\cos(2u) - \cos(2\alpha+2u)] \tag{11.3-53}$$

$$\bar{i}_{dg,\text{com}}^g = \frac{2\sqrt{3}}{\pi}\bar{i}_d \left[-\cos\left(u+\alpha-\frac{5\pi}{6}\right) + \cos\left(\alpha-\frac{5\pi}{6}\right)\right] \frac{3}{\pi}\frac{\sqrt{2}E}{l_c\omega_g}\cos\alpha$$

$$ \times [\sin(u+\alpha) - \sin\alpha] + \frac{1}{4}\frac{3}{\pi}\frac{\sqrt{2}E}{l_c\omega_g}[\sin(2u)$$

$$ - \sin(2\alpha+2u)] - \frac{3}{\pi}\frac{\sqrt{2}E}{l_c\omega_g}\frac{1}{2}u \tag{11.3-54}$$

Similarly, the conduction component of the average currents may be expressed by applying the reference-frame transformation to (11.3-44) with $\theta = \theta_g$ and integrating as set forth in (11.3-50) and (11.3-52):

$$\overline{i}_{qg,\text{cond}}^{g} = \frac{2\sqrt{3}}{\pi}\overline{i}_{d}\left[\sin\left(\alpha + \frac{7\pi}{6}\right) - \sin\left(\alpha + u + \frac{5\pi}{6}\right)\right] \tag{11.3-55}$$

$$\overline{i}_{dg,\text{cond}}^{g} = \frac{2\sqrt{3}}{\pi}\overline{i}_{d}\left[-\cos\left(\alpha + \frac{7\pi}{6}\right) + \cos\left(\alpha + u + \frac{5\pi}{6}\right)\right] \tag{11.3-56}$$

The relationships between the previous equations are conveniently summarized in the block diagram illustrated in Fig. 11.3-11, which represents a dynamic average-value model of the load-commutated converter. The inputs to this model include the firing angle α, the q- and d-axis components of the source voltage in the arbitrary reference frame, and the dc source voltage e_d. The outputs of the model include the fast average of the rectifier current $\overline{i}_d$ and the fast average of the q- and d-axis components of the ac currents in the arbitrary reference frame.

To illustrate the dynamic response that is established using the dynamic average-value model, it is assumed that the rated line-to-line source voltage is 208 V (rms). The commutating inductance l_c is 0.045 mH. Also, $r_{dc} = 0.5\,\Omega$, $L_{dc} = 1.33$ mH, and $e_d = 0$. In the following study, the dc and ac currents are initially zero, and rated voltages are suddenly applied at $t = 0$ with the delay angle α set to zero. The dynamic response is shown in Fig. 11.3-12 wherein the following variables are plotted: i_d, the dc current, i_{qg}^g the q-axis component of the ac current; and i_{dg}^g, the d-axis component of the ac current. The ac currents are expressed in the reference frame wherein $v_{dg}^g = 0$. The variables indicated with an overbar correspond to the average-value model in Fig. 11.3-11, whereas those that do not include the overbar correspond to the actual response. At the instant of time indicated in Fig. 11.3-11, the firing delay angle is stepped to 45°. As shown, the average-value model accurately portrays the fast-average dynamic response for the given study. The

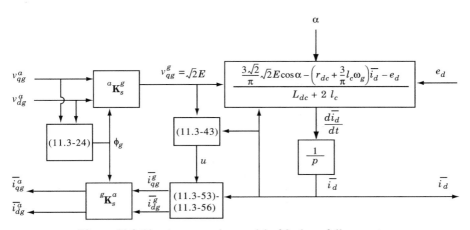

Figure 11.3-11   Average value model of 3-phase full converter.

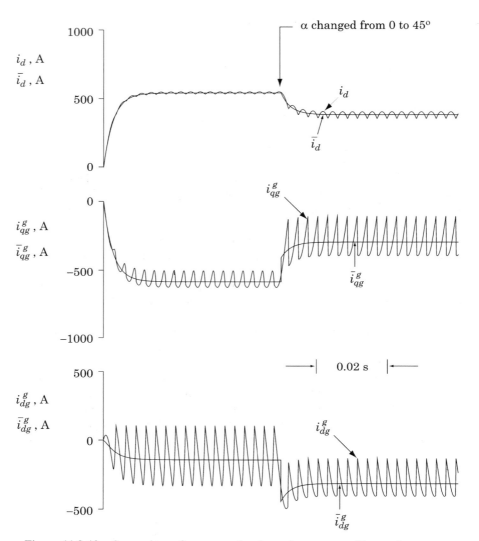

Figure 11.3-12   Comparison of average value dynamic response with actual response.

steady-state waveforms for $\alpha = 0$ and $\alpha = 45°$ are shown in Fig. 11.3-6a and 11.3-6b, respectively.

In the derivation of the average-value model, it was assumed that the commutating inductance l_c is constant. This is a reasonable assumption if the converter ac voltages are supplied by an electric utility. If, however, the ac voltages are supplied by a synchronous generator of comparable power rating, the commutating inductance will be a function of the rotor angular position. A detailed analysis of this configuration and corresponding average-value model are given in reference 5.

REFERENCES

[1] C. Adamson and N. G. Hingarani, *High Voltage Direct Current Power Transmission*, Garroway Limited, London, England, 1960.

[2] E. W. Kimbark, *Direct Current Transmission*, Vol. 1, John Wiley and Sons, New York, 1971.

[3] S. B. Dewan and A. Straughen, *Power Semiconductor Circuits*, John Wiley and Sons, New York, 1975.

[4] P. Wood, *Switching Power Converters*, Van Nostrand Reinhold Co., New York, 1981.

[5] S. D. Sudhoff, K. A. Corzine, H. J. Hegner, and D. E. Delisle, Transient and Dynamic Average-Value Modeling of Synchronous Machine Fed Load—Commutated Converters, *IEEE Transactions on Energy Conversion*, Vol. 11, No. 3, September 1996, pp. 508–514.

[6] M. H. Rashid, *SPICE for Power Electronics and Electric Power*, Prentice-Hall, Englewood Cliffs, NJ, 1993.

[7] P. W. Tuinenga, *SPICE: A Guide to Circuit Simulation and Analysis Using PSpice*, Prentice-Hall, Englewood Cliffs, NJ, 1995.

[8] *Saber Designer Reference Manual*, Analogy, Inc. 1998.

[9] J. Ogrodzki, *Circuit Simulation Methods and Algorithms*, CRC Press, 1994.

[10] L. T. Pillage, R. A. Rohrer, and C. Visweswariah, *Electronic Circuit and Systems Simulation Methods*, McGraw-Hill, New York, 1995.

[11] O. Wasynczuk and S. D. Sudhoff, Automated State Model Generation Algorithm for Power Circuits and Systems, *IEEE Transactions on Power Systems*, Vol. 11, No. 9, November 1996, pp. 1951–1956.

[12] *Power System Blockset for Use with Simulink*, User's Guide, Version 2, Hydro-Quebec TEQSIM International and The MathWorks Inc., 2000.

PROBLEMS

1 Using the average-value equations derived in Section 11.2, calculate the average dc voltage and current for each of the conditions in Fig. 11.2-7. Compare with the average values plotted in Fig. 11.2-7.

2 Using the average-value equations derived in Section 11.3, calculate the average dc voltage and current for each of the conditions in Fig. 11.3-6. Compare with the average values plotted in Fig. 11.3-6.

3 Assume that the ac source voltages applied to the 3-phase load commutated converter have an *acb* phase sequence. Relabel the variables in Fig. 11.3-4 and indicate the sequence in which the thyristors should be fired.

4 Derive (11.3-37).

5 For the parameters assumed in Fig. 11.3-12, calculate the steady-state values of V_d, I_d, I_{qg}^g, I_{dg}^g for $\alpha = 0$ and $\alpha = 45°$. Calculate the power delivered to the dc system for $\alpha = 0$ and $\alpha = 45°$.

6 For the parameters assumed in Fig. 11.3-12, calculate the real, reactive, and apparent power supplied by the ac source for $\alpha = 0$ and $\alpha = 45°$. Compare the real power with the dc power established in Problem 5.

Chapter 12

dc MACHINE DRIVES

12.1 INTRODUCTION

Due primarily to the advent of power electronic devices and the subsequent development of highly sophisticated brushless dc and induction machine drive systems, the traditional dc machine is not used as extensively as in the past. Nevertheless, the dc drive still plays an important role in drive applications and a treatment of converter–machine drive systems would be incomplete without including dc drives. As in Chapter 2, our focus will be on the permanent magnet and shunt machines. After a brief overview of the ac/dc and dc/dc converters which are used in dc drives, emphasis is then placed on the ac/dc converter dc drives. Here, we are able to draw upon the work in Chapter 11 where the operation of ac/dc converters are discussed and average-value models are developed for several types of converters. The operation of dc drives supplied from ac/dc converters is illustrated by computer traces, and the steady-state response is calculated using average-value models and compared to the computer traces. Next, one-, two-, and four-quadrant dc/dc converters are also considered, and steady-state and dynamic performance are illustrated. Dynamic average-value models are developed and a comparison is made with computer traces of the actual dc drive. Finally, speed and current controls are discussed, and time-domain block diagrams are given which can be used to analyze and design these types of drive systems.

12.2 SOLID-STATE CONVERTERS FOR dc DRIVE SYSTEMS

Numerous types of ac/dc and dc/dc converters are used in variable-speed drive systems to supply an adjustable dc voltage to the dc drive machine. In the case of ac/dc converters, half-wave, semi-, full, and dual converters are used depending upon the amount of power being handled and the application requirements, such as fast response time, regeneration, and reversible or nonreversible drives. In the case of dc/dc converters, one-, two-, and four-quadrant converters are common. Obviously, we cannot treat all types of converters and all important applications; instead, it is our objective in this section to present the widely used converters and to set the stage for the following sections wherein the analysis and performance of several common dc drive systems are set forth.

Single-Phase ac/dc Converters

Several types of single-phase phase-controlled ac/dc converters are shown in Fig. 12.2-1. Therein, the converters consist of silicon controlled rectifiers (SCRs) and diodes. The dc machine is illustrated in abbreviated form without showing the field winding and the resistance and inductance of the armature winding. The dc machines, which are generally used with ac/dc converters, are permanent magnet, shunt, or series machines. Half-wave, semi-, full, and dual converters are shown in Fig. 12.2-1. The operation of a single-phase, full converter with continuous dc current is analyzed in Chapter 11, where various modes of operation are illustrated.

The half-wave converter yields discontinuous armature current in all modes of operation, and only positive current flows on the ac side of the converter. Analysis of the operation of a dc drive with discontinuous armature current is quite involved [1] and not considered. The other converters shown in Fig. 12.2-1 can operate with either a continuous or discontinuous armature current. The half-wave converters and the semiconverters allow a positive dc voltage and unidirectional armature current; however, the semiconverter may be equipped with a diode connected across the terminals of the machine (free-wheeling diode) to dissipate energy stored in the armature inductance when the converter blocks current flow. The full and dual converters can regenerate; that is, the polarity of the motor voltage may be reversed. However, the current of the full converter is unidirectional. Although a reversing switch may be used to change the connection of the full converter to the machine and thereby reverse the current flow through the armature, bidirectional current flow is generally achieved with a dual converter. Consequently, dual converters are used extensively in variable-speed drives wherein it is necessary for the machine to rotate in both directions as in rolling mills and crane applications.

3-Phase ac/dc Converters

For drive applications requiring over 20- to 30-hp, 3-phase converters are generally used. Typical 3-phase converters are illustrated in Fig. 12.2-2. The machine current is continuous in most modes of operation of dc drives with 3-phase converters. The semi- and full converters are generally used except in reversible drives where the dual converter is more appropriate. Continuous current operation

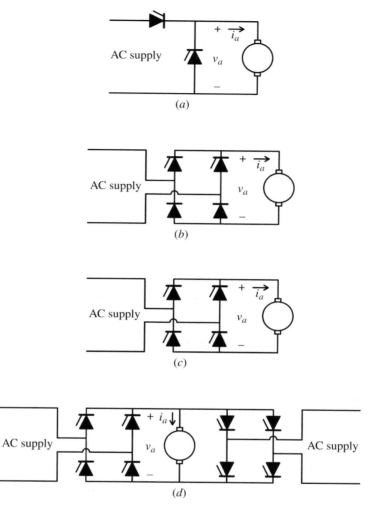

Figure 12.2-1 Typical single-phase, phase-controlled ac/dc converters. (*a*) Half-wave converter; (*b*) semiconverter; (*c*) full converter; (*d*) dual converter.

of a 3-phase, full converter is analyzed in Chapter 11, and several modes of operation are illustrated.

dc/dc Converters

The commonly used dc/dc converters in dc drive systems are shown in Fig. 12.2-3. Therein, the SCR or transistor is represented by a switch which can carry positive current only in the direction of the arrow. The one-quadrant converter (Fig. 12.2-3*a*)

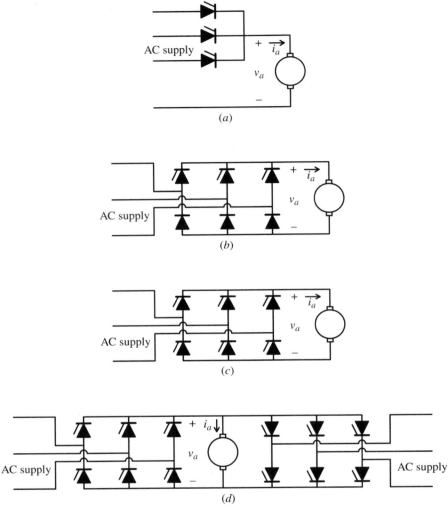

Figure 12.2-2 Typical 3-phase, phase-controlled ac/dc converters. (*a*) Half-wave converter; (*b*) semiconverter; (*c*) full converter; (*d*) dual converter.

is used extensively in low-power applications. Because the armature current will become discontinuous in some modes of operation, the analysis of the one-quadrant converter is somewhat involved. This analysis is set forth later in this chapter. The two- and four-quadrant converters are bidirectional with regard to current. In the four-quadrant converter, the polarity of the armature voltage can be reversed. All of these dc/dc converters will be considered later in this chapter.

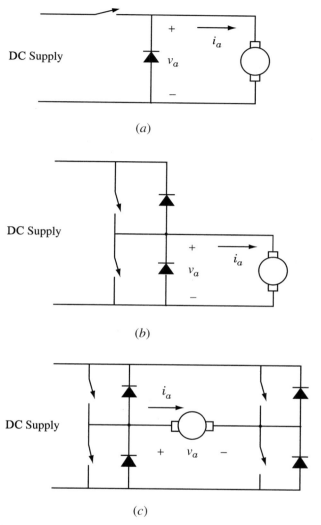

Figure 12.2-3 Typical dc/dc converters. (*a*) One-quadrant; (*b*) two-quadrant; (*c*) four-quadrant.

12.3 STEADY-STATE AND DYNAMIC CHARACTERISTICS OF ac/dc CONVERTER DRIVES

Although drive systems are often operated with a closed-loop speed control, there are several aspects of the ac/dc converter–machine combination without speed control (open-loop) which are of interest. In this section, the open-loop operation of a shunt machine supplied from single-phase and 3-phase full converters is considered.

We will then consider briefly the starting of a shunt motor supplied from a single-phase full converter in this section which employs a closed-loop current control.

Single-Phase Full-Converter dc Drive

A single-phase full-converter shunt-machine dc drive is shown in Fig. 12.3-1. Therein, e_{ga} is the ac source voltage, i_{ga} is the ac source current, and l_c is the inductance between the source and the converter which is commonly referred to as the commutating inductance. It is important to note that the field of the shunt machine is supplied from a diode rectifier that is also connected to the ac source through a commutating inductance L_c. Because the inductance of the armature circuit is small, the armature current i_a is likely to be discontinuous, especially at light loads unless external inductance is added to the armature circuit. On the other hand, the inductance of the field winding is large and the field current will be continuous with very little ripple content. Hence, for purposes of analysis, it is customary to disregard the presence of the diode rectifier supplying the field winding and assume that the field current is constant at the desired or rated value.

The computer traces shown in Figs. 12.3-2 and 12.3-3 depict steady-state operation of the 240-V 5-hp dc shunt machine supplied from a single-phase full converter with rated and 50% rated load torque, respectively. The parameters of the machine are given in Chapter 2 and repeated here for convenience; $R_f = 240\,\Omega$, $L_{FF} = 120\,\text{H}$, $L_{AF} = 1.8\,\text{H}$, $r_a = 0.6\,\Omega$, and $L_{AA} = 0.012\,\text{H}$. The inertia of the machine and connected load is $1\,\text{kg}\cdot\text{m}^2$. The voltage of the ac source is 280 V (rms) and l_c is 0.3 mH. It is assumed that the field current is 1 A (rated), and the ac current i_{ga} includes only the current flowing to the full converter supplying the armature. In these figures, e_{ga}, i_{ga}, v_a, and i_a are plotted. It is clear that T_e and i_a are related by a constant multiplier because i_f is constant. The converter is operating without phase delay; hence, whenever the forward voltage of the SCR becomes positive, it starts

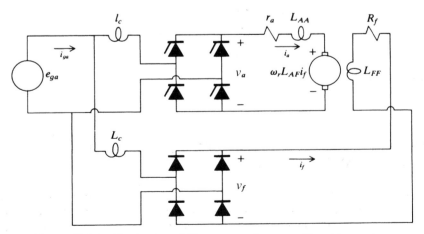

Figure 12.3-1 Single-phase full-converter, shunt–machine drive system.

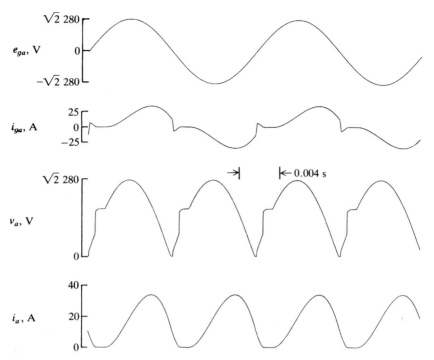

Figure 12.3-2 Operation of single-phase, full-converter 5-hp shunt–machine drive system with rated load torque; discontinuous armature current; $\omega_r = 138$ rad/s.

conducting just as if it were a diode (Chapter 11). Therefore, when the armature current is zero, all SCRs are blocking current flow due to the fact that they are all reverse biased. It is noted that the armature current is periodically zero (discontinuous) during the operating conditions depicted in Figs. 12.3-2 and 12.3-3. During this mode of operation, v_a is equal to the counter emf $(\omega_r L_{AF} i_f$ or $\omega_r k_v)$.

Let us consider the conduction of the SCRs during positive source current for the case where the load torque is 50% rated (Fig. 12.3-3). When the source voltage becomes more positive than the counter emf, the SCR in the upper left-hand part of the converter (Fig. 12.3-1) and the SCR in the lower right-hand part of the converter start to conduct. The current increases, lagging v_a (source voltage) due to the armature inductance. The lagging current causes the SCRs to continue to conduct current after the source voltage has decreased below the value of the counter emf. Once the terminal voltage (source voltage) has dropped below the counter emf sufficient to force the armature current to zero, the SCRs block current flow and the terminal voltage again becomes (jumps to) the counter emf (Fig. 12.3-3). It is clear that the magnitude of the "jump" in v_a is equal to the magnitude of $L_{AA}(di_a/dt)$ at the instant i_a becomes zero. Also, it is important to note that, in Fig. 12.3-3, the armature current reaches zero before v_a becomes zero. This is not the case in Fig. 12.3-2

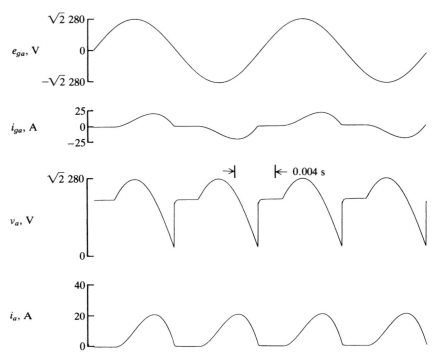

Figure 12.3-3 Operation of single-phase, full-converter 5-hp shunt–machine drive system with 50% rated load torque; discontinuous armature current; $\omega_r = 161$ rad/s.

(full-load operation) where the armature current is still positive when the terminal voltage becomes zero. Hence, commutation occurs simultaneously between the two SCRs in the upper and the two SCRs in lower part of the converter. The terminal voltage remains at zero during commutation (Chapter 11). However, after commutation is completed, the terminal voltage plus $L_{AA}(di_a/dt)$, which aids the terminal voltage because the current is decreasing, is still less than the counter emf; hence, the armature current continues to decrease and becomes zero before the terminal voltage has increased to the point at which it can maintain current flow. The armature current is blocked (discontinuous) and the terminal voltage becomes the counter emf until the source voltage again becomes larger than the counter emf.

The torque-versus-speed characteristics plotted in Fig. 2.4-3 for a typical shunt machine reveal that the speed of the machine changes very little for a relatively large change in load torque. This was illustrated in Fig. 2.5-3, where the rotor speed of the 5-hp shunt motor changed from approximately 128 rad/s to 131 rad/s when the load torque was decreased from rated to 50% rated. If the armature current is discontinuous and if the converter–machine system is operating without speed control, the change in rotor speed is much larger for a given change in load torque than when the current is continuous. For example, in Fig. 12.3-2 where the single-phase, full-converter shunt-machine dc drive is delivering rated torque load, the rotor speed is 138 rad/s compared to 161 rad/s for 50% rated (Fig. 12.3-3). Moreover, with the machine originally operating steadily at rated load, the load torque was stepped to 50% rated and it required between 4.5 and 5 s for the machine to establish steady-state operation at 50% rated load. It is apparent that when the current is discontinuous, a change in rotor speed has less effect on the change of the average current than when the current is continuous.

Continuous armature current may be achieved in a single-phase full-converter dc drive system by adding an external inductance in series with the armature circuit. Full-load operation of the example drive system is illustrated in Fig. 12.3-4 with a 0.108-H inductance connected in series with the armature circuit, making the total armature circuit inductance 10 times its original value; assuming that the inductance is ideal, the armature time constant τ_a is 10 times larger. As the source voltage changes polarity, the armature current is commutated from one SCR to the other simultaneously in the top and bottom parts of the converter. During this commutation period, the terminal voltage is zero. Once commutation is complete, v_a jumps to the value of the source voltage minus the $l_c(di_a/dt)$ voltage drop. However, due to the larger armature circuit inductance, the current does not subsequently become discontinuous as in Fig. 12.3-2.

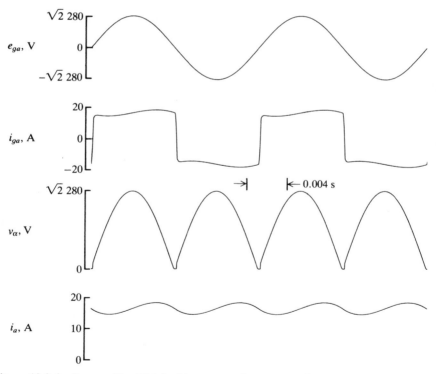

Figure 12.3-4   Same as Fig. 12.3-2 with armature time constant increased by 10; continuous armature current; $\omega_r = 134.9$ rad/s.

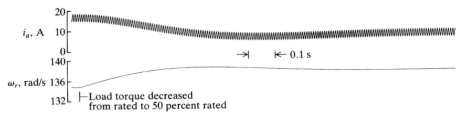

Figure 12.3-5   Dynamic performance of single-phase full-converter 5-hp shunt-machine drive following a decrease in load torque from rated to 50% rated; armature time constant as in Fig. 12.3-4.

The dynamic response (i_a or T_e and ω_r) of this system due to a change in load torque is shown in Fig. 12.3-5. The field current is fixed at 1 A. The steady-state speed changes from approximately 134.8 rad/s to 137.8 rad/s (2% change). It is clear that if we are to operate a single-phase full-converter drive system without closed-loop speed control, we should add inductance to the armature circuit to ensure continuous armature current over the range of loads anticipated; otherwise, large speed variations will occur with load changes.

Because, during continuous armature current operation, the 280-V source voltage yields a terminal voltage larger (maximum $\sqrt{2} \times 280 \times 0.636 = 252$ V) than the rated value of 240 V, the speed of the motor is higher for a given torque load. Compare 134.8 rad/s to 127.7 rad/s which is rated for full-load operation. The output voltage of the converter may be decreased by introducing a phase delay. For example, steady-state operation of the example 5-hp motor with continuous armature current and with the phase delay angle set to yield rated rotor speed (127.7 rad/s) is shown in Fig. 12.3-6. Here, α is 16.8°. It is interesting to note that if we use the voltage equation (11.2-15) for constant output current and if we set the output voltage equal to 240 V and the output current equal to 16.2 A, α is calculated to be 17.3°. Clearly, this approximate equation for the output voltage of the converter is adequate for most calculations if the armature current is continuous.

As mentioned earlier, a phase-controlled converter can also be used to reduce the terminal voltage during the starting period and thereby maintain the armature current at a safe level. This may be achieved by the elementary automatic control scheme shown in Fig. 12.3-7. Here, the armature current is maintained at a level below twice rated current over most of the acceleration period. The controller is generally an integral type where $G(s) = 1/\tau s$. A limit is incorporated in Fig. 12.3-7 to restrict the output of the controller when the motor is accelerated to a speed were the counter emf is large enough to maintain the armature current below twice rated. This method of starting is illustrated in Fig. 12.3-8. The single-phase full-converter shunt-machine drive is the example system with the armature time constant increased by 10. The armature current is continuous over the normal operating range. It is assumed, however, that a constant voltage of rated value is applied to the field winding at the instant voltage is applied to the armature circuit. In other words, the diode rectifier supplying the field winding, as shown in Fig. 12.3-1, is considered a

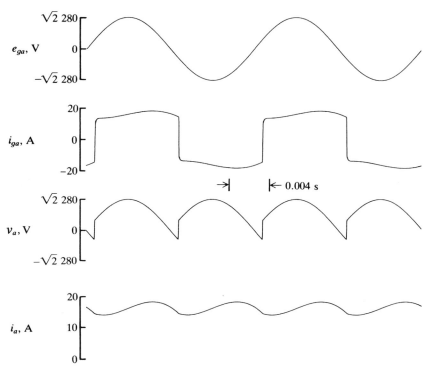

Figure 12.3-6 Same as Fig. 12.3-4 with a phase delay angle of $16.8°$; $\omega_r = 127.7$ rad/s (rated).

constant voltage source. The time constant of the integral controller is set at $\tau = 0.281$ s and the load torque (T_L) is set equal to $B_L\omega_r$, where B_L is selected as $0.2287\,\text{N}\cdot\text{m}\cdot\text{s/rad}$ so that rated load torque $(29.2\,\text{N}\cdot\text{m})$ occurs at rated speed $(127.7\,\text{rad/s})$. The damping coefficient, B_m, in (2.3-6) is set equal to zero. (It is clear

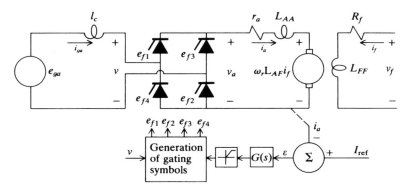

Figure 12.3-7 Single-phase, full-converter 5-hp shunt-machine drive with current control for starting purposes.

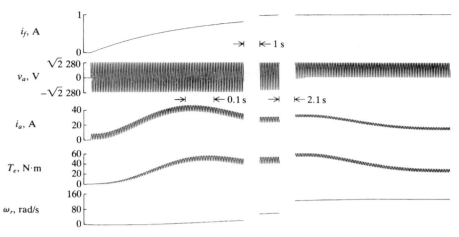

Figure 12.3-8 Starting characteristics of a single-phase full-converter 5-hp shunt-machine drive; armature time constant as in Fig. 12.3-4.

that we could have let $T_L = 0$ and set $B_m = B_L$.) Initially, the average armature current increases above the desired value of twice rated current (32.4 A). Following this initial overshoot, the phase delay of the converter is automatically controlled by decreasing the phase delay as the motor accelerates to maintain the current below twice rated current. When the speed of the machine reaches a value sufficient for the counter emf to limit the current below twice rated, the action of the control causes the converter to operate without phase delay.

3-Phase Full-Converter dc Drive

A 3-phase full-converter shunt-machine drive is shown in Fig. 12.3-9. The three-phase source voltages are e_{ga}, e_{gb}, and e_{gc} and the commutating inductance is denoted l_c. The circuit that could be used to obtain the field voltage is not shown as it is in Fig. 12.3-1; however, this could be achieved using either a single-phase or a 3-phase diode rectifier. The operation of a 3-phase full-converter with constant output current is analyzed in Chapter 11.

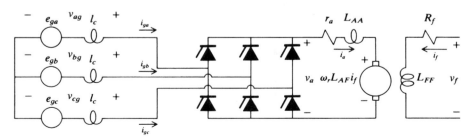

Figure 12.3-9 A 3-phase full-converter shunt-machine drive system.

Except at very light loads, the 3-phase full-converter dc drive operates with continuous armature current. Full-load operation of the example 200-hp 250-V dc shunt machine supplied from a 3-phase full converter is shown in Fig. 12.3-10. In this system, l_c is 0.045 mH and the 3-phase ac source is 208 V (rms), line to line. The parameters of the dc machine are given in Chapter 2 and repeated here for convenience; $R_f = 12\,\Omega$, $L_{FF} = 9\,\text{H}$, $r_a = 0.012\,\Omega$, and $L_{AA} = 0.35\,\text{mH}$. The inertia of the rotor and load is $30\,\text{kg} \cdot \text{m}^2$. The ac voltage e_{ga}, ac current i_{ga}, terminal voltage v_a, and armature current i_a are shown. The field current is constant at rated value (20.83 A). The converter is operating without phase delay; hence, the SCRs start conducting whenever they become forward biased. The commutation period is quite evident in v_a. The rotor speed is 69.6 rad/s, which is higher than the rotor speed of 64.6 rad/s when the terminal voltage is 250 V (Fig. 2.5-4). This higher speed occurs because the output voltage of the converter is larger than 250 V ($3\sqrt{6} \times 208\,\text{V}/ (\sqrt{3} \times \pi) = 280\,\text{V}$, if l_c is neglected).

The dynamic performance of the example 3-phase full-converter dc drive following a step decrease in load torque from rated (2375 N · m) to 50% rated is illustrated in Fig. 12.3-11. The dynamic response of this 200-hp motor with a constant

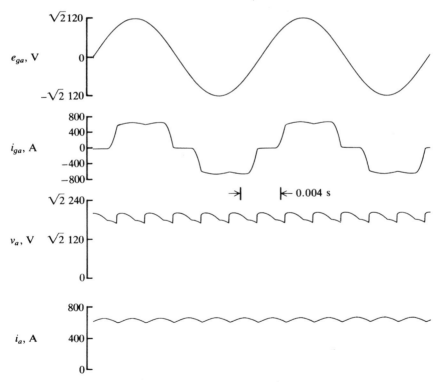

Figure 12.3-10 Rated load torque operation of 3-phase full-converter 200-hp shunt-machine drive system; $\omega_r = 69.5$ rad/s.

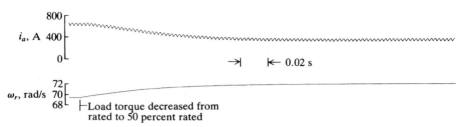

Load torque decreased from rated to 50 percent rated

Figure 12.3-11   Dynamic performance of 3-phase full-converter 200-hp shunt-machine drive following a decrease in load torque from rated to 50% rated.

terminal voltage of 250 V during the same type of load torque change is shown in Fig. 2.5-4. Therein, it is noted that the rotor speed exhibits a damped oscillation about the higher steady-state speed. This oscillation is not present in the speed response shown in Fig. 12.3-11 due to the fact that the commutating inductance appears as an added resistance to the armature circuit (Chapter 11) which increases the damping of rotor oscillations. From Fig. 12.3-11, the full-load speed is 69.6 rad/s and 71.9 rad/s at 50% rated load torque. This is approximately a 3.5% change in speed.

With 250 V applied to the 200-hp motor, the rotor speed is 64.6 rad/s when rated load torque is applied to the rotor (Fig. 2.5-4). This same operating speed may be achieved with the 3-phase converter by delaying the firing of the SCRs and thereby reducing the average voltage applied to the dc machine. This mode of operation is depicted in Fig. 12.3-12 wherein the delay angle is 21.6°. We may approximate this delay angle by using the expression for output average voltage given by (11. 3-16). We know from Section 2.5 that the current of this machine is 633 A with a load torque of 2375 N·m. If we appropriately substitute the values of the source voltage (208 V line to line), commutating inductance (0.045 mH), and armature current (633 A) into (11.3-16) and set this equal to 250 V, we can solve for the delay angle. This calculation yields $\alpha = 21.6°$, which is the same value obtained experimentally.

> **Example 12A**   In Problems 1 and 2 at the end of the chapter, you are asked to describe analytically the approximate dynamic response of the single- and 3-phase full-converter shunt-machine dc drives depicted in Figs. 12.3-5 and 12.3-11, respectively. It is instructive to set the stage for this work. In the case of the 5-hp shunt machine used with the single-phase converter, the armature circuit resistance is increased by the equivalent converter resistance due to the commutating inductance. From (11.2-10) or (11.2-15), the equivalent converter resistance, denoted here as r_c, is
>
> $$r_c = \frac{\omega_e l_c}{\pi} \tag{12A-1}$$
>
> with $l_c = 0.3$ mH, $r_c = 0.036\ \Omega$. Also, the total armature inductance is 0.012 H due to placing an ideal inductance of 0.108 H (denoted as L_x) in series with the

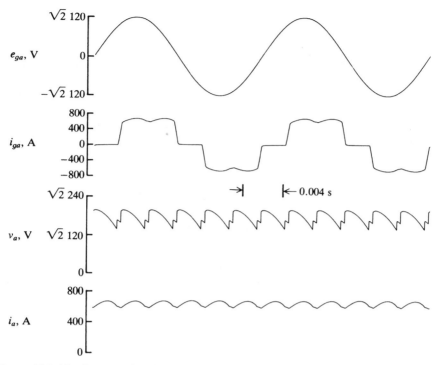

Figure 12.3-12   Same as Fig. 12.3-10 with a phase delay of $21.6°$; $\omega_r = 64.6\,\text{rad/s}$ (rated).

armature. Hence the armature circuit time constant τ_a becomes

$$\tau_a = \frac{L_{AA} + L_x}{r_a + r_c} = \frac{0.012 + 0.108}{0.6 + 0.036} = 0.189\,\text{s} \tag{12A-2}$$

$$\tau_m = \frac{J(r_a + r_c)}{(L_{AF}i_f)^2} = \frac{1(0.6 + 0.36)}{(1.8 \times 1)^2} = 0.196\,\text{s} \tag{12A-3}$$

Therefore, from (2E-6) to (2E-9) in Example 2E (Chapter 2),

$$\omega_n = 5.2\,\text{rad/s} \tag{12A-4}$$

$$\alpha = 2.65 \tag{12A-5}$$

$$D_r = 0.51 \tag{12A-6}$$

$$b_1, b_2 = -2.65 \pm j4.47 \tag{12A-7}$$

The addition of the ideal series inductance to achieve continuous armature current makes the system response (current and speed) underdamped. You will recall in Example 2E that the 5-hp machine was overdamped when operated from a dc source and without the additional inductance and equivalent converter resistance in the armature circuit.

The initial steady-state operating point may be approximated by assuming that the armature current is constant and the operation of the single-phase converter may be described by (11.2-15). Thus, with the load torque equal to 29.2 N·m (full load) and the field current of one ampere, I_a is 16.2 A. The ac source voltage is 280 V (rms); hence from (11.2-3), $V_{d0} = 252$ V. If we add the equivalent resistance of the converter to the resistance of the armature winding, then, because the delay angle is zero we obtain

$$V_{d0} = (r_a + r_c)I_a + \omega_r L_{AF} I_f \tag{12A-8}$$

Solving for ω_r and substituting the appropriate values yield

$$\omega_r = \frac{252 - 16.2(0.6 + 0.036)}{1.8 \times 1} = 134.3 \text{ rad/s} \tag{12A-9}$$

This value is essentially the same as that in Fig. 12.3-5.

In the case of the three-phase converter connected to the 200-hp shunt machine, the commutating inductance is 0.045 mH; thus

$$r_c = \frac{3\omega_e l_c}{\pi} = \frac{3 \times 377 \times 0.045 \times 10^{-3}}{\pi} = 0.016 \, \Omega \tag{12A-10}$$

Hence

$$\tau_a = 0.0125 \text{ s} \tag{12A-11}$$
$$\tau_m = 0.0597 \text{ s} \tag{12A-12}$$
$$\omega_n = 36.6 \text{ rad/s} \tag{12A-13}$$
$$\alpha = 40 \tag{12A-14}$$
$$D_r = 1.09 \tag{12A-15}$$
$$b_1, b_2 = -23.9, -56.1 \tag{12A-16}$$

Here, we see that the added resistance in the armature circuit, due to the commutating inductance, causes the armature current and speed responses to be slightly overdamped. In Example 2E, the 200-hp machine operating from a zero-impedance dc source had slightly underdamped current and speed responses.

With a load torque of 2375 N · m, the steady-state current is 633 A. From (11.3-3), V_{d0} is 280 V for a 208 line-to-line 3-phase ac source. Thus, from (12A-8), the initial steady-state speed is

$$\omega_r = \frac{280 - 633(0.012 + 0.016)}{0.18 \times 20.83} = 69.9 \text{ rad/s} \tag{12A-17}$$

This is essentially the same value of speed obtained in Fig. 12.3-11.

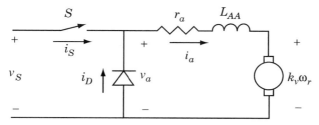

Figure 12.4-1   One-quadrant chopper drive system.

12.4   ONE-QUADRANT dc/dc CONVERTER DRIVE

In the previous section, we were able to refer to the analysis and the steady-state equivalent circuits derived in Chapter 11 to approximate the response of the shunt machine supplied from an ac/dc converter. We have not as yet analyzed dc/dc converters, commonly called choppers, which are used in dc drive systems. In this section, we will analyze the operation and establish the average-value model for a one-quadrant chopper drive. A brief word regarding nomenclature; dc/dc converter and chopper will be used interchangeably throughout the text.

One-Quadrant dc/dc Converter

A one-quadrant dc/dc converter is depicted in Fig. 12.4-1. The switch S is either a SCR with auxiliary turn-off circuitry or a transistor. It is assumed to be ideal. That is, if the switch S is closed, current is allowed to flow in the direction of the arrow; current is not permitted to flow opposite to the arrow. If S is open, current is not allowed to flow in either direction regardless of the voltage across the switch. If S is closed and the current is positive, the voltage drop across the switch is assumed to be zero. Similarly, the diode D is ideal. Therefore, if the diode current i_D is greater than zero, the voltage across the diode, v_a, is zero. The diode current can never be less than zero. In this analysis, it will be assumed that the dc machine is either a permanent magnet or a shunt with constant field current. Hence, k_v is used rather than $L_{AF} i_f$, and the field circuit will not be shown in any of the illustrations.

A voltage control scheme that is often used in dc drives is shown in Fig. 12.4-2. As illustrated in Fig. 12.4-2, a ramp generator provides a sawtooth waveform of

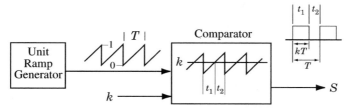

Figure 12.4-2   Switching logic for voltage control of one-quadrant chopper drive shown in Fig. 12.4-1.

period T which ramps from zero to one. This ramp is compared to k, which is referred to as the duty cycle control signal. As the name implies, k is often the output variable of an open- or closed-loop control. The switch S is controlled by the output of the comparator. The duty cycle control signal may vary between zero and one ($0 \leq k \leq 1$). From Fig. 12.4-2, we see that whenever k is greater than the ramp signal, the logic output of the comparator is high and S is closed. This corresponds to the time interval t_1 in Fig. 12.4-2. Now because the ramp signal (sawtooth waveform) varies between zero and one and since $0 \leq k \leq 1$, we can relate k, T, and t_1 as

$$t_1 = kT \qquad (12.4\text{-}1)$$

which may be written as

$$t_1 = \frac{k}{f_s} \qquad (12.4\text{-}2)$$

where f_s is the switching or chopping frequency of the chopper ($f_s = 1/T$).

When k is less than the ramp signal, the logic output is low and S is open. This corresponds to the time interval t_2. Thus, because

$$t_1 + t_2 = T \qquad (12.4\text{-}3)$$

we can write

$$t_2 = (1 - k)T = (1 - k)\frac{1}{f_s} \qquad (12.4\text{-}4)$$

It follows that if k is fixed at one, S is always closed; and if k is fixed at zero, S is always open.

The one-quadrant chopper is unidirectional; and as its name implies, the armature voltage v_a and the armature current i_a can only be positive or zero ($0 \leq v_a, 0 \leq i_a$). In the continuous-current mode of operation, $i_a > 0$. The discontinuous-current mode of operation occurs when i_a becomes zero either periodically or during a transient following a system disturbance.

Continuous-Current Operation

The continuous-current mode of operation for the one-quadrant dc drive is shown in Fig. 12.4-3. The armature current varies periodically between I_1 and I_2. This is considered steady-state operation because v_S, k_v or $\omega_r L_A i_f$, and ω_r are all considered constant. It may at first appear that ω_r cannot be constant because the armature current and thus the electromagnetic torque varies periodically. The rotor speed, however, is essentially constant because the switching frequency of the chopper is generally high so that the change in rotor speed due to the current switching is very small.

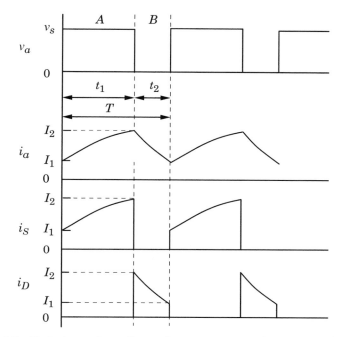

Figure 12.4-3 Typical waveforms for continuous-current steady-state operation of a one-quadrant chopper drive.

The period T in Fig. 12.4-3 is divided into interval A and interval B. During interval A or t_1 $(0 \leq t \leq kT)$, switch S is closed and the source voltage is applied to the armature circuit. During interval B or t_2 $(kT \leq t \leq T)$, switch S is open and the armature winding is short-circuited through the diode D.

During interval A, $v_a = v_S$, $i_a = i_S$, and $i_D = 0$. The armature voltage during this interval is

$$v_S = r_a i_a + L_{AA} \frac{di_a}{dt} + k_v \omega_r \qquad (12.4\text{-}5)$$

where r_a and L_{AA} are the resistance and inductance of the armature circuit, respectively. In (12.4-5), k_v is used to emphasize that the following analysis is for a permanent magnet or a shunt machine with a constant field current. During interval B, $v_a = 0$, $i_S = 0$, and $i_a = i_D$. For this interval we have

$$0 = r_a i_a + L_{AA} \frac{di_a}{dt} + k_v \omega_r \qquad (12.4\text{-}6)$$

Let us solve for the armature current for intervals A and B. From (12.4-5), for interval A we have

$$L_{AA} \frac{di_a}{dt} + r_a i_a = v_S - k_v \omega_r \qquad (12.4\text{-}7)$$

Because v_S and ω_r are assumed constant, the solution of (12.4-7) may be expressed in the form

$$i_a(t) = i_{a,ss} + i_{a,tr} \qquad (12.4\text{-}8)$$

where $i_{a,ss}$ is the steady-state current that would flow if the given interval were to last indefinitely. This current can be calculated by assuming $di_a/dt = 0$, whereupon from (12.4-7) we obtain

$$i_{a,ss} = \frac{v_s - k_v\omega_r}{r_a} \qquad (12.4\text{-}9)$$

The transient component $(i_{a,tr})$ of (12.4-7) is the solution of the homogeneous or force-free equation

$$L_{AA}\frac{di_a}{dt} + r_a i_a = 0 \qquad (12.4\text{-}10)$$

Thus,

$$i_{a,tr} = Ke^{-t/\tau_a} \qquad (12.4\text{-}11)$$

where $\tau_a = L_{AA}/r_a$. Thus, during interval A where $0 \le t \le t_1$ or $0 \le t \le kT$, the armature current may be expressed as

$$i_a = \frac{1}{r_a}(v_s - k_v\omega_r) + Ke^{-t/\tau_a} \qquad (12.4\text{-}12)$$

At $t = 0$ for interval A, $i_a(0) = I_1$ (Fig. 12.4-3), thus

$$I_1 = \frac{1}{r_a}(v_s - k_v\omega_r) + K \qquad (12.4\text{-}13)$$

Solving for K and substituting the result into (12.4-12) yields the following expression for i_a, which is valid for interval A:

$$i_a = I_1 e^{-t/\tau_a} + \frac{(v_s - k_v\omega_r)}{r_a}(1 - e^{-t/\tau_a}) \qquad (12.4\text{-}14)$$

If $T \ll \tau_a$, which is generally the case in most practical dc drives, (12.4-14) can approximated by using the first two terms of the Taylor series $e^x = 1 + x + (1/2)x^2 + \cdots$. In particular, (12.4-14) may be written as

$$i_a \approx I_1(1 - \frac{t}{\tau_a}) + \frac{v_s - k_v\omega_r}{L_{AA}}t \qquad \text{for } T \ll \tau_a \qquad (12.4\text{-}15)$$

At $t = t_1$ or kT, $i_a = I_2$; from (12.4-14) we obtain

$$I_2 = I_1 e^{-kT/\tau_a} + \frac{(v_s - k_v \omega_r)}{r_a} (1 - e^{-kT/\tau_a}) \qquad (12.4\text{-}16)$$

Equation (12.4-16) relates I_2, the current at the end of interval A, to I_1, the current at the beginning of interval A.

During interval B we have

$$L_{AA} \frac{di_a}{dt} + r_a i_a = -k_v \omega_r \qquad (12.4\text{-}17)$$

which is (12.4-6) rewritten. Solving for the steady-state current yields

$$i_{a,ss} = -\frac{k_v \omega_r}{r_a} \qquad (12.4\text{-}18)$$

and $i_{a,tr}$ is still (12.4-11). Thus

$$i_a = -\frac{k_v \omega_r}{r_a} + K e^{-t/\tau_a} \qquad (12.4\text{-}19)$$

For convenience of analysis, we will define a "new" time zero at the beginning of interval B. Thus at this new $t = 0$, we have $i_a = I_2$ and

$$I_2 = -\frac{k_v \omega_r}{r_a} + K \qquad (12.4\text{-}20)$$

Therefore, during interval B, with $t = 0$ at the start of interval B, we have

$$i_a = I_2 e^{-t/\tau_a} - \frac{k_v \omega_r}{r_a} (1 - e^{-t/\tau_a}) \qquad (12.4\text{-}21)$$

If $T \ll \tau_a$, then (12.4-21) may be approximated as

$$i_a \approx I_2 (1 - \frac{t}{\tau_a}) - \frac{k_v \omega_r}{L_{AA}} t \qquad \text{for } T \ll \tau_a \qquad (12.4\text{-}22)$$

Equation (12.4-14) defines the current during interval A, assuming that the initial current, I_1, for this interval is known, whereas (12.4-21) defines the current during interval B, assuming the initial current, I_2, for interval B is known. How can we establish these currents? Well, the initial current during interval B is the final current in interval A. That is, I_2 is calculated from (12.4-14) by setting $t = kT$, giving an expression for I_2 in terms of I_1 (12.4-16). But what determines the value of I_1? At the end of interval B, when $t = t_2$ in (12.4-21), the current i_a must return to I_1 for steady-state operation. Now, from (12.4-4), $t_2 = (1 - k)T$. In other words, $i_a = I_1$ when t in (12.4-21) is $(1 - k)T$. Thus,

$$I_1 = I_2 e^{-(1-k)T/\tau_a} - \frac{k_v \omega_r}{r_a}(1 - e^{-(1-k)T/\tau_a}) \tag{12.4-23}$$

Equations (12.4-16) and (12.4-23) can be used to solve for I_1 and I_2 in terms of v_s, k, T, ω_r, and the machine parameters. In particular, with some work, we can write

$$I_1 = \frac{v_s}{r_a}\left[\frac{e^{-T/\tau_a}(e^{kT/\tau_a} - 1)}{1 - e^{-T/\tau_a}}\right] - \frac{k_v \omega_r}{r_a} \tag{12.4-24}$$

$$I_2 = \frac{v_s}{r_a}\left[\frac{1 - e^{-kT/\tau_a}}{1 - e^{-T/\tau_a}}\right] - \frac{k_v \omega_r}{r_a} \tag{12.4-25}$$

If $T \ll \tau_a$, then (12.4-24) and (12.4-25) may be approximated as

$$I_1 \approx \frac{kv_S}{r_a}(1 - \frac{T}{\tau_a}) - \frac{k_v \omega_r}{r_a} \qquad \text{for } T \ll \tau_a \tag{12.4-26}$$

$$I_2 \approx \frac{kv_S}{r_a} - \frac{k_v \omega_r}{r_a} \qquad \text{for } (T \ll \tau_a) \tag{12.4-27}$$

The current ripple is defined as

$$\Delta I = I_2 - I_1 \tag{12.4-28}$$

Substituting (12.4-24) and (12.4-25) into (12.4-28) yields

$$\Delta I = \frac{v_s}{r_a}\left[\frac{1 - e^{-kT/\tau_a} + e^{-T/\tau_a} - e^{-(1-k)T/\tau_a}}{1 - e^{-T/\tau_a}}\right] \tag{12.4-29}$$

The value of k that produces maximum ripple in the current occurs when $d(\Delta I)/dk = 0$. This yields

$$e^{-kT/\tau_a} - e^{-(1-kT)/\tau_a} = 0 \tag{12.4-30}$$

Solving (12.4-30) for k gives $k = 0.5$ for maximum ripple, an expected result. With $k = 0.5$

$$\Delta I_{\max} = \frac{v_s}{r_a}\left[\frac{1 - 2e^{-T/2\tau_a} + e^{-T/\tau_a}}{1 - e^{-T/\tau_a}}\right] \tag{12.4-31}$$

After considerable work, (12.4-31) can be expressed as

$$\Delta I_{max} = \frac{v_s}{r_a} \tanh \frac{T}{4\tau_a} \qquad (12.4\text{-}32)$$

Note that when the switching period is much less than the armature time constant $(T \ll \tau_a)$, e^{-T/τ_a} approaches unity and ΔI from (12.4-29) approaches zero.

Discontinuous-Current Operation

Steady-state discontinuous-current operation of a one-quadrant chopper is shown in Fig. 12.4-4. In this case, the period T is divided into three intervals. In interval A, where $0 \leq t \leq t_1$ or $0 \leq t \leq kT$, the switch S is closed and the source voltage is applied to the armature circuit. The armature current increases from zero ($I_1 = 0$) to I_2, at which time the ramp signal (Fig. 12.4-2) becomes larger than k and S is opened. The armature current is instantaneously diverted through the diode. During internal B, the armature voltage v_a is zero and the armature current decays to zero, whereupon the diode stops conducting. As in the case of continuous operation, we will select a "new" time zero, for analysis purposes, at the beginning of interval B. Interval B continues for the time t_2. Thus, measuring from the new time zero,

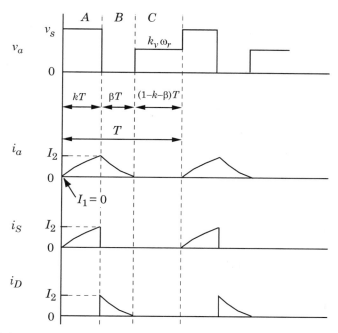

Figure 12.4-4 Typical waveforms for discontinuous-current steady-state operation of a one-quadrant chopper drive.

interval B occurs from $0 \leq t \leq t_2$ or $0 \leq t \leq \beta T$, where $t_2 = \beta T$. Interval C begins at the instant i_a becomes zero and continues to the end of the period T. This time interval is denoted t_3 in Fig. 12.4-4. During interval C, the armature winding is open-circuited and the counter emf $(k_v \omega_r)$ appears at the terminals. That is, during interval C, $v_a = k_v \omega_r$. From Fig. 12.4-4, it is clear that $T = t_1 + t_2 + t_3$. Because $t_1 = kT$ and $t_2 = \beta T$, t_3 may be expressed as $(1 - k - \beta)T$.

The armature current over the interval A can be expressed from (12.4-14) with $I_1 = 0$. Thus

$$i_a = \frac{(v_s - k_v \omega_r)}{r_a}(1 - e^{-t/\tau_a}) \tag{12.4-33}$$

If $T \ll \tau_a$, (12.4-33) may be approximated as

$$i_a \approx \frac{v_s - k_v \omega_r}{L_{AA}}t \qquad \text{for } T \ll \tau_a \tag{12.4-34}$$

During interval B, where, with our new time zero, $0 \leq t \leq \beta T$, we obtain

$$i_a = I_2 e^{-t/\tau_a} - \frac{k_v \omega_r}{r_a}(1 - e^{-t/\tau_a}) \tag{12.4-35}$$

If $T \ll \tau_a$, (12.4-35) may be approximated as

$$i_a \approx I_2(1 - \frac{t}{\tau_a}) - \frac{k_v \omega_r}{L_{AA}}t \qquad \text{for } T \ll \tau_a \tag{12.4-36}$$

When $t = kT$ in (12.4-33) with $i_a = I_2$, we obtain

$$I_2 = \frac{(v_s - k_v \omega_r)}{r_a}(1 - e^{-kT/\tau_a}) \tag{12.4-37}$$

When $t = \beta T$ in (12.4-35) with $i_a = 0$, we have

$$0 = I_2 e^{-\beta T/\tau_a} - \frac{k_v \omega_r}{r_a}(1 - e^{-\beta T/\tau_a}) \tag{12.4-38}$$

Substituting (12.4-37) into (12.4-38) and solving for β yields

$$\beta = \frac{\tau_a}{T} \ln \left[\frac{(v_s - k_v \omega_r)}{k_v \omega_r}(1 - e^{-kT/\tau_a}) + 1 \right] \tag{12.4-39}$$

If $T \ll \tau_a$, (12.4-39) may be approximated as

$$\beta \approx \frac{(v_s - k_v \omega_r)k}{k_v \omega_r} \qquad \text{for } T \ll \tau_a \tag{12.4-40}$$

Recall that $t_3 = (1 - k - \beta)T$; therefore discontinuous-current operation occurs only if

$$1 - k - \beta > 0 \tag{12.4-41}$$

If (12.4-40) is substituted into (12.4-41), discontinuous-current operation will occur approximately when

$$k_v \omega_r > k v_S \qquad \text{for } T \ll \tau_a \tag{12.4-42}$$

At first glance, we might be led to believe from (12.4-42) that the occurrence of discontinuous-current operation is independent of the switching period T. Clearly, this is not the case because increasing T will tend to increase the time of discontinuous-current operation. In this approximation, we have made T very small, whereupon (12.4-42) can be used to approximate when discontinuous-current operation will occur if $T \ll \tau_a$. Actually, (12.4-42) is quite logical once we realize that $k v_S$ is the average armature voltage over a period T if the current is continuous.

Average-Value Analysis

It may at first appear that the harmonics introduced by the switching of the chopper would drastically complicate the analysis of dc/dc converter drive systems. Fortunately, this is generally not the case. Because the switching period (T) of the chopper is generally much smaller than the time constant of the armature circuit (τ_a) and always much smaller than the time constant of the mechanical system (τ_m), it is possible to describe the performance of a dc/dc converter drive system with average-value variables. This can be accomplished by establishing a continuous average-value equivalent of the armature voltage v_a. Once this continuous average value of v_a, which we will denote $\bar{v}_a$, is determined, we will show that the voltage and torque equations that have previously been derived may be used directly to predict the average-value dynamic and steady-state responses of the dc drive system.

The goal of "average valuing" is to establish continuous averages of the waveforms (v_a, i_a, and ω_r) which predict the average-value performance of the dc drive during dynamic and steady-state operation. Our first step in establishing justification for replacing the switched and discontinuous variables with average-value equivalents is to follow a procedure similar to that used in reference 2. In particular, we will define the dynamic average of a variable as

$$f_{(da)} = \frac{1}{T} \int_{t-T}^{t} f(\xi) \, d\xi \tag{12.4-43}$$

where ξ is a dummy variable and, for our purpose, T is the switching period of the chopper and f can represent either v_a, i_a, or ω_r. The averaging process is achieved by continuously performing (12.4-43) on all variables as the averaging interval, $(t - T)$, moves or slides with time. Therefore, the result of this process is often referred to as

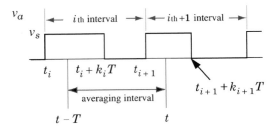

Figure 12.4-5 Averaging interval for $t_{i+1} \leq t \leq t_{i+1} + k_{i+1}T$; continuous-current operation of one-quadrant chopper drive.

the moving or sliding average rather than the dynamic average, which we will use here.

Our next step is to derive the dynamic average value of the armature voltage which would be valid for period-to-period changes in the duty cycle k with v_S constant. For this purpose, the change in k is depicted in Fig. 12.4-5 for continuous-current operation of a one-quadrant chopper. Therein, t_i is the beginning of the ith interval wherein the duty cycle is k_i. During the next interval, the ith + 1 interval, which begins at t_{i+1}, the duty cycle is denoted k_{i+1}. In Fig. 12.4-5, we see that $k_{i+1} < k_i$; hence the switch S in Fig. 12.4-1 is closed longer during the ith interval than during the ith + 1 interval. Superimposed upon the plot of v_a is the averaging interval, which is equal in length to the switching period T and which begins at $t - T$ and ends at t. In order to establish the dynamic average of v_a, we will let this averaging interval move from $t = t_{i+1}$ to $t = t_{i+1} + T$.

The averaging interval will yield the same dynamic average for v_a as it is moved to the right from $t = t_{i+1}$ until v_a becomes zero in the $i + 1$ interval at $t = t_{i+1} + k_{i+1}T$. This dynamic average of the armature voltage may be expressed as

$$
\begin{aligned}
v_{a(da)} &= \frac{1}{T} \left[\int_{t-T}^{t_i+k_iT} v_S \, d\xi + \int_{t_{i+1}}^{t} v_S \, d\xi \right] \\
&= \frac{1}{T} [v_S(t_i + k_iT - t + T) + v_S(t - t_{i+1})] \\
&= k_i v_S \qquad \text{for } t_{i+1} \leq t \leq t_{i+1} + k_{i+1}T \qquad (12.4\text{-}44)
\end{aligned}
$$

where ξ is a dummy variable of integration. In the evaluation of (12.4-44), we use the fact that $t_i + T = t_{i+1}$. The result that $v_{a(da)} = k_i v_S$ for $t_{i+1} \leq t \leq t_{i+1} + k_{i+1}T$ is expected. This of course would be the average taking over the interval T from t_i to t_{i+1}. Clearly, this average will not change as the averaging interval is moved to the right until a change is encountered at t that is not encountered at $t - T$ as is the case at $t = t_{i+1} + k_{i+1}T$.

Once t becomes larger than $t_{i+1} + k_{i+1}T$, the averaging interval is positioned as shown in Fig. 12.4-6. The leading edge of the sliding averaging interval includes $v_a = 0$ in the ith + 1 interval; the trailing edge value is v_S until $t - T = t_i + k_iT$ which may be written as $t = t_{i+1} + k_iT$. Thus, for the time interval from

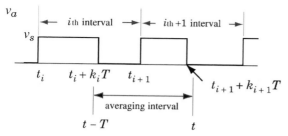

Figure 12.4-6   Averaging interval for $t_{i+1} + k_{i+1}T \leq t \leq t_{i+1} + k_iT$; continuous-current operation of one-quadrant chopper drive.

$t_{i+1} + k_{i+1}T \leq t \leq t_{i+1} + k_iT$, the dynamic average of v_a may be expressed as

$$v_{a(da)} = \frac{1}{T} \left[\int_{t-T}^{t_i+k_iT} v_S \, d\xi + \int_{t_{i+1}}^{t_{i+1}+k_{i+1}T} v_S \, d\xi + \int_{t_{i+1}+k_{i+1}T}^{t} 0 \, d\xi \right]$$

$$= \frac{1}{T} [v_S(t_i + k_iT - t + T) + v_S(t_{i+1} + k_{i+1}T - t_{i+1})]$$

$$= v_S \left[\frac{t_{i+1} - t}{T} + (k_i + k_{i+1}) \right] \quad \text{for } t_{i+1} + k_{i+1}T \leq t \leq t_{i+1} + k_iT \quad (12.4\text{-}45)$$

As the trailing edge of the averaging interval slides past $t - T = t_i + k_iT$, only the v_S, which appears in the ith $+1$ interval, is within the averaging interval. Thus, for $t_{i+1} + k_iT \leq t \leq t_{i+1} + T$, the dynamic average of v_a is $k_{i+1}v_S$. Let us show this as

$$v_{a(da)} = \frac{1}{T} \int_{t-T}^{t_{i+1}} 0 v_S \, d\xi + \frac{1}{T} \int_{t_{i+1}}^{t_{i+1}+k_{i+1}T} v_S \, d\xi + \frac{1}{T} \int_{t_{i+1}+k_{i+1}T}^{t} 0 \, d\xi$$

$$= \frac{1}{T} [v_S(t_{i+1} + k_{i+1}T - t_{i+1})]$$

$$= k_{i+1}v_S \quad \text{for } t_{i+1} + k_iT \leq t \leq t_{i+1} + T \quad (12.4\text{-}46)$$

After we have gone through the dynamic averaging process, we realize that we could have written the results given in (12.4-44) and in (12.4-46) by inspection and concern ourselves only with finding the dynamic average during the transition interval $t_{i+1} + k_{i+1}T \leq t \leq t_{i+1} + k_iT$.

A plot of the dynamic average of v_a as a function of time is piecewise linear as shown in Fig. 12.4-7. In general, $v_{a(da)}$ is a staircase with ramp changes occurring

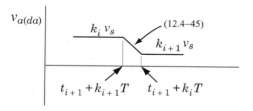

Figure 12.4-7 $v_{a(da)}$ versus t for v_a shown in Figs. 12.4-5 and 12.4-6.

once per switching cycle due to changes in k. If the change in duty cycle is small from one period to the next, $v_{a(da)}$ will appear essentially continuous, equal to kv_S. It is important to recall that if we took the average (not the sliding average) of v_a over the ith interval, we would get $k_i v_S$ and over the ith $+ 1$ interval we would get $k_{i+1}v_S$. In other words, the period-to-period average value of v_a, which we will denote as $\bar{v}_a$ and refer to as the average value of v_a, is

$$\bar{v}_a = kv_S \qquad (12.4\text{-}47)$$

Let us repeat. It appears that $\bar{v}_a$, which is continuous, is essentially equal to the dynamic average, $v_{a(da)}$, if the change in k is small from one period to the next, which is generally the case. Let us look more closely at this. For this purpose, it is interesting to compare the period-to-period average armature voltage (average-value), $\bar{v}_a$, and the dynamic average, $v_{a(da)}$, obtained by continuously performing the moving average (12.4-43). This is illustrated in Fig. 12.4-8. Here v_a, $v_{a(da)}$, and $\bar{v}_a$ are shown for $v_S = 10$ V and $T = 0.1$ ms, and k is expressed as

$$k = 0.5 + 0.5 \sin\left[r(t)t\right] \qquad (12.4\text{-}48)$$

where $r(t)$ is a ramp signal. In particular, $r(t) = k_r t$; it is left to the reader to determine k_r. Before proceeding, it is important to note the difference between $v_{a(da)}$ and $\bar{v}_a$ near time zero. At $t = 0$, $\bar{v}_a$ is 0.5 v_S; however, $v_{a(da)}$ is zero. Recall that at time zero the averaging interval is over the period before time zero ($t - T$ to t). Therefore $v_{a(da)}$ is zero until t increases from zero and the interval begins its slide with t, whereupon $v_{a(da)}$ ramps to slightly larger than 5 V in the first period T.

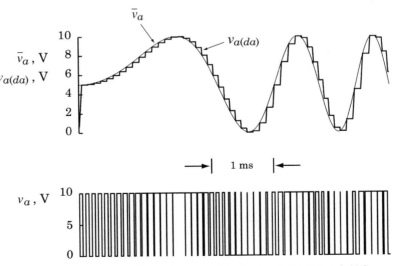

Figure 12.4-8   Comparison of $v_{a(da)}$ and $\bar{v}_a$ for $k = 0.5 + 0.5 \sin[r(t)t]$.

When the frequency of k (12.4-48) is small, changing slowly relative to the switching, $\bar{v}_a$ is an excellent approximation of $v_{a(da)}$. However, when the frequency of the variation in k approaches the switching frequency, the duty cycle k can change significantly from one switching interval to the next and $\bar{v}_a$ is less accurate in approximating $v_{a(da)}$. In most applications, k and v_S do not change significantly from one switching period to the next and $v_{a(da)}$ is sufficiently approximated by $\bar{v}_a$. Therefore, we need not use the sliding averaging interval to determine $v_{a(da)}$ for a given mode of converter operation. Instead we will evaluate the $\bar{v}_a$ over one period and assume that it will be a valid approximation for slow changes in k and v_S relative to the switching frequency. Let us apply this to determine $\bar{v}_a$ for discontinuous-current operation (Fig. 12.4-4). Here

$$
\bar{v}_a = \frac{1}{T} \left[\int_0^{kT} v_S \, d\xi + \int_{(k+\beta)T}^{T} k_v \bar{\omega}_r \, d\xi \right]
$$
$$
= k v_S + (1 - k - \beta) k_v \bar{\omega}_r \qquad (12.4\text{-}49)
$$

We have established that $\bar{v}_a$ can be used for $v_{a(da)}$ and we will now assume that $\bar{v}_a$ will give rise to $\bar{i}_a$ (or $\bar{T}_e$) and $\bar{\omega}_r$, which all together will portray the average-value dynamic and steady-state performance of the dc drive. This seems to be a logical assumption and we will omit its proof; however, for those who might question this, the details are given in reference 2. We can now predict the average-value dynamic and steady-state performance of the one-quadrant drive for continuous- and discontinuous-current mode of operation using the average-value time-domain block diagram shown in Fig. 12.4-9.

Operating Characteristics

It is instructive to observe the performance of a dc drive supplied from a one-quadrant chopper and to compare the detailed system response with that predicted by the average-value model that we have developed. The parameters of the fractional horsepower dc machine are $r_a = 3.8 \, \Omega$, $L_{AA} = 1.14 \, \text{mH}$, $k_v = 0.031 \, \text{V} \cdot \text{s/rad}$, $J = 1.41 \times 10^{-5} \, \text{N} \cdot \text{m} \cdot \text{s}^2$, $B_m = 2.82 \times 10^{-4} \, \text{N} \cdot \text{m} \cdot \text{s/rad}$, and $T_L = 0$. The source voltage, v_S, is 12 V. A computer study that depicts the one-quadrant chopper drive is

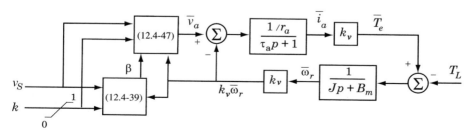

Figure 12.4-9 Average-value model for a one-quadrant chopper drive.

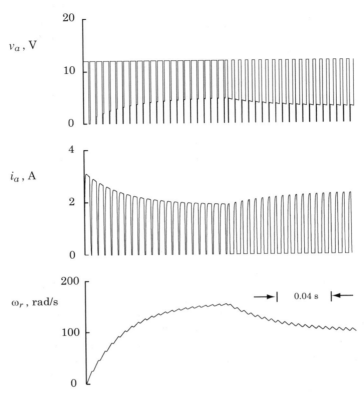

Figure 12.4-10 Operating characteristics of a one-quadrant chopper drive.

shown in Fig. 12.4-10, wherein v_a, i_a, and ω_r are plotted. Initially, the rotor speed is zero with k fixed at zero. The duty cycle is stepped to $k = 0.8$; the machine accelerates and approaches steady-state operation. The duty cycle is then stepped to $k = 0.4$, and the rotor speed decreases and again approaches steady-state operation. The last few cycles of this study are shown in Fig. 12.4-11 with an expanded time scale.

It is noted that discontinuous operation occurs throughout this study. This was done intentionally by selecting a switching frequency of 200 Hz which is at least an order of magnitude lower than normal. This makes the operation (switching) of the converter more clearly discerned, and the discontinuous operation provides the most difficult mode of operation for comparison purposes with the average-value model response that is shown in Fig. 12.4-12. This computer study was performed by simulating the time-domain block diagram of the one-quadrant chopper drive shown in Fig. 12.4-9. It appears that this average-value model adequately predicts the average performance of this system; however, there is something that can be done to make this comparison even more meaningful. It is difficult to compare $\bar{v}_a$ and $\bar{i}_a$ (Fig. 12.4-12) with v_a and i_a (Fig. 12.4-10). Fortunately, the dynamic averaging process can be used to aid with this comparison. If we perform the dynamic averaging

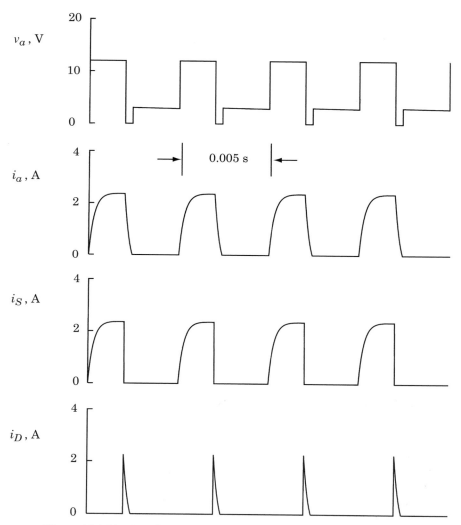

Figure 12.4-11   Last few cycles of Fig. 12.4-10 with expanded time scale.

(moving or sliding averaging) process as described by (12.4-43) upon each of the variables (v_a, i_a, and ω_r) shown in Fig. 12.4-10, we will obtain $v_{a(da)}$, $i_{a(da)}$ and $\omega_{r(da)}$ as shown in Fig. 12.4-13. The comparison of $\bar{v}_a$ and $\bar{i}_a$ with $v_{a(da)}$ and $i_{a(da)}$, respectively, establishes clearly the accuracy of the average-value model. Please realize that by selecting T much larger than normal to intentionally insure discontinuous-current operation for the machine parameters and load conditions, a very strict test has been imposed upon the accuracy of average-value modeling. From the comparison of Figs. 12.4-10, 12.4-12, and 12.4-13, it is clear that the

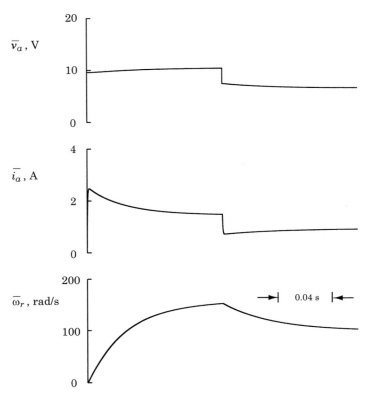

Figure 12.4-12 Same as Fig. 12.4-10 using the average-value model shown in Fig. 12.4-9.

average-value model can be used without reservation to analyze and design a one-quadrant chopper drive system for most modes of operation.

Example 12B A change in source voltage is depicted in Fig. 12B-1. During the ith interval the source voltage is v_{Si} and $v_{S(i+1)}$ during the ith $+ 1$ interval. The duty cycle k and the switching period are constant. Let us determine the dynamic average. We realize that for $t \leq t_{i+1}$ the dynamic average is kv_{Si} and for $t - T \geq t_i + kT$ the dynamic average is $kv_{S(i+1)}$. During the interval for $t_{i+1} \leq t \leq t_{i+1} + kT$, the dynamic average is

$$v_{a(da)} = \frac{1}{T} \left[\int_{t-T}^{t_i+kT} v_{Si} \, d\xi + \int_{t_{i+1}}^{t} v_{S(i+1)} \, d\xi \right]$$

$$= \frac{1}{T} \left[v_{Si}(t_i + kT - t + T) + v_{S(i+1)}(t - t_{i+1}) \right]$$

$$= kv_{Si} + (v_{Si} - v_{S(i+1)}) \left(\frac{t_{i+1} - t}{T} \right) \qquad \text{for } t_{i+1} \leq t \leq t_{i+1} + kT \quad (12B\text{-}1)$$

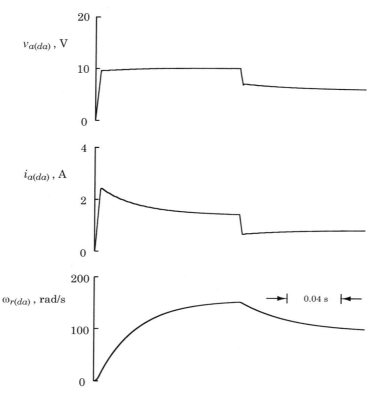

Figure 12.4-13   Dynamic average of variables shown in Fig. 12.4-10.

A plot of (12B-1) is shown in Fig. 12B-2. It is interesting to compare $v_{a(da)}$ and $\bar{v}_a$ for

$$v_S = 5 + 5 \sin r(t)t \qquad (12\text{B-}2)$$

The comparison is shown in Fig. 12B-3, wherein $T = 0.1$ ms and $k = 0.6$. The plots in Fig. 12B-3 depict a comparison of $v_{a(da)}$ and $\bar{v}_a$ similar to that noted for

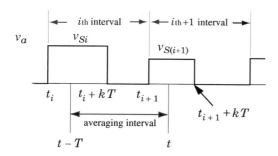

Figure 12B-1   Dynamic averaging of v_a for a change in source voltage v_S.

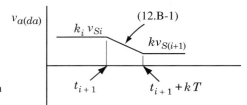

Figure 12B-2 $v_{a(da)}$ versus t for v_a shown in Fig. 12B-1.

a change in k (Fig. 12.4-8). The average voltage $\bar{v}_a$ closely approximates $v_{a(da)}$ until the frequency of the change in v_S approaches the switching frequency.

12.5 TWO-QUADRANT dc/dc CONVERTER DRIVE

A simplified schematic diagram of a two-quadrant chopper supplying a dc machine is shown in Fig. 12.5-1. Typical waveforms of the converter variables during steady-state operation are shown in Fig. 12.5-2. With a two-quadrant chopper, the armature voltage cannot be negative $(v_a \geq 0)$; however, the armature current can be positive

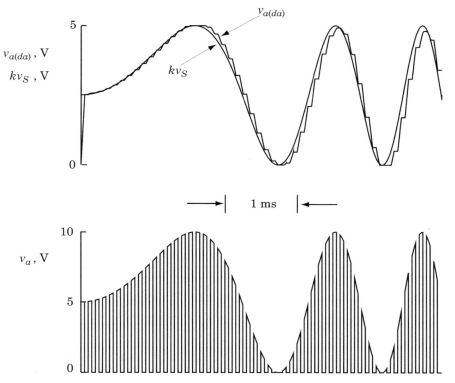

Figure 12B-3 Comparison of $v_{a(da)}$ and $\bar{v}_a$ for $v_S = 5 + 5\sin[r(t)t]$.

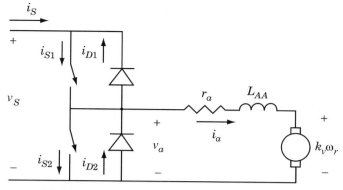

Figure 12.5-1   Two-quadrant chopper drive system.

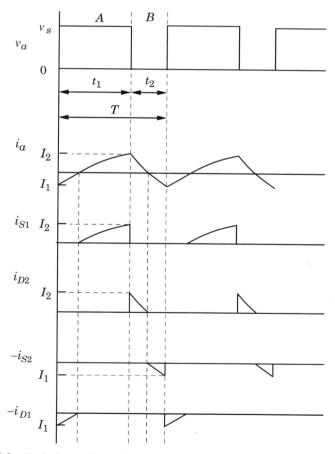

Figure 12.5-2   Typical waveforms for steady-state operation of a two-quadrant chopper drive.

or negative. That is, I_1 and I_2 (Fig. 12.5-2) can both be positive, whereupon the two-quadrant chopper is operating as a continuous-current one-quadrant chopper, or I_1 can be negative and I_2 positive, or I_1 and I_2 can both be negative. In Fig. 12.5-2, I_1 is negative and I_2 is positive and the average value of i_a is positive. Because the average value of v_a is positive, the mode of operation depicted is motor action if ω_r is positive (ccw).

In a two-quadrant chopper, there are two switches and two diodes arranged as shown in Fig. 12.5-1. As in the case of the one-quadrant chopper, we will assume that these devices are ideal. The switching logic is generated from the duty cycle k as shown in Fig. 12.4-2. When the comparator output signal is high, $S1$ is closed and $S2$ is open (interval A in Fig. 12.5-2); when it is low, $S1$ is open and $S2$ is closed (interval B in Fig. 12.5-2). There is a practical consideration that must be mentioned. Electronic switches have finite turn-off and turn-on times. The turn-off time is generally longer than the turn-on time. Therefore, the switching logic must be arranged so that the turn-on signal is delayed in order to prevent short-circuiting the source causing "shoot through." Although the delay is very short, it must be considered in the design; however, it does not make our analysis invalid wherein we will assume instant-on, instant-off operation.

It is important to discuss the mode of operation depicted in Fig. 12.5-2. During interval A, $S1$ is closed and $S2$ is open; and, at the start of interval A we have $i_a = I_1$, which is negative in Fig. 12.5-2. Because $S2$ is open, a negative $i_a(I_1)$ can only flow through $D1$. It is important to note that $-i_{D1}$ and $-i_{S2}$ are plotted in Fig. 12.5-2 to allow ready comparison with the waveform of i_a. Let us go back to the start of interval A. How did i_a become negative? Well during the interval B in the preceding period, $S2$ was closed with $S1$ open. With $S2$ closed, the armature terminals are short-circuited and the counter emf has driven i_a negative. Therefore, when $S1$ is closed and $S2$ is opened at the start of interval A, the source voltage has to contend with this negative I_1. We see from Fig. 12.5-2 that the average value of i_a is slightly positive. Therefore, v_S is larger than the counter emf; and at the start of interval A when v_S is applied to the machine, the armature current begins to increase toward zero from the negative value of I_1. Once i_a reaches zero, the diode $D1$ blocks the current flow. That is, i_{D1} cannot become negative; however, $S1$ has been closed since the start of interval A; and because i_{S1} can only be positive, $S1$ is ready to carry the positive i_a. The armature current, which is now i_{S1}, continues to increase until the end of interval A (I_2).

At the beginning of interval B, $S1$ is opened and $S2$ is closed; however, $S2$ cannot conduct a positive armature current. Therefore, the positive current (I_2) is diverted to diode $D2$, which is short-circuiting the armature terminals. Now, the counter emf has the positive current (I_2) with which to contend. It is clear that if the armature terminals were permanently short-circuited, the counter emf would drive i_a negative. At the start of interval B, the counter emf begins to do just that; however, when i_a becomes zero, diode $D2$ blocks i_{D2} and the negative armature current is picked up by $S2$, which has been closed since the beginning of interval B, waiting to be called upon to conduct a negative armature current. This continues until the end of interval B, whereupon we are back to where we started.

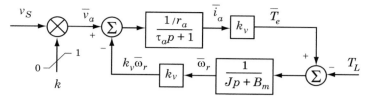

Figure 12.5-3   Average-value model of a two-quadrant chopper drive.

It is apparent that if the mode of operation is such that I_1 and I_2 are both positive, then the machine is acting as a motor with a substantial load torque if ω_r is positive (ccw). In this mode, either S1 or D2 will carry current during a switching period T. If both I_1 and I_2 are negative, the machine is operating as a generator, delivering power to the source if ω_r is positive (ccw). In this case, either S2 or D1 will carry current during a switching period.

From Fig. 12.5-2, we see that each switching period is divided into interval A and interval B. This is identical to that shown in Fig. 12.4-3 for a one-quadrant chopper in the continuous-current mode of operation. Hence, the relationships developed in the previous section for continuous-current operation of a one-quadrant chopper apply directly to the two-quadrant chopper. The average-value time-domain block diagram for the two-quadrant chopper drive system is shown in Fig. 12.5-3.

12.6   FOUR-QUADRANT dc/dc CONVERTER DRIVE

A simplified schematic diagram of a four-quadrant chopper drive system is shown in Fig. 12.6-1. Typical steady-state waveforms that depict the operation of the converter are shown in Fig. 12.6-2. As the name implies, four-quadrant operation (current versus voltage) is possible. That is, the instantaneous armature current i_a and the instantaneous armature voltage may be positive or negative. In fact, four-quadrant operation is depicted in each switching period in Fig. 12.6-2. In particular, I_1 is negative and I_2 is positive and v_a is v_S during interval A and $-v_S$ during interval B; however, the average v_a and the average i_a are positive. Therefore, from an average-value

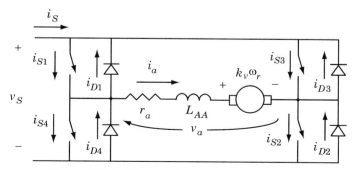

Figure 12.6-1   Four-quadrant chopper drive system.

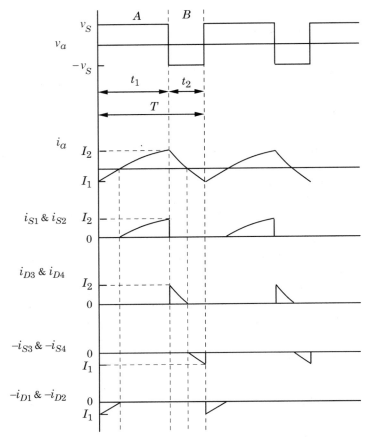

Figure 12.6-2 Typical waveforms for steady-state operation of a four-quadrant chopper drive.

point of view, the dc drive system depicted in Fig. 12.6-2 is operating as a motor if the rotor speed ω_r is positive (ccw). This is 1st quadrant operation of an average-current versus average-voltage plot even though four-quadrant operation of i_a versus v_a occurs each switching period. One must distinguish between four-quadrant operation during a period and four-quadrant average-value operation.

If v_a is positive and i_a is negative (average or instantaneous), then operation is in the 4th quadrant of an i_a versus v_a plot; and if ω_r is positive (ccw), the machine is operating as a generator. In the 2nd quadrant, v_a is negative and i_a is positive; and if ω_r is negative (cw), we have generator operation. In the 3rd quadrant, v_a is negative and i_a is negative; and if ω_r is negative (cw), we have motor operation. It is important to emphasize that one-, two-, and four-quadrant chopper operation has been defined from the plot of i_a versus v_a. However, the T_e versus ω_r plot is also used to define drive operation where the 1st and 3rd quadrants depict motor operation and the 2nd and 4th quadrants depict generator operation. At first glance, one might assume that the quadrants of the i_a versus v_a plot can be assigned the same modes of operation. Actually this is only the case if ω_r is positive (ccw) when v_a is positive and negative (ccw) when v_a is negative as stipulated above. In order to illustrate this, let us assume that v_a and i_a are both positive and ω_r is positive (ccw); the machine is operating as a motor, and operation is in the 1st quadrant of i_a versus v_a and in the 1st quadrant of T_e versus ω_r. If v_a and i_a are positive and ω_r is zero, the power $(v_a i_a)$ is being dissipated in r_a and the machine is neither a motor nor a generator. If, however, ω_r is made slightly negative by supplying an input torque, v_a and i_a can still both be positive (1st quadrant) and yet generator action is occurring because T_e is positive and ω_r is negative (2nd quadrant of T_e versus ω_r). Therefore, we must know the direction (sign) of ω_r when assigning motor or generator action to the four quadrants of the i_a versus v_a plot.

There are numerous switching strategies that might be used with a four-quadrant chopper. The switching depicted in Fig. 12.6-2 is perhaps one of the least involved. In this case, there are only two states. In the first state, which occurs over interval A, $S1$ and $S2$ are closed and $S3$ and $S4$ are open. The second state occurs over interval B, wherein the $S3$ and $S4$ are closed and $S1$ and $S2$ are open. As in the case of the previous dc/dc converters, we will consider the switches and diodes as being ideal.

During interval A, $S1$ and $S2$ are closed and $S3$ and $S4$ are open. At the beginning of the interval, i_a is negative (I_1) in Fig. 12.6-2. Because $S1$ and $S2$ cannot carry negative armature current, I_1 must flow through diodes $D1$ and $D2$. Note in Fig. 12.6-2 that $-i_{D1}$, $-i_{D2}$, $-i_{S3}$, and $-i_{S4}$ are plotted for the purpose of a direct comparison with i_a. During interval A, the armature voltage v_a is v_S; and because v_S is larger than the counter emf, the armature current increases from the negative value of I_1 toward zero. During this part of the interval, the source current is $-i_{D1}$, which is also $-i_{D2}$. When i_a reaches zero, $D1$ and $D2$ block positive armature current flow; however, $S1$ and $S2$ are closed ready to carry a positive i_a. Hence, the current increases from zero to I_2 through $S1$ and $S2$. During this part of the interval, the source current i_S is i_{S1}, which is also i_{S2}.

During interval B, v_a is $-v_S$ and are $S1$ and $S2$ are open with $S3$ and $S4$ closed. At the beginning of interval B, i_a is positive (I_2); however, $S3$ and $S4$ cannot conduct a positive armature current. Hence at the beginning of interval B the positive I_2 flows through diodes $D3$ and $D4$. This continues until i_a is driven to zero by $-v_S$. During this part of interval B, the source current i_S is $-i_{D3}$ or $-i_{D4}$. When i_a reaches zero, diodes $D3$ and $D4$ block negative i_a; thus, $S3$ and $S4$ carry the negative armature current to the end of interval B where $i_a = I_1$, which is negative. During this part of interval B, the source current i_S is i_{S3} or i_{S4}. We have completed a switching cycle.

Expressions for I_1 and I_2 can be derived by a procedure similar to that used in the case of the previous choppers. It can be shown that

$$I_1 = \frac{v_s}{r_a}\left[\frac{2e^{-(1-k)T/\tau_a} - e^{-T/\tau_a} - 1}{1 - e^{-T/\tau_a}}\right] - \frac{k_v \omega_r}{r_a} \tag{12.6-1}$$

$$I_2 = \frac{v_s}{r_a}\left[\frac{1 - 2e^{-kT/\tau_a} + e^{-T/\tau_a}}{1 - e^{-T/\tau_a}}\right] - \frac{k_v \omega_r}{r_a} \tag{12.6-2}$$

If k and v_S do not change significantly from one switching period to the next, the average armature voltage may be expressed as

$$
\begin{aligned}
\bar{v}_a &= \frac{1}{T}\left[\int_0^{kT} v_S \, d\xi + \int_{kT}^{T} -v_S \, d\xi\right] \\
&= \frac{1}{T}[kTv_S - (1-k)Tv_S] \\
&= (2k-1)v_S
\end{aligned}
\tag{12.6-3}
$$

Note that when $k = 0$, $v_a = -v_S$ and when $k = 1$, $v_a = v_S$. It is clear that the time-domain block diagram for the four-quadrant chopper drive is the same as that shown in Fig. 12.5-3 for the two-quadrant chopper drive with $v_a = kv_S$ replaced with $v_a = (2k-1)v_S$ which is (12.6-3).

12.7 MACHINE CONTROL WITH VOLTAGE-CONTROLLED dc/dcCONVERTER

Although we will not become involved in the design of closed-loop controls, it is important to discuss typical control systems and to set the stage for control design using the average-value models. When the dc/dc converter is used to control the voltage applied to the armature circuit, v_a, through the duty cycle, k, it is a voltage-controlled dc/dc converter. We will continue with this type of converter in this section. In the following section, we will introduce the so-called current-controlled dc/dc converter, which is generally implemented by appropriate control of a four-quadrant dc/dc converter.

Speed Control—Voltage-Controlled dc/dc Converter

A time-domain block diagram of a machine speed control with a voltage-controlled chopper is shown in Fig. 12.7-1. Therein, ω_r^* is the reference or commanded speed.

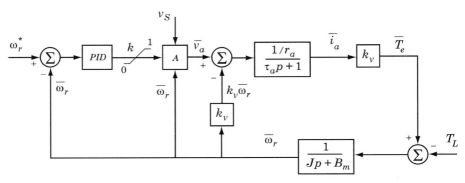

Figure 12.7-1   Speed control of dc machine with voltage-controlled dc/dc converter.

The error signal is supplied to a basic controller herein noted as a *PID* (proportional-integral-derivative) controller. The output of the *PID* controller is the duty cycle, which, along with the source voltage and the rotor speed, is an input to block *A*. The output of block *A* is the average armature voltage. The rest of the block diagram is self-explanatory.

It is clear that the calculation of the average armature voltage which occurs in block *A* will depend upon the type of chopper being used. For a one-quadrant chopper, the calculation performed in block *A* is

$$\bar{v}_a = kv_S + (1 - k - \beta)k_v\bar{\omega}_r \tag{12.7-1}$$

For a two-quadrant chopper we have

$$\bar{v}_a = kv_S \tag{12.7-2}$$

For a four-quadrant chopper we have

$$\bar{v}_a = (2k - 1)v_S \tag{12.7-3}$$

The average rotor speed $\bar{\omega}_r$ is shown as an input to block *A*. It is needed to calculate β for the one-quadrant chopper; however, it is not needed to calculate $\bar{v}_a$ for the two- and four-quadrant choppers. It is clear that only the four-quadrant chopper can be used for bidirectional speed control.

Speed Control—Voltage-Controlled dc/dc Converter with Feedforward Voltage Control

A time-domain block diagram of a machine speed control with a voltage-controlled chopper equipped with feedforward voltage control is shown in Fig. 12.7-2. This block diagram is identical to that shown in Fig. 12.7-1 except for the addition of block *B* between the *PID* controller and block *A*. In this case, the *PID* controll establishes the desired or commanded $\bar{v}_a^*$ from the speed error signal. The source voltage and average rotor speed are inputs to block *B* along with $\bar{v}_a^*$. The output of block *B* is the duty cycle, which is an input to block *A* as in Fig. 12.7-1.

The calculations to be performed in block *B* are obtained by replacing $\bar{v}_a$ with $\bar{v}_a^*$ in (12.7-1)–(12.7-3) and then solving each equation for the duty cycle *k*. In

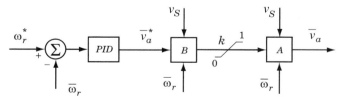

Figure 12.7-2 Same as Fig. 12.7-1 with voltage feedforward control.

particular, for a one-quadrature chopper the calculation performed in block B is

$$k = \frac{\bar{v}_a^* - k_v \bar{\omega}_r (1 - \beta)}{v_s - k_v \bar{\omega}_r} \qquad \text{with } 0 \leq \bar{v}_a^* \leq v_S \qquad (12.7\text{-}4)$$

For a two-quadrant chopper we have

$$k = \frac{\bar{v}_a^*}{v_S} \qquad \text{with } 0 \leq \bar{v}_a^* \leq v_s \qquad (12.7\text{-}5)$$

For the four-quadrant chopper we have

$$k = \frac{1}{2} + \frac{\bar{v}_a^*}{2v_S} \qquad \text{with } -v_S \leq \bar{v}_a^* \leq v_S \qquad (12.7\text{-}6)$$

As in the case of block A, the average rotor speed is not needed in block B for the two- and four-quadrant choppers.

The purpose of the feedforward (block B) is clear once we consider the response of the speed control shown in Fig. 12.7-1 to a change in source voltage, v_S. If, for example, v_S decreases, then $\bar{v}_a$, the output of block A (Fig. 12.7-1), decreases by the same amount because the duty cycle k will not change to compensate for the decrease in v_S until the rotor speed has decreased and the action of the PID controller increases k. With the feedforward voltage control, the calculation in block B is aware of the decrease in v_S and instantaneously increases k to compensate for this decrease. Hence, $\bar{v}_a$ remains essentially unaware of a change in v_S; consequently the average torque and thus the speed are not affected by a change in source voltage.

It should be mentioned that in the case of the one-quadrant chopper, the calculation in block B may not be that given by (12.7-4). Instead, changes in source voltage may not be completely compensated for during discontinuous-current operation. In particular, (12.7-4) may be replaced by (12.7-5), whereupon there would be some transient change in speed due to a change in source voltage whenever the chopper is operating in the discontinuous-current mode.

12.8 MACHINE CONTROL WITH CURRENT-CONTROLLED dc/dc CONVERTER

Current-control of dc machines is very desirable because by controlling the current, the torque is controlled. Moreover, current control can be used to prevent large damaging armature currents during start-up. Some form of current control or limiting can be achieved using the one- and two-quadrant choppers; however, the four-quadrant chopper is the most appropriate for current control because the polarity of the armature voltage can be reversed, thereby allowing the maximum possible current-control capability. It is not our purpose to become involved in the design, operation, and application of current-controlled dc/dc converters. Nevertheless, we will set forth the operation of the four-quadrant current-controlled chopper, from an ideal point of view, and, where appropriate, establish an average-value model that may be used for analysis and design purposes.

Current-Controlled Four-Quadrant dc/dc Converter

It is recalled from our discussion of the four-quadrant chopper (Fig. 12.6-1) that the source current equals i_a ($i_S = i_a$) where $S1$ and $S2$ are closed and $i_S = -i_a$ when $S3$ and $S4$ are closed. Because we wish to control i_a, it is necessary to measure it directly or establish it from a measurement of i_S. The latter is generally done, as shown in Fig. 12.8-1, by picking up a signal proportional to i_S from a resistive shunt placed in the i_S path on the low voltage (negative) side of the source voltage. That is, because the low side of the source voltage is grounded, the resistive shunt is connected in series with i_S between the bottom connection of $S4$ and $D4$ and ground. The demodulator shown in Fig. 12.8-1 is used to obtain a signal proportional to i_a (Ki_a) from i_S (Ki_S). As indicated in Fig. 12.8-1, this is achieved by switching low-power transistors with $S1$ ($S2$) and $S3$ ($S4$) unfolding Ki_S into Ki_a. The signal Ki_a is then compared to the commanded Ki_a^*, and the error signal is supplied to a comparator with hysteresis (Fig. 12.8-2). When the output of this comparator (L) is high, $S1$ and $S2$ are closed and $S3$ and $S4$ are open; when L is low, $S3$ and $S4$ are closed and $S1$ and $S2$ are open.

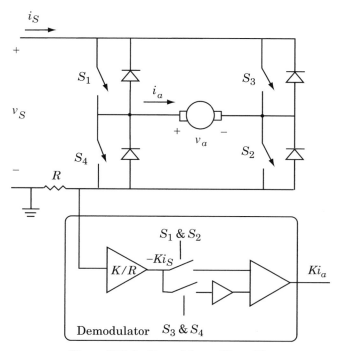

Figure 12.8-1   Demodulator; Ki_S to Ki_a.

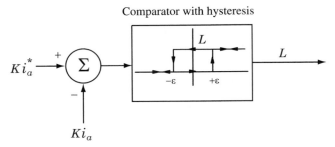

Figure 12.8-2 Hysteresis-type current controller.

The action of this type of current control is illustrated in Fig. 12.8-3. Therein, i_a^* is stepped from zero to a positive value. At the instant of the step increase, the error signal, which is the input to the comparator with hysteresis, is positive and greater than ε. The output of the comparator is high, and $S1$ and $S2$ are closed and $S3$ and $S4$ are open. The armature current increases until $i_a = i_a^* + \varepsilon$, at which time the comparator output steps low and $S1$ and $S2$ are opened and $S3$ and $S4$ are closed. The armature current decreases until $i_a = i_a^* - \varepsilon$, at which time the comparator output becomes high; $S1$ and $S2$ are closed and $S3$ and $S4$ are opened. This cycling will continue until the commanded current is changed. It is apparent from Figs. 12.8-2 and 12.8-3 that the period of this cycling or switching, which is denoted as T in Fig. 12.8-3, is directly related to the bounds ($\pm\varepsilon$) within the hysteresis comparator.

For analysis purposes, the switching period is divided into two intervals, t_1 and t_2. During t_1, $S1$ and $S2$ are closed and v_S is applied to the armature circuit; during t_2, $S3$ and $S4$ are closed and $-v_S$ is applied to the armature circuit. The rise time is also shown in Fig. 12.8-3. This is denoted t_r, and it is defined as the time for the current to increase from zero to the commanded current with the rotor speed equal to zero. If we set $I_1 = 0$ and $\omega_r = 0$ in (12.4-14), then

$$i_a = \frac{v_S}{r_a}(1 - e^{-t/\tau_a}) \tag{12.8-1}$$

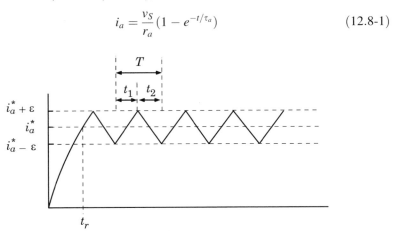

Figure 12.8-3 Typical response of a current-controlled four-quadrant chopper for a step increase in i_a^* from zero.

If we now let $i_a = i_a^*$ at $t = t_r$, (12.8-1) can be solved for t_r:

$$t_r = -\tau_a \ln\left(1 - \frac{r_a i_a^*}{v_S}\right) \tag{12.8-2}$$

It is now of interest to obtain expressions for t_1 and t_2 in terms of v_S, ε (the hysteresis limits), and machine parameters. It is clear that ε is generally made small in order to achieve minimum variation in i_a about i_a^*; and because r_a is generally small, i_a can be approximated by a straight line between i_a^* and $\pm\varepsilon$. The familiar voltage equation for the armature circuit is

$$v_a = r_a i_a + L_{AA} \frac{di_a}{dt} + \omega_r k_v \tag{12.8-3}$$

Our purpose is to obtain a straight-line approximation for $i_a^* - \varepsilon \leq i_a \leq i_a^* + \varepsilon$. If in (12.8-3) we let $r_a i_a = r_a i_a^*$ and if we assume that v_S and ω_r are constant over the switching period T, then with $S1$ and $S2$ closed we obtain

$$i_a = i_a^* - \varepsilon + \frac{1}{L_{AA}}(v_S - r_a i_a^* - k_v \omega_r)t \qquad \text{for } 0 \leq t < t_1 \tag{12.8-4}$$

Note that we have selected time zero at the beginning of the switching period T. During the interval when $S3$ and $S4$ are closed, with t set to 0 at the beginning of this interval, where we have

$$i_a = i_a^* + \varepsilon - \frac{1}{L_{AA}}(v_S + r_a i_a^* + k_v \omega_r)t \qquad \text{for } 0 \leq t < t_2 \tag{12.8-5}$$

If in (12.8-4) we set $t = t_1$, then $i_a = i_a^* + \varepsilon$ and t_1 becomes

$$t_1 = \frac{2L_{AA}\varepsilon}{v_S - r_a i_a^* - k_v \omega_r} \tag{12.8-6}$$

This expression for t_1, albeit an approximation, provides us with information that we would have expected if we would have thought about it. As $v_S - r_a i_a^* - k_v \omega_r$ approaches zero, t_1 approaches infinity. This says that as the rotor speed increases, the source voltage is not large enough ($S1$ and $S2$ fixed closed) to maintain the commanded current. In other words, the chopper is unable to provide the commanded current, and it has lost current tracking capability. We would expect this unless v_S could be increased without limit.

Now what about t_2? Well, if in (12.8-5) we set $t = t_2$, and $i_a = i_a - \varepsilon$ and solve for t_2, we obtain

$$t_2 = \frac{2L_{AA}\varepsilon}{v_S + r_a i_a^* + k_v \omega_r} \tag{12.8-7}$$

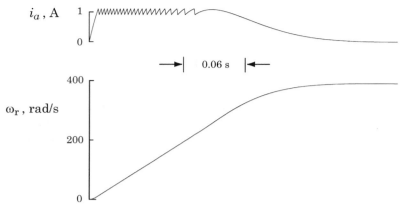

Figure 12.8-4 Acceleration of a dc drive from stall with current-controlled dc/dc converter.

Hence, t_2 approaches zero as ω_r approaches infinity. In other words, the higher the speed, the faster i_a decreases from $i_a^* + \varepsilon$ to $i_a^* - \varepsilon$. That seems logical. We can now express the switching period T. After some manipulation we obtain

$$T = t_1 + t_2 = \frac{4L_{AA}v_S\varepsilon}{v_S^2 - (r_a i_a^* + k_v\omega_r)^2} \tag{12.8-8}$$

Here, we again see that which (12.8-6) already told us; T approaches infinity as $v_S - r_a i_a^* - k_v\omega_r$ approaches zero. In other words, T becomes larger as the rotor speed increases.

The operation of a four-quadrant current-controlled chopper during the starting of a dc machine is shown in Fig. 12.8-4. Therein, the current and rotor speed are plotted. The dynamic torque-versus-speed characteristics for this acceleration period are shown in Fig. 12.8-5. The steady-state torque-versus-speed characteristic is also shown in this figure by a dashed line. The parameters of the fractional horsepower dc

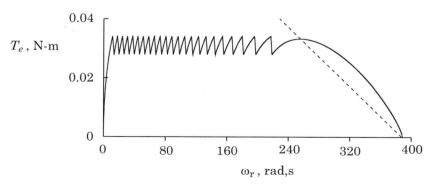

Figure 12.8-5 Torque versus speed for acceleration shown in Fig. 12.8-4.

machine and current controller are as follows: $r_a = 3.8\,\Omega$, $L_{AA} = 76\,\text{mH}$, $k_v = 0.031\,\text{V}\cdot\text{s/rad}$, $J = 1.41 \times 10^{-5}\,\text{kg}\cdot\text{m}^2$, $\varepsilon = 0.1\,\text{A}$, $i_a^* = 1\,\text{A}$, and $v_S = 12\,\text{V}$.

The plot of i_a (Fig. 12.8-4) illustrates the characteristics of the current control discussed earlier. In particular, t_1 and T increase as the speed increases while t_2 decreases. The commanded current i_a^* is set at 1 A. It appears from Figs. 12.8-4 and 12.8-5 that current tracking is lost when ω_r is near 250 rad/s, after which the machine accelerates to the steady-state speed with $v_a = v_S$ (S1 and S2 closed). Actually, we can approximate the speed at which current tracking will be lost. When S1 and S2 are closed and if we let $i_a = i_a^*$, the armature voltage equations becomes

$$v_S = r_a i_a^* + L_{AA} \frac{di_a^*}{dt} + \omega_r k_v \qquad (12.8\text{-}9)$$

Because di_a^*/dt is zero, we can approximate the maximum rotor speed at which v_S can maintain i_a^* with S1 and S2 fixed closed. In particular, for 1st quadrant operation

$$\omega_r = \frac{1}{k_v}\left(v_S - r_a i_a^*\right) \qquad (12.8\text{-}10)$$

This becomes $(1/0.031)(12 - 3.8 \times 1) = 264.5\,\text{rad/s}$ and should be compared to 250 rad/s, which was approximated from the conditions depicted in Figs. 12.8-4 and 12.8-5.

It is difficult to portray the action of a current-controlled converter with an average-value model when current tracking is lost. In Fig. 12.8-4, current tracking occurs in most of the acceleration period; however, current tracking does not occur during the initial rise of the current and during the final acceleration period. It is clear that one continuous average-value model cannot predict all modes of operation. Switching or high gain clamping circuits must be incorporated. Such models are beyond the scope of this treatment of dc drives. The interested reader is referred to recent research in this area described in reference 3.

Speed Control: Current-Controlled dc/dc Converter

An average-value model that can be used for speed control of a dc machine being supplied from a current-controlled four-quadrant dc/dc converter is shown in Fig. 12.8-6. This average-value model is valid only when i_a^* is being maintained. When current tracking is lost, this average-value model is not valid. It is clear that it may be appropriate to include a safeguard to prevent its use when current tracking is lost. Perhaps incorporating a version of (12.8-10) to limit ω_r^* might be sufficient. In any event, the average-value model is valid for analysis and design purposes during current tracking.

In order to demonstrate speed control of a dc drive and to verify the accuracy of the average value model shown in Fig. 12.8-6, let us consider the fractional horsepower dc machine used in Figs. 12.8-4 and 12.8-5. In this case, however, the

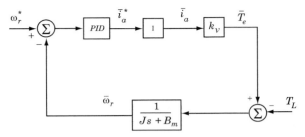

Figure 12.8-6 Speed control of dc machine with current-controlled dc/dc converter: current tracking operation.

armature circuit time constant is selected as $\tau_a = 2$ ms, which is actually more practical than that $(\tau_a = 76 \times 10^{-3}/3.8 = 0.02\,\text{s})$ used in Figs. 12.8-4 and 12.8-5. For this study the *PID* controller is a simple proportional control with a gain of unity, and i_a^* is limited to ± 2 A with $\varepsilon = 0.1$ A. The load torque, T_L, is $k_L\omega_r^2$ times the sign of ω_r, where k_L is selected so that $T_L = 0.039\,\text{N}\cdot\text{m}$ when $\omega_r = 100$ rad/s.

The operation of this speed control system is shown in Figs. 12.8-7 and 12.8-8. Initially, $\omega_r^* = 100$ rad/s and the drive is operating in the steady state; ω_r^* is then stepped to -100 rad/s. The armature current and speed are shown in Fig. 12.8-7 and T_e versus ω_r in Fig. 12.8-8. Once ω_r^* is stepped to -100 rad/s, i_a^* steps to the limit of -2 A. Switches $S3$ and $S4$ are closed and the armature current decrease to $-(2\,\text{A} + \varepsilon)$ whereupon the current control maintains i_a at $-2\,\text{A} \pm \varepsilon$. Once the speed has decreased to near -100 rad/s, i_a^* increases to the value $(-1\,\text{A})$ necessary to maintain $\omega_r = -100$ rad/s. It is interesting to point out that initially T_e and ω_r are positive (motor action); however, when ω_r^* is stepped to

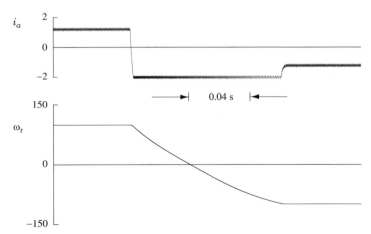

Figure 12.8-7 Speed control with four-quadrant chopper: ω_r^* stepped from 100 rad/s to -100 rad/s with i_a^* limited to ± 2 A.

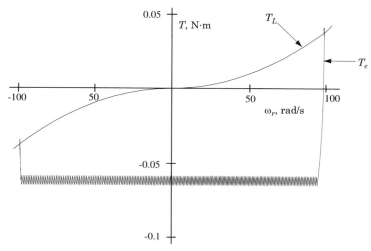

Figure 12.8-8 T_e versus ω_r for Fig. 12.8-7.

-100 rad/s, T_e is negative and ω_r is positive. Generator action (regeneration) occurs until ω_r reaches zero. When ω_r becomes negative, the dc machine is again operating as a motor.

The computer traces shown in Figs. 12.8-9 and 12.8-10 are the modes of operation shown in Figs. 12.8-7 and 12.8-8, respectively, using the average-value model shown in Fig. 12.8-6. It appears that the average-value model is sufficiently accurate to predict the operation of this type of dc drive system even though current tracking was lost briefly when ω_r^* was stepped to -100 rad/s and later when i_a^* increased to -1 A from its negative limit.

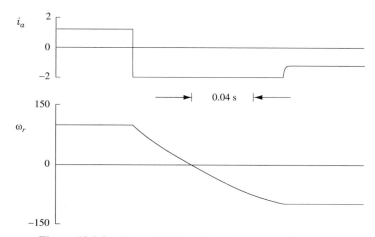

Figure 12.8-9   Figure 12.8-7 using the average-value model.

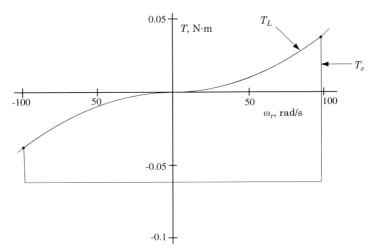

Figure 12.8-10   Figure 12.8-8 using the average-value model.

REFERENCES

[1] P. C. Sen, *Thyristor DC Drives*, John Wiley and Sons, New York, 1981.

[2] J. G. Kassakian, M. F. Schlecht, and G. C. Verghese, *Principles of Power Electronics*, Addison-Wesley, Reading, MA, 1991.

[3] E. A. Walters, Automated Averaging Techniques for Power Electronic-Based Systems, Ph.D. Thesis, Purdue University, 1999.

[4] M. H. Rashid, *Power Electronics: Circuits, Devices, and Applications*, Prentice-Hall, Englewood Cliffs, NJ, 1993.

[5] N. Mohan, T. M. Undeland, W. P. Robbins, *Power Electronics: Converters, Applications, and Design*, John Wiley and Sons, New York, 1995.

[6] P. T. Krein, *Elements of Power Electronics*, Oxford University Press, New York, 1998.

[7] S. E. Lyshevski, *Electromechanical Systems, Electric Machines, and Applied Mechatronics*, CRC Press, Boca Raton, FL, 1999.

PROBLEMS

1 The average-value steady-state equivalent circuit for the single-phase load commutated full converter is derived in Chapter 11 (Fig. 11.2-6). Consider the 5-hp shunt machine described in Section 12.3 and in Example 12A. Assume that this average-value equivalent circuit can be used with the dynamic equations that describe the dc machine. Show that when the load torque is stepped from rated to 50% of rated value, the average rotor speed and armature current may be expressed as

$$\omega_r = 137.2 - 3.2\, e^{-t/0.377} \sin(4.47t + 118.6°)$$
$$i_a = 8.1 + 9.4 e^{-t/0.377} \sin(4.47 + 59.3°)$$

Compare these responses with those shown in Fig. 12.3-5.

2   Repeat Problem 1 for the 200-hp dc shunt machine described in Section 12.3 and considered in Example 12A. In this case, the machine is being supplied from a three-phase full converter. Using the average-value relationship (11.3-16), show that the transient response to a 50% step decrease in load torque from rated are

$$\omega_r = 72.3 - 2.9\, e^{-t/0.042} + 0.5\, e^{-t/0.018}$$

$$i_a = 316.7 + 551.1 e^{-t/0.042} - 234.8\, e^{-t/0.018}$$

Compare these responses with those shown in Fig. 12.3-11.

3   Consider the one-quadrant chopper shown in Fig. 12.4-1. Let $v_S = 100\,\text{V}$, $L_{AA} = 5\,\text{mH}$, $r_a = 10\,\Omega$, $k_v = 1\,\text{V} \cdot \text{s/rad}$, $\omega_r = 0$, and $f_s = 10\,\text{kHz}$. Calculate I_1 and I_2 for steady-state operation at $k = 0.5$. Sketch i_a, v_a, i_S, i_D, and the instantaneous electric power supplied to the motor, $P_e = v_a i_a$.

4   Calculate the average values for the plots of i_a, v_a, i_S, i_D, and P_e shown in Problem 3. Make the simplifying approximation that the current waveform is made up of straight-line segments. Compare average P_e with average v_a times average i_a and the average power P_S supplied by the source.

5   Consider the one-quadrant chopper given in Problem 3. Assume $\omega_r = 80\,\text{rad/s}$. Determine the minimum value of k where continuous-current operation will occur.

6   Approximate ΔI (12.4-29) for $T \ll \tau_a$ using (a) the first two terms of Taylor's series and (b) the first three terms.

7   Consider the waveform of v_a in Fig. 12.P-1 where the value of v_S changes from v_{S1} to v_{S2} at $t_i + \frac{k}{2}T$. Assume T and k are constant. Determine and plot the dynamic average of the armature voltage, $v_{a(da)}$.

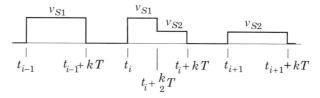

Figure 12.P-1 v_a with a change in source during a switching cycle.

8   Obtain β for $T \ll \tau_a$ [see (12.4-40)].

9   A one-quadrant chopper is driving a permanent-magnet dc motor. The parameters are $r_a = 2\,\Omega$ and $k_v = 0.05\,\text{V} \cdot \text{s/rad}$. The source voltage v_S is 20 V and the switching period T is 0.1 ms. The steady-state armature current increases (straight-line) from zero at the start of the switching period to 2 A in 0.05 ms. It decreases back to zero in 0.03 ms and remains at zero for 0.02 ms, which ends the switching period. (a) Sketch v_a. (b) Calculate $\bar{v}_a$. (c) Calculate $\bar{\omega}_r$. (d) Approximate τ_a.

10   Plot the source current i_S and the variables shown in Fig. 12.5-2 for the two-quadrant chopper operating with (a) I_1 and I_2 positive and (b) I_1 and I_2 negative.

11   Consider the two-quadrant chopper shown in Fig. 12.5-1. Let $v_S = 12\,\text{V}$, $f_s = 10\,\text{kHz}$, $L_{AA} = 1.14\,\text{mH}$, $r_a = 3.8\,\Omega$, $k_v = 0.031\,\text{V} \cdot \text{s/rad}$.   If $\omega_r = 100\,\text{rad/s}$ and

$T_e = 2.82 \times 10^{-2}\,\text{N} \cdot \text{m}$, (a) calculate k, (b) determine I_1 and I_2, and (c) sketch i_a, v_a, i_{S1}, i_{D1}, i_S. Also, sketch P_S and P_e, the instantaneous electric power supplied from the source and to the machine, respectively.

12   Refer to Problem 11. Make simplifying straight-line approximations to calculate the average i_a, v_a, i_{S1}, i_{D1}, i_S, P_S, and P_e.

13   A two-quadrant chopper is driving a permanent-magnet dc motor. The parameters are $r_a = 1\,\Omega$, $L_{AA} = 1\,\text{mH}$, and $k_v = 0.02\,\text{V} \cdot \text{s/rad}$. The source voltage, v_S, is 25 V and $T = 0.2\,\tau_a$. Both T_L and B_m are zero. The steady-state armature current, i_a, increases as a straight line from $-I$ at the start of each switching period, T, to I at $T/2$. It then decreases from I to $-I$ completing the switching period. Assume the speed is constant. (a) Calculate $\bar{v}_a$. (b) Calculate $\bar{\omega}_r$ and I.

14   Plot the source current, i_S, and the variables shown in Fig. 12.6-2 for steady-state operation of a four-quadrant chopper drive system with ω_r negative (cw) and T_e positive.

15   Plot i_a versus v_a for the modes of operation depicted in Figs. 12.4-3, 12.4-4, 12.5-2, and 12.6-2. Show I_1 and I_2, the intervals A and B, and the direction of transversing each i_a versus v_a trajectory. Assuming ω_r is positive (ccw) in all cases, indicate in each quadrant the type of operation, either motor or generator.

16   Consider the i_a versus v_a plots for the modes of operation depicted in Figs. 12.4-3, 12.4-4, 12.5-2, and 12.6-2. Determine for each the quadrant of operation average i_a versus average v_a. If ω_r is positive (ccw), indicate the type of operation (motor or generator) for each figure.

17   Obtain (12.6-1) and (12.6-2).

18   Use a straight-line approximation to determine t_r in Fig. 12.8-4. In the first approximation, neglect the $i_a r_a$ voltage drop; for the second approximation include an average $i_a^* r_a$ voltage drop during the rise time.

19   Obtain (12.8-8) from (12.8-6) and (12.8-7).

20   A four-quadrant current-controlled chopper is used to drive a permanent-magnet dc motor. Initially, the dc motor is stalled ($\omega_r = 0$) and i_a is zero. The current command is stepped to i_a^*. Assume current tracking occurs and $T_L = 0$. Express ω_r in terms of k_v, i_a^*, B_m, and J.

21   A four-quadrant current-controlled chopper is used to supply a permanent-magnet dc motor. When the speed is fixed at zero, the armature current increases from zero to I in time t_1 and decreases from I back to zero in time t_2 where $T = t_1 + t_2$. Assume the waveforms are straight lines. (a) Express t_1 and t_2 in terms of T. (b) Express i_a^* and ε in terms of I. (c) Plot i_{S1} and i_{S4} (Fig. 12.6-1).

22   In Figs. 12.8-7 and 12.8-8, i_a^* was limited to ± 2 A. Assume that τ_a was decreased by decreasing the armature circuit inductance and that $r_a = 3.8\,\Omega$. (a) Approximate the maximum limits that could have been used for i_a^* without losing current tracking. (b) Superimpose upon Fig. 12.8-8 the steady-state T_e versus ω_r plot with $S1$ and $S2$ (Fig. 12.6-1) fixed closed. (c) With $S3$ and $S4$ fixed closed.

23   Consider the 240-V 5-hp dc shunt machine that has been used in Chapter 2 and this chapter. The parameters are as follows: $r_a = 0.6\,\Omega$, $L_{AA} = 0.012\,\text{H}$, and $L_{AF} = 1.8\,\text{H}$. Assume T_L is $k_L \omega_r^2$ times the sign of ω_r. The field current i_f is maintained at 1 A, and T_L

is 80% of rated torque when $\omega_r = \pm 120\,$rad/s. Design a speed control over the range of $-120\,$rad/s $\le \omega_r \le 120\,$rad/s using a current-controlled four-quadrant chopper. Select (a) v_S, (b) k_L, (c) i_a^* limits, and (d) ε. Justify each selection. (e) Assume a proportional controller and sketch the T_e versus ω_r response for a step change when ω_r^* is stepped from $-120\,$rad/s to 120 rad/s. (f) Superimpose upon the plot in (e) the steady-state T_e versus ω_r plots for S1 and S2 (Fig. 12.6-1) fixed closed and then S3 and S4 fixed closed.

24 Refer to Problem 23 wherein the parameters of a 5-hp dc shunt machine are given along with specifications on i_f and T_L. Design a speed control over the range of $-120\,$rad/s $\le \omega_r \le 120\,$rad/s using a voltage-controlled four-quadrant chopper with feedforward voltage control. Select (a) v_S, (b) k_L, (c) A in Fig. 12.7-1, and (d) the switching frequency. (e) Why would it not be advisable to implement large changes in ω_r^*? (f) Suggest an approximate means of changing ω_r^* to prevent the problem associated with step changes in ω_r^*.

Chapter 13

FULLY CONTROLLED 3-PHASE
BRIDGE CONVERTERS

13.1 INTRODUCTION

In our study of induction, synchronous, and brushless dc machines we set forth control strategies that assumed that the machine was driven by a 3-phase variable frequency voltage or current source, without mention of how a source is actually obtained, or what its characteristics might be. In this chapter, the operation of the 3-phase fully controlled bridge converter is set forth. It is shown that by suitable control this device can be used to achieve either a 3-phase controllable voltage source or a 3-phase controllable current source, as was assumed to exist in previous chapters.

13.2 THE 3-PHASE BRIDGE CONVERTER

The converter topology that serves as the basis for nearly all 3-phase variable speed drive systems is shown in Fig. 13.2-1. This type of converter is comprised of six controllable switches or valves labeled T1–T6. Physically, bipolar junction transistors (BJTs), metal-on-silicon field-effect transistors (MOSFETs), insulated-gate bipolar junction transistors (IGBTs), and metal-on-silicon controlled thyristors (MCTs) are just a few of the devices that can be used as switches. Across each switch is an antiparallel diode used to ensure that there is a path for inductive current in the event that a switch that would normally conduct current of that direction is turned off.

In Fig. 13.2-1, v_{dc} denotes the dc voltage applied to the converter bridge, and i_{dc} designates the dc current flowing into the bridge. The bridge is divided into three legs, one for each phase of the load. The line-to-ground voltage of the a-, b-, and c-phase legs of the converter are denoted v_{ag}, v_{bg}, and v_{cg}, respectively. In this

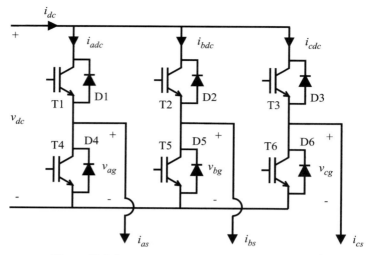

Figure 13.2-1 The 3-phase bridge converter topology.

text, the load current will generally be the stator current into a synchronous, induction, or brushless dc machine; therefore i_{as}, i_{bs}, and i_{cs} are used to represent the current into each phase of the load. Finally, the dc currents from the upper rail into the top of each phase leg are designated i_{adc}, i_{bdc}, and i_{cdc}.

To understand the operation of this basic topology, it must first be understood that none of the semiconductor devices shown are ever intentionally operated in the active region of their i–v characteristic. Their operating point is either in the saturated region (on) or in the cutoff region (off). If the devices were operated in their active region, then by applying a suitable gate voltage to each device the line-to-ground voltage of each leg could be continuously varied from 0 to v_{dc}. At first, such control appears advantageous because each leg of the converter could be used as a controllable voltage source. The disadvantage of this strategy is that if the switching devices are allowed to operate in their active region, there will be both a voltage across and current through each semiconductor device resulting in power loss. On the other hand, if each semiconductor is either on or off, then either there is a current through the device but no voltage, resulting in no power loss, or there is a voltage across the device but no current, again resulting in no power loss. Of course, in a real device there will be some power losses due to (a) the small voltage drop that occurs even when the device is in saturation (on) and (b) losses that are associated with turning the switching devices on or off (switching losses); nevertheless, inverter efficiencies greater than 95% are readily obtained.

In this study of the operation of the converter bridge, it will be assumed that either the upper switch or lower switch of each leg is gated on, except during switching transients (the result of turning one switch on while turning another off). Ideally, the leg-to-ground voltage of a given phase will be v_{dc} if the upper switch is on and the lower switch is turned off, or 0 if the lower switch is turned on and the upper switch is off. This assumption is often useful for analysis purposes as well as for computer simulation of systems in which the dc supply voltage is much greater than the semiconductor voltage drops. If a more detailed analysis or computer simulation is desired (hence the voltage drops across the semiconductors are not neglected), then the line-to-ground voltage is determined both by the switching devices turned on and by the direction of the phase current.

To illustrate this, consider the diagram of one leg of the bridge as is shown in Fig. 13.2-2. Therein, x can be a, b, or c, to represent the a, b, or c phase, respectively. Figure 13.2-3a illustrates the effective equivalent circuit shown in Fig. 13.2-2 if the upper transistor is on and the current i_{xs} is positive. For this condition it can be seen that the line-to-ground voltage v_{xg} will be equal to the dc supply voltage v_{dc} less the voltage drop across the switch v_{sw}. The voltage drop across the switch is generally in the range of 0.7–3.0 V. Although the voltage drop is actually a function of the switch current, it can often be represented as a constant. From Fig. 13.2-3a, the dc current into the bridge, i_{xdc}, is equal to the phase current i_{xs}.

If the upper transistor is on and the phase current is negative, then the equivalent circuit is as shown in Fig. 13.2-3b. In this case, the dc current into the leg i_{xdc} is again equal to the phase current i_{xs}. However, because the current is now flowing through the diode, the line-to-ground voltage v_{xg} is equal to the dc supply voltage v_{dc} plus the diode forward voltage drop v_d. If the upper switch is on and the phase current is zero, it seems reasonable to assume that the line-to-ground voltage is equal to the supply voltage as indicated in Fig. 13.2-3c. Although other estimates could be argued (such as averaging the voltage from the positive and negative current conditions), it must be remembered that this is a rare condition so a small inaccuracy will not have a perceptible effect on the results.

The positive, negative, and zero phase current equivalent circuits that represent the phase leg when the lower switching device is on and the upper switching device is off are illustrated in Fig. 13.2-3d to Fig. 13.2-3f, respectively. The situation is entirely analogous to the case in which the upper switch is on.

One final possibility is the case in which neither transistor is turned on. As stated previously, it is assumed that in the drives considered herein, either the upper or

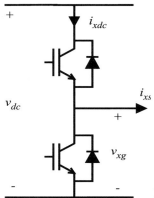

Figure 13.2-2 One-phase leg.

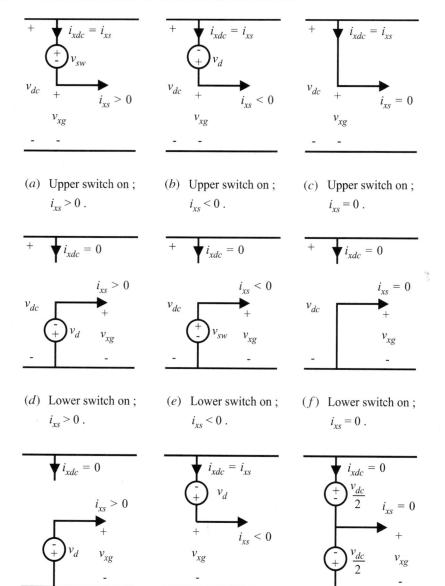

(a) Upper switch on ;
    $i_{xs} > 0$.

(b) Upper switch on ;
    $i_{xs} < 0$.

(c) Upper switch on ;
    $i_{xs} = 0$.

(d) Lower switch on ;
    $i_{xs} > 0$.

(e) Lower switch on ;
    $i_{xs} < 0$.

(f) Lower switch on ;
    $i_{xs} = 0$.

(g) Neither switch on ;
    $i_{xs} > 0$.

(h) Neither switch on ;
    $i_{xs} < 0$.

(i) Neither switch on ;
    $i_{xs} = 0$.

Figure 13.2-3   Phase leg equivalent circuits.

lower transistor is turned on. However, there is a delay between the time a switch is commanded to turn off and the time it actually turns off as well as a delay between the time a switch is commanded to turn on and the time it actually turns on. Sophisticated semiconductor device models are required to predict the exact voltage and current waveforms associated with the turn-on and turn-off transients of the switching devices [1–5]. However, as a first-cut representation it can be assumed that a device turns on with a delay T_{on} after the control logic commands it to turn on, and turns off after a delay T_{off} after the control logic commands it to turn off. The turn-off time is generally longer than the turn on-time. Unless the turn-on time and turn-off time are identical, there will be an interval in which either no device in a leg is turned on or both devices in a leg are turned on. The latter possibility is known as "shoot-through" and is extremely undesirable; therefore an extra delay is added into the control logic such that the device being turned off will do so before the complementary device is turned on (see Problem 10). Therefore it may be necessary to represent the condition in which neither device of a leg is turned on.

If neither device of a phase leg is turned on and the current is positive, then the situation is as in Fig. 13.2-3g. Because neither switching device is conducting, the current must flow through the lower diode. Thus, the line-to-ground voltage v_{xg} is equal to $-v_d$, and the dc current into the leg i_{xdc} is zero. Conversely, if the phase current is negative, then the upper diode must conduct as is indicated in Fig. 13.2-3h. In this case, the line-to-ground voltage is $v_{dc} + v_d$, and the dc current into the leg i_{xdc} is equal to phase current into the load i_{xs}. In the event that neither transistor is on, and that the phase current into the load is zero, it is difficult to identify what the line-to-ground voltage will be because it will become a function of the back emf of the machine to which the converter is connected. If, however, it is assumed that the period during which neither switching device is gated on is brief (on the order of a microsecond), then assuming that the line-to-ground voltage is $v_{dc}/2$ is an acceptable approximation. Note that this approximation cannot be used if the period during which neither switching device is gated on is extended. An example of the type of analysis that must be conducted if both the upper and lower switching devices are off for an extended periods appears in references 6–8.

Table 13.2-1 summarizes the calculation of line-to-ground voltage and dc current into each leg of the bridge for each possible condition. Once each of the line-to-ground voltages are found, the line-to-line voltages may be calculated. In particular,

$$v_{abs} = v_{ag} - v_{bg} \tag{13.2-1}$$

$$v_{bcs} = v_{bg} - v_{cg} \tag{13.2-2}$$

$$v_{cas} = v_{cg} - v_{ag} \tag{13.2-3}$$

and from Fig. 13.2-1, the total dc current into the bridge is given by

$$i_{dc} = i_{adc} + i_{bdc} + i_{cdc} \tag{13.2-4}$$

Because machines are often wye-connected, it is useful to derive equations for the line-to-neutral voltages produced by the 3-phase bridge. If the converter of

Table 13.2-1   Converter Voltages and Currents

Switch On	Current Polarity	v_{xg}	i_{xdc}
Upper	Positive	$v_{dc} - v_{sw}$	i_{xs}
	Negative	$v_{dc} + v_d$	i_{xs}
	Zero	v_{dc}	i_{xs}
Lower	Positive	$-v_d$	0
	Negative	v_{sw}	0
	Zero	0	0
Neither	Positive	$-v_d$	0
	Negative	$v_{dc} + v_d$	i_{xs}
	Zero	$v_{dc}/2$	0

Fig. 13.2-1 is connected to a wye-connected load, then the line-to-ground voltages are related to line-to-neutral voltages and the neutral-to-ground voltage by

$$v_{ag} = v_{as} + v_{ng} \tag{13.2-5}$$

$$v_{bg} = v_{bs} + v_{ng} \tag{13.2-6}$$

$$v_{cg} = v_{cs} + v_{ng} \tag{13.2-7}$$

Summing (13.2-5)–(13.2-7) and rearranging yields

$$v_{ng} = \frac{1}{3}\left(v_{ag} + v_{bg} + v_{cg}\right) - \frac{1}{3}\left(v_{as} + v_{bs} + v_{cs}\right) \tag{13.2-8}$$

The final term in (13.2-8) is recognized as the zero-sequence voltage of the machine; thus

$$v_{ng} = \frac{1}{3}\left(v_{ag} + v_{bg} + v_{cg}\right) - v_{0s} \tag{13.2-9}$$

For a balanced wye-connected machine such as a synchronous machine, induction machine, or brushless dc machine, summing the line-to-neutral voltage equations indicates that the zero-sequence voltage is zero. However, if the machine is unbalanced, this would not be the case. Another practical example of a case in which the zero-sequence voltage is not identically equal to zero is a brushless dc machine with a squarewave or trapezoidal back emf, in which case the sum of the 3-phase back emfs does not sum to zero. However, for the machines considered in this text in which the zero-sequence voltage must be zero, (13.2-8) reduces to

$$v_{ng} = \frac{1}{3}\left(v_{ag} + v_{bg} + v_{cg}\right) \tag{13.2-10}$$

Substitution of (13.2-10) into (13.2-5 - 13.2-7) and solving for the line-to-neutral voltages yields

$$v_{as} = \frac{2}{3} v_{ag} - \frac{1}{3} v_{bg} - \frac{1}{3} v_{cg} \tag{13.2-11}$$

$$v_{bs} = \frac{2}{3} v_{bg} - \frac{1}{3} v_{ag} - \frac{1}{3} v_{cg} \tag{13.2-12}$$

$$v_{cs} = \frac{2}{3} v_{cg} - \frac{1}{3} v_{ag} - \frac{1}{3} v_{bg} \tag{13.2-13}$$

13.3 180° VOLTAGE SOURCE OPERATION

In the previous section, the basic voltage and current relationships needed to analyze the 3-phase bridge were set forth with no discussion as to how the bridge would enable operation of a 3-phase ac machine from a dc supply. In this section, a basic method of accomplishing the dc to ac power conversion is set forth, a method known as 180° voltage source operation. In this mode of operation, each switch-flyback diode of each phase leg conducts for 180 electrical degrees per cycle.

   The operation of the 3-phase bridge in the 180° voltage source mode is shown in Fig. 13.3-1. Therein, the first three traces illustrate switching signals applied to the power electronic devices as a function of θ_c, the converter angle. The definition of the converter angle is dependent upon the type of machine the given converter is driving. For the present the converter angle can be taken to be $\omega_c t$, where t is time and ω_c is the radian frequency of the 3-phase output. In subsequent chapters, the converter angle will be related to the electrical rotor position or the position of the synchronous reference frame, depending upon the type of machine. Referring to Fig. 13.3-1, the logical complement of the switching command to the lower device of each leg is shown for convenience because this signal is equal to the switch command of the upper device if switching times are neglected. For purposes of explanation, it is further assumed that the diode and switching devices are ideal—that is, that they are perfect conductors when turned on or perfect insulators when turned off. With these assumptions the line-to-ground voltages are as shown in the central three traces of Fig. 13.3-1. From the line-to-ground voltages, the line-to-line voltages may be calculated from (13.2-1)–(13.2-3), which are illustrated in the final three traces. Because the waveforms are square waves rather than sine waves, the three-phase bridge produces considerable harmonic content in the ac output when operated in this mode. In particular, using Fourier series techniques, the a- to b-phase line-to-line voltage may be expressed as

$$v_{abs} = \frac{2\sqrt{3}}{\pi} v_{dc} \cos\left(\theta_c + \frac{\pi}{6}\right) + \frac{2\sqrt{3}}{\pi} v_{dc} \left(\sum_{j=1}^{\infty} \left(-\frac{1}{6j-1} \cos\left((6j-1)\left(\theta_c + \frac{\pi}{6}\right)\right) \right. \right.$$

$$\left. \left. + \frac{1}{6j+1} \cos\left((6j+1)\left(\theta_c + \frac{\pi}{6}\right)\right) \right) \right) \tag{13.3-1}$$

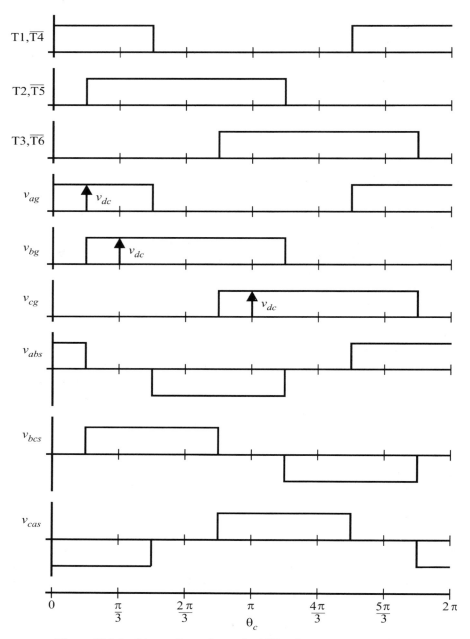

Figure 13.3-1 Line-to-line voltages for 180° voltage source operation.

From (13.3-1) it can be seen that the line-to-line voltage contains a fundamental component as well as the 5th, 7th, 11th, 13th, 17th, 19th... harmonic components. There are no even harmonics or odd harmonics that are a multiple of three. The effect of harmonics depends on the machine. In the case of a brushless dc with a sinusoidal back emf, the harmonics will result in torque harmonics but will not have any effect on the average torque. In the case of the induction motor, torque harmonics will again result; however, in this case the average torque will be affected. In particular, it can be shown that the $6j - 1$ harmonics form an *acb* sequence that will reduce the average torque, while the $6j + 1$ harmonics form an *abc* sequence that increases the average torque. The net result is usually a small decrease in average torque. In all cases, harmonics will result in increased machine losses.

Figure 13.3-2 again illustrates 180° voltage source operation, except that the formulation of the line-to-neutral voltages is considered. From the line-to-ground voltage, the neutral-to-ground voltage v_{ng} is calculated using (13.2-10). The line-to-neutral voltages are calculated using the line-to-ground voltages and line-to-neutral voltage from (13.2-5)-(13.2-7). From Fig. 13.3-2, the a-phase line-to-neutral voltage may be expressed as a Fourier series of the form

$$v_{as} = \frac{2}{\pi} v_{dc} \cos \theta_c + \frac{2}{\pi} v_{dc} \sum_{j=1}^{\infty} \left(\frac{(-1)^{j+1}}{6j - 1} \cos((6j - 1)\theta_c) + \frac{(-1)^j}{6j + 1} \cos((6j + 1)\theta_c) \right)$$

$$(13.3-2)$$

Relative to the fundamental component, each harmonic component of the line-to-neutral voltage waveform has the same amplitude as in the line-to-line voltage. The frequency spectrum of both the line-to-line and line-to-neutral voltages is illustrated in Fig. 13.3-3.

The effect of these harmonics on the current waveforms is illustrated in Fig. 13.3-4. In this study, a 3-phase bridge supplies a wye-connected load consisting of a 2-Ω resistor in series with a 1-mH inductor in each phase. The dc voltage is 100 V, and the frequency is 100 Hz. The a-phase voltage has the waveshape depicted in Fig. 13.3-2, and the impact of the a-phase voltage harmonics on the a-phase current is clearly evident. Because of the harmonic content of the waveforms, the power going into the 3-phase load is not constant, which implies that the power into the converter, and hence the dc current into the converter, is not constant. As can be seen, the dc current waveform repeats every 60 electrical degrees; this same pattern will also be shown to be evident in q- and d-axis variables.

Because the analysis of electric machinery is based upon reference frame theory, it is convenient to determine q- and d-axis voltages produced by the converter. To do this, we will define the converter reference frame to be a reference frame in which θ of (3.3-4) is equal to θ_c. In this reference frame the average q-axis voltage is equal to the peak value of the fundamental component of the applied line-to-neutral voltage and the average d-axis voltage is zero. This transformation will be designated K_s^c.

Usually, the converter reference frame will be the rotor reference frame in the case of a brushless dc machine or the synchronously rotating reference frame in the case of an induction motor. Deriving expressions analogous to (13.3-2) for the b- and c-phase line-to-neutral voltages and transforming these voltages to the converter

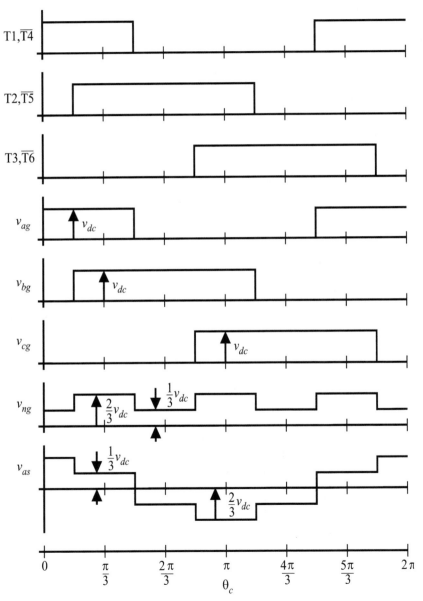

Figure 13.3-2 Line-to-neutral voltage for 180° voltage source operation.

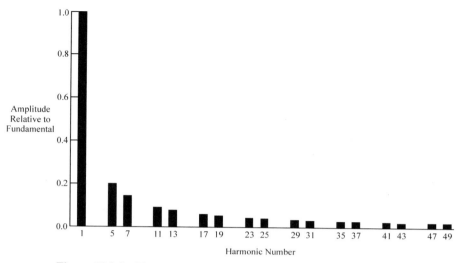

Figure 13.3-3   Frequency spectrum of 180° voltage source operation.

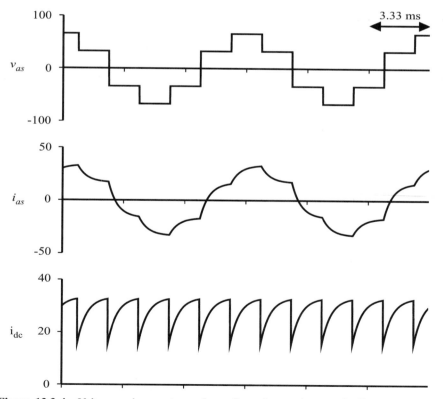

Figure 13.3-4   Voltage and current waveforms for a six-step inverter feeding an RL load.

reference frame yields

$$v_{qs}^c = \frac{2}{\pi} v_{dc} - \frac{2}{\pi} v_{dc} \sum_{j=1}^{\infty} \frac{2(-1)^j}{36j^2 - 1} \cos(6j\theta_c) \tag{13.3-3}$$

$$v_{ds}^c = \frac{2}{\pi} v_{dc} \sum_{j=1}^{\infty} \frac{12j}{36j^2 - 1} \sin(6j\theta_c) \tag{13.3-4}$$

From (13.3-3)–(13.3-4) it can be seen that the q- and d-axis variables will contain a dc component in addition to multiples of the 6th harmonic. In addition to being evident in qd variables, the 6th harmonic is also apparent in the torque waveforms of machines connected to converters operated in the 180° voltage source mode.

For the purposes of machine analysis, it is often convenient to derive an average-value model of the machine in which harmonics are neglected. From (13.3-3)–(13.3-4) the average q- and d-axis voltage may be expressed as

$$\bar{v}_{qs}^c = \frac{2}{\pi} v_{dc} \tag{13.3-5}$$

$$\bar{v}_{ds}^c = 0 \tag{13.3-6}$$

where the line above the variables denotes average value.

It is interesting to compare the line-to-neutral voltage to the q- and d-axis voltage. Such a comparison appears in Fig. 13.3-5. As can be seen, the q- and d-axis voltages repeat every 60 electrical degrees, which is consistent with the fact that these waveforms only contain a dc component and harmonics that are a multiple of six. The qd currents, qd flux linkages, and electromagnetic torque also possess the property of repeating every 60 electrical degrees.

In order to calculate the average dc current into the inverter, note that the instantaneous power into the inverter is given by

$$P_{in} = i_{dc} v_{dc} \tag{13.3-7}$$

The power out of the inverter is given by

$$P_{out} = \frac{3}{2}(v_{qs}i_{qs} + v_{ds}i_{ds}) \tag{13.3-8}$$

Neglecting inverter losses, the input power must equal the output power; therefore

$$i_{dc} = \frac{\frac{3}{2}(v_{qs}i_{qs} + v_{ds}i_{ds})}{v_{dc}} \tag{13.3-9}$$

Equation (13.3-9) is true on an instantaneous basis in any reference frame. Therefore it is also true on average, thus

$$\bar{i}_{dc} = \frac{3}{2}\left(\frac{\overline{v_{qs}i_{qs} + v_{ds}i_{ds}}}{v_{dc}}\right) \tag{13.3-10}$$

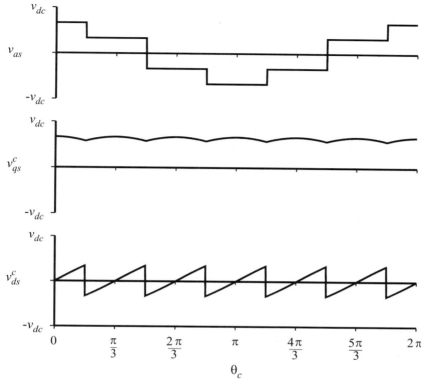

Figure 13.3-5 Comparison of a-phase voltage to q- and d-axis voltage.

If the use of (13.3-10) is restricted to a frame of reference in which the fundamental components of the applied voltages are constant and if the power transmitted through the bridge via the harmonics of the voltage and current waveforms is neglected, (13.3-10) may be approximated as

$$\bar{i}_{dc} = \frac{3}{2}\frac{\bar{v}_{qs}\bar{i}_{qs} + \bar{v}_{ds}\bar{i}_{ds}}{\bar{v}_{dc}} \tag{13.3-11}$$

It should be emphasized that (13.3-11) is only valid in a reference frame in which the variables are constant in the steady state (the converter reference frame, rotor reference frame of a synchronous or brushless dc machine, or synchronous reference frame) and when the harmonic power can be neglected.

Example 3A Suppose a six-step bridge converter drives a three-phase RL load. The system parameters are as follows: $v_{dc} = 100$ V, $r = 1.0\ \Omega$, $l = 1.0$ mH, $\omega_c = 2\pi 100$ rad/s. Estimate the average dc current into the inverter. From (13.3-5)–(13.3-6) we have that $\bar{v}_{qs}^c = 63.7$ V and $\bar{v}_{ds}^c = 0.0$ V. From the steady-state equations representing the RL circuit in the converter reference frame, we obtain

$$
\begin{bmatrix} \bar{i}_{qs}^c \\ \bar{i}_{ds}^c \end{bmatrix} = \begin{bmatrix} r & -\omega_c l \\ \omega_c l & r \end{bmatrix}^{-1} \begin{bmatrix} \bar{v}_{qs}^c \\ \bar{v}_{ds}^c \end{bmatrix}
\tag{3A-1}
$$

from which we obtain $\bar{i}_{qc} = 45.6$ A and $\bar{i}_{ds}^c = 28.7$ A. From (13.3-11) we have that $\bar{i}_{dc} = 43.6$ A. It is instructive to do this calculation somewhat more accurately by including the harmonic power. In particular, from (13.3-2) the harmonic content of the voltage waveform can be calculated, which can then be used to find the total power being supplied by the load as

$$
P_{out} = \frac{3}{2}r \left(\left| \frac{\frac{2}{\pi}v_{dc}}{r + j\omega_c l} \right|^2 + \sum_{i=1}^{\infty} \left(\left| \frac{\frac{2}{\pi(6k-1)}v_{dc}}{r + j(6k-1)\omega_c l} \right|^2 + \left| \frac{\frac{2}{\pi(6k+1)}v_{dc}}{r + j(6k+1)\omega_c l} \right|^2 \right) \right)
\tag{3A-2}
$$

This yields $P_{out} = 4389$ W, which requires an average dc current of 43.9 A. Thus, at least for this load, the approximations made in deriving (13.3-11) are valid.

The 180° voltage source mode of operation just considered is the simplest strategy for controlling the 3-phase bridge topology so as to synthesize a 3-phase ac voltage source from a single-phase dc voltage source. By varying ω_c, variable frequency operation is readily achieved. Nevertheless, there are two distinct disadvantages of this type of operation. First, the only way that the amplitude of the fundamental component can be achieved is by varying v_{dc}. Although this is certainly possible by deriving the dc supply from a phase-controlled rectifier, appropriate control of the power electronic switches can also be used which allows the use of a less expensive uncontrolled dc supply voltage. Such a method is considered in the following section. Secondly, the harmonic content inevitably lowers the machine efficiency. An appropriate switching strategy can substantially alleviate this problem. Thus, although the control strategy just considered is simple, more sophisticated methods of control are generally preferred. The one advantage of the method besides its simplicity is that the amplitude of the fundamental component is the largest possible with the topology considered. For this reason, many other control strategies effectively approach 180° voltage source operation as the desired output voltage increases.

13.4 PULSE-WIDTH MODULATION

In this section, the next level of refinement over 180° voltage source operation is presented. In particular, the pulse-width modulation control strategy set forth in this section allows the amplitude of the fundamental component to be readily controlled.

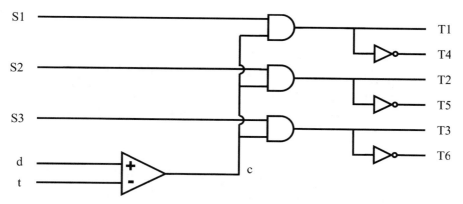

Figure 13.4-1 Pulse-width modulation control schematic (deadtime logic not shown).

Figure 13.4-1 illustrates the logic control strategy and line-to-ground voltages for this pulse-width modulation strategy. Therein, the logic signals S1–S3 are the same as the switching signals $T1$–$T3$ of the 180° voltage source inverter strategy. The control input to the converter is the duty cycle d which may be varied from 0 to 1. The signal t is a triangle waveform that also varies between 0 and 1. The duty cycle d and triangle wave t are inputs of a comparitor, the output of which will be denoted c. The comparitor output is logically added with S1–S3 to yield the switching signals into the actual devices.

Operation of this control circuit is illustrated in Fig. 13.4-2. As alluded to previously, the signals S1–S3 in the pulse-width modulated strategy are identical to $T1$–$T3$ in the 180° voltage-source mode. The duty cycle, d, is assumed to be constant or to vary slowly relative to the sawtooth waveform. The frequency of the triangle wave is the switching frequency f_{sw} (the number of times each switching device is turned on per second), which should be much greater than the frequency of the fundamental component of the output. The output of the comparitor, c, is a square wave whose average value is d. When c is high, the switching signals to the transistors $T1$–$T3$, and hence the voltages, are all identical to those of those of 180° voltage source operation. When c is low, all the voltages are zero.

In order to analyze this pulse-width modulation strategy, it is convenient to make use of the fact the voltages produced by this control strategy are equal to voltages applied in the 180° voltage source operation multiplied by the output of the comparitor. Using Fourier series techniques, the comparitor output may be expressed as

$$c = d + 2d \sum_{k=1}^{\infty} \text{sinc}(kd) \cos k\theta_{sw} \qquad (13.4\text{-}1)$$

where θ_{sw} is the switching angle defined by

$$p\theta_{sw} = \omega_{sw} \qquad (13.4\text{-}2)$$

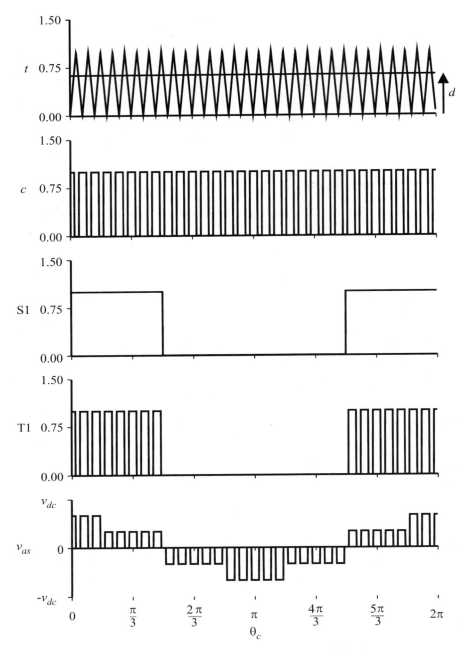

Figure 13.4-2   Pulse-width modulation control signals.

where $\omega_{sw} = f_{sw}/2\pi$. Multiplying (13.4-1) by (13.3-2) yields a Fourier series expression for the a-phase line-to-neutral voltage:

$$v_{as} = \frac{2dv_{dc}}{\pi}\left(\cos\theta_c + \sum_{j=1}^{\infty}\left(\frac{(-1)^{j+1}}{6j-1}\cos((6j-1)\theta_c) + \frac{(-1)^j}{6j+1}\cos(6j+1)\theta_c\right)\right)$$

$$+\frac{2d}{\pi}v_{dc}\sum_{k=1}^{\infty}\mathrm{sinc}(kd)\cos(k\theta_{sw}-\theta_c)$$

$$+\frac{2dv_{dc}}{\pi}\sum_{k=1}^{\infty}\mathrm{sinc}(kd)\sum_{j=1}^{\infty}\left(\frac{(-1)^{j+1}}{6j-1}\cos(k\,\theta_{sw}-(6j-1)\theta_c)\right.$$

$$\left.+\frac{(-1)^j}{6j+1}\cos(k\theta_{sw}-(6j+1)\theta_c)\right)$$

$$+\frac{2d}{\pi}v_{dc}\sum_{k=1}^{\infty}\mathrm{sinc}(kd)\cos(k\theta_{sw}+\theta_c)$$

$$+\frac{2dv_{dc}}{\pi}\sum_{k=1}^{\infty}\mathrm{sinc}(kd)\sum_{j=1}^{\infty}\left(\frac{(-1)^{j+1}}{6j-1}\cos(k\theta_{sw}+(6j-1)\theta_c)\right.$$

$$\left.+\frac{(-1)^j}{6j+1}\cos(k\theta_{sw}+(6j+1)\theta_c)\right) \tag{13.4-3}$$

As can be seen, (13.4-3) is quite involved. The first line indicates that the PWM drive will produce all the harmonics produced by the 180° voltage source inverter, except that all components, including the fundamental, will be scaled by the duty cycle. The next two lines represent the spectrum of the 180° voltage source inverter as projected onto the lower side band of the fundamental and harmonics of the switching frequency. The final two lines represent the spectrum of the 180° voltage source inverter as projected onto the upper side band of the fundamental and harmonics of the switching frequency. Although the high-frequency harmonic components are not of direct interest for machine analysis, the location of these harmonics is important in identification of acoustic and electromagnetic noise.

From (13.4-3) it is apparent that the fundamental component of the applied voltage is given by

$$|v_{as}|_{fund} = d\frac{2}{\pi}v_{dc}\cos\theta_c \tag{13.4-4}$$

From (13.4-3), it follows that the average q- and d-axis voltages are given by

$$\bar{v}_{qs}^c = \frac{2}{\pi}dv_{dc} \tag{13.4-5}$$

$$\bar{v}_{ds}^c = 0 \tag{13.4-6}$$

Thus, by varying the duty cycle, the amplitude of the fundamental component of the inverter voltage is readily achieved with a fixed dc supply voltage.

Figure 13-4.3 illustrates the voltage and current waveforms obtained using the pulse-width modulation control strategy. The system parameters are the same as for Fig. 13.3-4, except that the duty cycle is 0.628, and the switching frequency is 3000 Hz. As can be seen, the a-phase current waveform is approximately 0.628 times the current waveform in Fig. 13.3-4, if the higher frequency components of the a-phase current are neglected.

Although this PWM control strategy allows the fundamental component of the applied voltage to be readily controlled, the disadvantage of this method is that the low-frequency harmonic content adversely affects the performance of the drive. The next modulation scheme considered, sine-triangle modulation, also allows for the control of the applied voltage. However, in this case, there is relatively little low-frequency harmonic content, resulting in nearly ideal machine performance.

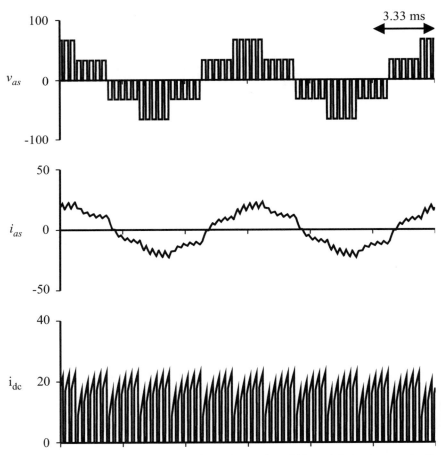

Figure 13.4-3   Voltage and current waveforms for a pulse-width modulation inverter feeding an RL load.

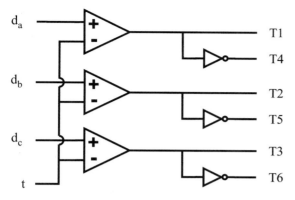

Figure 13.5-1 Pulse-width modulation control schematic (deadtime logic not shown).

13.5 SINE-TRIANGLE MODULATION

In the previous section, a method to control the amplitude of the applied voltages was set forth. Although quite simple, considerable low-frequency harmonics were generated. The sine-triangle modulation strategy illustrated in Fig. 13.5-1 does not share this drawback. Therein, the signals d_a, d_b, and d_c represent duty cycles that vary in a sinusoidal fashion, and t is a triangle wave that varies between -1 and 1 with a period T_{sw}. In practice, each of these variables is typically scaled such that the actual voltage levels make the best use of the hardware on which they are implemented.

Figure 13.5-2 illustrates the triangle wave, a-phase duty cycle, and resulting a-phase line-to-ground voltage. Therein, the a-phase duty cycle is shown as being constant even though it is in fact sinusoidal. This is because the triangle wave is assumed to be of a much higher switching frequency than the duty cycle signals,

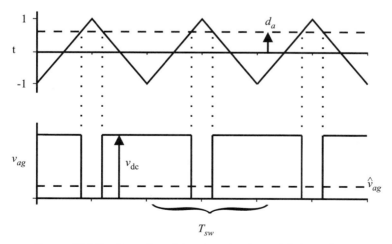

Figure 13.5-2 Operation of the sine-triangle modulation scheme.

so that on the time scale shown the a-phase duty cycle appears to be constant. For the purposes of analysis, it is convenient to define the "fast average" of a variable—that is, the average value over of a period of time T_{sw}—as

$$\hat{x}(t) = \frac{1}{T_{sw}} \int_{t-T_{sw}}^{t} x(\tau) \, d\tau \tag{13.5-1}$$

From Fig. 13.5-2 and (13.5-1) it can be shown that

$$\hat{v}_{ag} = \frac{1}{2}(1 + d_a)v_{dc} \tag{13.5-2}$$

Similarly, it can be shown that

$$\hat{v}_{bg} = \frac{1}{2}(1 + d_b)v_{dc} \tag{13.5-3}$$

$$\hat{v}_{cg} = \frac{1}{2}(1 + d_c)v_{dc} \tag{13.5-4}$$

If d_a, d_b, and d_c form a balanced three-phase set, then these three signals must sum to zero. Making use of this fact, substitution of (13.5-2)–(13.5-4) into (13.2-11)–(13.2-13) yields

$$\hat{v}_{as} = \frac{1}{2}d_a v_{dc} \tag{13.5-5}$$

$$\hat{v}_{bs} = \frac{1}{2}d_b v_{dc} \tag{13.5-6}$$

$$\hat{v}_{cs} = \frac{1}{2}d_c v_{dc} \tag{13.5-7}$$

Although it has been assumed that the duty cycles are sinusoidal, (13.5-5)–(13.5-7) hold whenever the sum of the duty cycles is zero. If the duty cycles are specified as

$$d_a = d\cos\theta_c \tag{13.5-8}$$

$$d_b = d\cos\left(\theta_c - \frac{2\pi}{3}\right) \tag{13.5-9}$$

$$d_c = d\cos\left(\theta_c + \frac{2\pi}{3}\right) \tag{13.5-10}$$

it follows from (13.5-5)–(13.5-7) that

$$\hat{v}_{as} = \frac{1}{2}d v_{dc} \cos\theta_c \tag{13.5-11}$$

$$\hat{v}_{bs} = \frac{1}{2}d v_{dc} \cos\left(\theta_c - \frac{2\pi}{3}\right) \tag{13.5-12}$$

$$\hat{v}_{cs} = \frac{1}{2}d v_{dc} \cos\left(\theta_c + \frac{2\pi}{3}\right) \tag{13.5-13}$$

Recall that the ^ denotes the fast average value. Thus, assuming that the frequency of the triangle wave is much higher than the frequency of the desired waveform, the sine-triangle modulation strategy does not produce any low-frequency harmonics. Transforming (5.5-11)–(5.5-13) to the converter reference frame yields

$$\hat{v}_{qs}^c = \frac{1}{2} d v_{dc} \tag{13.5-14}$$

$$\hat{v}_{ds}^c = 0 \tag{13.5-15}$$

Equations (13.5-14)–(13.5-15) serve as both average-value and (because there are no low-frequency harmonics) fast-average-value expressions.

Figure 13.5-3 illustrates the performance of a sine-triangle modulated converter feeding an RL load. The system parameters are identical to the study in Fig. 13.4-3, except that $d = 0.4$, which results in the voltage waveform with the same fundamental component as in Fig. 13.4-3. Comparing Fig. 13.5-3 to Fig. 13.4-3, it is evident

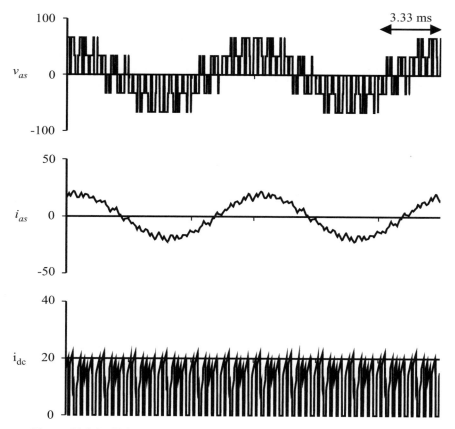

Figure 13.5-3 Voltage and current waveforms using sine-triangle modulation.

that the sine-triangle modulation strategy results in greatly reduced low-frequency current harmonics. This is even more evident as the switching frequency is increased.

From (13.5-11)–(13.5-13) or (13.5-14)–(13.5-15) it can be seen that if d is limited to values between 0 and 1, then the amplitude of the applied voltage varies from 0 to $v_{dc}/2$, whereas in the case of simple duty cycle modulation the amplitude varies between 0 and $2v_{dc}/\pi$. The maximum amplitude produced by the sine-triangle modulation scheme can be increased to the same value as for duty cycle modulation by increasing d to a value greater than one, a mode of operation known as overmodulation.

Figure 13.5-4 illustrates overmodulated operation. In the upper trace, the two lines indicate the envelope of the triangle wave. The action of the comparitors given the value of the duty cycle relative to the envelope of the triangle wave in the upper trace of Fig. 13.5-4 results in the following description of the fast average of the a-phase line-to-ground voltage:

$$\hat{v}_{ag} = \begin{cases} v_{dc} & d_a > 1 \\ \frac{1}{2}(1+d_a)v_{dc} & -1 < d_a < 1 \\ 0 & d_a < 1 \end{cases} \tag{13.5-16}$$

This is illustrated in the second trace of Fig. 13.5-4, wherein the angles θ_1 and θ_2 mark the points at which the a-phase duty cycle is equal to 1 and -1, respectively. Using Fourier analysis, v_{ag} may be expressed in terms of its average value and fundamental component as

$$|v_{ag}|_{avg+fund} = \frac{v_{dc}}{2} + \frac{2v_{dc}}{\pi}f(d)\cos\theta_c \tag{13.5-17}$$

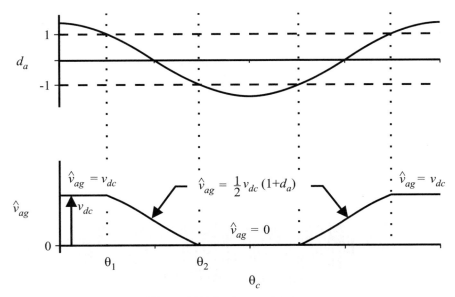

Figure 13.5-4 Overmodulation.

where

$$f(d) = \frac{1}{2}\sqrt{1 - \left(\frac{1}{d}\right)^2} + \frac{1}{4}d\left(\pi - 2\arccos\left(\frac{1}{d}\right)\right) \tag{13.5-18}$$

and d must be greater than unity (overmodulated). The b- and c-phase voltages may be similarly expressed by subtracting and adding $120°$ from θ_c in (13.5-17), respectively, whereupon (13.2-11)–(13.2-13) may be used to express the line-to-neutral voltages. In particular,

$$|v_{as}|_{fund} = \frac{2v_{dc}}{\pi}f(d)\cos(\theta_c) \tag{13.5-19}$$

As d varies from one to infinity, $f(d)$ varies from $\pi/4$ to 1. Thus the amplitude of the fundamental component increases as the duty cycle becomes greater than 1. However, this increase is at a cost; low-frequency harmonics will be present, and they will increase with duty cycle. In particular, at a duty cycle of 1, no harmonics will be present, but at $d = \infty$ the harmonics are equal to those produced by the six-step inverter.

Expressing the b- and c-phase voltages analogously to (13.5-19) and transforming to the converter reference frame yields

$$\bar{v}_{qs}^c = \frac{2v_{dc}}{\pi}f(d) \qquad d \geq 1 \tag{13.5-20}$$

$$\bar{v}_{ds}^c = 0 \qquad d \geq 0 \tag{13.5-21}$$

It is interesting to observe the performance of the overmodulated sine-triangle controlled bridge. Figure 13.5-5 illustrates system performance for the same conditions as illustrated in Fig. 13.5-3, except that the d has been increased to 2. As can be seen, the fundamental component of the voltage and current waveforms has increased; however, this is at the expense of an increase in the low-frequency harmonics. As the duty cycle is further increased, the relative both the voltage and current waveforms will approach those shown in Fig. 13.3-4.

13.6 THIRD-HARMONIC INJECTION

One of the chief limitations of sine-triangle modulation is that the peak value of the fundamental component of the line-to-neutral voltage is limited to $v_{dc}/2$. As it turns out, this limit can be increased by changing the duty-cycle waveforms from the expression given by (13.5-8)–(13.5-10) to the following:

$$d_a = d\cos(\theta_c) - d_3\cos(3\theta_c) \tag{13.6-1}$$

$$d_b = d\cos\left(\theta_c - \frac{2\pi}{3}\right) - d_3\cos(3\theta_c) \tag{13.6-2}$$

$$d_c = d\cos\left(\theta_c + \frac{2\pi}{3}\right) - d_3\cos(3\theta_c) \tag{13.6-3}$$

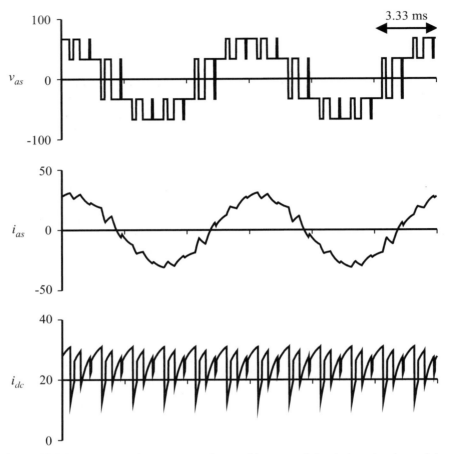

Figure 13.5-5 Voltage and current waveforms with overmodulated sine-triangle modulation.

In order to understand why this can be used to increase the maximum fundamental component of the line-to-neutral voltage, note that applying the fast average definition to (13.2-10) yields

$$\hat{v}_{ng} = \frac{1}{3}\left(\hat{v}_{ag} + \hat{v}_{bg} + \hat{v}_{cg}\right) \tag{13.6-4}$$

Applying the same fast-average definition to (13.2-5)–(13.2-7) and solving for the line-to-neutral voltage yields

$$\hat{v}_{as} = \hat{v}_{ag} - \hat{v}_{ng} \tag{13.6-5}$$

$$\hat{v}_{bs} = \hat{v}_{bg} - \hat{v}_{ng} \tag{13.6-6}$$

$$\hat{v}_{cs} = \hat{v}_{cg} - \hat{v}_{ng} \tag{13.6-7}$$

Substitution of (13.6-1)–(13.6-3) into (13.5-2)–(13.5-4) and then substituting the resulting expressions for $\hat{v}_{ag}$, $\hat{v}_{bg}$, and $\hat{v}_{cg}$ into (13.6-4) and then (13.6-5)–(13.6-7) yields

$$\hat{v}_{as} = \frac{1}{2} d\hat{v}_{dc} \cos(\theta_c) \tag{13.6-8}$$

$$\hat{v}_{bs} = \frac{1}{2} d_b\hat{v}_{dc} \cos\left(\theta_c - \frac{2\pi}{3}\right) \tag{13.6-9}$$

$$\hat{v}_{cs} = \frac{1}{2} d\hat{v}_{dc} \cos\left(\theta_c + \frac{2\pi}{3}\right) \tag{13.6-10}$$

This is the same result as was obtained for sine-triangle modulation in the previous section [see (13.5-11)–(13.5-13)] and, like the previous result, is valid provided that $|d_a|$, $|d_b|$, and $|d_c|$ are less than unity for all θ_c. The difference is that this requirement on $|d_a|$, $|d_b|$, and $|d_c|$ is met. In particular, in the case of sine-triangle modulation, ensuring that $|d_a|$, $|d_b|$, and $|d_c|$ are all less than unity is met by requiring $|d| < 1$, which forces the fundamental component of the line-to-neutral voltage to be limited to $v_{dc}/2$. In the case of third harmonic modulation, the requirement that $|d_a|$, $|d_b|$, and $|d_c|$ are all less than unity can be met with $d > 1$ because the third harmonic term can be used to reduce the peak value of the phase duty-cycle waveforms.

It remains to establish what value should be used for d_3 and what the maximum value of d which can be used is. Because of symmetry, these quantities can be determined by considering just the a-phase over the range $0 \le \theta_c \le \pi/6$. Note that over this range, the effect of the third harmonic term is to reduce the magnitude of d_a (provided that d is positive). However, at the point $\theta_c = \pi/6$, $\cos(3\theta_c)$ is zero and so the amount of the reduction is zero. Evaluating (13.6-1) at $\theta_c = \pi/6$ leads to the requirement that

$$d\cos(\pi/6) \le 1 \tag{13.6-11}$$

which means that

$$d \le \frac{2}{\sqrt{3}} \tag{13.6-12}$$

for the strategy to work correctly. The next step is to establish the value of d_3. To derive this value, require that (13.6-1) have a peak value less than unity for all θ_c when d is its maximum value of $2/\sqrt{3}$. Solving this problem numerically yields $d_3 = d/6$. This answer is unique; any other value will result in overmodulation when $d = 2/\sqrt{3}$.

The primary advantage of this strategy is the increase in available voltage that can be obtained. In particular, substituting $d = 2/\sqrt{3}$ into (13.6-8) the fundamental component of the line-to-neutral voltage is increased to $v_{dc}/\sqrt{3}$; a 15% increase in amplitude over sine-triangle modulation. In regard to the average value modeling of this strategy, (13.5-14)–(13.5-15) are valid provided that $d \le 2/\sqrt{3}$.

13.7   SPACE-VECTOR MODULATION

Another method of achieving 3-phase voltage waveforms that are devoid of low-frequency harmonic content is space-vector modulation [9]. This modulation strategy is particularly designed to work with voltage commands expressed in terms of qd variables. In particular, in this strategy, voltage commands expressed in a stationary reference frame (v_{qs}^{s*} and v_{ds}^{s*}) are sampled at the beginning of each switching cycle, and then the inverter semiconductors are switched in such a way that the fast average of the actual q- and d-axis voltages in the stationary reference frame ($\hat{v}_{qs}^{s}$ and $\hat{v}_{ds}^{s}$) are obtained over the ensuing switching period.

When describing the space-vector modulator algorithm, it is convenient to define the q- and d-axis modulation indexes as the q- and d-axis voltages in the stationary reference frame normalized to the dc voltage:

$$m_q^s = \hat{v}_q^s / v_{dc} \qquad (13.7\text{-}1)$$

$$m_d^s = \hat{v}_d^s / v_{dc} \qquad (13.7\text{-}2)$$

It is likewise convenient to define the commanded modulation indexes as

$$m_q^{s*} = v_q^{s*} / v_{dc} \qquad (13.7\text{-}3)$$

$$m_d^{s*} = v_d^{s*} / v_{dc} \qquad (13.7\text{-}4)$$

Assuming that the dc voltage is constant or at least slowly varying compared to the switching frequency, it is apparent that the fast average of the q- and d-axis voltage will be equal to the commanded voltages if the fast average of the q- and d-axis modulation index is equal to the commanded modulation index.

The space-vector modulation strategy can now be explained in terms of the space-vector diagram illustrated in Fig. 13.7-1. Therein, the q- and d-axis modulation index

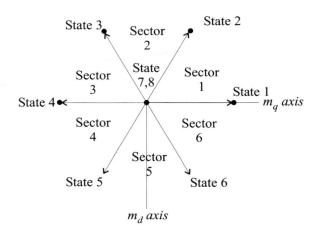

Figure 13.7-1   Space-vector diagram.

Table 13.7-1 Modulation Indexes Versus State

State	$T1/\overline{T4}$	$T2/\overline{T5}$	$T3/\overline{T6}$	$m_{q,x}$	$m_{d,x}$
1	1	0	0	$2/3\cos(0°)$	$-2/3\sin(0°)$
2	1	1	0	$2/3\cos(60°)$	$-2/3\sin(60°)$
3	0	1	0	$2/3\cos(120°)$	$-2/3\sin(120°)$
4	0	1	1	$2/3\cos(180°)$	$-2/3\sin(180°)$
5	0	0	1	$2/3\cos(240°)$	$-2/3\sin(240°)$
6	1	0	1	$2/3\cos(300°)$	$-2/3\sin(300°)$
7	1	1	1	0	0
8	0	0	0	0	0

vector corresponding to each of the eight possible switching states of the converter is shown. The numerical values of the q- and d-axis modulation index corresponding to the ith state, $m_{q,x}$ and $m_{d,x}$, respectively, along with on/off status of the inverter transistors corresponding to that state, are listed in Table 13.7-1. In order to determine the sequence of states required in order to achieve the desired modulation index for a switching cycle, the following steps are performed. First, given the q- and d-axis voltage command in the stationary reference frame, the q- and d-axis modulation index command is calculated using (13.7-3)–(13.7-4). The next step is to limit the magnitude of the modulation index command to reflect the voltage limitation applied to the converter. The magnitude of the modulation index command is defined as

$$m^* = \sqrt{(m_q^{s*})^2 + (m_d^{s*})^2} \qquad (13.7\text{-}5)$$

In the stationary reference frame, the modulation index command vector has a magnitude of m^* and rotates in the qd plane at the desired electrical frequency. The largest magnitude that can be achieved without introducing low-frequency harmonics corresponds to the radius of the largest circle that can be circumscribed within the boundaries of the hexagon connecting the switching state vectors in Fig. 13.7-1. This radius is given by

$$m_{\max} = \frac{1}{\sqrt{3}} \qquad (13.7\text{-}6)$$

The limited modulation index command is next found as follows. First, the magnitude of the raw command is computed using (13.7-5). Then the condition modulation index commands are calculated as follows:

$$m_q^{**} = \begin{cases} m_q^* & m^* \leq m_{\max} \\ m_{\max}\dfrac{m_q^*}{|m^*|} & m^* > m_{\max} \end{cases} \qquad (13.7\text{-}7)$$

$$m_d^{**} = \begin{cases} m_d^* & m^* \leq m_{\max} \\ m_{\max}\dfrac{m_d^*}{|m^*|} & m^* > m_{\max} \end{cases} \qquad (13.7\text{-}8)$$

The next step is to compute the sector of the conditioned modulation command. This is readily calculated from

$$Sector = \text{ceil}\left(\frac{\text{angle}(m_q^{**} - jm_d^{**})3}{\pi}\right) \qquad (13.7\text{-}9)$$

where angle( ) returns the angle of its complex argument and has a range of 0 to 2π, and ceil( ) returns next greatest integer.

Once the sector has been determined, the sequence of states used in the ensuing switching cycle are as set forth in Table 13.7-2. This sequence consists of four states consisting of the initial state denoted α, the second state denoted β, the third state denoted γ, and the final state denoted δ. The initial state is always 7 or 8, and the final state will be 8 if the initial state is 7 and will be 7 if the initial state is 8. Therefore the switching state always begins and ends in a state in which the instantaneous modulation indexes are zero. Another property of the listed state sequence is that only the three states (with states 7 and 8 counted as a single state because they produce identical voltages) with modulation indexes spatially closest to the desired modulation index are used. It is also interesting to observe that with the state sequence listed, the transition between each state and the following state is always achieved by switching the semiconductors in a single converter leg. This is an important feature because it minimizes switching frequency.

After the state sequence has been determined, the time to be spent in each state has to be determined. It can be shown that the fast average of the modulation index is given by

$$\hat{m}_q = \frac{t_\beta}{T_{sw}} m_{q,\beta} + \frac{t_\gamma}{T_{sw}} m_{q,\gamma} \qquad (13.7\text{-}10)$$

$$\hat{m}_d = \frac{t_\beta}{T_{sw}} m_{d,\beta} + \frac{t_\gamma}{T_{sw}} m_{d,\gamma} \qquad (13.7\text{-}11)$$

Table 13.7-2   State Sequence

Sector	Initial State (α)	2nd State (β)	3rd State (γ)	Final State (δ)
1	7	2	1	8
2	7	2	3	8
3	7	4	3	8
4	7	4	5	8
5	7	6	5	8
6	7	6	1	8
1	8	1	2	7
2	8	3	2	7
3	8	3	4	7
4	8	5	4	7
5	8	5	6	7
6	8	1	6	7

where t_β and t_γ denote the amount of time spent in the 2nd and 3rd states of the sequence, β and γ denote index (1 through 6, see Table 13.7-1) of the 2nd and 3rd states of the sequence as determined from Table 13.7-2, and T_{sw} denotes the switching period. Setting the fast average modulation indexes equal to the limited modulation index commands and solving (13.7-9)–(13.7-10) for the switching times yields

$$t_\beta = T_{sw}(m_{d,\gamma}m_q^{**} - m_{q,\gamma}m_d^{**})/D \tag{13.7-12}$$

$$t_\gamma = T_{sw}(-m_{d,\beta}m_q^{**} + m_{q,\beta}m_d^{**})/D \tag{13.7-13}$$

where

$$D = m_{q,\beta}m_{d,\gamma} - m_{q,\gamma}m_{d,\beta} \tag{13.7-14}$$

Once t_β and t_γ have been found, the last step is to determine the instants at which the state transitions will occur. To this end, it is convenient to define $t = 0$ as the beginning of the switching cycle and to define t_A, t_B, and t_C as the times at which the transition from state α to β, β to γ, and γ to δ, respectively, are made. These times are determined in accordance with

$$t_A = (T_{sw} - t_\beta - t_\gamma)/2 \tag{13.7-15}$$

$$t_B = t_A + t_\beta \tag{13.7-16}$$

$$t_C = t_B + t_\gamma \tag{13.7-17}$$

In summary, the space-vector modulator operates as follows. At the beginning of a switching cycle, the commanded modulation indexes are calculated using (13.7-3)–(13.7-4). Next, the conditioned modulation index commands are limited using (13.7-5)–(13.7-8) in order to reflect the voltage limitations of the converter. Next, the sector of the modulation command is determined using (13.7-9) from which the state sequence is established using Table 13.7-2. At this point, (13.7-12)–(13.7-14) are used to determine the amount of time in which each state is spent, and then (13.7-15)–(13.7-17) are used to calculate the actual transition times.

The modeling of this switching algorithm is quite straightforward. In particular, neglecting deadtime and voltage drops, it may be assumed that the output voltage in the stationary reference frame may be expressed as

$$\hat{v}_{qs}^s = m_q^{**}v_{dc} \tag{13.7-18}$$

$$\hat{v}_{ds}^s = m_d^{**}v_{dc} \tag{13.7-19}$$

It is interesting to note that because of the limitation on the magnitude of the modulation index (13.7-6), the limit on the peak value of the fundamental component of the line-to-neutral voltage that can be produced is $v_{dc}/\sqrt{3}$, which is identical to that of third harmonic injection.

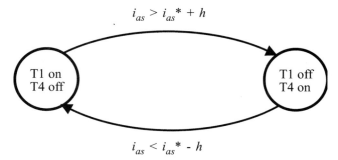

Figure 13.8-1   State transition diagram.

13.8 HYSTERESIS MODULATION

Thus far, all the bridge control strategies considered have resulted in a 3-phase controllable voltage source. However, it is also possible for the bridge to be controlled such that it appears to be a current source. In particular, let i_{as}^*, i_{bs}^*, and i_{cs}^* denote the desired machine or load currents. In order that the actual a-phase current be maintained within a certain tolerance of the desired a-phase currents, the control strategy depicted in Fig. 13.8-1, known as a hysteresis current regulator, is used. As can be seen, if the a-phase current becomes greater than the reference current plus the hysteresis level, the lower transistor of the a-phase leg is turned on, which tends to reduce the current. If the a-phase current becomes less than the reference current minus the hysteresis level, the upper transistor is turned on, which tends to increase the a-phase current. The b- and c-phases are likewise controlled. The net effect is that the a-phase current is within the hysteresis level h of the actual current, as is illustrated in Fig. 13.8-2. As can be seen, the a-phase current tends to wander back and forth between the two error bands. However, note that the slope of the a-phase currents has inflections even when the current is not against one of the error bands; these are due to the switching in the other phase legs.

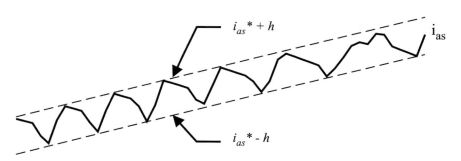

Figure 13.8-2   Allowable current bound.

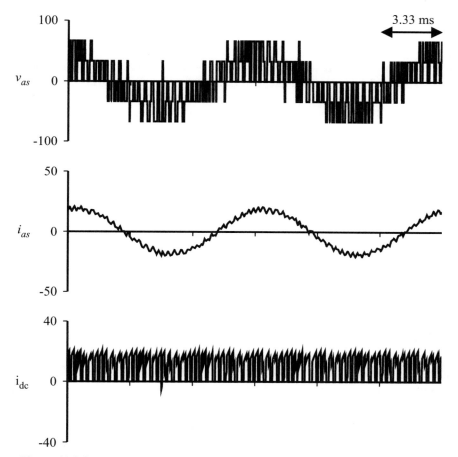

Figure 13.8-3   Voltage and current waveforms using a hysteresis current modulator.

The performance of the a-current-regulated bridge is illustrated in Fig. 13.8-3 for the same conditions are illustrated in Fig. 13.4-3 and Fig. 13.5-3. In this case the commanded a-phase reference current is

$$i_{as}^* = 19.1 \cos{(\theta_c - 17.4°)}$$

and the b- and c-phase reference currents lag the a-phase reference currents by 120° and 240°, respectively. This current command is the fundamental component of the current obtained in Figs. 13.4-3 and 13.5-3. The hysteresis level is set at 2 A. As can be seen, as in the case of the sine-triangle modulated converter, relatively little low-frequency harmonic content is generated.

Although the concept of having a controllable current source is attractive in that it allows us to ignore the stator dynamics, there are several limitations of the current regulated modulation. First, there is a limit on the range of currents which can actually be commanded. In particular, assume that for a given current command the peak line-to-neutral terminal voltage is v_{pk}. Because the peak line-to-neutral voltage the bridge can supply is $2v_{dc}/3$, it is apparent that v_{pk} must be less than $2v_{dc}/3$ if the commanded current is to be obtained. There is another constraint, which is that the peak line-to-line voltage $\sqrt{3}\,v_{pk}$ must be less than the peak line-to-line voltage the converter can achieve, which is equal to v_{dc}. This requirement is more restrictive and defines the steady-state range over which we can expect the currents to be tracked. In particular,

$$v_{pk} < \frac{v_{dc}}{\sqrt{3}} \tag{13.8-1}$$

Note that the maximum voltage achieved using current regulated modulation is greater than that which is achieved using sine-triangle modulation, but equal to that of third-harmonic or space-vector modulation.

In addition to the steady-state limitation on whether the commanded currents will be tracked, there is also a dynamic limitation. In particular, because the stator currents of a machine are algebraically related to the state variables, they cannot be changed instantaneously. Therefore, current tracking will be lost during any step change in commanded currents. When the current command is being changed in a continuous fashion, then current tracking will be maintained provided that the peak line-to-neutral voltage necessary to achieve the commanded currents does not exceed (13.8-1).

One disadvantage of the hysteresis controlled modulation scheme is that the switching frequency cannot be directly controlled. Indirectly, it can be controlled by setting h to an appropriate level—making h smaller increases the switching frequency and making h larger decreases the switching frequency; however, once h is set, the switching frequency will vary depending on the machine parameters and the operating point. For this reason, current source inverters are often synthesized by using suitable control of a voltage source inverter with current feedback.

In regard to average value modeling, the most straightforward approach is to assume that the actual currents are equal to the commanded currents. Because this involves neglecting the dynamics associated with the load, such an approach constitutes a reduced-order model. When taking this approach, a check should be conducted to make sure that sufficient voltage is available to actually achieve the current command because such a modeling approach is not valid if sufficient voltage is not present. In the event that a more sophisticated model is required, the reader is referred to references 10 and 11, which describe how to include dynamics of hysteresis modulation and how to model the effects of loss of current tracking due to insufficient inverter voltage, respectively.

13.9 DELTA MODULATION

An alternate method to achieve current source operation is through the use of delta modulation. This strategy has an advantage over hysteresis modulation in that a maximum switching frequency is set. The disadvantage is that there is no guarantee on how closely the actual current will track the commanded current.

In this strategy, the current error of each phase is calculated in accordance with

$$e_{xs} = i_{xs}^* - i_{xs} \tag{13.9-1}$$

Every T_{sw} seconds (the switching period), the current error is sampled. If the current error is positive, the upper switch is turned on; if it is negative, the lower switch is turned on. Clearly, as the switching period is decreased, the actual current will track the desired current more closely. It should be observed that because the sign of the error does not necessarily change from one sampling to the next, the phase leg involved will not necessarily switch at every sampling. Therefore, the actual switching frequency is considerably lower than $1/T_{sw}$.

There are two variations of this strategy. In the first, the three phase legs are sampled and switched simultaneously. In the second, the switching between phases is staggered. The second method is preferred because it provides slightly higher bandwidth and is more robust with respect to electromagnetic compatibility concerns because the switching in one phase will not interfere with the switching in another. This robustness, coupled with its extreme simplicity in regard to hardware implementation, make the strategy very attractive.

As in the case of hysteresis modulation, there are limitations on how well and under what conditions a current waveform can be achieved. The limitations arising from available voltage are precisely the same as for hysteresis modulation, and so no further discussion will be given in this regard. However, in the case of delta modulation there is an additional limitation in that there is no guarantee on how closely the waveform will track the reference. This must be addressed through careful selection of the switching frequency. Trading off waveform quality versus the switching frequency, while keeping in mind that the actual switching frequency will be lower than the set switching frequency, is a tradeoff best made through the use of a detailed computer simulation of the converter-machine system.

13.10 OPEN-LOOP VOLTAGE AND CURRENT CONTROL

In the previous sections, a variety of modulation strategies were set forth which achieve voltages or currents of a certain magnitude and frequency. For each of these, a method to predict the fast average of the q- and d-axis voltages or currents in the converter reference frame was set forth. In this section, we examine the inverse problem—that of obtaining the appropriate duty cycle(s) and the converter reference frame position in order to achieve a desired fast average synchronous reference frame q- and d-axis voltage or current. The approach in this section is to provide a basic open-loop algorithm. In the next section, closed-loop control algorithms that ensure that commands are exactly obtained, at least in an average value sense, are set forth.

The first modulation strategy considered in this chapter which was capable of implementing a q- and d-axis voltage command was duty-cycle modulation. In order to see how the variables associated with this modulation strategy are related to a voltage command, observe that from (3.6-7) we have that

$$\begin{bmatrix} v_{qs}^e \\ v_{ds}^e \end{bmatrix} = \begin{bmatrix} \cos\theta_{ce} & \sin\theta_{ce} \\ -\sin\theta_{ce} & \cos\theta_{ce} \end{bmatrix} \begin{bmatrix} v_{qs}^c \\ v_{ds}^c \end{bmatrix} \qquad (13.10\text{-}1)$$

where θ_{ce} is angular displacement of the converter reference frame from the synchronous reference frame, that is,

$$\theta_{ce} = \theta_c - \theta_e \qquad (13.10\text{-}2)$$

Replacing v_{qs}^e with the commanded value v_{qs}^{e*}, v_{ds}^e with the commanded value v_{ds}^{e*}, and v_{qs}^c and v_{ds}^c with the average values expressions given by (13.3-5) and (13.3-6) in (13.10-1) yields

$$\begin{bmatrix} v_{qs}^{e*} \\ v_{ds}^{e*} \end{bmatrix} = \begin{bmatrix} \cos\theta_{ce} & \sin\theta_{ce} \\ -\sin\theta_{ce} & \cos\theta_{ce} \end{bmatrix} \begin{bmatrix} \frac{2}{\pi} d v_{dc} \\ 0 \end{bmatrix} \qquad (13.10\text{-}3)$$

From (13.10-3) we obtain

$$d = \frac{\pi}{2v_{dc}} \sqrt{(v_{qs}^{e*})^2 + (v_{ds}^{e*})^2} \qquad (13.10\text{-}4)$$

$$\theta_{ce} = angle(v_{qs}^{e*} - j v_{ds}^{e*}) \qquad (13.10\text{-}5)$$

Together (13.10-4) and (13.10-5) suggest the control strategy illustrated in Fig. 13.10-1. Therein the inputs are the q- and d-axis voltage commands in the synchronous reference frame v_{qs}^{e*} and v_{ds}^{e*}, the dc input voltage to the inverter v_{dc}, and the position of the synchronous reference frame θ_e. The outputs are the duty cycle d and the position of the converter reference frame θ_c, as required by the switch level

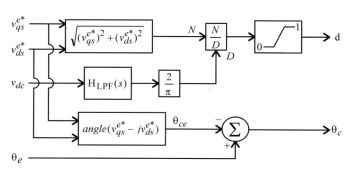

Figure 13.10-1   Voltage control for duty-cycle modulator.

control defined by Fig. 13.4-1 (in which S1, S2, and S3 are defined in the same way as T1, T2, and T3 in Fig. 13.5-1). As can be seen in Fig. 13.10-1, the duty cycle is essentially calculated in accordance with (13.10-4), with the exception that the dc voltage is filtered through a transfer function $H_{LPF}(s)$ to eliminate noise and for the purposes of stability. In addition, a limit is placed on the duty cycle d. The position of the converter reference frame is established by simply adding θ_{ce}, as set forth in (13.10-5), to the position of the synchronous reference frame θ_{ce}.

The next modulation strategy considered was sine-triangle modulation. However, sine-triangle modulation is rarely used in its pure form; it is normally utilized in conjunction with the third-harmonic injection technique because this yields the potential for a greater ac voltage for a given dc voltage than sine-triangle modulation. The development of a strategy to generate the duty cycle and the position of the converter reference frame from the q- and d-axis voltage command is nearly identical to the case for duty cycle modulation except that (13.5-14) and (13.5-15) replace (13.3-5) and (13.3-6) in the development, which results in a change in the gain following the low-pass filter output from $2/\pi$ to $1/2$, the change of the limit on the duty cycle from 1 to $2/\sqrt{3}$, and the introduction of the duty cycle d_3. These modifications are reflected in Fig. 13.10-2. Using the output of this block, the gating of the transistors is readily determined as explained in Sections 13.3 and 13.5.

In the case of space vector modulation, the situation is more straightforward since this switching algorithm is based on a q- and d-axis voltage command, albeit in the stationary reference frame. In this case, the q- and d-axis voltage in the stationary reference frame is calculated from the q- and d-axis command in the stationary reference frame using the frame to frame transformation; in particular, this yields

$$v_{qs}^{s*} = v_{qs}^{e*} \cos \theta_e + v_{ds}^{e*} \sin \theta_e \tag{13.10-6}$$

$$v_{ds}^{s*} = -v_{qs}^{e*} \sin \theta_e + v_{ds}^{e*} \cos \theta_e \tag{13.10-7}$$

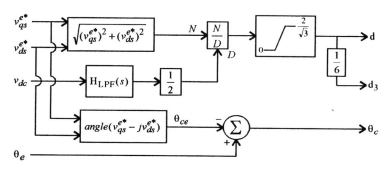

Figure 13.10-2   Voltage control for third-harmonic modulator.

The final modulation strategies discussed in this chapter were hysteresis and delta modulation. Both of these strategies are based on an *abc* variable current command, which is readily computed in terms of a *q*- and *d*-axis current command in the synchronous reference, and the position of the synchronous reference frame, θ_e, using the inverse transformation. In particular, this yields

$$i^*_{abcs} = K^{e^{-1}}_s i^{e*}_{qd0s} \qquad (13.10\text{-}8)$$

13.11 CLOSED-LOOP VOLTAGE AND CURRENT CONTROLS

In the previous section, several strategies for obtaining *q*- and *d*-axis voltage and current commands were discussed. However, each of these methods was open loop. In the case of the voltage control strategies, errors will arise because of logic propagation delays, switching deadtime, and the voltage drop across the semiconductors. In the case of current control, even if the inverter is operated in an ideal sense there will still be a deviation between the actual and commanded current which will have the net effect that the average *q*- and *d*-axis current obtained will not be equal to the commanded values. In this section, closed-loop methods of obtaining *q*- and *d*-axis voltage controls are set forth. These methods are based upon the synchronous regulator concept set forth in reference 12. This concept is based upon the observation that integral feedback loop is most effective if implemented in the synchronous reference frame. Because of the integral feedback, there will be no error for dc terms, provided that the inverter can produce the required voltage. In other words, the average value of the voltages or currents (as expressed in the synchronous reference frame) will be exactly achieved. Because the average value in the synchronous reference frame corresponds to the fundamental component in *abc* variables, it can be seen that integral feedback loop implemented in a synchronous reference frame will ensure that the desired fundamental component of the applied voltages or currents is precisely achieved.

Figure 13.11-1 illustrates a method whereby integral feedback can be introduced into either a voltage control or current control system. Therein, *f* can denote either

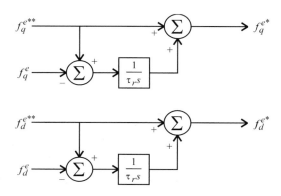

Figure 13.11-1 Synchronous regulator.

voltage v or current i. Furthermore, the superscript $**$ designates a physically desired value, whereas the superscript $*$ designates the inverter command (which will be used in accordance with one of the modulation strategies described in Section 13.10). Note that the strategy is dependent upon the measured value of voltage or current in the synchronous reference frame, f_q^e and f_d^e. These variables are obtained by measuring the abc voltages or currents and transforming them to the synchronous reference frame.

For the purposes of analysis, it is sufficient to only consider the q-axis loop (as the d-axis will yield identical results), whereupon it is convenient to assume that q-axis quantity f_q^e will be equal to the q-axis inverter command f_q^{e*} plus an error term; in particular,

$$f_q^e = f_q^{e*} + f_{q,err}^e \qquad (13.11\text{-}1)$$

Incorporating (13.11-1) into Fig. 13.11-1, it is straightforward to show that the transfer function between the q-axis quantity f_q^e, the command f_q^{e**}, and the error $f_{q,err}^e$ is given by

$$f_q^e = f_q^{e**} + \frac{\tau_r s}{\tau_r s + 1} f_{q,err}^e \qquad (13.11\text{-}2)$$

From (13.12-1), it is readily seen that in the steady state, the average value of the q-axis quantity f_q^e will be equal to the q-axis command f_q^{e**}. It is also possible to see that from the perspective of (13.11-2), it is desirable to make τ_r as small as possible because this decreases the frequency range and extent to which $f_{q,err}^e$ can corrupt f_q^e. However, there is a constraint on how small τ_r can be made. In particular, again using (13.11-1) in conjunction with Fig. 13.11-1 it can be shown that

$$f_q^{e*} = f_q^{e**} - \frac{1}{\tau_r s + 1} f_{q,err}^e \qquad (13.11\text{-}3)$$

As this point, it is important to keep in mind that f_{qs}^{e*} should be relatively free from harmonic content or otherwise distortions in the switching pattern will result. Because $f_{q,err}^e$ contains a considerable high-frequency switching component, τ_r must be large enough that significant switching harmonics are not present in f_q^{e*}. The selection of τ_r is somewhat a function of the modulation strategy. For example, if this regulator is used in conjunction with obtaining a current command, then selecting

$$\tau_r \approx \frac{5}{2\pi f_{sw,est}} \qquad (13.11\text{-}4)$$

where $f_{sw,est}$ is the estimated switching frequency (which can be determined through a computer simulation), should normally produce adequate attenuation of the switching ripple in the inverter command. However, if the regulator is being used in conjunction of sine-triangle modulation with third-harmonic injection or space-vector modulation, then there will be considerable voltage error ripple, where-upon selecting

$$\tau_r \approx \frac{20}{2\pi f_{sw}}$$ (13.11-5)

(where f_{sw} is the switching frequency) is more appropriate. Finally, for voltage source operation using duty cycle modulation, the presence of low-frequency harmonics necessitates an even larger time constant, perhaps on the order of

$$\tau_r \approx \frac{20}{2\pi 5 f_{min}}$$ (13.11-6)

where f_{min} denotes the minimum frequency of the fundamental component of the applied waveform which will be used (this can require a very long time constant and implies poor transient performance).

The regulator shown in Fig. 13.11-1 is designed as a trimming loop wherein voltage source modulation strategy is used to achieve a voltage source command, or in which a current source modulation strategy is used to achieve a current source command. However, it is often the case that a voltage source modulation strategy will be used to obtain a current command. The advantage of this approach to obtaining a current command is that it allows a fixed switching frequency modulation strategy to be used.

One approach to achieving a voltage-source-based current regulator is depicted in Fig. 13.11-2. Inputs to this control are the q- and d-axis current commands i_{qs}^{e*} and i_{ds}^{e*}, the measured q- and d-axis currents i_{qs}^{e} and i_{ds}^{e} (obtained by measuring the abc currents and transforming to the synchronous reference frame), and finally the speed of the synchronous reference frame ω_e. The outputs of the control are the q- and d-axis voltage commands in a synchronous reference frame v_{qs}^{e*} and v_{ds}^{e*}, which are achieved using one of the open-loop control strategies discussed in Section 13.10 (closed-loop voltage control is not necessary because errors in the currents will be

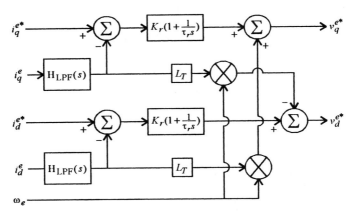

Figure 13.11-2   Voltage-source-based current control.

mitigated by the integral feedback present in the current control loop). Parameters associated with this strategy are the regulator gain K_r, time constant τ_r, and a Thevenin equivalent inductance of the load L_T. The low-pass filter $H_{LPF}(s)$ is designed to have unity gain at dc with a cutoff frequency somewhat below the switching frequency.

In order to gain insight into the operation of this control loop, let us assume that the actual q- and d-axis voltages v_{qs}^e and v_{ds}^e are equal to their commanded values, v_{qs}^{e*} and v_{ds}^{e*}, and that the low-pass filter has dynamics which are appreciably faster than those of this regulator so that they may be ignored for the purpose of designing this control loop, and that on the time scale where this control loop operates (which is much faster than the typical fundamental component of the waveforms but slower than the switching frequency), the load on the inverter may be approximated as

$$v_{qs}^e = \omega_e L_T i_{ds}^e + L_T p i_{qs}^e + e_{qT} \tag{13.11-7}$$

$$v_{ds}^e = -\omega_e L_T i_{qs}^e + L_T p i_{ds}^e + e_{dT} \tag{13.11-8}$$

where e_{qT} and e_{dT} are slowly varying quantities. In essence, this is the model of a voltage behind inductance load. Many machines, including brushless dc machines (see Problem 19) and induction machines (see Problem 20), can have their stator equation approximated by this form for fast transients. Incorporating these assumptions into (13.11-7) through (13.11-8) yields

$$i_{qs}^e = \frac{K_r(\tau_r s + 1)i_{qs}^{e*} + \tau_r s e_{qT}^e}{L_T \tau_r \left(s^2 + \frac{K_r}{L_T} s + \frac{K_v}{L_T \tau_v} \right)} \tag{13.11-9}$$

A similar result can be derived for the d axis. Inspection of (13.11-9) reveals that there will be no steady-state error and that there is no interaction between the q and d axis. This interaction was eliminated by the L_T term in the control. Of course, if this term is not used, or if the value used is not equal to the Thevenin equivalent inductance, then interaction between the q and d axis will exist and can be quite pronounced.

The gain K_r and time constant τ_r may be readily chosen using pole-placement techniques. In particular, if is desired that the pole locations be at $s = -s_1$ and $s = -s_2$, wherein s_1 and s_2 are chosen to be fast as possible, subject to the constraint that the two poles will be considerably slower than the low pass filter and the switching frequency, then the gain and time constant may be readily expressed as

$$K_r = L_T(s_1 + s_2) \tag{13.11-10}$$

$$\tau_r = \frac{1}{s_1} + \frac{1}{s_2} \tag{13.11-11}$$

In utilizing (13.11-10)–(13.11-11), one choice is to make the system critically damped and choose

$$s_1 = s_2 \approx \frac{\pi f_{sw}}{5} \tag{13.12-12}$$

where f_{sw} is the switching frequency. A numerical example in applying this design procedure to the design of the current control loops of a large induction motor drive is set forth in reference 13, and the application of the same general technique to a ac power supply is set forth in reference 14; this latter reference includes an excellent discussion of the decoupling mechanism.

Any of the techniques used in this section will guarantee that provided enough inverter voltage is present, the desired fundamental component of the applied voltage or current will be exactly obtained. Of course, low levels of low-frequency harmonics (including negative sequence terms, 5th and 7th harmonics, etc) and high switching frequency harmonics will still be present. A method of eliminating low-frequency harmonics is set forth in references 15 and 16.

REFERENCES

[1] J. G. Kassakian, M. F. Schlecht, and G. C. Verghese, *Principals of Power Electronics*, Addison-Wesley, Reading, MA, 1991.

[2] Mohan, Undeland, and Robbins, *Power Electronics*, - 2nd ed., John Wiley and Sons / IEEE Press, New York, 1995.

[3] M. H. Rashid, *Power Electronics*, 2nd ed., Prentice-Hall, Englewood Cliffs, NJ, 1993.

[4] R. S. Ramshaw, *Power Electronics Semiconductor Switches*, 2nd ed., Chapman and Hall, London, 1993.

[5] J. T. Tichenor, S. D. Sudhoff, and J. L. Drewniak, Behavioral IGBT Modeling for Prediction of High Frequency Effects in Motor Drives, in *Proceedings of the 1997 Naval Symposium on Electric Machines*, July 28–31, 1997, Newport, RI, pp. 69–75.

[6] R. R. Nucera, S. D. Sudhoff, and P. C. Krause, Computation of Steady-State Performance of an Electronically Commutated Motor, *IEEE Transactions on Industry Applications*, Vol. 25, November/December 1989, pp. 1110–1117.

[7] S. D. Sudhoff and P. C. Krause, Average-Value Model of the Brushless DC 120° Inverter System, *IEEE Transactions on Energy Conversion*, Vol. 5, September 1990, pp. 553–557.

[8] S. D. Sudhoff and P. C. Krause, Operating Modes of the Brushless DC Motor with a 120° Inverter, *IEEE Transactions on Energy Conversion*, Vol. 5, September 1990, pp. 558–564.

[9] H. W. van der Broek, H.-Ch. Skudelny, and G. Stanke, *Analysis and Realization of a Pulse Width Modulator Based on Voltage Space Vectors*, IEEE Industry Applications Society, 1986.

[10] S. F. Glover, S. D. Sudhoff, H. J. Hegner, and H. N. Robey, Jr., Average Value Modeling of a Hysteresis Controlled DC/DC Converter for Use in Electrochemical System Studies, in *Proceedings of the 1997 Naval Symposium on Electric Machines*, July 28–31, 1997, Newport, RI, pp. 77–84.

[11] K. A. Corzine, S. D. Sudhoff, and H. J. Hegner, Analysis of a Current-Regulated Brushless DC Drive, *IEEE Transactions on Energy Conversion*, Vol. 10, No. 3, September 1995, pp. 438–445.

[12] T. M. Rowan and R. J. Kerkman, A New Synchronous Current Regulator and an Analysis of Current-Regulated Inverters, *IEEE Transactions on Industry Applications*, Vol. IA-22, No. 4, 1986, pp. 678–690.

[13] S. D. Sudhoff, J. T. Alt, H. J. Hegner, and H. N. Robey, Jr., and Control of a 15-Phase Induction Motor Drive System, in *Proceedings of the 1997 National Symposium on Electric Machines*, July 28–31, 1997, Newport, RI, pp. 103–110.

[14] O. Wasynczuk, S. D. Sudhoff, T. D. Tran, D. H. Clayton, and H. J. Hegner, A Voltage Control Strategy for Current-Regulated PWM Inverters, *IEEE Trans. on Power Electronics*, Vol. 11, No. 1, Jan. 1996, pp. 7–15.

[15] P. L. Chapman and S. D. Sudhoff, A Multiple Reference Frame Synchronous Estimator/ Regulator, *IEEE Transactions on Energy Conversion*, 1999.

[16] P. L. Chapman and S. D. Sudhoff, Optimal Control of Permanent-Magnet AC Drives with a Novel Multiple Reference Frame Synchronous Estimator/Regulator, in *Proceedings of the 34th Industry Applications Society Annual Meeting*, 1999.

PROBLEMS

1 Show that v_{0s} is zero for a balanced 3-phase induction motor.

2 Show that v_{0s} is zero for a balanced 3-phase synchronous machine.

3 Show that v_{0s} is zero for a balanced 3-phase brushless dc machine with a sinusoidal back emf.

4 Figure 13.P-1 illustrates the a-phase line-to-ground voltage of a 3-phase bridge converter. Determine the diode and transistor forward voltage drops.

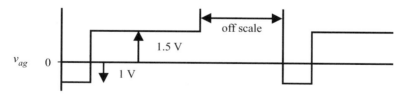

Figure 13.P-1   The a-phase line-to-ground voltage of a 3-phase bridge converter.

5 From Fig. 13.3-1 derive (13.3-1).

6 From (13.3-1) deduce analogous expressions for v_{bcs} and v_{cas}.

7 From Fig. 13.3-2 derive (13.3-2).

8 From (13.3-2) deduce analogous expressions for v_{bs} and v_{cs}.

9 Consider a 3-phase bridge supplying a wye-connected load in which the a-phase, b-phase, and c-phase resistances are 2 Ω, 4 Ω, and 4 Ω, respectively. Given that the dc supply voltage is 100 V and the control strategy is 180° voltage source operation, sketch the a-phase line-to-neutral voltage waveform.

10 Figure 13.P-2 illustrates a circuit that can be used to avoid shoot-through. If 5-V logic is used, the gate threshold turn-on voltage is 3.4 V, and the resistor is 1 kΩ, compute the capacitance necessary to ensure that gate turn-off of one transistor of a pair is gated off 1.5 μs before the second transistor of the pair is gated on.

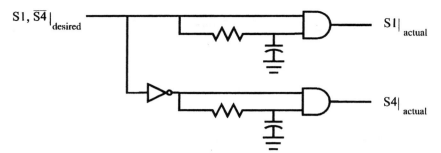

Figure 13.P-2   Circuit that can be used to avoid shoot-through.

11   Consider the 3-hp induction motor whose parameters are listed in Table 4.10-1. Plot the torque–speed and dc current–speed curves if it is being fed from a 3-phase bridge in 180° operation, assuming that the dc voltage is 560 V and the frequency is 120 Hz. Neglect harmonics.

12   Consider the system discussed in Problem 11. Compute the effect of the fifth and seventh harmonics on the average torque if the machine is operating at a slip of 0.025 relative to the fundamental component of the applied voltages.

13   A duty-cycle modulation drive with a dc voltage of 600 V and a duty cycle of 0.75 is used to drive a brushless dc machine. At a certain operating speed, the fundamental component of the stator frequency is 300 Hz. If the switching frequency is 10 kHz, compute the amplitude of the strongest two harmonics in the region of 50 kHz.

14   A brushless dc machine is to be operated from the 3 phase bridge using the PWM control strategy. The dc voltage is 100 V, and the desired q- and d-axis voltages are $v_{qs}^r = 50$ V and $v_{ds}^r = 10$ V. Specify the duty-cycle d and the relationship between θ_c and θ_r such that these voltages are obtained.

15   Derive (13.5-17) and (13.5-18) from Fig. 13.5-4.

16   Consider the 3-hp induction motor if Table 3.1-1. The machine is being fed from a 3-phase bridge operated using sine-triangle modulation with $v_{dc} = 280$ V. If the machine is being operated at a speed of 1710 rpm and the frequency of the fundamental component of the applied voltages is 60 Hz, plot the torque versus duty cycle as the duty-cycle d is varied from 0 to 5.

17   A 3-pole brushless dc machine with the parameters $r_s = 2.99\ \Omega$, $L_{ss} = 11.35$ mH, and $\lambda_m' = 0.156$ Vs/rad is operated from a current regulated inverter with $v_{dc} = 140$ V. If it is being operated at 2670 rpm, plot the locus of points in the q-axis current command versus d-axis current command plane that describes the limits of the region over which the current command can be expected to be obtained.

18   Rederive (13.11-9)–(13.11-11) if a resistive term is included in the load model (13.11-7)–(13.11-8).

19   Ignoring stator resistance and taking the synchronous reference frame to be the rotor reference frame, express L_T, e_{qT}, and e_{dT} in terms of electrical rotor speed for a surface-mounted (nonsalient) permanent magnet synchronous machine.

20 Ignoring stator resistance and assuming that the rotor flux linkages in the synchronous reference frame are constants, express L_T, e_{qT}, and e_{dT} for an induction machine in terms of the q- and d-axis rotor flux linkages and the electrical rotor speed. As an aside, because the rotor windings are shorted, their time derivative tends to be small, which leads to this approximation; it is akin to putting the model in subtransient form in the case of synchronous machines.

Chapter 14

INDUCTION MOTOR DRIVES

14.1 INTRODUCTION

The objective of this chapter is to explore the use of induction machines in variable-speed drive systems. There are three strategies that will be considered herein. The first, volts-per-hertz control, is designed to accommodate variable speed commands by using the inverter to apply a voltage of correct magnitude and frequency so as to approximately achieve the commanded speed without the use of speed feedback. The second strategy is constant slip control. In this control, the drive system is designed so as to accept a torque command input, and therefore speed control requires an additional feedback loop. Although this strategy requires the use of a speed sensor, it has been shown to be highly robust with respect to changes in machine parameters and results in high efficiency of both the machine and inverter. One of the disadvantages of this strategy is that in closed-loop speed control situations the response can be somewhat sluggish. The third strategy considered is field-oriented control. In this method, nearly instantaneous torque control can be obtained, allowing the drive to act as a torque transducer. The disadvantage of this strategy is that in its direct form the sensor requirements are significant, and in its indirect form it is sensitive to parameter variations unless on-line parameter estimation or other steps are taken.

14.2 VOLTS-PER-HERTZ CONTROL

Perhaps the simplest and least expensive induction motor drive strategy is constant volt-per-hertz control. This is a speed control strategy that is based on two observations. The first of these is that the torque–speed characteristic of an induction machine is normally quite steep in the neighborhood of synchronous speed and so the electrical rotor speed will be near to the electrical frequency. Thus, by controlling the frequency one can approximately control the speed. The second observation is based upon the a-phase voltage equation that may be expressed as

$$v_{as} = r_s i_{as} + p\lambda_{as} \tag{14.2-1}$$

For steady-state conditions at intermediate to high speeds wherein the flux linkage term dominates the resistive term in the voltage equation, the magnitude of the applied voltage is related to the magnitude of the stator flux linkage by

$$V_s = \omega_e \Lambda_s \tag{14.2-2}$$

which suggests that in order to maintain constant flux linkage (to avoid saturation), the stator voltage magnitude should be proportional to frequency.

Figure 14.2-1 illustrates one possible implementation of a constant volts-per-hertz drive. Therein, the speed command denoted ω_{rm}^* acts as input to a slew rate limiter (SRL) that acts to reduce transients by limiting the rate of change of the speed command to values between α_{min} and α_{max}. The output of the SRL is multiplied by $P/2$, where P is the number of poles in order to arrive at the electrical rotor speed command ω_r^* to which the radian electrical frequency ω_e is set. The electrical frequency is then multiplied by the volts-per-hertz ratio V_b/ω_b, where V_b is rated voltage and ω_b is rated radian frequency in order to form a rms line-to-neutral voltage command V_s. The rms voltage command V_s is then multiplied by $\sqrt{2}$ in order to obtain a q-axis voltage command v_{qs}^{e*} (the voltage is arbitrarily placed in the q axis). The d-axis voltage command is set to zero. In a parallel path, the electrical frequency ω_e is integrated to determine the position of a synchronous reference frame θ_e. The integration to determine θ_e is periodically reset by an integer multiple of 2π in order to keep θ_e bounded. Together, the q- and d-axis voltage commands may then be passed to any one of a number of modulation strategies in order to achieve the commanded voltage as discussed in Chapter 13. The advantages of this control are that it is simple and that it is relatively inexpensive by virtue of being entirely open loop; speed can be controlled (at least to a degree) without feedback. The principal drawback of this type of control is that because it is open loop, some measure of error will occur, particularly at low speeds.

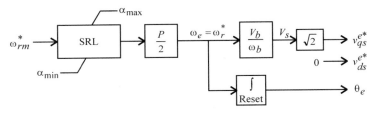

Figure 14.2-1   Elementary volts-per-hertz drive.

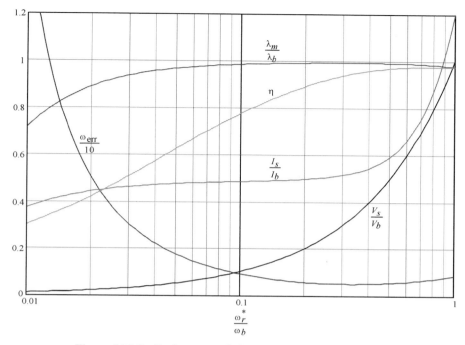

Figure 14.2-2   Performance of elementary volts-per-hertz drive.

Figure 14.2-2 illustrates the steady-state performance of the voltage-per-hertz drive strategy shown in Fig. 14.2-1. In this study, we used a 50-Hp, 4-pole, 1800-rpm, 460-V (line-to-line, rms) machine with the parameters $r_s = 72.5 \, \text{m}\Omega$, $L_{ls} = 1.32 \, \text{mH}$, $L_M = 30.1 \, \text{mH}$, $L'_{lr} = 1.32 \, \text{mH}$, and $r'_r = 41.3 \, \text{m}\Omega$, and the load torque is assumed to be of the form

$$T_L = T_b \left(0.1 S(\omega_{rm}) + 0.9 \left(\frac{\omega_{rm}}{\omega_{bm}} \right)^2 \right) \qquad (14.2\text{-}3)$$

where $S(\omega_{rm})$ is a stiction function that varies from 0 to 1 as ω_{rm} goes from 0 to 0^+. Figure 14.2-3 illustrates the percent error in speed ω_{err} $100(\omega^*_{rm} - \omega_{rm})/\omega^*_{rm}$, normalized voltage V_s/V_b, normalized current I_s/I_b, efficiency η, and normalized air gap flux linkage λ_m/λ_b versus normalized speed command $\omega^*_{rm}/\omega_{bm}$. The base for the air-gap flux linkage is taken to be the no-load air-gap flux linkage that is obtained at rated speed and rated voltage.

As can be seen, the voltage increases linearly with speed command while the rms current remains approximately constant until about 0.5 pu and then rises to approximately 1.2 pu at a speed command of 1 pu. Also, it is evident that the percent speed error remains less than 1% for speeds from 0.1 pu to 1 pu; however, the speed error becomes quite large for speeds less than 0.1 pu. The reason for this is the fact that the

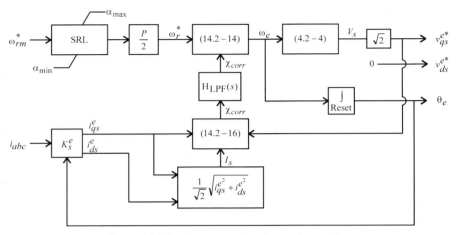

Figure 14.2-3   Compensated volts-per-hertz drive.

magnetizing flux drops to zero as the speed command goes to zero due to the fact that the resistive term dominates the flux-linkage term in (14.2-1) at low speeds. As a result, the torque–speed curve loses its steepness about synchronous speed, resulting in larger percentage error between commanded and actual speed.

The low-speed performance of the drive can be improved by increasing the voltage command at low frequencies in such a way as to make up for the resistive drop. One method of doing this is based on the observation that the open-loop speed regulation becomes poorer at low speeds, because the torque–speed curve becomes decreasingly steep as the frequency is lowered if the voltage is varied in accordance with (14.2-2). To prevent this, it is possible to vary the rms voltage in such a way that the slope of the torque–speed curve at synchronous speed becomes independent of the electrical frequency. Taking the derivative of torque with respect to rotor speed in (4.9-19) at about synchronous speed for an arbitrary electrical frequency and setting it equal to the same derivative about base electrical frequency yields

$$V_s = V_b \sqrt{\frac{r_{s,est}^2 + \omega_e^2 L_{ss,est}^2}{r_{s,est}^2 + \omega_b^2 L_{ss,est}^2}} \qquad (14.2\text{-}4)$$

where $r_{s,est}$ and $L_{ss,est}$ are the estimated value of r_s and L_{ss}, respectively. The block diagram of this version of volts-per-hertz control is identical to that shown in Fig. 14.2-1 with the exception that (14.2-4) replaces (14.2-2).

Several observations are in order. First, it can be readily shown that varying the voltage in accordance with (14.2-4) will yield the same air-gap flux at zero frequency as is seen for no load conditions at rated frequency; thus the air-gap flux does not fall to zero at low frequency as it does when (14.2-2) is used. It is also interesting to observe that (14.2-4) reduces to (14.2-2) at a frequency such that $\omega_e L_{ss,est} \gg r_{s,est}$.

In order to further increase the performance of the drive, one possibility is to utilize the addition of current feedback in determining the electrical frequency command. Although this requires at least one (but more typically two) current sensor(s) that will increase cost, it is often the case that a current sensor(s) will be utilized in any case for overcurrent protection of the drive. In order to derive an expression for the correct feedback, first note that near synchronous speed, the electromagnetic torque may be approximated as

$$T_e = K_{tv}(\omega_e - \omega_r) \tag{14.2-5}$$

where

$$K_{tv} = -\left.\frac{\partial T_e}{\partial \omega_r}\right|_{\omega_r = \omega_e} \tag{14.2-6}$$

If (14.2-4) is used, we obtain

$$K_{tv} = \frac{3\left(\frac{P}{2}\right) L_M^2 \, r_r' \, V_b^2}{r_r'^2 (r_s^2 + \omega_b^2 L_{ss}^2)} \tag{14.2-7}$$

regardless of synchronous speed. Next note that from (4.6-4), torque may be expressed as

$$T_e = \frac{3}{2}\frac{P}{2}(\lambda_{ds}^e \, i_{qs}^e - \lambda_{qs}^e \, i_{ds}^e) \tag{14.2-8}$$

From (4.5-10)–(4.5-11), for steady-state conditions the stator flux linkage equations may be expressed as

$$\lambda_{ds}^e = \frac{v_{qs}^e - r_s \, i_{qs}^e}{\omega_e} \tag{14.2-9}$$

and

$$\lambda_{qs}^e = -\frac{v_{ds}^e - r_s \, i_{ds}^e}{\omega_e} \tag{14.2-10}$$

Approximating v_{qs}^e by its commanded value of v_{qs}^{e*}, approximating v_{ds}^e by its commanded value of zero in (14.2-9) and (14.2-10), and substituting the results into (14.2-8) yields

$$T_e = \frac{3}{2}\frac{P}{2}\frac{1}{\omega_e}(v_{qs}^{e*} \, i_{qs}^e - 2r_s \, I_s^2) \tag{14.2-11}$$

where

$$I_s = \frac{1}{\sqrt{2}} \sqrt{i_{qs}^{e2} + i_{ds}^{e2}} \qquad (14.2\text{-}12)$$

Equating (14.2-7) and (14.2-11) and solving for ω_e yields

$$\omega_e = \frac{\omega_r^* + \sqrt{\omega_r^{*2} + 3P\left(v_{qs}^{e*}\, i_{qs}^{e} - 2r_s\, I_s^2\right)/K_{tv}}}{2} \qquad (14.2\text{-}13)$$

In practice, (14.2-13) is implemented as

$$\omega_e = \frac{\omega_r^* + \sqrt{\max\left(0, \omega_r^{*2} + \mathrm{X}_{corr}\right)}}{2} \qquad (14.2\text{-}14)$$

where

$$\mathrm{X}_{corr} = H_{LPF}(s)\chi_{corr} \qquad (14.2\text{-}15)$$

and where

$$\chi_{corr} = 3P(v_{qs}^{e*}\, i_{qs}^{e} - 2r_s\, I_s^2)/K_{tv} \qquad (14.2\text{-}16)$$

In (14.2-15), $H_{LPF}(s)$ represents the transfer function of a low-pass filter which is required for stability and to remove noise from the measured variables. This filter is often simply a first-order lag filter. The resulting control is depicted in Fig. 14.2-3.

Figure 14.2-4 illustrates the steady-state performance of the compensated voltage-per-hertz drive for the same operating conditions as those of the study depicted in Fig. 14.2-2. Although in many ways the characteristic shown in Fig. 14.2-4 are similar to those of Fig. 14.2-2, there are two important differences. First the air-gap flux does not go to zero at low-speed commands. Second, the speed error is dramatically reduced over the entire operating range of drive. In fact, the speed error using this strategy is less than 0.1% for speed commands ranging from 0.1 to 1.0 pu—without the use of a speed sensor.

In practice, Fig. 14.2-4 is over optimistic for two reasons. First, the presence of a large amount of stiction can result in reduced low-speed performance (the machine will simply stall at some point). Second, it is assumed in the development that the desired voltage is applied. At extremely low commanded voltages, semiconductor voltage drops and the effects of dead time can become important and result in reduced control fidelity. In this case it is possible to use either closed-loop (such as discussed in Section 13.11) or open-loop compensation techniques to help ensure that the desired voltages are actually obtained.

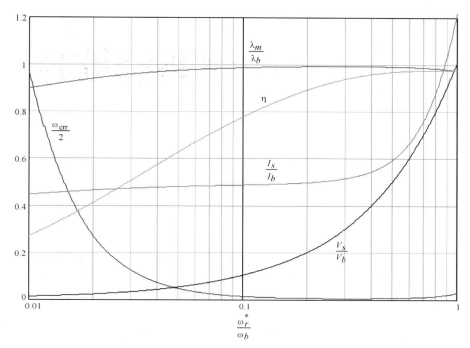

Figure 14.2-4   Performance of compensated volts-per-hertz drive.

Figure 14.2-5 illustrates the start-up performance of the drive for the same conditions as Fig. 14.2-2. In this study, the total mechanical inertia is taken to be $0.82\,\text{N}\cdot\text{m}\cdot\text{s}^2$, and the low-pass filter used to calculate X_{corr} was taken to be a first-order lag filter with a 0.1-s time constant. The acceleration limit, α_{max}, was set to 75.4 rad/s^2. Variables depicted in Fig. 14.2-5 include the mechanical rotor speed ω_{rm}, the electromagnetic torque T_e, the peak magnitude of the air-gap flux linkage $\lambda_m = \sqrt{\lambda_{qm}^2 + \lambda_{dm}^2}$, and finally the a-phase current i_{as}. Initially, the drive is completely off; approximately 0.6 s into the study, the mechanical rotor speed command is stepped from 0 to 188.5 rad/s. As can be seen, the drive comes to speed in roughly 3 s, and the buildup in speed is essentially linear (following the output of the slew rate limit). The air-gap flux takes some time to reach rated value; however, after approximately 0.5 s it is close to its steady-state value. The a-phase current is very well behaved during start-up, with the exception of an initial (negative) peak—this was largely the result of the initial dc offset. Although the drive could be brought to rated speed more quickly by increasing the slew rate, this would have required a larger starting current and therefore a larger and more costly inverter. There are several other compensation techniques set forth in the literature [1,2].

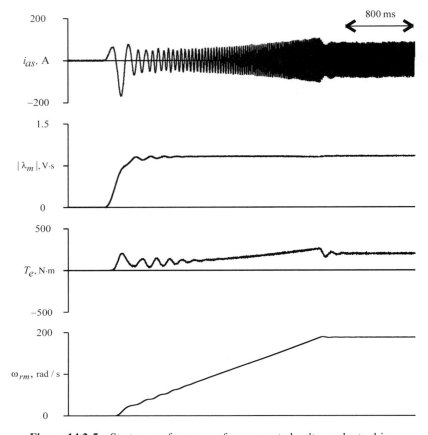

Figure 14.2-5 Start-up performance of compensated volts-per-hertz drive.

14.3 CONSTANT-SLIP CURRENT CONTROL

Although the three-phase bridge inverter is fundamentally a voltage source device, by suitable choice of modulation strategy (such as be hysteresis or delta modulation) it is possible to achieve current source based operation. One of the primary disadvantages of this approach is that it requires phase current feedback (and its associated expense); however, at the same time this offers the advantage that the current is readily limited, making the drive extremely robust, and, as a result, enabling the use of less conservatism when choosing the current ratings of the inverter semiconductors.

One of the simplest strategies for current control operation is to utilize a fixed slip frequency, defined as

$$\omega_s = \omega_e - \omega_r \tag{14.3-1}$$

By appropriate choice of the radian slip frequency, ω_s, several interesting optimizations of the machine performance can be obtained, including achieving the optimal torque for a given value of stator current (maximum torque per amp) as well as the maximum efficiency [3,4].

In order to explore these possibilities, it is convenient to express the electromagnetic torque as given by (4.9-16) in terms of slip frequency that yields

$$T_e = \frac{3\left(\frac{P}{2}\right)\omega_s L_M^2 I_s^2 r_r'}{(r_r')^2 + (\omega_s L_{rr}')^2} \tag{14.3-2}$$

From (14.3-2) it is apparent that in order to achieve a desired torque T_e^* utilizing a slip frequency ω_s, the rms value of the fundamental component of the stator current should be set in accordance with

$$I_s = \sqrt{\frac{2|T_e^*|(r_{r,est}'^2 + (\omega_s L_{rr,est}')^2)}{3P|\omega_s| L_{M,est}^2 r_{rr,est}'}} \tag{14.3-3}$$

In (14.3-3), the parameter subscripts in (14.3-2) have been augmented with 'est' in order to indicate that this relationship will be used in a control system in which the parameter values reflect estimates of the actual values.

As alluded to previously, the development here points toward control in which the slip frequency is held constant at a set value $\omega_{s,set}$. However, before deriving the value of slip frequency to be used, it is important to establish when it is reasonable to use a constant slip frequency. The fundamental limitation that arises in this regard is magnetic saturation. In order to avoid overly saturating the machine, a limit must be placed on the flux linkages. A convenient method of accomplishing this is to limit the rotor flux linkage. From the steady-state equivalent circuit, the a-phase rotor flux linkage may be expressed as

$$\tilde{\lambda}_{ar} = L_{lr}\tilde{I}_{ar}' + L_M(\tilde{I}_{as} + \tilde{I}_{ar}') \tag{14.3-4}$$

From the steady-state equivalent circuit it is also clear that

$$\tilde{I}_{ar}' = -\tilde{I}_{as}\frac{j\omega_e L_M}{j\omega_e L_{rr}' + r_r'/s} \tag{14.3-5}$$

Substitution of (14.3-5) into (14.3-4) yields

$$\tilde{\lambda}_{ar}' = \tilde{I}_{as} L_M \frac{r_r'}{j\omega_s L_{rr}' + r_r'} \tag{14.3-6}$$

Taking the magnitude of both sides of (14.3-6) yields

$$\lambda_r' = I_s L_M \frac{r_r'}{\sqrt{\omega_s^2 L_{rr}'^2 + r_r'^2}} \tag{14.3-7}$$

where λ_r and I_s are the rms value of the fundamental component of the referred a-phase rotor flux linkage and a-phase stator current, respectively. Combining (14.3-7) with (14.3-2) yields

$$T_e = 3\frac{P}{2}\frac{\omega_s \lambda_r^2}{r_r'} \tag{14.3-8}$$

Now, if a constant slip frequency $\omega_{s,set}$ is used and the rotor flux is limited to $\lambda_{r,max}$, then the maximum torque that can be achieved in such an operating mode, denoted $T_{e,thresh}$, is

$$T_{e,thresh} = 3\frac{P}{2}\frac{\omega_{s,set} \lambda_{r,max}^2}{r_{r,est}'} \tag{14.3-9}$$

From (14.3-8), for torque commands in which $|T_e^*| > T_{e,thresh}$, the slip must be varied with in accordance with

$$\omega_s = \frac{2T_e^* r_{r,est}'}{3P \lambda_{r,max}^2} \tag{14.3-10}$$

Figure 14.3-1 illustrates the combination of the ideas into a coherent control algorithm. As can be seen, based on the magnitude of the torque command, the magnitude of the slip frequency ω_s is either set equal to the set point value $\omega_{s,set}$ or to the value arrived at from (14.3-10); and the result is given the sign of T_e^*. The slip frequency ω_s and torque command T_e^* are together used to calculate the rms magnitude of the fundamental component of the applied current I_s, which is scaled by $\sqrt{2}$ in order to arrive at a q-axis current command i_{qs}^{e*}. The d-axis current command i_{ds}^{e*} is set to zero. Of course, the placement of the current command into the q-axis was completely arbitrary; it could have just as well been put in the d axis or any combination of the two, provided that the proper magnitude is obtained. In addition to being used to determine I_s, the slip frequency ω_s is added to the electrical rotor speed ω_r in order to arrive at the electrical frequency ω_e, which is in turn integrated in order to yield the position of the synchronous reference frame θ_e. There are a variety of ways to achieve the commanded q- and d-axis currents as discussed in Chapter 13. Finally, it should also be observed that the control depicted in Fig. 14.3-1 is a torque rather than speed control system; speed control is readily achieved through a separate control loop in which the output is a torque command. Using this approach, it is important that the speed control loop be set to be slow relative to the speed of the torque control, which can be shown to have a dynamic response on the order of the rotor time constant.

One remaining question is the selection of the slip frequency set point $\omega_{s,set}$. Herein, two methods of selection are considered; the first will maximize torque for a given stator current, and the second will maximize the machine efficiency.

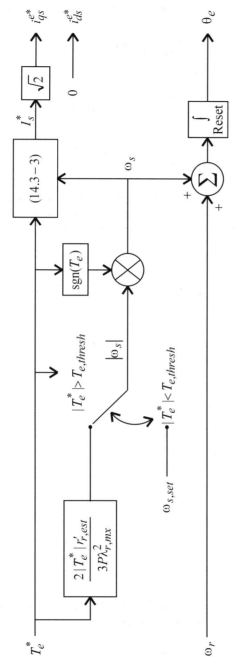

Figure 14.3-1 Constant slip frequency drive.

In order to maximize torque for a given stator current, note that by setting $\omega_s = \omega_{s,set}$ in (14.3-1), torque is maximized for a given stator current by maximizing the ratio

$$\frac{T_e}{I_s^2} = \frac{3\left(\frac{P}{2}\right)\omega_{s,set} L_M^2 r_r'}{(r_r')^2 + (\omega_{s,set} L_r')^2} \tag{14.3-11}$$

Setting the derivative of the right-hand side of (14.3-11) with respect to $\omega_{s,set}$ equal to zero and solving for $\omega_{s,set}$ yields the value of $\omega_{s,set}$ which maximizes the torque for a given stator current. This yields

$$\omega_{s,set} = \frac{r_{r,est}'}{L_{rr,est}'} \tag{14.3-12}$$

In order to obtain an expression for slip frequency that will yield maximum efficiency, it is convenient to begin with an expression for the input power of the machine. With $\tilde{I}_{as} = I_s$, the input power may be expressed as

$$P_{in} = 3I_s \, \mathrm{Re}(\tilde{V}_{as}) \tag{14.3-13}$$

Using the induction motor equivalent circuit model, it is possible to expand (14.3-13) to

$$P_{in} = 3r_s I_s^2 + \frac{3I_s^2 \, \omega_e \, L_M^2 \, \omega_s \, r_r'}{r_r'^2 + (\omega_s L_{rr}')^2} \tag{14.3-14}$$

Comparison of (14.3-14) to (14.3-2) yields

$$P_{in} = 3r_s I_s^2 + \frac{2}{P}\omega_e T_e \tag{14.3-15}$$

Noting that $\omega_e = \omega_s + \omega_r$ and that

$$P_{out} = \frac{2}{P}\omega_r T_e \tag{14.3-16}$$

(14.3-15) may be expressed as

$$P_{in} - P_{out} = 3r_s I_s^2 + \frac{2}{P}T_e \, \omega_s \tag{14.3-17}$$

Substitution of (14.3-2) into (14.3-17) yields an expression for the power losses in terms of torque and slip frequency; in particular,

$$P_{loss} = \frac{2}{P}T_e \left[\frac{r_r' \, r_s}{\omega_s \, L_m^2} + \frac{\omega_s \, r_s \, L_{rr}^2}{r_r' \, L_m^2} + \omega_s \right] \tag{14.3-18}$$

Setting $T_e = T_e^*$ and $\omega_s = \omega_{s,set}$ in (14.3-18) and then minimizing the right-hand side with respect to $\omega_{s,set}$ yields a slip frequency set point of

$$\omega_{s,set} = \frac{r'_{r,est}}{L'_{rr,est}} \frac{1}{\sqrt{\dfrac{L^2_{m,est}}{L'^2_{rr,est}}\dfrac{r_{s,est}}{r'_{r,est}} + 1}} \qquad (14.3\text{-}19)$$

Assuming that $L_{m,est} \approx L'_{rr,est}$, and that $r_{s,est} \approx r'_{r,est}$, it is apparent that the slip frequency for maximum efficiency is lower than the slip frequency for maximum torque per amp by a factor of roughly $1/\sqrt{2}$.

The steady-state performance of a constant slip control drive is depicted in Figure 14.3-2, wherein $\omega_{s,set}$ is determined using (14.3-12), and in Fig. 14.3-3, wherein $\omega_{s,set}$ is determined using (14.3-13). In these studies, the parameters are those of the 50-Hp induction motor discussed in Section 14.2, the maximum rotor flux allowed is set to be the value obtained for no-load operation at rated speed and rated voltage, and the estimated values of the parameters are assumed to be correct. It is assumed that the speed in this study is equal to the commanded speed (the assumption being the drive is used in the context of a closed-loop speed control because rotor position feedback is present). As can be seen, this drive results in appreciably lower losses for low-speed operation than in the case of the volts-per-hertz

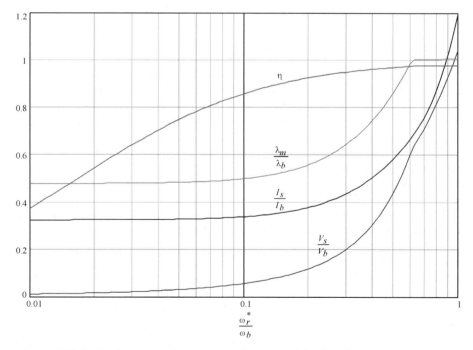

Figure 14.3-2   Performance of constant slip frequency drive (maximum torque per amp).

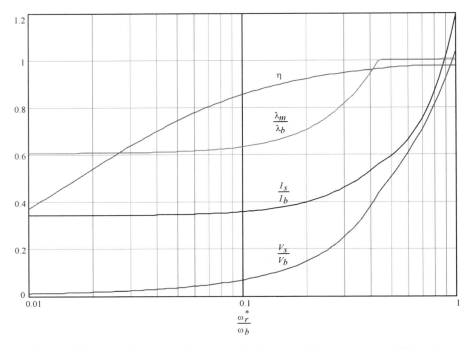

Figure 14.3-3   Performance of constant slip frequency drive (maximum efficiency).

drives discussed in the previous section. Because core losses are not included in Figs. 14.3-2 and 14.3-3, the fact that these strategies utilize reduced flux levels will further accentuate the difference between constant slip and volts-per-hertz controls. In comparing Fig. 14.3-2 to Fig. 14.3-3, it is interesting to observe that setting the slip frequency to achieve maximum torque per amp performance yields nearly the same efficiency as setting the slip frequency to minimize losses. Because inverter losses go up with current, this suggests that setting the slip to optimize torque per amp may yield higher overall efficiency than setting the slip to minimize machine losses—particularly in view of the fact that the lower flux level in maximum torque per amp control will reduce core losses relative to maximum efficiency control.

Another question that arises in regard to the control is the effect that errors in the estimated value of the machine parameters will have on the effectiveness of the control. As it turns out, this algorithm is very robust with respect to parameter estimation, because the optimums being sought (maximum torque per amp or maximum efficiency) are broad. An extended discussion of this is set forth in references 3 and 4.

The use of the constant slip control in the context of a speed control system is depicted in Fig. 14.3-4. Initially the system is at zero speed. Approximately 2 s into the study, the speed command is stepped to 188.5 rad/s. In this study, the machine and load are identical to those in the study shown in Fig. 14.2-4. However,

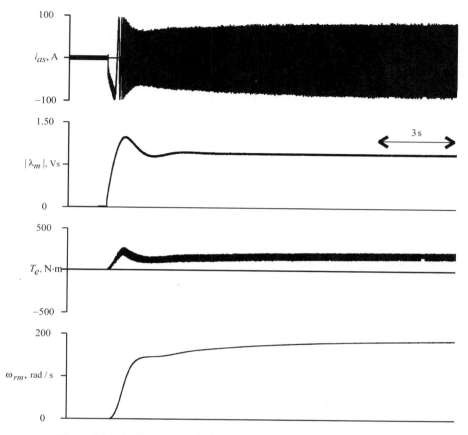

Figure 14.3-4   Start-up performance of constant slip controlled drive.

because the constant-slip control is a torque input control, a speed control is necessary for speed regulation. For the study shown in Fig. 14.3-4, the torque command is
calculated in accordance with the speed control shown in Fig. 14.3-5. This is a relatively simple PI control with a limited output, and anti-windup integration that prevents the integrator from integrating the positive (negative) speed error whenever the
maximum (minimum) torque limit is invoked. For the purposes of this study, the
maximum and minimum torque commands were taken to 218 N · m (1.1 pu) and
0 N · m, respectively, while K_{sc} and τ_{sc} were selected to be 1.64 N · m · s/rad and
2 s, respectively. It can be shown that if $T_e = T_e^*$ and if the machine were unloaded,
this would result in a transfer function between the actual and commanded speeds
with two critically damped poles with 1-s time constants. Also used in conjunction
with the control system was a synchronous current regulator in order to precisely
achieve the current command output of the constant-slip control. To this end, the
synchronous current regulator depicted in Fig. 13.11-1 was used. The time constant
of the regulator was set to 16.7 ms.

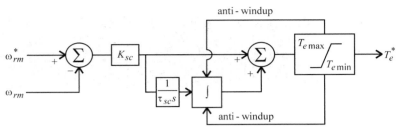

Figure 14.3-5   Speed control.

As can be seen, the start-up performance using the constant-slip control is much slower than using the constant volts-per-hertz control; this is largely because of the fact that the speed control needed to be fairly slow in order to accommodate the sluggish torque response. However, one point of interest is that the stator current, by virtue of the tight current regulation, is very well behaved; in fact the peak value is only slightly above the steady-state value.

14.4   FIELD-ORIENTED CONTROL

In many motor drive systems, it is desirable to make the drive act as a torque transducer wherein the electromagnetic torque can nearly instantaneously be made equal to a torque command. In such a system, speed or position control is dramatically simplified because the electrical dynamics of the drive become irrelevant to the speed or position control problem. In the case of induction motor drives, such performance can be achieved using a class of algorithms collectively known as field-oriented control. There are a number of permutations of this control—stator flux-oriented, rotor flux-oriented, and air-gap flux-oriented—and of these types there are direct and indirect methods of implementation. This text will consider the most prevalent types, which are direct rotor flux-oriented control and indirect rotor field-oriented control. For discussions of the other types, the reader is referred to texts entirely devoted to field-oriented control, such as references 5 and 6.

The basic premise of field-oriented control may be understood by considering the current loop in a uniform flux field shown in Fig. 14.4-1. From the Lorenz force equation, it is readily shown that the torque acting on the current loop is given by

$$T_e = -2BiNLr \sin \theta \qquad (14.4\text{-}1)$$

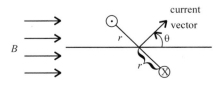

Figure 14.4-1   Torque on a current loop.

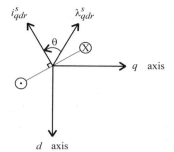

d axis   **Figure 14.4-2** Torque production in an induction motor.

where B is the flux density, i is the current, N is the number of turns, L is the length of the coil into the page, and r is the radius of the coil. Clearly, the magnitude of the torque is maximized when the current vector (defined perpendicular to the surface of the winding forming the current loop and in the same direction as the flux produced by that loop). The same conclusion is readily applied to an induction machine. Consider Fig. 14.4-2. Therein, qd-axis rotor current and flux linkage vectors, $i'_{qdr} = [\, i'_{qr} \quad i'_{dr}\,]^T$ and $\lambda'_{qdr} = [\, \lambda'_{qr} \quad \lambda'_{dr}\,]^T$, respectively, are shown at some instant of time. Repeating (4.6-3), we obtain

$$T_e = \frac{3}{2}\frac{P}{2}(\lambda'_{qr}\, i'_{dr} - \lambda'_{dr}\, i'_{qr}) \tag{14.4-2}$$

which may be expressed as

$$T_e = -\frac{3}{2}\frac{P}{2}|\lambda'_{qdr}||i'_{qdr}|\sin\theta \tag{14.4-3}$$

which is analogous to (14.4-1). Again, for a given magnitude of flux linkage, torque is maximized when the flux linkage and current vectors are perpendicular.

Thus, it is desirable to keep the rotor flux linkage vector perpendicular to the rotor current vector. As it turns out, this is readily accomplished in practice. In particular, in the steady state the rotor flux linkage vector and rotor current vector are always perpendicular for all singly fed induction machines. To see this, consider the rotor voltage equations (4.5-13) and (4.5-14). With the rotor circuits short-circuited and using the synchronous reference frame, it can be shown that the rotor currents may be expressed as

$$i^e_{qr} = -\frac{1}{r'_r}(\omega_e - \omega_r)\lambda^e_{dr} \tag{14.4-4}$$

$$i^e_{dr} = \frac{1}{r'_r}(\omega_e - \omega_r)\lambda^e_{qr} \tag{14.4-5}$$

The dot product of the rotor flux linkage and rotor current vectors may be expressed as

$$\lambda'^e_{qdr} \cdot i'^e_{qdr} = \lambda'^e_{qr}\, i'^e_{qr} + \lambda'^e_{dr}\, i'^e_{dr} \tag{14.4-6}$$

Substitution of (14.4-4) and (14.4-5) into (14.4-6) reveals that this dot product is zero whereupon it may be concluded that the rotor flux and rotor current vectors, as expressed in the synchronous reference frame, are perpendicular. Furthermore, if they are perpendicular in the synchronous reference frame, they are perpendicular in every reference frame. In this sense, in the steady state every singly excited induction machine operates with an optimal relative orientation of the rotor flux and rotor current vectors. However, the defining characteristic of a field-oriented drive is that this characteristic is maintained during transient conditions as well. It is this feature which results in the high transient performance capabilities of this class of drive.

In both direct and indirect field-oriented drives, the method to achieve the condition that the rotor flux and rotor current vectors are always perpendicular is twofold. The first part of the strategy is to ensure that

$$\lambda_{qr}^{\prime e} = 0 \tag{14.4-7}$$

and the second to is to ensure that

$$i_{dr}^{\prime e} = 0 \tag{14.4-8}$$

Clearly, if (14.4-7) and (14.4-8) hold during transient conditions, then by (14.4-6) the rotor flux linkage and rotor current vectors are perpendicular during those same conditions. By suitable choice of θ_e on an instantaneous basis, (14.4-7) can always be satisfied by choosing the position of the synchronous reference frame to put all of the rotor flux linkage in the d axis. Satisfying (14.4-8) can be accomplished by forcing the d-axis stator current to remain constant. To see this, consider the d-axis rotor voltage equation (with zero rotor voltage):

$$0 = r_r^{\prime} i_{dr}^{\prime e} + (\omega_e - \omega_r)\lambda_{qr}^{\prime e} + p\lambda_{dr}^{\prime e} \tag{14.4-9}$$

By suitable choice of reference frame, (14.4-7) is achieved; therefore $\lambda_{qr}^{\prime e}$ can be set to zero in (14.4-9) to yield

$$0 = r_r^{\prime} i_{dr}^{\prime e} + p\lambda_{dr}^{\prime e} \tag{14.4-10}$$

Next, substitution of the d-axis rotor flux linkage equation (4.5-20) into (14.4-10) and rearranging yields

$$p i_{dr}^{\prime e} = -\frac{r_r^{\prime}}{L_{rr}} i_{dr}^{\prime e} - \frac{L_M}{L_{rr}} p i_{ds}^{e} \tag{14.4-11}$$

Equation (14.4-11) can be viewed as a stable first-order differential equation in $i_{dr}^{\prime e}$ with $p i_{ds}^{e}$ as input. Therefore, if i_{ds}^{e} is held constant, then $i_{dr}^{\prime e}$ will go to, and stay at, zero, regardless of other transients that may be taking place.

Before proceeding further, it is motivational to explore some of the other implications of (14.4-7) and (14.4-8) being met. First, combining (14.4-8) with (4.5-17) and (4.5-20), respectively, it is clear that

$$\lambda_{ds}^e = L_{ss} i_{ds}^e \qquad (14.4\text{-}12)$$

and that

$$\lambda_{dr}^{\prime e} = L_M i_{ds}^e \qquad (14.4\text{-}13)$$

Clearly, the d-axis flux levels are set solely by the d-axis stator current. Combining (14.4-2) with (14.4-7), it can be seen that torque may be expressed as

$$T_e = -\frac{3}{2}\frac{P}{2}\lambda_{dr}^{\prime e} i_{qr}^{\prime e} \qquad (14.4\text{-}14)$$

Furthermore, from (14.4-7) and (4.5-19) it can be shown that

$$i_{qr}^{\prime e} = -\frac{L_M}{L_{rr}} i_{qs}^e \qquad (14.4\text{-}15)$$

Combining (14.4-14) and (14.4-15), we obtain

$$T_e = \frac{3}{2}\frac{P}{2}\frac{L_M}{L_{rr}}\lambda_{dr}^{\prime e} i_{qs}^e \qquad (14.4\text{-}16)$$

Together, (14.4-13) and (14.4-16) suggest the "generic" rotor flux-oriented control depicted in Fig. 14.4-3. Therein, variables of the form x^*, $\bar{x}$, and $\hat{x}$ denote commanded, measured, and estimated, respectively; in the case of parameters an addition of a ",est" to the subscript indicates the estimated value. As can be seen, a dc source supplies an inverter driving an induction machine. Based on a torque command T_e^*, the assumed values of the parameters, and the estimated value of the d-axis rotor flux $\hat{\lambda}_{dr}^{\prime e*}$, (14.4-13) is used to formulate a q-axis stator current command i_{qs}^{e*}. The d-axis stator current command i_{ds}^{e*} is calculated such as to achieve

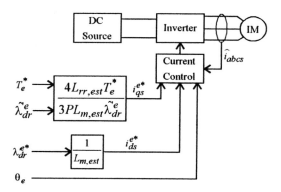

Figure 14.4-3 Generic rotor flux-oriented control.

a rotor flux command (which is typically maintained constant or varied only slowly) $\lambda_{dr}^{\prime e*}$ based on (14.4-10). The q- and d-axis stator current command is then achieved using any one of a number of current source current controls as discussed in Section 13.11. However, this diagram of the rotor flux-oriented field-oriented control is incomplete in two important details: the determination of $\hat{\lambda}_{dr}^{\prime e}$ and the determination of θ_e. The difference in direct and indirect field-oriented control is in how these two variables are established.

14.5 DIRECT ROTOR-ORIENTED FIELD-ORIENTED CONTROL

In direct field-oriented control, the position of the synchronous reference frame based upon the value of the q- and d-axis rotor flux linkages in the rotor reference frame. From (3.6-7), upon setting the position of the stationary reference frame to be zero, we have that

$$\begin{bmatrix} \lambda_{qr}^{\prime e} \\ \lambda_{dr}^{\prime e} \end{bmatrix} = \begin{bmatrix} \cos\theta_e & -\sin\theta_e \\ \sin\theta_e & \cos\theta_e \end{bmatrix} \begin{bmatrix} \lambda_{qr}^{\prime s} \\ \lambda_{dr}^{\prime s} \end{bmatrix} \tag{14.5-1}$$

In order to achieve $\lambda_{qr}^{\prime e} = 0$, from (14.5-1) it is sufficient to define the position of the synchronous reference frame in accordance with

$$\theta_e = angle(\lambda_{qr}^{\prime s} - j\lambda_{dr}^{\prime s}) + \frac{\pi}{2} \tag{14.5-2}$$

whereupon it can be shown that

$$\lambda_{dr}^{\prime e} = \sqrt{(\lambda_{qr}^{\prime s})^2 + (\lambda_{dr}^{\prime s})^2} \tag{14.5-3}$$

The difficulty in this approach is that $\lambda_{qr}^{\prime s}$ and $\lambda_{dr}^{\prime s}$ are not directly measurable quantities. However, they can be estimated using direct measurement of the air-gap flux. In this method, hall-effect sensors (or some other means) are placed in the air-gap and used to measure the air-gap flux in the q and d axis of the stationary reference frame (because the position of the sensors is fixed in a stationary reference frame). The net effect is that λ_{qm}^s and λ_{dm}^s may be regarded as measurable. In order to establish $\lambda_{qr}^{\prime s}$ and $\lambda_{dr}^{\prime s}$ from λ_{qm}^s and λ_{dm}^s, note that

$$\lambda_{qm}^s = L_M(i_{qs}^s + i_{qr}^{\prime s}) \tag{14.5-4}$$

Therefore,

$$i_{qr}^{\prime s} = \frac{\lambda_{qm}^s - L_M i_{qs}^s}{L_M} \tag{14.5-5}$$

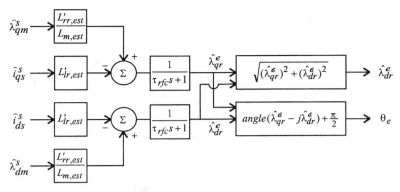

Figure 14.5-1   Rotor flux calculator.

Recall that the q-axis rotor flux linkages may be expressed as

$$\lambda_{qr}^{\prime s} = L_{lr}\, i_{qr}^{\prime s} + L_M(i_{qs}^s + i_{qr}^{\prime s}) \tag{14.5-6}$$

Substitution of (14.5-5) into (14.5-6) yields

$$\lambda_{qr}^{\prime s} = \frac{L_{rr}^{\prime}}{L_M}\lambda_{qm}^s - L_{lr}^{\prime}i_{qs}^s \tag{14.5-7}$$

Performing an identical derivation for the d axis yields

$$\lambda_{dr}^{\prime s} = \frac{L_{rr}^{\prime}}{L_M}\lambda_{dm}^s - L_{lr}^{\prime}\, i_{ds}^s \tag{14.5-8}$$

This suggests the rotor flux calculator shown in Fig. 14.5-1, which calculates both the position of the synchronous reference frame and the d-axis flux linkage. This is based directly upon (14.5-2), (14.5-3), (14.5-7), and (14.5-8), with the

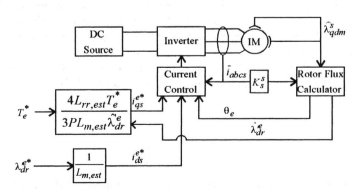

Figure 14.5-2   Direct field-oriented control.

addition of two low-pass filters in order to prevent switching frequency noise from affecting the control (the time constant τ_{rfc} must be set small enough that this transfer function has no effect on the highest-frequency fundamental component that will be utilized) and that, as in Fig. 14.4-1, a more careful distinction is made between measured and estimated values. Figure 14.5-2 depicts the incorporation of the rotor flux calculator into the direct field-oriented control. This will be important in future analysis of the effects of using parameter values in the control algorithms which are not equal to those of the machine (which are highly operating point dependent).

14.6 ROBUST DIRECT FIELD-ORIENTED CONTROL

One of the problems of the control strategy presented is that it is a function of the parameters of the machine. Because of magnetic nonlinearities and the distributed nature of the machine windings (particularly the rotor windings), the model is not particularly accurate. The machine resistances and inductances are highly operating point dependent. In order to understand the potential sources of error, let us first consider the rotor flux observer. From Fig. 14.5-1, recall the rotor flux vector is estimated as

$$\hat{\lambda}_{qdr}^{\prime e} = \frac{L_{rr,est}^{\prime}}{L_{M,est}} \hat{\lambda}_{qd,m}^{s} - L_{lr,est}^{\prime} i_{qds}^{s} \tag{14.6-1}$$

Assuming that the measured flux and measured current are accurate, (14.6-1) is relatively insensitive to parameter variation. To see this, let us first consider the first term on the right-hand side of (14.6-1). The term is a function of $L_{rr,est}/L_{M,est}$. However, note that because the rotor leakage inductance is much less that magnetizing inductance, this ratio will be close to unity regardless of the actual value of the parameters. Hence, this term will not be a strong function of the parameters of the machine. The second term in (14.6-1) is a strong function of the leakage inductance. However, the second term as a whole is considerably smaller than first because the first term represents the air-gap flux and the second has a magnitude equal to the rotor leakage flux. Thus, as a whole, (14.6-1) and the rotor flux estimator are relatively insensitive to the machine parameters.

Another key relationship used in the direct field-oriented control which is a function of the parameters of the machine is the calculation of the q-axis current; in particular,

$$i_{qs}^{e*} = \frac{T_e^*}{\frac{3}{2} \frac{P}{2} \frac{L_{M,est}}{L_{rr,est}} \hat{\lambda}_{dr}^{\prime e}} \tag{14.6-2}$$

Again, because the ratio of $L_{M,est}$ to $L_{rr,est}$ is close to unity for the normal range of parameters, this relationship is again relatively insensitive to parameters.

However, this is not the case for the calculation of the d-axis current, which is calculated in accordance with

$$i_{ds}^{e*} = \frac{\lambda_{dr}^{e*}}{L_{M,est}} \tag{14.6-3}$$

As can be seen, this relationship is highly sensitive to $L_{M,est}$. An error in the d-axis current command will result in an incorrect value of rotor flux linkages. Because the rotor flux linkages can be estimated using the rotor flux estimator shown in Fig. 14.5-1, this error can be readily eliminated by introducing a rotor flux feedback loop shown in Fig. 14.6-1. The basis of this loop is (14.6-3). However, integral feedback is utilized to force the d-axis rotor flux linkage to be equal to its commanded value. For the purposes of design of this feedback loop, it is convenient to assume that $\lambda_{dr}^e = L_M i_{ds}^{e*}$ and that $\hat{\lambda}_{dr}^{\prime e} = \lambda_{dr}^{\prime e}$, whereupon it can be shown that the transfer function between the actual and commanded flux linkages is given by

$$\frac{\lambda_{dr}^{\prime e}}{\lambda_{dr}^{\prime e*}} = \frac{\tau_\lambda s + 1}{\tau_\lambda \frac{L_{M,est}}{L_M} s + 1} \tag{14.6-5}$$

From the form of this transfer function, it can be seen that in the steady-state the rotor flux will be equal to the commanded value. Furthermore, note that if $L_{M,est} = L_M$ the transfer function between the commanded and actual rotor flux is unity. The value of τ_f is chosen so that that $\tau_f 2\pi f_{sw} L_{M,est}/L_M \gg 1$; as a worst-case estimate, $L_{M,est}/L_M$ can be taken to be 0.7 or so in this process.

Although this approach goes along way in making the direct field-oriented control robust with respect to parameter variations, the design can be made even more robust by adding a torque calculator and feedback loop. From (4.6-4), recall that torque may be expressed as

$$T_e = \frac{3}{2} \frac{P}{2} (\lambda_{ds}^s i_{qs}^s - \lambda_{qs}^s i_{ds}^s) \tag{14.6-6}$$

Furthermore, the stator flux may be expressed as

$$\lambda_{qds}^s = L_{ls} i_{qds}^s + \lambda_{qdm}^s \tag{14.6-7}$$

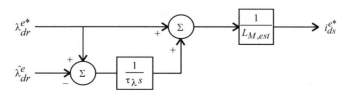

Figure 14.6-1   Flux control loop.

Substitution of (14.6-7) into (14.6-6) yields

$$T_e = \frac{3}{2}\frac{P}{2}(\lambda^s_{dm} i^s_{qs} - \lambda^s_{qm} i^s_{ds})$$ (14.6-8)

which suggests that an estimate for torque can be calculated as

$$\hat{T}_e = \frac{3}{2}\frac{P}{2}(\hat{\lambda}^s_{dm} \hat{i}^s_{qs} - \hat{\lambda}^s_{qm} \hat{i}^s_{ds})$$ (14.6-9)

With the torque calculator present, it is possible to introduce a torque feedback loop shown in Fig. 14.6-2. For the purposes of analysis of this loop, it is convenient to define

$$K_{t,est} = \frac{3}{2}\frac{P}{2}\frac{L_{M,est}}{L'_{rr,est}}\lambda^{\prime e*}_{dr}$$ (14.6-10)

which will be treated as a constant parameter for the purposes of torque loop design. For the purpose of gaining intuition about the performance of the flux loop, it is convenient to assume that

$$T_e = K_t i^{e*}_{qs}$$ (14.6-11)

where

$$K_t = \frac{3}{2}\frac{P}{2}\frac{L_M}{L'_{rr}}\lambda^{\prime e}_{dr}$$ (14.6-12)

Under these conditions, it is readily shown that transfer function between actual and commanded torque is given by

$$\frac{T_e}{T^*_e} = \frac{\tau_t s + 1}{\tau_t \frac{K_{t,est}}{K_t}s + 1}$$ (14.6-13)

Upon inspection of (14.6-13) it is clear that at dc there will be no error between the actual and commanded torque in the steady state (at least if the error in the

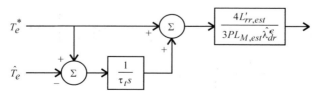

Figure 14.6-2 Torque control loop.

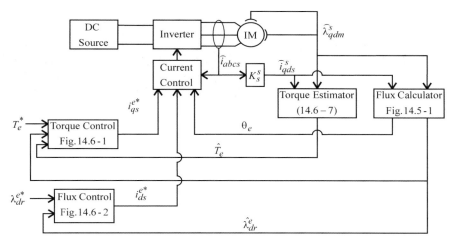

Figure 14.6-3 Robust direct rotor field-oriented control.

current and flux sensors is ignored). Furthermore if K_t and $K_{t,est}$ are equal, the transfer function will be unity, whereupon it would be expected that the actual torque would closely tract the commanded torque even during transients. The time constant τ_t is chosen as small as possible subject to the constraint that $2\pi f_{sw}\tau_t K_{t,est}/K_t \gg 1$ so that switching frequency noise does not enter into the torque command.

Incorporating the rotor and torque feedback loops into the direct field oriented control yields the robust field-oriented control depicted in Fig. 14.6-3. Therein, the use of a flux estimator, torque calculator, and closed-loop torque and flux controls yields a drive which is highly robust with respect to deviations of the parameters from their anticipated values.

The start-up performance of the direct field oriented control is depicted in Fig. 14.6-4. Therein, the machine, load, and speed controls are the same as the study depicted in Fig. 14.3-4, with the exception that the parameters of the speed control were changed to $K_{sc} = 16.4\,\text{N} \cdot \text{m} \cdot \text{s/rad}$ and $\tau_{sc} = 0.2\,\text{s}$ in order to take advantage of the nearly instantaneous torque response characteristic of field-oriented drives. Parameters of the field-oriented controller were $\tau_{rfc} = 100\,\mu\text{s}$, $\tau_\lambda = 50\,\text{ms}$, and $\tau_t = 50\,\text{ms}$. The current commands were achieved using a synchronous current regulator (Fig. 13.11-1) in conjunction with a delta-modulated current control. The synchronous current regulator time constant τ_{scr} and delta modulator switching frequency were set to 16.7 ms and 10 kHz, respectively. Initially, the drive is operating at zero speed, when, approximately 250 ms into the study, the mechanical speed command is stepped from 0 to 188.5 rad/s. The electromagnetic torque steps to the torque limit (which was set to 218 N · m) for approximately 1.5 s, after which the torque command begins to decrease as the speed approaches the commanded value. The drive reaches steady-state conditions within 2 s, and at the same time the peak current utilized was only slightly larger than the steady-state value. In

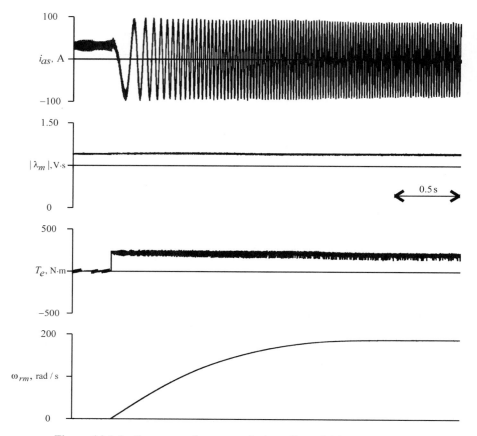

Figure 14.6-4   Start-up performance of robust direct field-oriented drive.

this context, it can be seen that although the control is somewhat elaborate, it can be used to achieve a high degree of dynamic performance with minimal inverter requirements. It is also interesting to observe that the magnitude of the air-gap flux was essentially constant throughout the entire study.

14.7   INDIRECT ROTOR FIELD-ORIENTED CONTROL

Although direct field-oriented control can be made fairly robust with respect to variation of machine parameters, the sensing of the air-gap flux linkage (typically) using hall-effect sensors is somewhat problematic (and expensive) in practice. This has led to considerable interest in indirect field-oriented control methods that are more sensitive to knowledge of the machine parameters but do not require direct sensing of the rotor flux linkages.

In order to establish an algorithm for implementing field-oriented control without knowledge of the rotor flux linkages, it is useful to first establish the electrical frequency that is utilized in direct field-oriented control. From the q-axis rotor voltage equation, we obtain

$$0 = r'_r i'^e_{qr} + (\omega_e - \omega_r)\lambda'^e_{dr} + p\lambda'^e_{qr} \tag{14.7-1}$$

Because $\lambda'^e_{qr} = 0$ for the direct field-oriented control (14.7-1) becomes

$$\omega_e = \omega_r - r'_r \frac{i'^e_{qr}}{\lambda'^e_{dr}} \tag{14.7-2}$$

Using (14.4-12) to express i'^e_{qr} in terms of i^e_{qs} and using (14.4-10) to express λ'^e_{dr} in terms of i^e_{ds}, (14.7-2) becomes

$$\omega_e = \omega_r + \frac{r'_r}{L'_{rr}} \frac{i^e_{qs}}{i^e_{de}} \tag{14.7-3}$$

This raises an interesting question. Suppose that instead of establishing θ_e utilizing the rotor flux calculator in Fig. 14.5-1, it is instead calculated by integrating ω_e, where ω_e is established by

$$\omega_e = \omega_r + \frac{r'_r}{L'_{rr}} \frac{i^{e*}_{qs}}{i^{e*}_{ds}} \tag{14.7-4}$$

As it turns out, this is sufficient to satisfy the conditions for field-oriented control $\lambda'^e_{qr} = 0$ and $i'^e_{dr} = 0$, provided that i^{e*}_{ds} is held constant. To show this, first consider the rotor voltage equations

$$0 = r'_r i'^e_{qr} + (\omega_e - \omega_r)\lambda'^e_{dr} + p\lambda'^e_{qr} \tag{14.7-5}$$

$$0 = r'_r i'^e_{dr} - (\omega_e - \omega_r)\lambda'^e_{qr} + p\lambda'^e_{dr} \tag{14.7-6}$$

Substitution of (14.7-4) into (14.7-5) and (14.7-6) yields

$$0 = r'_r i'^e_{qr} + \frac{r'_r}{L'_{rr}} \frac{i^{e*}_{qs}}{i^{e*}_{de}} \lambda'^e_{dr} + p\lambda'^e_{qr} \tag{14.7-7}$$

$$0 = r'_r i'^e_{dr} - \frac{r'_r}{L'_{rr}} \frac{i^{e*}_{qs}}{i^{e*}_{de}} \lambda'^e_{qr} + p\lambda'^e_{dr} \tag{14.7-8}$$

The next step is to substitute the rotor flux linkage equations into (14.4-7) and (14.4-8), which, upon making the assumption the stator currents are equal to their commanded values, yields

$$0 = r_r' \left[\frac{\lambda_{qr}^{le} - L_M\, i_{qs}^{e*}}{L_{rr}'} \right] i_{qr}^{le} + \frac{r_r'}{L_{rr}'} \frac{i_{qs}^{e*}}{i_{de}^{e*}} \left[L_{rr}'\, i_{dr}^{le} + L_M\, i_{ds}^{e*} \right] + p\lambda_{qr}^{le} \qquad (14.7\text{-}9)$$

$$0 = r_r'\, i_{dr}^{le} - \frac{r_r'}{L_{rr}'} \frac{i_{qs}^{e*}}{i_{de}^{e*}} \lambda_{qr}^{le} + p\left[L_{rr}'\, i_{qr}^{le} + L_M\, i_{ds}^{e*} \right] \qquad (14.7\text{-}10)$$

Noting that $pi_{ds}^{e*} = 0$ and rearranging (14.7-9) and (14.7-10) yields

$$p\lambda_{qr}^{le} = -\frac{r_r'}{L_{rr}'} \lambda_{qr}^{le} - r_r' \frac{i_{qs}^{e*}}{i_{ds}^{e*}} i_{dr}^{le} \qquad (14.7\text{-}11)$$

$$pi_{dr}^{le} = -\frac{r_r'}{L_{rr}'} i_{dr}^{le} + \frac{r_r'}{(L_{rr}')^2} \frac{i_{qs}^{e*}}{i_{ds}^{e*}} \lambda_{qr}^{le} \qquad (14.7\text{-}12)$$

Provided that $pi_{ds}^{e*} = 0$, (14.7-11) and (14.7-12) constitute a set of asymptotically stable differential equations with an equilibrium point of $\lambda_{qr}^{le} = 0$ and $i_{dr}^{le} = 0$. The conclusion is that λ_{qr}^{le} and i_{dr}^{le} will go to and stay at zero, thereby satisfying the conditions for field-oriented control.

Figure 14.7-1 depicts the block diagram of the indirect rotor field-oriented control, which is based upon (14.6-2), (14.6-3), and (14.7-3). As can be seen, it is considerably simpler than the direct field-oriented control—though it is much more susceptible to performance degradation as a result of error in estimating the effective machine parameters.

The start-up performance of the indirect field-oriented drive is depicted in Fig. 14.7-2. Therein, the parameters of the induction machine, speed control, inverter, and current regulator are all identical to those of the corresponding study shown in Fig. 14.6-4. In fact, comparison of Fig. 14.6-4 to Fig. 14.7-2 reveals that the two

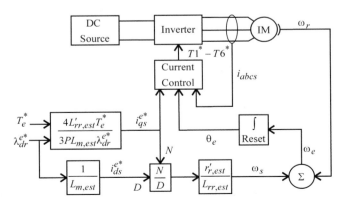

Figure 14.7-1 Indirect rotor field-oriented control.

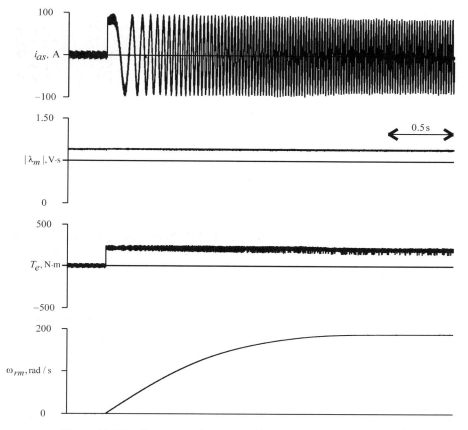

Figure 14.7-2   Start-up performance of indirect field-oriented drive.

controls give identical results. This is largely the result of the fact that the estimated parameters were taken to be the parameter of the machine, and that the machine was assumed to behave in accordance with the machine model described in Chapter 4. However, in reality the machine parameter can vary significantly. Because of the feedback loops, in the case of the direct field-oriented control parameter, variations will have relatively little effect on performance. In the case of the indirect field-oriented drive, significant degradation of the response can result. This is illustrated in Fig. 14.7-3, which is identical to Fig. 14.7-2 with the exception that an error in the estimated parameters is included in the analysis; in particular, $L_{M.est} = 1.25L_m$ and $r'_{r.est} = 0.75r'_r$. As can be seen, although the speed control still achieves the desired speed, the transient performance of the drive is compromised as can be seen by the variation in air-gap flux linkages and electromagnetic torque. This degradation is particularly important at low-speeds where instability in the speed or position controls can result.

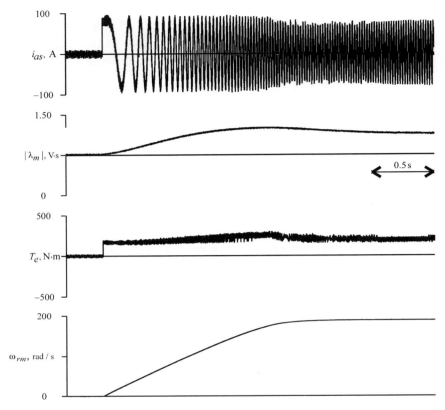

Figure 14.7-3 Start-up performance of indirect field-oriented drive with errors in estimated parameter values.

14.8 CONCLUSIONS

In this chapter, a variety of induction motor drive schemes have been explored including volts-per-hertz control, compensated volts-per-hertz control, constant slip-control, and rotor flux-oriented control. Although most induction motor drives fall into these classes, there are other schemes as well, such as direct torque control [7,8]. Much of the research in the induction motor drives area is focused on sensorless methods of speed control, and methods of obtaining robust indirect flux-oriented drives through the use of parameter estimation and other schemes. To this end, the reader is referred to reference 9.

REFERENCES

[1] A. Muñoz-García, T. A. Lipo, and D. W. Novotny, A New Induction Motor *V/f* Control Method Capable of High-Performance Regulation at Low Speeds, *IEEE Transactions on Industry Applications*, Vol. 34, No. 4, July/August 1998, pp. 813–821.

[2] P. P. Waite and G. Pace, Performance Benefits of Resolving Current in Open-Loop AC Drives, in *Proceedings of the 5th European Conference on Power Electronics and Applications*, Brighton, UK, September 13–16, 1993, pp. 405–409.

[3] O. Wasynczuk, S. D. Sudhoff, K. A. Corzine, J. L. Tichenor, P. C. Krause, I. G. Hansen, and L. M. Taylor, A Maximum Torque Per Ampere Control Strategy for Induction Motor Drives, *IEEE Transactions on Energy Conversion*, Vol. 13, No. 2, June 1998, pp. 163–169.

[4] O. Wasynczuk and S. D. Sudhoff, Maximum Torque Per Ampere Induction Motor Drives— An Alternative to Field-Oriented Control, *1998 SAE Transactions, Journal of Aerospace*, Section 1, pp. 85–93.

[5] A. M. Trzynadlowski, *The Field Orientation Principle in Control of Induction Motors*, Kluwer Academic Publishers, Hingham, MA, 1994.

[6] I. Boldea and S. A. Nasar, *Vector Control of AC Drives*, CRC Press, Boca Raton, FL ,1992.

[7] I. Takahashi and T. Noguchi, A New Quick-Response and High-Efficiency Control Strategy of an Induction Motor, *IEEE Transactions on Industry Applications*, Vol. 22, Sept./Oct. 1986, pp. 820–827.

[8] J. Maes and J. Melkebeek, Discrete Time Direct Torque Control of Induction Motors using Back-EMF measurement, *Conference Record—IAS Annual Meeting (IEEE Industry Applicaions Society)*, Vol. 1, 1998 pp. 407–414.

[9] K. Rajashekara, A. Kawamura, and K. Matsuse, *Sensorless Control of AC Motor Drives— Speed and Position Sensorless Operation*, IEEE Press, Piscataway, NJ, 1996 (selected reprints).

PROBLEMS

1 Derive (14.2-4) and (14.2-7).

2 Calculate the characteristics shown in Fig. 14.2-4 if (*a*) $r_{s,est} = 0.75r_s$ and (*b*) $L_{ss,est} = 1.1L_{ss}$.

3 Consider the 50-hp induction machine used in the studies in this chapter. Suppose the combined inertia of the machine and load is $2\text{N} \cdot \text{m} \cdot \text{s}^2$. Compute the minimum value of $\alpha_{\max}$ of a slew rate limiter by assuming that there is no-load torque and that the rated electrical torque is obtained.

4 Using the parameters of the 50-hp induction motor set forth in this chapter, plot the ratio of power loss divided by torque [see (14.3-18)] and corresponding value of the magnitude of the air-gap flux as a function of slip frequency ω_s.

5 Repeat the study depicted in Fig. 14.3-2 if (*a*) $r'_{r,est} = 0.75r'_r$ and (*b*) $r'_{r,est} = 1.25r'_r$.

6 Derive the transfer function between commanded and actual speed if the control used in Fig. 14.2-5 is used. Assume that the electromagnetic torque is equal to its commanded value, that the load torque is zero, and that the combined inertia of the electric machine and load is J.

7 Suppose it is desired that the rms value of the fundamental component of the rotor flux, λ_r, in the constant-slip control is limited to the value that would be obtained at rated speed, at rated frequency, and rated voltage for no-load conditions. Compute the numerical value of λ_r. If maximum torque per amp control is used, at what percentage of base torque does the control change from constant slip to constant flux?

8 Using the same criterion as in Problem 6, compute λ_{dr}^{e*} for field-oriented control.

9 At moderate and high speeds, it is possible to measure the applied voltages, and currents and based on this information form an estimate of λ_{qs}^s and λ_{ds}^s. Draw a block diagram of a control that could achieve this. Given λ_{qs}^s and λ_{ds}^s, devise flux and torque control loops that could be used to add robustness to the indirect field-oriented controller. Why won't this method work at low-speeds?

10 Derive 14.6-5.

11 Derive 14.6-13.

12 Derive an indirect field-oriented control strategy in which $\lambda_{dr}^{\prime e} = 0$ and $i_{qr}^{\prime e} = 0$.

Chapter 15

BRUSHLESS dc MOTOR DRIVES

15.1 INTRODUCTION

For the purposes of this text, a brushless dc motor drive is defined as a permanent-magnet synchronous machine used in the context of a drive system with rotor position feedback. There are a great variety of brushless dc motor drives. Generally, these drives may be described by the block diagram in Fig. 15.1-1. Therein, the brushless dc drive is seen to consist of four main parts: a power converter, a permanent-magnet synchronous machine (PMSM), sensors, and a control algorithm. The power converter transforms power from the source (such as the local utility or a dc supply bus) to the proper form to drive the permanent-magnet synchronous machine, which, in turn, converts electrical energy to mechanical energy. One of the salient features of the brushless dc drive is the rotor-position sensor (or at least an observer). Based on the rotor position and a command signal(s), which may be a torque command, voltage command, speed command, and so on, the control algorithms determine the gate signal to each semiconductor in the power electronic converter.

The structure of the control algorithms determines the type of brushless dc drive, of which there are two main classes: voltage-source-based drives and current-source (or regulated) drives. Both voltage- and current-source-based drives may be used with permanent-magnet synchronous machines with either sinusoidal or nonsinusoidal back emf waveforms. Machines with sinusoidal back emfs may be controlled so as to achieve nearly constant torque; however, machines with a nonsinusoidal back emf offer reduced inverter sizes and reduced losses for the same power level. The discussion in this chapter will focus on the machines with sinusoidal back emfs; for information on the nonsinusoidal drives, the reader is referred to references 1–3.

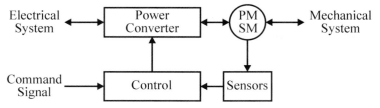

Figure 15.1-1 Brushless dc drive.

In this chapter, a variety of voltage- and current-source-based drives featuring machines with sinusoidal back emf waveforms are analyzed. For each drive considered, detailed simulations will be used to demonstrate performance. Next, average-value models for each drive are set forth, along with a corresponding linearized model for control synthesis. Using these models, the steady-state, harmonic, transient, and dynamic performance of each drive considered will be set forth. Extended design examples will be used to illustrate the performance of the drive within the context of a control system environment.

15.2 VOLTAGE-SOURCE INVERTER DRIVES

Figure 15.2-1 illustrates a voltage-source inverter-based brushless dc drive. Power is supplied from the utility through a transformer, which is depicted as an equivalent voltage behind inductance. The transformer output is rectified, and the rectifier output is connected to the dc link filter, which may be simply an LC filter (L_{dc}, C_{dc}) but which may include a stabilizing filter (L_{st}, r_{st}, C_{st}) as well. The filtered rectifier output is used as a dc voltage source for the inverter, which drives the permanent-magnet synchronous machine. As can be seen, rotor position is an input to the controller. Based upon rotor position and other inputs, the controller determines the switching states of each of the inverter semiconductors. The command signal to the controller may be quite varied depending upon the structure of the controls in the system in which the drive is embedded; it will often be a torque command. Other inputs to the control algorithms may include rotor speed, dc link voltage, and rectifier voltage. Other outputs may include gate signals to the rectifier thyristors if the rectifier is phase-controlled.

Variables of particular interest in Fig. 15.2-1 include the utility supply voltages, v_{au}, v_{bu}, and v_{cu}, the utility currents into the rectifier, i_{au}, i_{bu}, and i_{cu}, the rectifier output voltage, v_r, the rectifier current, i_r, the stabilizing filter current, i_{st}, the stabilizing filter capacitor voltage, v_{st}, the inverter voltage, v_{dc}, the inverter current, i_{dc}, the three-phase currents into the machine, i_{as}, i_{bs}, and i_{cs}, the machine line-to-neutral voltages, v_{as}, v_{bs}, and v_{cs}, and the electrical rotor position, θ_r.

Even within the context of the basic system shown in Fig. 15.2-1, there are many possibilities for control, depending upon whether or not the rectifier is phase-controlled and the details of the inverter modulation strategy. Regardless of the

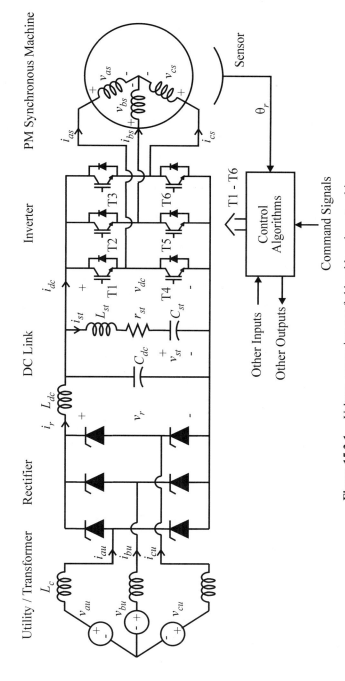

Figure 15.2-1 Voltage-source inverter-fed brushless dc motor drive.

control strategy, it is possible to relate the operation of the converter back to the idealized analysis set forth in Chapter 6, which will be the starting point for our investigation into voltage-source inverter-fed brushless dc drive systems.

15.3 EQUIVALENCE OF VSI SCHEMES TO IDEALIZED SOURCE

In order to make use of the idealized analysis of voltage-source operation of the brushless dc machine set forth in Chapter 6, it is necessary to relate the voltage-source inverter to an ideal source. This relationship is a function of the type of modulation strategy used. In this section, the equivalencies of 180° voltage-source operation, duty-cycle modulation, sine-triangle modulation, third-harmonic injection modulation, and space-vector modulation strategies to idealized sources are established.

The 180° voltage-source inverter-based brushless dc drive is the simplest of all the topologies to be considered in terms of generating the signals required to control the inverter. It is based on the use of relatively inexpensive Hall-effect rotor position sensors. For this reason, the 180° voltage-source inverter drive is a relatively low-cost drive. Furthermore, because the frequency of the switching of the semiconductors corresponds to the speed of the machine, fast semiconductor switching is not important, and switching losses will therefore be negligible. However, the inverter does produce considerable harmonic content that will result in increased machine losses.

In the 180° voltage-source inverter drive, the on or off status of each of the semiconductors is directly tied to electrical rotor position, which is accomplished through the use of the Hall-effect sensors. These sensors can be configured to have a logical high output when they are under a south magnetic pole and a logic zero when they are under a north magnetic pole of the permanent magnet, and they are arranged on the stator of the brushless dc machine as illustrated in Fig. 15.3-1, where ϕ_h denotes

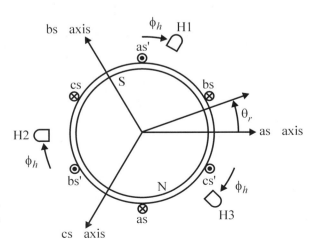

Figure 15.3-1 Electrical diagram of a brushless dc machine.

the shift angle of the Hall-effect sensors. The gate signals for T1, T2, and T3 are set equal to the logical outputs of sensors H1, H2, and H3, respectively. The gate signals T4, T5, and T6 are the logical complements of T1, T2, and T3.

Comparing the gating signals shown in Fig. 15.3-2 to those illustrated in the generic discussion of 180° voltage-source inverters in Chapter 13 (see Fig 13.3-1), it can be seen that the two sets of waveforms are identical provided that the converter angle, θ_c, is related to rotor position, θ_r, and the Hall-effect sensor position, ϕ_h, by

$$\theta_c = \theta_r + \phi_h \tag{15.3-1}$$

From Section 13.3, we know that the fast-average value of the q- and d-axis voltages in the converter reference frame may be expressed as

$$\bar{v}_{qs}^c = \frac{2}{\pi} \bar{v}_{dc} \tag{15.3-2}$$

$$\bar{v}_{ds}^c = 0 \tag{15.3-3}$$

In (15.3-2) and (15.3-3) and throughout this chapter, v_{dc} is taken to be a fast-average value, as indicated by the overbar notation. There are a number of subtle points in doing this, however. The most important of these points is that while we found the expressions for the q- and d-axis voltage using the fast-average definition with a period tied to the converter angle, the average of v_{dc} will often be determined

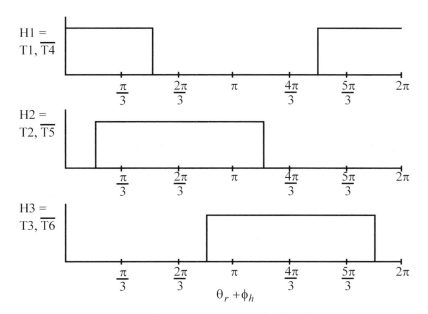

Figure 15.3-2   Semiconductor switching signals.

using the fast-average definition with a period based on the fundamental frequency of the input to the rectifier forming the dc supply. One method of managing this potential inconsistency in a formal sense is through the notion of double averaging [4]. However, in this text the averages will be treated as if they can simply be used together. For the purposes of the analysis considered in this text, this admittedly heuristic practice yields excellent results.

From (15.3-1) the difference in the angular position between the converter reference frame and rotor reference frame is the Hall-effect phase-delay angle, ϕ_h. Using this information, the fast-average of the stator voltages may be determined in the rotor reference frame using the frame-to-frame transformation ${}^cK_s^r$ (see Section 3.6), which yields

$$\bar{v}_{qs}^r = \frac{2}{\pi}\bar{v}_{dc}\cos\phi_h \tag{15.3-4}$$

$$\bar{v}_{ds}^r = -\frac{2}{\pi}\bar{v}_{dc}\sin\phi_h \tag{15.3-5}$$

Comparison of (15.3-4) and (15.3-5) indicates that at least in terms of the fundamental component, the operation of the brushless dc machine from a 180° inverter is identical to that of a brushless dc machine fed by an ideal three-phase variable-frequency voltage source with an rms amplitude of

$$v_s = \frac{1}{\sqrt{2}}\frac{2}{\pi}\bar{v}_{dc} \tag{15.3-6}$$

and a phase delay of

$$\phi_v = \phi_h \tag{15.3-7}$$

Figure 15.3-3 illustrates the steady-state performance of a six-step voltage-source inverter. In this study, the inverter voltage v_{dc} is regulated at 125 V and the mechanical rotor speed is 200 rad/s. The machine parameters are $r_s = 2.98\,\Omega$, $L_q = L_d = 11.4$ mH, $\lambda_m' = 0.156$ V·s, and $P = 4$. There is no phase advance. As can be seen, the nonsinusoidal a-phase voltage results in time varying q- and d-axis voltages. The effect of the harmonics is clearly evident in the a-phase current waveform as well as in the q- and d-axis current waveforms. Also apparent are the low-frequency torque harmonics (six times the fundamental frequency) that result. The current harmonics do not contribute to the average torque; therefore the net effect of the harmonics is to increase machine losses. On the other hand, because the inverter is switching at a relatively low-frequency (six times the electrical frequency of the fundamental component of the applied voltage), switching losses are extremely low.

This drive system is extremely easy to implement in hardware; however, at the same time it is difficult to utilize in a speed control system, because the fundamental component of the applied voltage cannot be regulated unless a phase-controlled rectifier is used. Although this is certainly possible, and has often been done in

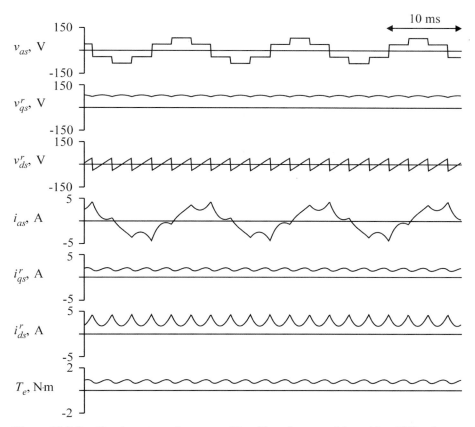

Figure 15.3-3   Steady-state performance of brushless dc motor drive with a 180° voltage-source inverter.

the past, it is generally advantageous to control the applied voltages with the inverter rather than rectifier, because this minimizes the total number of controlled power semiconductor devices.

In order to control the amplitude of the fundamental component of the applied voltage, duty-cycle PWM may be used, as is discussed in Section 13.4. In this case, the gate drive signals T1–T6 are modulated in order to control the amplitude of the applied voltage. Recall from Section 13.4 that for duty-cycle PWM the fast-average q- and d-axis voltages are given by

$$\bar{v}_{qs}^c = \frac{2}{\pi} d\bar{v}_{dc} \tag{15.3-8}$$

and

$$\bar{v}_{ds}^c = 0 \tag{15.3-9}$$

Using (15.3-1) to relate the positions of the converter and rotor reference frames, the frame-to-frame transformation may be used to express the q- and d-axis voltages in the rotor reference frame. In particular,

$$\bar{v}_{qs}^r = \frac{2}{\pi} d\bar{v}_{dc} \cos \phi_h \tag{15.3-10}$$

$$\bar{v}_{ds}^r = -\frac{2}{\pi} d\bar{v}_{dc} \sin \phi_h \tag{15.3-11}$$

From (15.3-10)–(15.3-11), it is clear that the rms value of the fundamental component of the applied voltage is

$$v_s = \frac{1}{\sqrt{2}} \frac{2}{\pi} d\bar{v}_{dc} \tag{15.3-12}$$

The phase delay given by (15.3-7) is applicable to the duty-cycle modulated PWM drive in addition to the 180° voltage-source inverter.

Figure 15.3-4 illustrates the performance of a duty-cycle modulated PWM drive. For this study, the parameters are identical to those of the study depicted in Fig. 15.3-3, with the exception of the modulation strategy. The duty cycle is 0.9, the switching frequency is 5 kHz, and the dc voltage is 138.9 V, which yields the same fundamental component of the applied voltage as in the previous study. As can be seen, the voltage waveforms possess an envelope similar in shape to that of the six-step case; however, they are rapidly switching within that envelope. The current waveforms are virtually identical to the previous study, although there is a small high-frequency component.

By utilizing PWM, the amplitude of the applied voltage is readily varied. However, due to the increased switching frequency, the switching losses in the converter are increased. The losses in the machine will be similar to those in the previous study.

Like duty-cycle modulation, sine-triangle modulation may also be used to control the amplitude of the voltage applied to the brushless dc machine. However, in this case, Hall-effect sensors are not generally adequate to sense rotor position. Recall from Section 13.5 that phase-leg duty cycles are continuous functions of converter angle, which implies that they will be continuous functions of rotor position. For this reason, a resolver or an optical encoder must be used as the rotor position sensor. Although this increases the cost of the drive, and also increases the switching losses of the power electronics devices, the sine-triangle modulated drive does have an advantage in that the low-frequency harmonic content of the machine currents are greatly reduced, thereby reducing losses in machines with sinusoidal back emfs as well as reducing acoustic noise and torque ripple.

In the case of the sine-triangle PWM inverter, the angular position used to determine the phase-leg duty cycles, the converter angle, is equal to the electric rotor position plus an offset, that is,

$$\theta_c = \theta_r + \phi_v \tag{15.3-13}$$

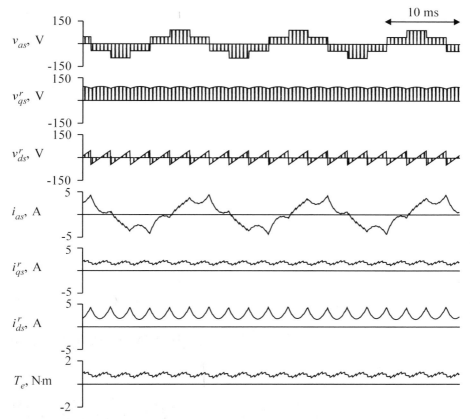

Figure 15.3-4   Steady-state performance of a duty-cycle modulated brushless dc motor drive.

From Section 13.5,

$$\bar{v}_{qs}^c = \begin{cases} \frac{1}{2}d\bar{v}_{dc} & 0 < d \le 1 \\ \frac{2}{\pi}\bar{v}_{dc}f(d) & d > 1 \end{cases} \tag{15.3-14}$$

$$\bar{v}_{ds}^c = 0 \tag{15.3-15}$$

where

$$f(d) = \frac{1}{2}\sqrt{1 - \left(\frac{1}{d}\right)^2} + \frac{1}{4}d\left(\pi - 2\arccos\left(\frac{1}{d}\right)\right) \qquad d > 1 \tag{15.3-16}$$

Using (15.3-13) to compute the angular difference of the locations of the converter and rotor reference frames, we may express the fast average of the q- and d-axis stator voltages as

$$\bar{v}_{qs}^r = \begin{cases} \frac{1}{2}\bar{v}_{dc}\, d\cos\phi_v & d \le 1 \\ \frac{2}{\pi}\bar{v}_{dc}\, f(d)\cos\phi_v & d > 1 \end{cases} \tag{15.3-17}$$

$$\bar{v}_{ds}^r = \begin{cases} -\frac{1}{2}\bar{v}_{dc}\, d\sin\phi_v & d \le 1 \\ -\frac{2}{\pi}\bar{v}_{dc}\, f(d)\sin\phi_v & d > 1 \end{cases} \tag{15.3-18}$$

Figure 15.3-5 illustrates the performance of a sine-triangle modulated drive. The parameters and operating conditions are identical to those in the previous study, with

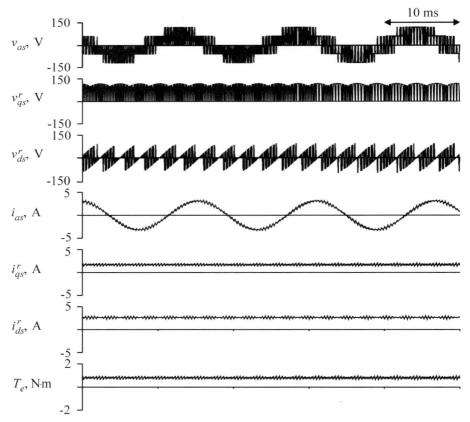

Figure 15.3-5   Steady-state performance of sine-triangle modulated brushless dc motor drive.

the exception that the dc voltage has been increased to 176.8 V (the duty cycle is 0.9 and the switching frequency is 5 kHz). This yields the same fundamental component of the applied voltage as in the previous two studies. Although the amplitude of the fundamental component of the voltage waveforms is similar to the duty-cycle modulated case, the harmonic content of the waveform has been significantly altered. This is particularly evident in the current waveforms, which no longer contain significant harmonic content. As a result, the torque waveform is also devoid of low-frequency harmonics. Like duty-cycle modulation, this strategy allows the fundamental component of the applied voltage to be changed. In addition, the phase can be readily changed, and low-frequency current and torque harmonics are eliminated. However, the price for these benefits is that rotor position must be known on a continuous basis, which requires either an optical encoder or resolver, sensors that are considerably more expensive than Hall-effect devices. Methods of eliminating the need for the encoder or resolver have been set forth, but are beyond the scope of this text [5,6].

In Chapter 13, the next modulation strategy considered was third-harmonic modulation. Recall that for this strategy the analysis is the same as for sine-triangle modulation, with the exception that the amplitude of the duty-cycle d may be increased to $2/\sqrt{3}$ before overmodulation occurs. Therefore, we have that

$$\bar{v}_{qs}^r = \frac{1}{2}d\bar{v}_{dc}\cos\phi_v \qquad 0 \le d \le 2/\sqrt{3} \qquad (15.3\text{-}19)$$

$$\bar{v}_{ds}^r = -\frac{1}{2}d\bar{v}_{dc}\sin\phi_v \qquad 0 \le d \le 2/\sqrt{3} \qquad (15.3\text{-}20)$$

The final voltage-source modulation strategy considered in Chapter 13 was space-vector modulation. This strategy is designed to accept a q- and d-axis voltage command in the stationary reference frame as an input (which is obtained by transforming the voltage command in the rotor reference frame and utilizing the frame-to-frame transformation), and the fast-average of the output q- and d-axis voltages are equal to the input, provided that the peak commanded line-to-neutral input voltage magnitude is less than $v_{dc}/\sqrt{3}$. If this limit is exceeded, the q- and d-output voltage vector retains its commanded direction, but its magnitude is limited. Thus we have that

$$\bar{v}_{qs}^r = \begin{cases} v_{qs}^{r*} & v_{spk}^* < v_{dc}/\sqrt{3} \\[2mm] \frac{\bar{v}_{dc}}{\sqrt{3}}\frac{v_{qs}^{r*}}{v_{spk}^*} & v_{spk}^* \ge v_{dc}/\sqrt{3} \end{cases} \qquad (15.3\text{-}21)$$

$$\bar{v}_{ds}^r = \begin{cases} v_{ds}^{r*} & v_{spk}^* < v_{dc}/\sqrt{3} \\[2mm] \frac{\bar{v}_{dc}}{\sqrt{3}}\frac{v_{ds}^{r*}}{v_{spk}^*} & v_{spk}^* \ge v_{dc}/\sqrt{3} \end{cases} \qquad (15.3\text{-}22)$$

where

$$v_{spk}^* = \sqrt{\left(v_{qs}^{r*}\right)^2 + \left(v_{ds}^{r*}\right)^2} \qquad (15.3\text{-}23)$$

In order to summarize the results of this section, notice that in each case the fast-average q- and d-axis voltages may be expressed as

$$\bar{v}_{qs}^r = \bar{v}_{dc} m \cos \phi_v \qquad (15.3\text{-}24)$$

$$\bar{v}_{ds}^r = -\bar{v}_{dc} m \sin \phi_v \qquad (15.3\text{-}25)$$

where

$$m = \begin{cases} \frac{2}{\pi} & 180° \text{ voltage-source modulation} \\ \frac{2}{\pi}d & \text{duty-cycle modulation} \\ \frac{1}{2}d & \text{sine-triangle modulation } (d \leq 1) \\ \frac{2}{\pi}f(d) & \text{sine-triangle modulation } (d \geq 1) \\ \frac{1}{2}d & \text{third-harmonic modulation } (d \leq 2/\sqrt{3}) \\ \frac{v_{spk}^*}{v_{dc}} & \text{space-vector modulation } (v_{spk}^* \leq v_{dc}/\sqrt{3}) \\ \frac{1}{\sqrt{3}} & \text{space-vector modulation } (v_{spk}^* \geq v_{dc}/\sqrt{3}) \end{cases} \qquad (15.3\text{-}26)$$

In the case of space-vector modulation, observe that ϕ_v is defined as

$$\phi_v = angle(v_{qs}^{r*} - jv_{ds}^{r*}) \qquad (15.3\text{-}27)$$

15.4 AVERAGE-VALUE ANALYSIS OF VSI DRIVES

The average-value model of basic voltage-source inverters drive consist of four parts: (i) the rectifier model, (ii) the dc link and stabilizing filter model, (iii) the inverter model, and (iv) the machine model. From Chapter 11, recall that the fast-average rectifier voltage is given by

$$\bar{v}_r = v_{ro} \cos \alpha - r_r \bar{i}_r - l_r p \bar{i}_r \qquad (15.4\text{-}1)$$

where v_{r0}, r_r, and l_r are given by

$$v_{r0} = \begin{cases} \frac{3\sqrt{3}}{\pi}\sqrt{2}E & \text{three phase rectifier} \\ \frac{2}{\pi}\sqrt{2}E & \text{single phase rectifier} \end{cases} \qquad (15.4\text{-}2)$$

$$r_r = \begin{cases} \frac{3}{\pi}\omega_{eu} L_c & \text{three phase rectifier} \\ \frac{2}{\pi}\omega_{eu} L_c & \text{single phase rectifier} \end{cases} \qquad (15.4\text{-}3)$$

$$l_r = \begin{cases} 2L_c & \text{three phase rectifier} \\ L_c & \text{single phase rectifier} \end{cases} \qquad (15.4\text{-}4)$$

In (15.4-2)–(15.4-4), ω_{eu} is the radian electrical frequency of the source feeding the rectifier (not the fundamental frequency being synthesized by the drive) E is the rms line-to-neutral utility voltage (line-to-line voltage in single-phase applications), and L_c is the commutating inductance. In the typical case wherein a transformer/rectifier is used, E and L_c reflect the utility voltage and transformer leakage impedance referred to the secondary (drive) side of the transformer.

The rectifier electrical dynamics may be expressed as

$$L_{dc}pi_r = v_r - v_{dc} - r_{dc}i_r \tag{15.4-5}$$

where r_{dc} is the resistance associated with the inductor L_{dc}. Treating the variables in (15.4-5) as fast-average values yields

$$L_{dc}p\bar{i}_r = \bar{v}_r - \bar{v}_{dc} - r_{dc}\bar{i}_r \tag{15.4-6}$$

In (15.4-5) the rectifier voltage is given by (15.4-1); however, that expression for the rectifier voltage involves the time derivative of i_r. Hence (15.4-1) and (15.4-5) should be combined into a single differential equation. In particular,

$$p\bar{i}_r = \frac{v_{r0}\cos\alpha - \bar{v}_{dc} - r_{rl}\bar{i}_r}{L_{rl}} \tag{15.4-7}$$

where

$$r_{rl} = r_r + r_{dc} \tag{15.4-8}$$

$$L_{rl} = L_r + L_{dc} \tag{15.4-9}$$

Finally, using Kirchhoff's laws the dc voltage, stabilizing filter current, and stabilizing filter voltage are governed by

$$p\bar{v}_{dc} = \frac{\bar{i}_r - \bar{i}_{st} - \bar{i}_{dc}}{C_{dc}} \tag{15.4-10}$$

$$p\bar{i}_{st} = \frac{\bar{v}_{dc} - \bar{v}_{st} - r_{st}\bar{i}_{st}}{L_{st}} \tag{15.4-11}$$

and

$$p\bar{v}_{st} = \frac{\bar{i}_{st}}{C_{st}} \tag{15.4-12}$$

respectively. Because the rectifier current must be positive, (15.4-7) is only valid for this condition. If the rectifier current is zero and the derivative given by (15.4-7) is negative, then the derivative should be set to zero because the diodes or thyristors will be reverse biased. From (13.3-11) the dc current into the converter may be approximated as

$$\bar{i}_{dc} = \frac{3}{2} \frac{\bar{v}_{qs}^r \bar{i}_{qs}^r + \bar{v}_{ds}^r \bar{i}_{ds}^r}{\bar{v}_{dc}} \qquad (15.4\text{-}13)$$

Substituting (15.3-19) and (15.3-20) into (15.4-13) and simplifying yields

$$\bar{i}_{dc} = \frac{3}{2} m (\bar{i}_{qs}^r \cos \phi_v - \bar{i}_{ds}^r \sin \phi_v) \qquad (15.4\text{-}14)$$

The next step in developing the average-value model for the voltage-source inverter drive is the incorporation of the electrical dynamics of the machine in average-value form. Taking the fast-average of (6.3-10) and (6.3-11) and rearranging yields

$$p\bar{i}_{qs}^r = \frac{\bar{v}_{qs}^r - r_s\bar{i}_{qs}^r - \omega_r L_d \bar{i}_{ds}^r - \omega_r \lambda_m'}{L_q} \qquad (15.4\text{-}15)$$

$$p\bar{i}_{ds}^r = \frac{\bar{v}_{ds}^r - r_s\bar{i}_{ds}^r + \omega_r L_q \bar{i}_{qs}^r}{L_d} \qquad (15.4\text{-}16)$$

Note that in (15.4-15) and (15.4-16) the electrical rotor speed is not given an average-value designation. Because the rotor speed varies slowly compared to the time scale of the fast average, it can generally be considered a constant as far as the fast-average procedure is concerned. However, there are instances when this approximation may not be completely accurate. For example, inaccuracies may occur in the case of 180° voltage-source modulated brushless dc drive with an exceptionally low inertia during the initial part of the start-up transient. Normally, however, the approximation works well in practice.

From Chapter 6, the expression for instantaneous electromagnetic torque is given by

$$T_e = \frac{3}{2} \frac{P}{2} (\lambda_m' i_{qs}^r + (L_d - L_q) i_{qs}^r i_{ds}^r) \qquad (15.4\text{-}17)$$

Upon neglecting the correlation between the q-axis current harmonics and the d-axis current harmonics, (15.4-17) may be averaged to yield

$$\bar{T}_e = \frac{3}{2} \frac{P}{2} (\lambda_m' \bar{i}_{qs}^r + (L_d - L_q) \bar{i}_{qs}^r \bar{i}_{ds}^r) \qquad (15.4\text{-}18)$$

This approximation (that is, assuming that the average of the product term is equal to the product of the averages) is very accurate in the case of sine-triangle modulation wherein there is minimal low-frequency harmonic content. However, in the case of the 180° or duty-cycle modulation, some error arises from this simplification in salient machines. In the case of nonsalient machines in which L_d is equal to L_q (15.4-18) is exact regardless of the modulation scheme.

To complete the average-value model of the drive, it only remains to include the mechanical dynamics. In particular,

$$p\omega_r = \frac{P}{2}\frac{\bar{T}_e - \bar{T}_l}{J} \qquad (15.4\text{-}19)$$

and, if rotor position is of interest,

$$p\theta_r = \omega_r \qquad (15.4\text{-}20)$$

Equations (15.4-19) and (15.4-20) complete the average-value model of the voltage-source inverter drive. It is convenient to combine these relationships and express them in matrix-vector form. This yields

$$
p\begin{bmatrix} \bar{i}_r \\ \bar{v}_{dc} \\ \bar{i}_{st} \\ \bar{v}_{st} \\ \bar{i}^r_{qs} \\ \bar{i}^r_{ds} \\ \omega_r \end{bmatrix} =
\begin{bmatrix}
-\frac{r_{rl}}{L_{rl}} & -\frac{1}{L_{rl}} & 0 & 0 & 0 & 0 & 0 \\
\frac{1}{C_{dc}} & 0 & -\frac{1}{C_{dc}} & 0 & 0 & 0 & 0 \\
0 & \frac{1}{L_{st}} & -\frac{r_{st}}{L_{st}} & -\frac{1}{L_{st}} & 0 & 0 & 0 \\
0 & 0 & \frac{1}{C_{st}} & 0 & 0 & 0 & 0 \\
0 & 0 & 0 & 0 & -\frac{r_s}{L_q} & 0 & -\frac{\lambda'_m}{L_q} \\
0 & 0 & 0 & 0 & 0 & -\frac{r_s}{L_d} & 0 \\
0 & 0 & 0 & 0 & \frac{3}{2}\left(\frac{P}{2}\right)^2\frac{1}{J}\lambda'_m & 0 & 0
\end{bmatrix}
\begin{bmatrix} \bar{i}_r \\ \bar{v}_{dc} \\ \bar{i}_{st} \\ \bar{v}_{st} \\ \bar{i}^r_{qs} \\ \bar{i}^r_{ds} \\ \omega_r \end{bmatrix}
$$

$$
+
\begin{bmatrix}
\frac{1}{L_{rl}}v_{r0}\cos\alpha \\
-\frac{3}{2}\frac{m}{C_{dc}}\left(\cos\phi_v\, \bar{i}^r_{qs} - \sin\phi_v\, \bar{i}^r_{ds}\right) \\
0 \\
0 \\
\frac{1}{L_q}\bar{v}_{dc}\, m\cos\phi_v - \frac{L_d}{L_q}\omega_r\, \bar{i}^r_{ds} \\
-\frac{1}{L_d}\bar{v}_{dc}\, m\sin\phi_v + \frac{L_q}{L_d}\omega_r\, \bar{i}^r_{qs} \\
\frac{P}{2}\frac{1}{J}\left(\frac{3}{2}\frac{P}{2}(L_d - L_q)\bar{i}^r_{qs}\, \bar{i}^r_{ds} - T_L\right)
\end{bmatrix}
\qquad (15.4\text{-}21)
$$

15.5 STEADY-STATE PERFORMANCE OF VSI DRIVES

In the previous section, a large-signal average-value model of a voltage-source inverter-fed brushless dc motor drive was set forth. Before using this model to explore the transient behavior of the drive, it is appropriate to first consider the steady-state performance. From the work presented in Chapter 13, it is clear that given the modulation strategy and $\bar{V}_{dc}$ the fast average of the q- and d-axis voltages may be obtained whereupon the work set forth in Chapter 6 may be used to calculate any quantity of interest. Therefore, the primary goal of this section will be primarily to establish an expression for $\bar{v}_{dc}$.

The differential equations that govern the fast-average value performance of the drive have inputs that are constants in the steady state; therefore the solution of these equations is also constant in the steady state, assuming that a stable solution exists. Therefore, the steady-state solution may be found by setting the derivative terms equal to zero. Thus, for steady-state conditions, the rectifier voltage equation (15.4-7) necessitates that

$$0 = v_{r0} \cos \alpha - \bar{V}_{dc} - r_{rl} \bar{I}_r \tag{15.5-1}$$

Similarly, substitution of (15.4-14) into (15.4-10) and setting the time derivative to zero yields

$$0 = \bar{I}_r - \bar{I}_{st} - \frac{3}{2}m(\bar{I}_{qs}^r \cos \phi_v - \bar{I}_{ds}^r \sin \phi_v) \tag{15.5-2}$$

Due to the series capacitance in the stabilizing filter, the fast average of the stabilizing filter current must be equal to zero. Therefore, (15.5-2) reduces to

$$0 = \bar{I}_r - \frac{3}{2}m(\bar{I}_{qs}^r \cos \phi_v - \bar{I}_{ds}^r \sin \phi_v) \tag{15.5-3}$$

Combining (15.5-3) with (15.5-1) yields

$$\bar{V}_{dc} = v_{r0} \cos \alpha - \frac{3}{2}r_{rl}m(\bar{I}_{qs}^r \cos \phi_v - \bar{I}_{ds}^r \sin \phi_v) \tag{15.5-4}$$

The next step in the development is to eliminate the q- and d-axis stator currents from (15.5-4). To this end, setting the time derivatives in (15.4-15) and (15.4-16) equal to zero, we obtain

$$0 = \bar{V}_{dc} m \cos \phi_v - r_s \bar{I}_{qs}^r - \omega_r L_d \bar{I}_{ds}^r - \omega_r \lambda_m' \tag{15.5-5}$$

$$0 = -\bar{V}_{dc} m \sin \phi_v - r_s \bar{I}_{ds}^r + \omega_r L_q \bar{I}_{qs}^r \tag{15.5-6}$$

Solving for (15.5-5) and (15.5-6) simultaneously in terms of $\bar{V}_{dc}$, m, and ω_r yields

$$\bar{I}_{qs}^r = \frac{r_s(\bar{V}_{dc}m \cos \phi_v - \omega_r \lambda_m') + \omega_r L_d \bar{V}_{dc}m \sin \phi_v}{r_s^2 + \omega_r^2 L_d L_q} \tag{15.5-7}$$

$$\bar{I}_{ds}^r = \frac{\omega_r L_q(\bar{V}_{dc}m \cos \phi_v - \omega_r \lambda_m') - r_s \bar{V}_{dc}m \sin \phi_v}{r_s^2 + \omega_r^2 L_d L_q} \tag{15.5-8}$$

Finally, substituting (15.5-7) and (15.5-8) into (15.5-4) and solving for $\bar{V}_{dc}$, we have

$$\bar{V}_{dc} = \frac{(r_s^2 + \omega_r^2 L_d L_q) v_{r0} \cos \alpha + \frac{3}{2} r_{rl} m \, \omega_r \, \lambda'_m (r_s \cos \phi_v - \omega_r L_q \sin \phi_v)}{r_s^2 + \omega_r^2 L_d L_q + \frac{3}{2} m^2 \, r_{rl} \, r_s + \frac{3}{4} r_{rl} \, \omega_r (L_d - L_q) \, m^2 \sin 2\phi_v} \quad (15.5\text{-}9)$$

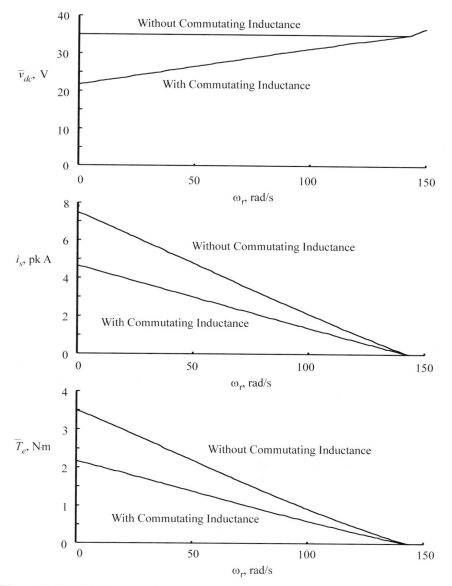

Figure 15.5-1 Steady-state voltage-source inverter-based brushless dc drive characteristics with and without commutating inductance.

Because (15.4-1) is only valid for rectifier currents greater than zero, it follows that (15.5-9) is only valid when it yields a dc supply voltage such that the rectifier current is positive. In the event that it is not, the rectifier appears as an open circuit, and all the diodes or thyristors are reverse-biased. In this case the average dc link current must be equal to zero. It follows from (15.4-14) that

$$\bar{I}_{qs}^r \cos\phi_v - \bar{I}_{ds}^r \sin\phi_v = 0 \qquad (15.5\text{-}10)$$

Substitution of (15.5-7) and (15.5-8) into (15.5-10) yields

$$\bar{V}_{dc}\big|_{\bar{I}_{dc}=0} = \frac{\omega_r \lambda_m' (r_s \cos\phi_v - \omega_r L_q \sin\phi_v)}{m\left(r_s + \frac{1}{2}\omega_r (L_d - L_q) \sin 2\phi_v\right)} \qquad (15.5\text{-}11)$$

Thus, as long as (15.5-9) yields a value of $\bar{V}_{dc}$ such that positive rectifier current is obtained, it is a valid expression. In the event that (15.5-9) yields negative rectifier current, (15.5-11) should be used.

The steady-state performance characteristics of a brushless dc drive are illustrated in Fig. 15.5-1. Therein the dc inverter voltage, the peak amplitude of the fundamental component of stator current, and the average electromagnetic torque are illustrated versus speed for the same parameters that were used in generating Fig. 15.3-3. In this case, however, the machine is connected to a transformer rectifier such that $v_{r0} = 35V$ and $r_r = 3.0\,\Omega$. Superimposed on each characteristic is the same trace that would be obtained if $\bar{V}_{dc}$ were held constant. As can be seen, the amplitudes of the stator current, the electromagnetic torque, and dc voltage are all considerably reduced due to the voltage drop that occurs due to the commutating reactance, although the difference decreases with speed. It is interesting to observe that above 145 rad/s, the dc voltage increases. This is due to fact that "rectified" machine voltage is greater than the rectified utility voltage; hence these diodes become reverse-biased. It should be noted that due to the voltage drop across the rectifier, the optimum phase delay given by (6.4-24) is no longer applicable. The derivation of an appropriate expression is left as an exercise for the reader.

15.6 TRANSIENT AND DYNAMIC PERFORMANCE OF VSI DRIVES

In this section the transient (large disturbance) and dynamic (small disturbance) behavior of voltage-source inverter based drives is examined. To this end, consider the drive system illustrated in Fig. 15.2-1. The parameters for this drive system are $E = 85.5$ V, $\omega_{eu} = 2\pi60$ rad/s, $L_c = 5$ mH, $L_{dc} = 5$ mH, and $C_{dc} = 1000\,\mu$F. The rectifier is uncontrolled, and the inverter is sine-triangle modulated. The machine parameters are identical to those of the machine considered in Section 15.3, and the load torque is equal to 0.005 N·m·s/rad times the mechanical rotor speed.

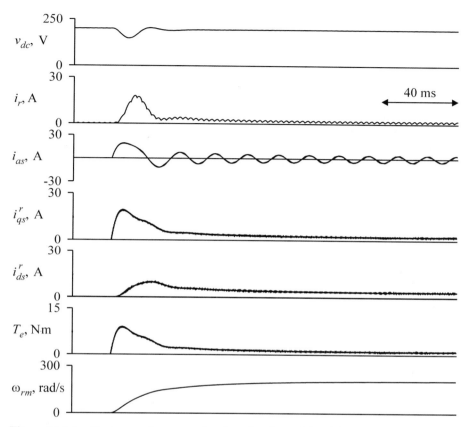

Figure 15.6-1   Start-up performance of a sine-triangle modulated brushless dc motor drive as calculated using a detailed computer model.

Figure 15.6-1 illustrates the start-up performance as the duty-cycle is stepped from 0 to 0.9. As can be seen, there is a large inrush of current on start-up because initially the impedance of the machine consists solely of the stator resistance and because there is no back emf. This results in a large initial torque, so the machine rapidly accelerates. Note that the large inrush current causes a substantial droop in the inverter voltage. Although the inrush current results in a large initial torque, this is generally an undesirable effect, because the initial current is well over the rated current of the machine (3.68 A, peak). In addition, if provision is not made to avoid these overcurrents, then both the inverter and the rectifier will have to be sized to ensure that the semiconductors are not damaged. Because the cost of the semiconductors is roughly proportional to the voltage rating times the current rating and because the overcurrent is five times rated current, the cost the inverter semiconductors will be five times that required had the inverter only been required to conduct steady-state current. Fortunately, by suitable control of the duty cycle, the overcurrent can be minimized.

It is interesting to compare the detailed portrayal of the drive's startup response to that predicted by the average-value model (15.4-21), which is illustrated in Fig. 15.6-2. Comparing the two figures, it is evident that the average-value model captures the salient features of the start-up with the exception of the harmonics, which were neglected in the averaging procedure. In addition to being considerably easy to code, the computation time using the average-value representation is approximately 120 times faster than the computation time required by a detailed representation in which the switching of all the semiconductors is taken into account, making it an ideal formulation for control system analysis and synthesis.

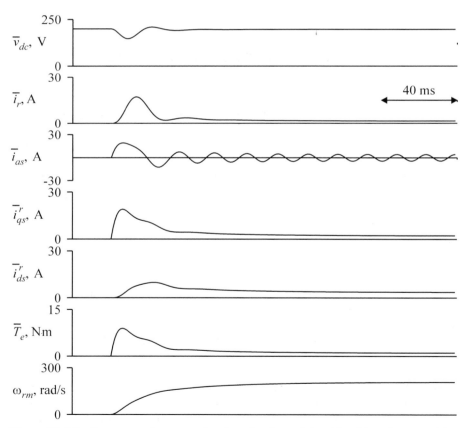

Figure 15.6-2 Start-up performance of a sine-triangle modulated brushless dc motor drive as calculated using an average-value model.

Because many control algorithms are based upon linear control theory, it is convenient to linearize the average value model. Linearizing (15.4-21) yields

$$p\left[\Delta \bar{i}_r \quad \Delta \bar{v}_{dc} \quad \Delta \bar{i}_{st} \quad \Delta \bar{v}_{st} \quad \Delta \bar{i}_{qs}^r \quad \Delta \bar{i}_{ds}^r \quad \Delta \omega_r\right]^T =$$

$$
\begin{bmatrix}
-\frac{r_{rl}}{L_{rl}} & -\frac{1}{L_{rl}} & 0 & 0 & 0 & 0 & 0 \\[4pt]
\frac{1}{C_{dc}} & 0 & -\frac{1}{C_{dc}} & 0 & -\frac{3}{2}\frac{m_0}{C_{dc}}\cos\phi_{v0} & \frac{3}{2}\frac{m_0}{C_{dc}}\sin\phi_{v0} & 0 \\[4pt]
0 & \frac{1}{L_{st}} & -\frac{r_{st}}{L_{st}} & -\frac{1}{L_{st}} & 0 & 0 & 0 \\[4pt]
0 & 0 & \frac{1}{C_{st}} & 0 & 0 & 0 & 0 \\[4pt]
0 & \frac{m_0}{L_q}\cos\phi_{v0} & 0 & 0 & -\frac{r_s}{L_q} & -\frac{L_d}{L_q}\omega_{r0} & -\frac{L_d}{L_q}\bar{i}_{ds0}^r-\frac{\lambda_m'}{L_q} \\[4pt]
0 & -\frac{m_0}{L_d}\sin\phi_{v0} & 0 & 0 & +\frac{L_q}{L_d}\omega_{r0} & -\frac{r_s}{L_d} & \frac{L_q}{L_d}\bar{i}_{qs0}^r \\[4pt]
0 & 0 & 0 & 0 & \frac{3}{2}\left(\frac{P}{2}\right)^2\frac{1}{J}(\lambda_m'+ & \frac{3}{2}\left(\frac{P}{2}\right)^2\frac{1}{J}+ & 0 \\
 & & & & (L_d-L_q)\bar{i}_{ds0}^r) & (L_d-L_q)\bar{i}_{qs0}^r
\end{bmatrix}
\begin{bmatrix}
\Delta \bar{i}_r \\ \Delta \bar{v}_{dc} \\ \Delta \bar{i}_{st} \\ \Delta \bar{v}_{st} \\ \Delta \bar{i}_{qs}^r \\ \Delta \bar{i}_{ds}^r \\ \Delta \omega_r
\end{bmatrix}
$$

$$
+
\begin{bmatrix}
\frac{v_{r0}}{L_{rl}} & \frac{\cos\alpha_0}{L_{rl}} & 0 & 0 & 0 \\[4pt]
0 & 0 & -\frac{3}{2}\frac{(\cos\phi_{v0}\bar{i}_{qs0}^r-\sin\phi_{v0}\bar{i}_{ds0}^r)}{C_{dc}} & \frac{3}{2}\frac{m_0(\sin\phi_{v0}\bar{i}_{qs0}^r+\cos\phi_{v0}\bar{i}_{ds0}^r)}{C_{dc}} & 0 \\[4pt]
0 & 0 & 0 & 0 & 0 \\[4pt]
0 & 0 & 0 & 0 & 0 \\[4pt]
0 & 0 & \frac{\bar{v}_{dc0}}{L_q}\cos\phi_{v0} & -\frac{\bar{v}_{dc0}m_0}{L_q}\sin\phi_{v0} & 0 \\[4pt]
0 & 0 & -\frac{\bar{v}_{dc0}}{L_d}\sin\phi_{v0} & -\frac{\bar{v}_{dc0}m_0}{L_d}\cos\phi_{v0} & 0 \\[4pt]
0 & 0 & 0 & 0 & -\frac{P}{2}\frac{1}{J}
\end{bmatrix}
\begin{bmatrix}
\Delta\cos\alpha \\ \Delta v_{r0} \\ \Delta m \\ \Delta\phi_v \\ \Delta T_L
\end{bmatrix}
$$

$$(15.6\text{-}1)$$

In (15.6-1), the addition of a subscript zero designates the initial equilibrium point about which the equations are linearized and Δ denotes a change in a variable. Thus

$$x = x_0 + \Delta x \qquad (15.6\text{-}2)$$

where x is any state, input variable, or output variable.

Figure 15.6-3 illustrates the start-up response as predicted by the average-value model linearized about the initial operating point. In this figure, (15.6-2) has been used to calculate each variable shown based on the initial operating point and the perturbation in the state variable given by (15.6-1). As can be seen, there are many discrepancies between the prediction of the linearized model and the performance of the drive illustrated in Fig. 15.6-1. In particular, the linearized model does not predict any perturbation to the dc voltage or the existence of rectifier current. In addition, the linearized model predicts a significantly higher q-axis current than is observed but fails to predict any d-axis current. The linearized model also significantly overestimates the peak torque and the final speed. Thus, this study illustrates

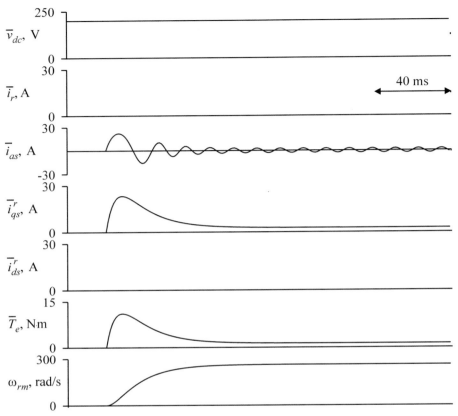

Figure 15.6-3 Start-up performance of a sine-triangle modulated brushless dc motor drive as calculated using a linearized model.

the hazards involved in using the linearized model to predict large disturbance transients.

Although the linearized model cannot be used to predict large-signal transients, it can be used for dynamic analysis such as operating point stability. To illustrate this, Figs. 15.6-4 and 15.6-5 depict the performance of the drive as predicted by a detailed computer simulation and the linearized model as the duty cycle is changed from 0.9 to 1. The linearized model has been determined about the initial operating point. As can be seen, in this case the linearized model accurately portrays the transient.

15.7 CONSIDERATION OF STEADY-STATE HARMONICS

Six-step and duty-cycle modulated drives exhibit considerable current and torque ripple. The steady-state harmonic current and torque harmonics may be calculated

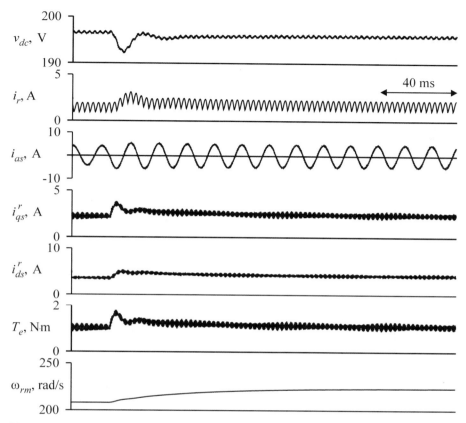

Figure 15.6-4 Response of a sine-triangle modulated brushless dc motor drive to a step change in duty cycle as calculated using a detailed computer model.

by employing the principle of superposition. To this end, it is observed that the dominant harmonics will be multiples of the 6th. The $6h$th harmonic (where h is in the set of natural numbers) of the q- and d-axis voltages may be expressed in the converter reference frame as

$$v_{qs}^c|_{6h} = \mathrm{Re}(\sqrt{2}\tilde{V}_{qs}^c \, e^{j6h\,\theta_c}) \tag{15.7-1}$$

$$v_{ds}^c|_{6h} = \mathrm{Re}(\sqrt{2}\tilde{V}_{ds}^c e^{j6h\theta_c}) \tag{15.7-2}$$

Recall that from (13.3-3) and (13.3-4) we have

$$v_{qs}^c|_{6h} = \frac{2}{\pi} dV_{dc}\frac{2(-1)^h}{36h^2-1}\cos{(6h\theta_c)} \tag{15.7-3}$$

$$v_{ds}^c|_{6h} = -\frac{2}{\pi} dV_{dc}\frac{12h}{36h^2-1}\sin{(6h\theta_c)} \tag{15.7-4}$$

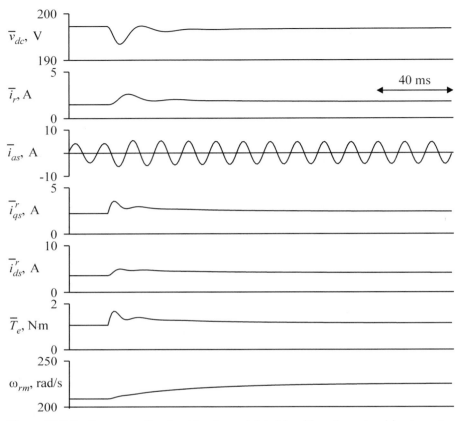

Figure 15.6-5 Response of a sine-triangle modulated brushless dc motor drive to a step change as calculated using a linearized model.

Comparing (15.7-1) to (15.7-3) and (15.7-2) to (15.7-4) it can be shown that

$$\tilde{V}_{qs6h}^{c} = \frac{\sqrt{2}}{\pi} dV_{dc} \frac{2(-1)^{h}}{36h^{2} - 1} \tag{15.7-5}$$

$$\tilde{V}_{ds6h}^{c} = j\frac{\sqrt{2}}{\pi} dV_{dc} \frac{12h}{36h^{2} - 1} \tag{15.7-6}$$

The transformation between the converter reference frame and the rotor reference frame is linear for steady-state conditions with a constant ϕ_{v}. Thus, it applies to phasor quantities in addition to instantaneous quantities. Transforming (15.7-5) and (15.7-6) to the rotor reference frame yields

$$\tilde{V}_{qs6h}^{r} = \tilde{V}_{qs6h}^{c} \cos \phi_{v} + \tilde{V}_{ds6h}^{c} \sin \phi_{v} \tag{15.7-7}$$

$$\tilde{V}_{ds6h}^{r} = -\tilde{V}_{qs6h}^{c} \sin \phi_{v} + \tilde{V}_{ds6h}^{c} \cos \phi_{v} \tag{15.7-8}$$

When considering each harmonic of the applied voltage separately, (14.3-10) and (14.3-11) may be written as

$$\tilde{V}^r_{qs6h} = (r_s + j6h\omega_r L_q)\,\tilde{I}^r_{qs6h} + \omega_r L_d\,\tilde{I}^c_{ds6h} \tag{15.7-9}$$

$$\tilde{V}^r_{ds6h} = (r_s + j6h\omega_r L_d)\,\tilde{I}^r_{ds6h} - \omega_r L_q\,\tilde{I}^c_{qs6h} \tag{15.7-10}$$

Solving for I^r_{qs6h} and I^r_{ds6h}, we obtain

$$I^r_{qs6h} = \frac{Z_{d6h}}{Z_{qd6h}}\,\tilde{V}^r_{qs6h} - \frac{\omega_r L_d}{Z_{qd6h}}\,\tilde{V}^r_{ds6h} \tag{15.7-11}$$

$$I^r_{ds6h} = \frac{\omega_r L_d}{Z_{qd6h}}\,\tilde{V}^r_{qs6h} + \frac{Z_{q6h}}{Z_{qd6h}}\,\tilde{V}^r_{ds6h} \tag{15.7-12}$$

where

$$Z_{q6h} = r_s + j6h\omega_r L_q \tag{15.7-13}$$

$$Z_{d6h} = r_s + j6h\omega_r L_d \tag{15.7-14}$$

$$Z_{qd6h} = r_s^2 - 36h^2\omega_r^2 L_q L_d + j6hr_s\,\omega_r\,(L_q + L_d) \tag{15.7-15}$$

Once each of the harmonic currents has been found using (15.7-11) and (15.7-12), the total steady-state q- and d-axis currents may be expressed as

$$I^r_{qs} = \bar{I}^r_{qs} + \sqrt{2}\sum_{h=1}^{\infty}\mathrm{Re}(\tilde{I}^r_{qs6h})\cos 6h\theta_r - \mathrm{Im}(\tilde{I}^r_{qs6h})\sin 6h\theta_r \tag{15.7-16}$$

$$I^r_{ds} = \bar{I}^r_{ds} + \sqrt{2}\sum_{h=1}^{\infty}\mathrm{Re}(\tilde{I}^r_{ds6h})\cos 6h\theta_r - \mathrm{Im}(\tilde{I}^r_{ds6h})\sin 6h\theta_r \tag{15.7-17}$$

Substitution of (15.7-16) and (15.7-17) into the expression for electromagnetic torque (14.3-14) yields

$$
\begin{aligned}
T_e ={}& \frac{3}{2}\frac{P}{2}\bar{I}^r_{qs}\left(\lambda'_m + (L_d - L_q)\,\bar{I}^r_{ds}\right) \\[4pt]
&+ \frac{3}{2}\frac{P}{2}\lambda'_m\left(\sum_{h=1}^{\infty}\sqrt{2}\mathrm{Re}(\tilde{I}^r_{qs6h})\cos 6h\,\theta_r + \sqrt{2}\mathrm{Re}(-j\tilde{I}^r_{qs6h})\sin 6h\,\theta_r\right) \\[4pt]
&+ \frac{3}{2}\frac{P}{2}(L_d - L_q)\bar{I}^r_{qs}\left(\sum_{H=1}^{\infty}\sqrt{2}\mathrm{Re}(\tilde{I}^r_{ds6H})\cos 6H\,\theta_r + \sqrt{2}\mathrm{Re}(-j\tilde{I}^r_{ds6H})\sin 6H\,\theta_r\right) \\[4pt]
&+ \frac{3}{2}\frac{P}{2}(L_d - L_q)\bar{I}^r_{ds}\left(\sum_{h=1}^{\infty}\sqrt{2}\mathrm{Re}(\tilde{I}^r_{qs6h})\cos 6h\,\theta_r + \sqrt{2}\mathrm{Re}(-j\tilde{I}^r_{qs6h})\sin 6h\,\theta_r\right) \\[4pt]
&+ \frac{3}{2}\frac{P}{2}(L_d - L_q)\left(\sum_{h=1}^{\infty}\sum_{H=1}^{\infty}\mathrm{Re}(\tilde{I}^r_{qs6h}\tilde{I}^r_{ds6H})\cos 6(h-H)\theta_r + \mathrm{Re}(\tilde{I}^r_{qs6h}\tilde{I}^r_{ds6H})\cos 6(h+H)\theta_r\right) \\[4pt]
&+ \frac{3}{2}\frac{P}{2}(L_d - L_q)\left(\sum_{h=1}^{\infty}\sum_{H=1}^{\infty}\mathrm{Re}(-j\tilde{I}^r_{qs6h}\tilde{I}^{r*}_{ds6H})\sin 6(h-H)\theta_r \right. \\[4pt]
&\qquad\qquad\qquad\qquad \left. + \mathrm{Re}(-j\tilde{I}^r_{qs6h}\tilde{I}^{r*}_{ds6H})\sin 6(h+H)\theta_r\right)
\end{aligned}
\tag{15.7-17}
$$

Example 15A. Using the techniques just set forth, let us attempt to estimate the peak-to-peak torque ripple for the study illustrated in Fig. 15.3-4. To this end, it will be assumed that the torque ripple may be adequately represented by its sixth-harmonic component. Because $L_d = L_q$ from (15.7-17), it is apparent that the sixth harmonic of the torque is only a function of the sixth harmonic of the q-axis current. To calculate this current, from (15.7-5) and (15.7-6) we have

$$\tilde{V}^c_{qs6} = 3.22 \angle 180° \tag{15A-1}$$

$$\tilde{V}^c_{ds6} = 19.3 \angle 90° \tag{15A-2}$$

Because $\phi_v = 0$, from (15.7-7) and (15.7-8) we have that $\tilde{V}^r_{qs6} = \tilde{V}^c_{qs6}$ and $\tilde{V}^r_{ds6} = \tilde{V}^c_{ds6}$. The sixth-harmonic q- and d-axis currents may be found from (15.7-11)–(15.7-12), which yields

$$I^c_{qs6} = 0.239 \angle 99.7° \tag{15A-3}$$

When we use (15A-3) and (15.7-17), the zero-to-peak amplitude of the sixth harmonic torque is 0.158 N·m, so that the peak-to-peak torque ripple is 0.316, which is in good agreement with Fig. 15.3-4.

15.8 CASE STUDY: VOLTAGE-SOURCE INVERTER-BASED SPEED CONTROL

Now that the basic analytical tools to analyze voltage-source inverter based brushless dc motor drives have been set forth, it is appropriate to consider the use of these tools in control system synthesis. To this end, consider a sine-triangle modulated voltage-source-based drive with the parameters listed in Table 15.8-1. It is desired to use this drive to achieve speed control of an inertial load. The design requirements are as follows: (i) There shall be no steady-state error, and (ii) the phase margin will be 60° (a common design requirement for well-behaved transient response) when the drive is operated at the nominal operating speed of 200 rad/s (mechanical). The design requirement of zero steady-state error necessitates integral feedback. Thus, a proportional-plus-integral controller would be appropriate. A block diagram of this control in a system context is illustrated in Fig. 15.8-1. Therein, the s

Table 15.8-1   Drive System Parameters

E	85.5 V	C_{dc}	1000 μF	L_d	11.4 mH
ω_{eu}	377 rad/s	J	5 mN·m·s²	λ'_m	0.156 V·s
L_c	5 mH	r_s	2.98 Ω	P	4
L_{dc}	5 mH	L_q	11.4 mH		

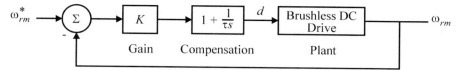

Figure 15.8-1   Speed control system.

represents the derivative operator in Laplace notation, which is typically used for control synthesis. In the time domain the control law is of the form

$$d = K(\omega_{rm}^* - \omega_{rm}) + \frac{K}{\tau} \int (\omega_{rm}^* - \omega_{rm})\, dt \qquad (15.8\text{-}1)$$

For the purpose of design, we will make use of a linearized model of the brushless dc drive, in which the system is linearized about an operating speed of 200 rad/s. The linearized model can be calculated using (15.6-1) or it can be calculated by

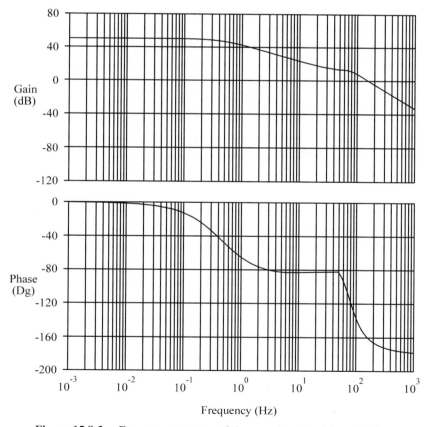

Figure 15.8-2   Frequency response of the open-loop brushless dc drive.

automatic linearization of a nonlinear average-value model, a feature common to many simulation languages.

Figure 15.8-2 illustrates the open-loop Bode plot of the brushless dc drive, wherein the output is the mechanical rotor speed and the input is the duty cycle. Because the Bode characteristic is based upon a linearized model, strictly speaking it is only valid about the operating point about which it was linearized (200 rad/s). From Fig. 15.8-2 we see that although the gain margin is infinite, the phase margin is only 20°. A phase margin of 30° is often considered to be the minimum acceptable.

The design process begins by selection of τ. The integral feedback will decrease the phase by 90° at frequencies much less than $1/(2\pi\tau)$. Because this will decrease the already small phase margin, it is important to pick τ so that the breakpoint frequency of the compensator is considerably less than the frequency at which the phase of the plant begins to decrease from zero. Selecting the breakpoint frequency of the compensator to be at 0.01 Hz yields a τ of 16 s.

The Bode characteristic of the compensated plant is depicted in Fig. 15.8-3. As can be seen, the phase margin is still 20°. The next step is to select K so as to obtain

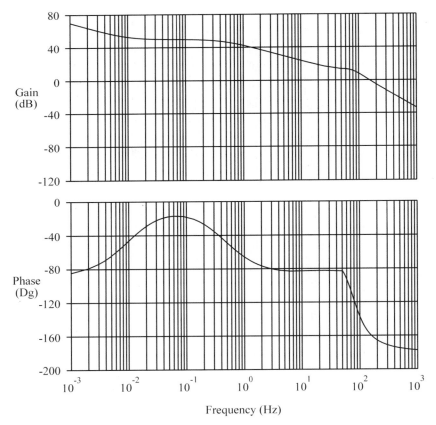

Figure 15.8-3 Frequency response of the compensated brushless dc drive.

the desired phase margin, which can be accomplished choosing the gain such that the phase at the gain crossover frequency is $-120°$. From Fig. 15.8-3, it can be seen that the phase of the compensated plant is $-120°$ when the gain of the compensated plant is 12 dB. Thus, choosing $K = 0.25 (-12$ dB) will result in the desired phase margin.

Figure 15.8-4 illustrates the Bode characteristic of the closed-loop plant. As can be seen, the bandwidth of the system is on the order of 100 Hz, and the resonant peak is not overly pronounced. However, the closed-loop frequency response cannot be used as a sole judge of the system's performance because the actual system is non-linear. For this, the simplest usable model is an average-value model. Figure 15.8-5 illustrates the system performance during a step change in commanded speed from 0 to 200 rad/s. As can be seen, the transient performance with regard to speed is quite well-behaved. Nevertheless, the reader might be surprised by Fig. 15.8-5 in several ways. First, it can be seen that the duty cycle, which is normally 0–1, is nearly 50 in the initial part of the study. Thus, the drive will be overmodulated and we can expect the current to exhibit considerable low-frequency harmonics on start-up (these are not apparent in Fig. 15.8-5 because an average-value model was used). Because the

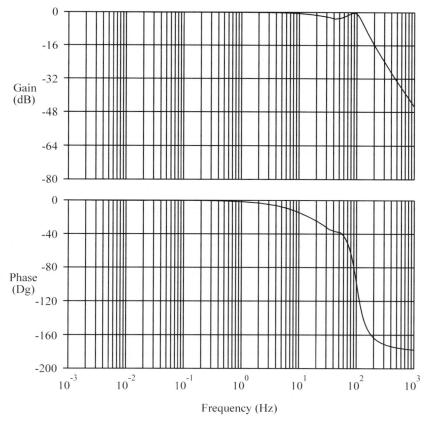

Figure 15.8-4   Frequency response of the closed-loop brushless dc drive.

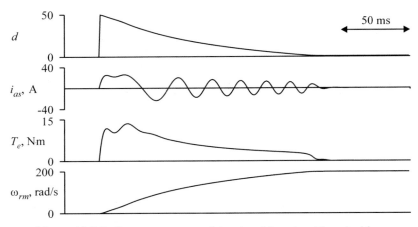

Figure 15.8-5   Start-up response of the closed-loop brushless dc drive.

applied voltage was effectively much lower than expected, the bandwidth for this large disturbance is not nearly the 100 Hz indicated in Fig. 15.8-4. Finally, the rated current for the machine is only 2.6 A, rms. Although the machine could probably withstand the temporary overcurrent, the inverter probably could not, and thus either the bandwidth should be reduced so as to alleviate the overcurrent or the duty cycle should be limited as a function of the current. Finally, close inspection reveals that at the end of the study, the speed is not 200 rad/s and does not even appear to be rapidly increasing. This is because the bandwidth of the compensation was chosen to be quite low, and as a result a small error in rotor speed will persist for some time, although it will eventually go to zero. A second design iteration addressing these issues is left to the reader as an exercise.

15.9 CURRENT-REGULATED INVERTER DRIVES

Sections 15.1–15.8 explored the performance of drives in which the machine is controlled through suitable regulation of the applied voltages. In the remainder of this chapter an alternate strategy is considered: control of the machine through the regulation of the stator currents. The hardware configuration for current-regulated inverter drives is identical to that of voltage-source inverter drives, as illustrated in Fig. 15.2-1. The only difference is in the way in which the gate signals to the individual semiconductors are established.

Current-regulated inverters have several distinctive features. First, because torque is a function of the machine current, the torque may be controlled with the same bandwidth by which the stator currents are controlled. In fact, it is often the case that for practical purposes the torque control is essentially instantaneous. A second feature of current-regulated drives is robustness with regard to changes in machine parameters. For example, current-regulated inverter drives are insensitive to

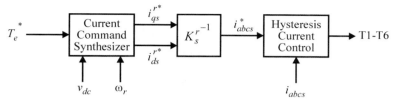

Figure 15.9-1 Hysteresis current-regulated drive.

parameter variations in the stator leakage inductance or stator resistance. Current-regulated drives are also robust in regard to faults. In the event of a winding-to-winding short circuit within the machine the currents are automatically limited, which prevents damage to the inverter. The currents are also automatically limited during start-up.

Figure 15.9-1 illustrates the control of the current-regulated drive. Therein, the q- and d-axis current commands, i_{qs}^r and i_{ds}^r, are formulated based on the commanded torque, T_e^*, electrical rotor speed, ω_r, and the inverter voltage, v_{dc}. Using the inverse transformation, the corresponding abc variable current command i_{abcs}^* is determined. Finally, based on the abc variable current command and the actual currents, the on and off status of each of the inverter semiconductors (T1–T6) is determined using the hysteresis current control strategy set forth in Section 13.8. An immediate question that arises is how the q- and d-axis current commands are generated to begin with; this question is addressed in detail in a following section. For the present it suffices to say that the command is determined in such a way that if the commanded currents are obtained, the commanded torque will also be obtained.

Figure 15.9-2 illustrates the steady-state performance of a hysteresis current-regulated brushless dc motor drive. Therein, the operating conditions are identical to those portrayed in Fig. 15.3-5 except for the modulation strategy. The q- and d-axis current commands are set to 1.73 A and 2.64 A, respectively, so that the fundamental component of the commanded current is identical to that in Fig. 15.3-5. As can be seen, although the modulation strategies are different, the waveforms produced by the sine-triangle modulation strategies and the hysteresis current strategy are very similar.

A second method to implement a current-regulated inverter drive is to utilize a current-control loop on a voltage-source inverter drive. This is illustrated in Fig. 15.9-3. Therein, the current command synthesizer serves the same function as in Fig. 15.9-1. Based on the commanded q- and d-axis currents and the measured q- and d-axis currents (determined by transforming the measured abc variable currents), the q- and d-axis voltage commands (v_{qs}^{r*} and v_{ds}^{r*}) are determined. The q- and d-axis voltage command is then converted to an abc variable voltage command v_{abcs}^*, which is scaled in order to determine the instantaneous duty cycles, d_a, d_b, and d_c of the sine-triangle modulation strategy. Based on these duty cycles, T1–T6 are determined as described in Section 13.5. There are several methods of developing the current control, such as a synchronous current regulator [7]. An example of the design of a feedback linearization-based controller is considered in Example 15B.

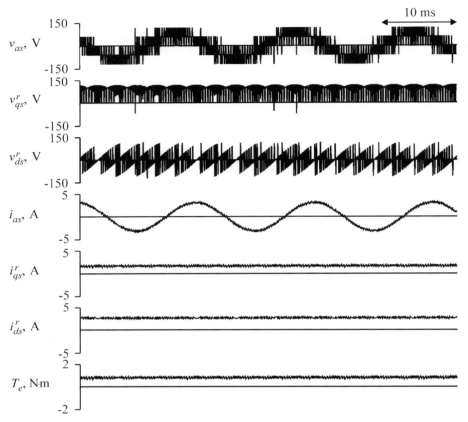

Figure 15.9-2   Steady-state performance of a hysteresis-modulated brushless dc motor drive.

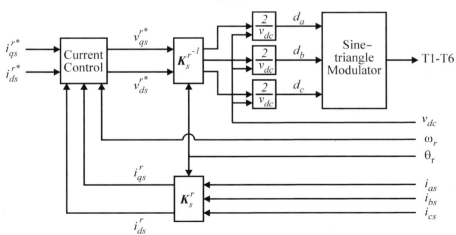

Figure 15.9-3   A sine-triangle-based current regulator.

Example 15B. Let us consider the design of a current regulator for a non-salient brushless dc motor. The goal is to determine the q- and d-axis voltage command so that the actual currents become equal to the commanded currents. Let us attempt to accomplish this goal by specifying the voltage commands as

$$v_{qs}^{r*} = \omega_r(L_{ss}i_{ds}^r + \lambda_m') + \left(K_p + \frac{K_i}{s}\right)(i_{qs}^{r*} - i_{qs}^r) \tag{15B-1}$$

$$v_{ds}^{r*} = -\omega_r L_{ss}i_{qs}^r + \left(K_p + \frac{K_i}{s}\right)(i_{ds}^{r*} - i_{ds}^r) \tag{15B-2}$$

where s denotes the Laplace operator, K_p is the proportional gain, and K_i is the integral gain. This control algorithm contains feedback terms that cancel the nonlinearities in the stator voltage equations, feedforward terms that cancel

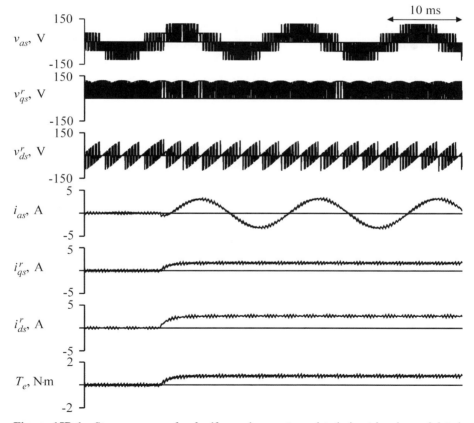

Figure 15B-1 Step response of a feedforward current-regulated sine-triangle modulated brushless dc motor drive.

the effect of the back emf, and a proportional-plus-integral (PI) control loop to drive the error to zero. Assuming that the actual q- and d-axis voltages are equal to the commanded q- and d-axis voltages, it can be shown that the transfer function between the commanded and actual q-axis currents is given by

$$\frac{i_{qs}^r(s)}{i_{qs}^{r*}(s)} = \frac{\frac{K_p}{L_{ss}}\left(s + \frac{K_i}{K_p}\right)}{s^2 + \frac{(r_s + K_p)}{L_{ss}}s + \frac{K_i}{L_{ss}}} \tag{15B-3}$$

The transfer function relating the d-axis current to the commanded d-axis current is identical. Assuming the same machine parameters as in the study illustrated in Fig. 15.9-2, and selecting pole locations of $s = -200$ and $s = -1000$ (note that the poles may be arbitrarily placed), we have that

$$K_i = 2280\,\Omega/s \tag{15B-4}$$

$$K_p = 10.7\,\Omega \tag{15B-5}$$

Figure 15B-1 illustrates the response of the brushless dc drive as the current command is stepped from zero to $i_{qs}^{r*} = 1.73\,\mathrm{A}$ and $i_{ds}^{r*} = 2.64\,\mathrm{A}$. All operating conditions are as in Fig. 15.9-2. As can be seen, the machine performance is extremely well behaved and is dominated by the pole at $s = -200$.

15.10 VOLTAGE LIMITATIONS OF CURRENT-SOURCE INVERTER DRIVES

As alluded to previously, assuming that the current control loop is sufficiently fast, the current-regulated drive can be thought of as an ideal current source. However, there are some limitations on the validity of this approximation. In particular, eventually the back emf of the machine will rise to the point where the inverter cannot supply the current commands due to the fact that the back emf of the machine becomes too large. Under such conditions, the machine is said to have lost current tracking.

In order to estimate the operating region over which current tracking is obtained, note that if it is obtained we have

$$\bar{i}_{qs}^r = i_{qs}^{r*} \tag{15.10-1}$$

$$\bar{i}_{ds}^r = i_{ds}^{r*} \tag{15.10-2}$$

Substituting (15.10-1) and (15.10-2) into the stator voltage equations and neglecting the stator dynamics yields

$$\bar{v}_{qs}^r = r_s\, i_{qs}^{r*} + \omega_r L_d\, i_{ds}^{r*} + \omega_r \lambda_m' \tag{15.10-3}$$

$$\bar{v}_{ds}^r = r_s\, i_{ds}^{r*} - \omega_r L_q\, i_{qs}^{r*} \tag{15.10-4}$$

Recall that the rms value of the fundamental component of the applied voltage is given by

$$v_s = \frac{1}{\sqrt{2}} \sqrt{\left(\bar{v}^r_{qs}\right)^2 + \left(\bar{v}^r_{ds}\right)^2} \tag{15.10-5}$$

Substitution of (15.10-3) and (15.10-4) into (15.10-5) yields

$$v_s = \frac{1}{\sqrt{2}} \sqrt{\left(r_s i^{r*}_{qs} + \omega_r L_d i^{r*}_{ds} + \lambda'_m \omega_r\right)^2 + \left(r_s i^{r*}_{ds} - \omega_r L_q i^{r*}_{qs}\right)^2} \tag{15.10-6}$$

Recall from Section 13.8 that for the hysteresis-modulated current-regulated inverters, the maximum rms value of the fundamental component of the applied voltage which can be obtained without low-frequency harmonics is given by

$$v_s = \frac{1}{\sqrt{6}} \bar{v}_{dc} \tag{15.10-7}$$

If low-frequency harmonics are tolerable and a synchronous regulator is used, either in the context of a sine-triangle voltage-source inverter-based current loop or with the hysteresis-modulated current regulator, then the maximum rms value of the fundamental component becomes

$$v_s = \frac{\sqrt{2}}{\pi} \bar{v}_{dc} \tag{15.10-8}$$

In the event that for a given current command and speed (15.10-8) cannot be satisfied, then it is not possible to obtain stator currents equal to the commanded currents. If (15.10-8) can be satisfied, but (15.10-7) cannot be satisfied, then it is possible to obtain stator currents that have the same fundamental component as the commanded currents provided that integral feedback in the rotor reference frame is present to drive the current error to zero. However, low-frequency harmonics will be present.

Figure 15.10-1 illustrates the effects of loss of current tracking. Initially, operating conditions are identical to those portrayed in Fig. 15.9-2. However, approximately 20 ms into the study, the dc inverter voltage is stepped from 177 V to 124 V, which results in a loss of current tracking. As can be seen, the switching of the hysteresis current regulator is such that some compensation takes place; nevertheless current tracking is lost. As a result, harmonics appear in the a-phase and q- and d-axis current waveforms, as well as in the electromagnetic torque.

15.11 CURRENT COMMAND SYNTHESIS

It is now appropriate to address the question as to how to determine the current command. Normally, when using a current-regulated inverter, the input to the

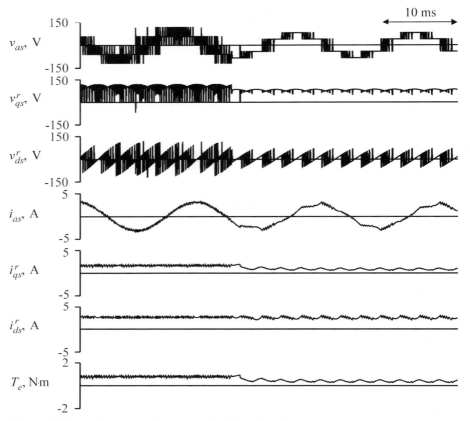

Figure 15.10-1   Response of hysteresis-modulated brushless dc motor drive to step decrease in dc inverter voltage.

controller is a torque command. Thus, the problem may be reformulated as the determination of the current command from the torque command. To answer this question, let us first consider a nonsalient machine in which $L_q = L_d$. From (6.3-14), torque may be expressed as

$$T_e = \frac{3}{2}\frac{P}{2}\lambda'_m i^r_{qs} \qquad (15.11\text{-}1)$$

Therefore, the commanded q-axis current may be expressed in terms of the commanded torque as

$$i^{r*}_{qs} = \frac{2}{3}\frac{2}{P}\frac{1}{\lambda'_m}T^*_e \qquad (15.11\text{-}2)$$

Clearly, if the desired torque is to be obtained, then (15.11-2) must be satisfied. The d-axis current does not affect average torque, and so its selection is somewhat arbitrary. Because d-axis current does not affect the electromagnetic torque, but does result in additional stator losses, it is often selected to be zero, that is,

$$i_{ds}^{r*} = 0 \tag{15.11-3}$$

This selection of d-axis current minimizes the current amplitude into the machine, thus maximizing torque-per-amp and at the same time maximizing the efficiency of the machine by minimizing the stator resistive losses.

Although (15.11-3) has several distinct advantages, there is one reason to command a nonzero d-axis current. To see this reason, consider (15.10-6) for the nonsalient case:

$$v_s = \frac{1}{\sqrt{2}} \sqrt{\left(r_s\, i_{qs}^{r*} + \omega_r L_{ss}\, i_{ds}^{r*} + \lambda_m'\, \omega_r \right)^2 + \left(r_s\, i_{ds}^{r*} - \omega_r L_{ss}\, i_{qs}^{r*} \right)^2} \tag{15.11-4}$$

From (15.11-4) we see that the required inverter voltage goes up with speed and with q-axis current (which is proportional to torque). However, when we examine the first squared term in (15.11-4), it can be seen that at positive speeds the required inverter voltage can be reduced by injecting negative d-axis current. In fact, by solving (15.11-4) for d-axis current in terms of the q-axis current command and speed, we have that the d-axis current injection that allows us to operate within the voltage constraint is given by

$$i_{ds}^{r*} = \frac{-\lambda_m' L_{ss}\, \omega_r^2 + \sqrt{2z^2 v_s^2 - \left(r_s \omega_r \lambda_m' + z^2\, i_{qs}^{r*} \right)^2}}{z^2} \tag{15.11-5}$$

where

$$z = \sqrt{r_s^2 + \omega_r^2 L_{ss}^2} \tag{15.11-6}$$

Thus, a logical current control strategy is to command zero d-axis current as long as the inverter voltage requirements are not exceeded, and to inject the amount of d-axis current specified by (15.11-5) if they are. Note that there are limitations on d-axis current injection in that (i) (15.11-5) may not have a solution, (ii) excessive d-axis current injection may result in demagnetization of the permanent magnet, and (iii) excessive d-axis current injection can result in exceeding the current limit of the machine or inverter. In addition, the use of (15.11-5) requires accurate knowledge of the dc inverter voltage (to determine the peak v_s), the rotor speed, and all of the machine parameters. A means of implementing such a control without knowledge of the dc inverter voltage, speed, and machine parameters is set forth in reference 8.

The process for determining the current command in salient machines, which typically are constructed using buried-magnet technology, is somewhat more involved than in the nonsalient case. Let us first consider the problem of computing the q- and d-axis current commands so as to maximize torque-per-amp performance.

In the case of the salient machine, from Chapter 6 the expression for electromagnetic torque is given by

$$T_e = \frac{3}{2}\frac{P}{2}(\lambda'_m i^r_{qs} + (L_d - L_q)i^r_{qs} i^r_{ds})$$ (15.11-7)

Solving (15.11-7) for d-axis current command in terms of the q-axis current command and in terms of the commanded torque yields

$$i^{r*}_{ds} = \frac{T_e}{\frac{3}{2}\frac{P}{2}(L_d - L_q)}\frac{1}{i^{r*}_{qs}} - \frac{\lambda'_m}{L_d - L_q}$$ (15.11-8)

In terms of the q- and d-axis current commands, the rms value of the fundamental component of the commanded current is given by

$$i_s = \frac{1}{\sqrt{2}}\sqrt{(i^{r*}_{qs})^2 + (i^{r*}_{ds})^2}$$ (15.11-9)

Substitution of (15.11-8) into (15.11-9) yields an expression for the magnitude of the stator current in terms of the commanded torque and q-axis current command. Setting the derivative of the resulting expression with respect to the q-axis current command equal to zero yields the following transcendental expression for the q-axis current command:

$$(i^{r*}_{qs})^4 + \frac{T_e \lambda'_m i^{r*}_{qs}}{\frac{3}{2}\frac{P}{2}(L_d - L_q)^2} - \left(\frac{T_e}{\frac{3}{2}\frac{P}{2}(L_d - L_q)}\right)^2 = 0$$ (15.11-10)

The solution of (15.11-10) yields the value of q-axis current that maximizes torque-per-amp performance. Once the q-axis current command is determined by solving (15.11-10), the d-axis current command may be found by solving (15.11-8). From the form of (15.11-10), it is apparent that the solution for the q-axis current must be accomplished numerically. For this reason, when implementing this control with a microprocessor, the q- and d- axis current commands are often formulated through a table look-up that has been constructed by solving (15.11-8) and (15.11-10) offline.

Once the q- and d-axis current commands have been formulated, it is necessary to check whether or not the inverter is capable of producing the required voltage. If it is not, it is necessary to recalculate the commanded q- and d-axis currents such that the required inverter voltage does not exceed that obtainable by the converter. This calculation can be conducted by solving (15.10-6) and (15.11-8) simultaneously for the q- and d-axis current command.

Figure 15.11-1 illustrates the graphical interpretation of the selection of the commanded q- and d-axis currents for a machine in which $r = 0.2\ \Omega$, $L_q = 20$ mH, $L_d = 10$ mH, and $\lambda'_m = 0.07$ V·s. The machine is operating at a speed of

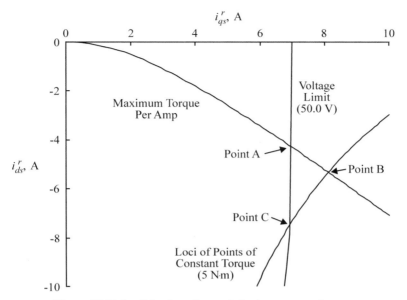

Figure 15.11-1   Selection of q- and d-axis current trajectory.

500 rad/s (electrical) and $v_s = 50$ V. Illustrated therein are the trajectory of the maximum torque-per-amp characteristic, the loci of points in the q- and d-axis current plane at which the electromagnetic torque of 5 N · m is obtained, and the loci of points representing the voltage limit imposed by (15.10-6). For a given electromagnetic torque command, the q- and d-axis current command is formulated using the maximum torque-per-amp trajectory, provided that this point is inside the voltage limit. However, q- and d-axis currents on this trajectory corresponding to torques greater than that obtainable at point A cannot be achieved if the q- and d-axis current is confined to the maximum torque-per-amp trajectory. Suppose a torque of 5 N · m is desired. Point B represents the point on this maximum torque-per-amp trajectory which has the desired torque. Unfortunately, point B is well outside of the limit imposed by the available voltage. However, note that any point on the constant torque locus will satisfy the desired torque. Thus, in this case the current command is chosen to correspond to point C.

15.12   AVERAGE-VALUE MODELING OF CURRENT-REGULATED INVERTER DRIVES

In this section, an average-value model of current-regulated and current-source inverter drives is formulated in much the same way as the average-value model of voltage-source inverter drives. Because the topology of the converter has not changed, it follows that the expressions for the time derivatives of the rectifier current, the dc link voltage, the stabilizing filter current, and the stabilizing filter voltage given by (15.4-6)–(15.4-9) are valid. Furthermore, the change in control strategy does not affect the mechanical dynamics, thus (15.4-14) and (15.4-15) may still be used to represent the machine. However, the change in control strategy will change the formulation of the expression for the dc link currents, the stator dynamics, and the expression for electromagnetic torque.

In order to formulate an expression for the dc link current, it is convenient to assume that the actual machine currents are equal to the commanded machine currents whereupon

$$\bar{i}_{qs}^r = i_{qs}^{r*} \tag{15.12-1}$$

$$\bar{i}_{ds}^r = i_{ds}^{r*} \tag{15.12-2}$$

Of course, this assumption is only valid when the dc link voltage is such that the desired current is actually obtained. An average-value model of brushless dc drives in which the current tracking is not obtained is set forth in reference 9. Assuming that the actual currents are equal to the commanded currents, the stator currents are no longer state variables. Neglecting the stator dynamics, the q- and d-axis voltages may be expressed as

$$\bar{v}_{qs}^{r*} = r_s i_{qs}^{r*} + \omega_r L_d i_{ds}^{r*} + \lambda_m' \omega_r \tag{15.12-3}$$

$$\bar{v}_{ds}^r = r_s i_{ds}^{r*} - \omega_r L_q i_{qs}^{r*} \tag{15.12-4}$$

The instantaneous power into the machine is given by

$$P = \frac{3}{2}[r_s(i_{qs}^{r*} + i_{ds}^{r*})^2 + \omega_r(L_d - L_q)i_{qs}^{r*} i_{ds}^{r*} + \omega_r \lambda_m' i_{qs}^r] \tag{15.12-5}$$

Assuming that no power is lost into the inverter, if follows that the dc link current is given by

$$\bar{i}_{dc} = \frac{P}{\bar{V}_{dc}} \tag{15.12-6}$$

Combining (15.12-5) with (15.12-6) yields

$$\bar{i}_{dc} = \frac{3}{2}\frac{1}{\bar{v}_{dc}}[r_s(i_{qs}^{r*} + i_{ds}^{r*})^2 + \omega_r(L_d - L_q)i_{qs}^{r*} i_{ds}^{r*} + \omega_r \lambda_m' i_{qs}^r] \tag{15.12-7}$$

The other expression affected by the change from a voltage-source inverter to a current-source or current-regulated inverter will be the expression for torque. In particular, from (15.4-12) and again assuming that the actual stator currents are equal to the commanded currents, we obtain

$$T_e = \frac{3}{2}\frac{P}{2}(\lambda_m' i_{qs}^{r*} + (L_d - L_q)i_{qs}^{r*} i_{ds}^{r*}) \tag{15.12-8}$$

As can be seen from (15.12-8), if it is assumed that the actual currents are equal to the commanded currents, then any desired torque may be instantaneously obtained.

Combining (15.4-7), (15.4-10), (15.4-11), (15.4-12), (15.4-19), (15.12-7), and (15.12-8) yields

$$
p\begin{bmatrix} \bar{i}_r \\ \bar{v}_{dc} \\ \bar{i}_{st} \\ \bar{v}_{st} \\ \omega_r \end{bmatrix} = \begin{bmatrix} -\frac{r_{rl}}{L_{rl}} & -\frac{1}{L_{rl}} & 0 & 0 & 0 \\ \frac{1}{C_{dc}} & 0 & -\frac{1}{C_{dc}} & 0 & 0 \\ 0 & \frac{1}{L_{st}} & -\frac{r_{st}}{L_{st}} & -\frac{1}{L_{st}} & 0 \\ 0 & 0 & \frac{1}{C_{st}} & 0 & 0 \\ 0 & 0 & 0 & 0 & 0 \end{bmatrix} \begin{bmatrix} \bar{i}_r \\ \bar{v}_{dc} \\ \bar{i}_{st} \\ \bar{v}_{st} \\ \omega_r \end{bmatrix}
$$
$$
+ \begin{bmatrix} \frac{v_{r0}\cos\alpha}{L_{rl}} \\ -\frac{1}{C_{dc}}\frac{3}{2}\frac{1}{V_{dc}}[r_s(i_{qs}^{r*}+i_{ds}^{r*})^2 + \omega_r(L_d-L_q)i_{ds}^{r*}\,i_{qs}^{r*} + \omega_r\lambda_m' \,i_{qs}^{r*}] \\ 0 \\ 0 \\ \frac{P}{2}\frac{1}{J}[\frac{3}{2}\frac{P}{2}(\lambda_m'\,i_{qs}^{r*} + (L_d-L_q)i_{qs}^{r*}\,i_{ds}^{r*}) - T_L] \end{bmatrix}
$$
$$(15.12\text{-}9)$$

15.13 CASE STUDY: CURRENT-REGULATED INVERTER-BASED SPEED CONTROLLER

The control of current-regulated inverter-based drives is considerably simpler than its voltage-source-based counterpart, due to the fact that when designing the speed or position control algorithms, the inverter and machine may be modeled as a nearly ideal torque transducer (neglecting the stator dynamics of the machine). To illustrate this, let us reconsider the speed control system discussed in Section 15.8. Assuming that a current command synthesizer and current regulator can be designed with sufficiently high bandwidth, the speed control algorithm may be designed by assuming that the drive will produce an electromagnetic torque equal to the desired torque,

$$T_e = T_e^* \qquad (15.13\text{-}1)$$

In order to ensure that there will be no steady-state error, let's propose a PI control law in accordance with (15.12-2):

$$T_e^* = K_p\left(1 + \frac{1}{\tau s}\right)(\omega_{rm}^* - \omega_{rm}) \qquad (15.13\text{-}2)$$

wherein ω_{rm}^* represents the speed command. Combining (15.13-1), (15.13-2), and the mechanical dynamics of the drive, it can be shown that the resulting transfer function between the actual and commanded rotor speed is given by

$$\frac{\omega_{rm}}{\omega_{rm}^*} = \frac{\frac{K}{J\tau}(\tau s + 1)}{s^2 + \frac{K}{J}s + \frac{K}{J\tau}} \qquad (15.13\text{-}3)$$

Because (15.13-3) represents a second-order system and there are two free parameters, the poles of (15.13-3) may be arbitrarily placed. However, some restraint should be exercised because it is important that the current regulator be much faster than the mechanical system if (15.13-1) and hence (15.13-3) are to remain valid. Placing the poles at $s = -5$ and $s = -50$ yields $K = 0.257$ N·m·s/rad and $\tau = 0.22$ s. The pole at $s = -5$ will dominate the response.

In order to complete the design, a current command synthesizer (to determine what the current command should be to achieve the desired torque) and a current regulation control strategy need to be designed. For this example, let us assume a simple current command synthesizer in which all of the current is injected into the q-axis, and let us use the sine-triangle-based current regulator set forth in Example 15B as a current regulator. Recall that the poles of the current regulator are at $s = -200$ and $s = -1000$, which are much faster than those of the mechanical system.

Practically speaking, there are two important refinements that can be made to this control system. First, the q-axis current command generated by the current command synthesizer should be limited to ±3.68 A in order to limit the current to the rated value of the machine. In order to avoid the current limit from causing windup of the speed control integrator, the contribution of the $K/(\tau s)$ portion of the speed control (that is the integral portion of the control) should be limited. Herein, the portion of the torque command contributed by the integral term will be limited so that the overshoot under worst-case conditions is limited to an acceptable value (some trial and error using dynamic simulations would be used to determine the exact number).

Figure 15.13-1 illustrates the interactions of the various controllers. Based on the speed error, the PI speed control determines a torque command T_e^* (the limit on the integral feedback is not shown). Then the current command synthesizer determines the q-axis current required to obtain the desired torque, subject to the q-axis current limit. In this controller, the d-axis current is set at zero. Based on the commanded q- and d-axis currents, the electrical rotor speed, the actual currents, and the dc

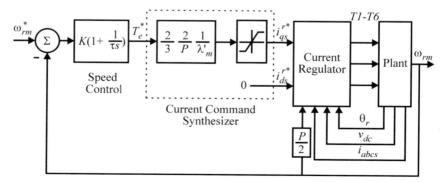

Figure 15.13-1   Current-regulated inverter-based speed control.

supply voltage of the current regulator determines the on or off status of each of the semiconductors in the inverter (T1–T6).

Figure 15.13-2 illustrates the performance of the speed control system. Initially the system is in the steady-state. However, 50 ms into the study the speed command is stepped from 0 to 200 rad/s. As can be seen, the torque command immediately jumps to the value that corresponds to the maximum q-axis current command. Because the electromagnetic torque is constant, the speed increases linearly with time. As can be seen, the magnitude of the ac current into the rectifier and the rectifier current both increase linearly with speed. This is due to the fact that the power going into the machine increases linearly with speed. The increasing rectifier current results in a dc inverter voltage that decreases linearly with time. Note that the dc link voltage initially undergoes a sudden dip of 5 V that we might not have expected; this is due to the fact that initially the rectifier was under no-load condition and hence charged the dc link capacitor to peak rather than average value of the rectifier voltage. Eventually, the machine reaches the desired speed. At this point the torque command falls off because the load is inertial. As a result, the electromagnetic torque, stator current, and rectifier current all decrease to their original values, and the dc inverter voltage increases to its original value. Comparing Fig. 15.13-2 to Fig. 15.8-5, the reader will observe that the current-based speed control system is

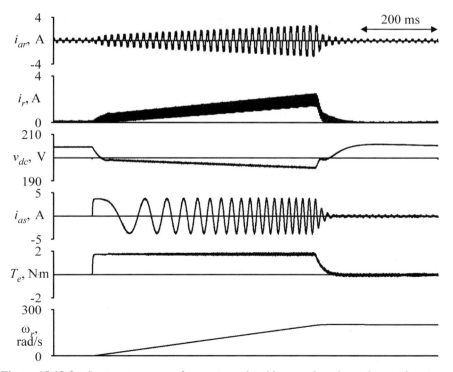

Figure 15.13-2 Start-up response of current-regulated inverter-based speed control system.

considerably more sluggish than the voltages-source-based speed control system. However, this is a result of the fact that the current-regulated inverter-based system did not exceed the current limits of the machine. In fact, the current-regulated inverter-based system brought the machine to the commanded speed as fast as possible subject to the limitation of the stator current.

REFERENCES

[1] P. L. Chapman, S. D. Sudhoff, and C. Whitcomb, Multiple Reference Frame Analysis of Non-Sinusoidal Brushless DC Drives, *IEEE Transactions on Energy Conversion*, Vol. 14., No. 3, September 1999, pp 440–446.

[2] P. L. Chapman, S. D. Sudhoff, and C. A. Whitcomb, Optimal Current Control Strategies for Non-Sinusoidal Permanent-Magnet Synchronous Machine Drives, *IEEE Transactions on Energy Conversion*, Vol. 14, No. 3, December 1999, pp. 1043–1050.

[3] P. L. Chapman and S. D. Sudhoff, A Multiple Reference Frame Synchronous Estimator/Regulator, *IEEE Transactions on Energy Conversion*, Vol. 15, No. 2. June 2000, pp. 197–202.

[4] B. T. Kuhn, S. D. Sudhoff, and C. A. Whitcomb, Performance Characteristics and Average-Value Modeling of Auxiliary Resonant Commutated Pole Converter Based Induction Motor Drives, *IEEE Transactions on Energy Conversion*, Vol. 14. No. 3, September 1999, pp. 493–499.

[5] H. G. Yeo, C. S. Hong, J. Y. Yoo, H. G. Jang, Y. D. Bae, and Y. S. Park, Sensorless Drive for Interior Permanent Magnet Brushless DC Motors, *IEEE International Electric Machines and Drives Conference Record*, May 18–21, 1997, pp. TD1-3.1–TD1-3.3.

[6] K. A. Corzine and S. D. Sudhoff, A Hybrid Observer for High Performance Brushless DC Drives, *IEEE Transactions on Energy Conversion*, Vol. 11, No. 2, June 1996, pp. 318–323.

[7] T. M. Rowan and R. J. Kerkman, A New Synchronous Current Regulator and an Analysis of Current-Regulated Inverters, *IEEE Transactions on Industry Applications*, Vol. IA-22, No. 4, 1986, pp. 678–690.

[8] S. D. Sudhoff, K. A. Corzine, and H. J. Hegner, A Flux-Weakening Strategy for Current-Regulated Surface-Mounted Permanent-Magnet Machine Drives, *IEEE Transactions on Energy Conversion*, Vol. 10, No. 3, September 1995, pp. 431–437.

[9] K. A. Corzine, S. D. Sudhoff, and H. J. Hegner, Analysis of a Current-Regulated Brushless DC Drive, *IEEE Transactions on Energy Conversion*, Vol. 10, No. 3, September 1995, pp. 438–445.

PROBLEMS

1 Consider the brushless dc drive whose characteristics are depicted in Fig. 15.5-1. Plot the characteristics if the phase delay is in accordance with (6.4-24).

2 Repeat Problem 1, except at each speed calculate the phase delay which maximizes torque [which is *not* given by (14.4-7)].

3 Consider the drive system whose parameters are given in Table 15.8-1. If the phase delay is to be set equal to zero, compute the turns ratio of the transformer with the minimum secondary voltage that would be required if the drive is to supply a 1.72-N·m load at a mechanical rotor speed of 200 rad/s. Assume that the primary of the transformer is connected to a 230-V source (rms line-to-line) and that the effective series leakage reactance will be 0.05 pu. Further assume that the VA rating of the transformer is 1.5 times the mechanical output power.

4 Consider the speed control system considered in Section 15.8. Plot the closed-loop frequency response of the system about a nominal operating speed of 20 rad/s (mechanical).

5 Consider the speed control system considered in Section 15.8. Estimate the bandwidth of the closed-loop plant that could be designed if the current is to be restricted to the rated value of 2.6 A, rms.

6 Assuming that the drive discussed in Example 15B is operating at an electrical rotor speed of 200 rad/s, compute the pole locations if the linearizing feedback terms are not used in making up the command voltages.

7 Consider a current-regulated buried magnet brushless dc motor drive in which stator resistances is negligible. Sketch the locus of obtainable q- and d-axis currents in terms of the maximum fundamental component of the applied voltage, the electrical rotor speed, the q- and d-axis inductances, and λ'_m.

8 A four-pole brushless dc motor drive has the following parameters: $r_s = 0.3\,\Omega$, $L_{ss} = 20$ mH, and $\lambda'_m = 0.2$ V·s. The machine is to deliver 10 N·m at a mechanical rotor speed of 200 rad/s. Compute the q- and d-axis current commands such that the power factor is maximized. What is the rms voltage and current applied to the machine, and what is the efficiency?

9 Repeat Problem 8, except choose the current command so as to minimize the required dc voltage.

10 Repeat Problem 8, except choose the current command so as to minimize the commanded current.

11 Compute the locations of points A, B, and C on Fig. 15.12-1.

Appendix A

TRIGONOMETRIC RELATIONS, CONSTANTS AND CONVERSION FACTORS, AND ABBREVIATIONS

TRIGONOMETRIC RELATIONS

$$\cos^2 x + \cos^2\left(x - \frac{2\pi}{3}\right) + \cos^2\left(x + \frac{2\pi}{3}\right) = \frac{3}{2}$$

$$\sin^2 x + \sin^2\left(x - \frac{2\pi}{3}\right) + \sin^2\left(x + \frac{2\pi}{3}\right) = \frac{3}{2}$$

$$\sin x \cos x + \sin\left(x - \frac{2\pi}{3}\right)\cos\left(x - \frac{2\pi}{3}\right) + \sin\left(x + \frac{2\pi}{3}\right)\cos\left(x + \frac{2\pi}{3}\right) = 0$$

$$\cos x + \cos\left(x - \frac{2\pi}{3}\right) + \cos\left(x + \frac{2\pi}{3}\right) = 0$$

$$\sin x + \sin\left(x - \frac{2\pi}{3}\right) + \sin\left(x + \frac{2\pi}{3}\right) = 0$$

$$\sin x \cos y + \sin\left(x - \frac{2\pi}{3}\right)\cos\left(y - \frac{2\pi}{3}\right) + \sin\left(x + \frac{2\pi}{3}\right)\cos\left(x + \frac{2\pi}{3}\right) = \frac{3}{2}\sin(x - y)$$

$$\sin x \sin y + \sin\left(x - \frac{2\pi}{3}\right)\sin\left(y - \frac{2\pi}{3}\right) + \sin\left(x + \frac{2\pi}{3}\right)\sin\left(y + \frac{2\pi}{3}\right) = \frac{3}{2}\cos(x - y)$$

$$\cos x \sin y + \cos\left(x - \frac{2\pi}{3}\right)\sin\left(y - \frac{2\pi}{3}\right) + \cos\left(x + \frac{2\pi}{3}\right)\sin\left(y + \frac{2\pi}{3}\right) = -\frac{3}{2}\sin(x - y)$$

$$\cos x \cos y + \cos\left(x - \frac{2\pi}{3}\right)\cos\left(y - \frac{2\pi}{3}\right) + \cos\left(x + \frac{2\pi}{3}\right)\cos\left(y + \frac{2\pi}{3}\right) = \frac{3}{2}\cos(x - y)$$

$$\sin x \cos y + \sin\left(x + \frac{2\pi}{3}\right)\cos\left(y - \frac{2\pi}{3}\right) + \sin\left(x - \frac{2\pi}{3}\right)\cos\left(y + \frac{2\pi}{3}\right) = \frac{3}{2}\sin(x + y)$$

$$\sin x \sin y + \sin\left(x + \frac{2\pi}{3}\right)\sin\left(y - \frac{2\pi}{3}\right) + \sin\left(x - \frac{2\pi}{3}\right)\sin\left(y + \frac{2\pi}{3}\right) = -\frac{3}{2}\cos(x + y)$$

$$\cos x \sin y + \cos\left(x + \frac{2\pi}{3}\right)\sin\left(y - \frac{2\pi}{3}\right) + \cos\left(x - \frac{2\pi}{3}\right)\sin\left(y + \frac{2\pi}{3}\right) = \frac{3}{2}\sin(x + y)$$

$$\cos x \cos y + \cos\left(x + \frac{2\pi}{3}\right)\cos\left(y - \frac{2\pi}{3}\right) + \cos\left(x - \frac{2\pi}{3}\right)\cos\left(y + \frac{2\pi}{3}\right) = \frac{3}{2}\cos(x + y)$$

CONSTANTS AND CONVERSION FACTORS

Permeability of free space . . . . $\mu_0 = 4\pi \times 10^{-7}$ Wb/At m
Permittivity (capacitivity) of free space . . . . $\varepsilon_0 = 8.854 \times 10^{-12}$ C^2/N · m^2
Acceleration of gravity . . . . $g = 9.807$ m/s^2
Length . . . . 1 m = 3.281 ft
                                                 = 39.37 in
Mass. . . . . 1 kg = 0.0685 slug
                                                 = 2.205 lb (mass)
Force . . . . 1 N = 0.225 lb
                                                 = 7.23 pdl
Torque . . . . 1 N · m = 0.738 lb · ft
Energy . . . . 1 J (W · s) = 0.738 lb · ft
Power . . . . 1 W = 1.341 × 10^{-3} hp
Moment of inertia . . . . 1 kg · m^2 = 0.738 slug · ft^2
                                                 = 23.7 lb · ft^2
Magnetic flux . . . . 1 Wb = 10^8 Mx (lines)
Magnetic flux density . . . . 1 Wb/m^2 = 10,000 G
                                                 = 64.5 klines/in^2
Magnetizing force . . . . 1 At/m = 0.0254 At/in
                                                 = 0.0126 Oe

ABBREVIATIONS

alternating current	ac	megawatt	MW
ampere	A	meter	m
ampere-turn	At	microfarad	µF
coulomb	C	millihenry	mH
direct current	dc	newton	N
foot	ft	newton meter	N·m
gram	g	oersted	Oe
henry	H	pound	lb
hertz	Hz	poundal	pdl
horsepower	hp	power factor	PF
inch	in	pulse-width modulation	PWM
joule	J	radian	rad
kilogram	kg	revolution per minute	r/min (rpm)
kilovar	kvar	second	s
kilovolt	kV	voltampere reactive	var
kilovoltampere	kVA	volt	V
kilowatt	kW	voltampere	VA
magnetomotive force	MMF	watt	W
maxwell	Mx	weber	Wb

INDEX

Abbreviations, 604
Ac/dc converter:
  single-phase, 428, 432–438
  3-phase, 428–429, 438–442
ac machines, 9
ac-to-dc converters, 89
Alternating-current (ac) machines, 1
Ampere's law, 4, 40, 44
Arbitrary reference frame(s), *see* Reference
    frame(s)
balanced steady-state:
  phasor relationships, 127–130
  voltage equations, 130–133
defined, 110
stationary circuit variables transformation:
  capacitive elements, 119–123
  inductive elements, 117–119
  resistive elements, 115–117
  3-phase, 133–137
stationary 3-phase series RL circuit variables,
    133–136
3-phase load commutated converter, 423
3-phase symmetrical induction machines:
  computer simulation, 184–187
  torque equations, 153–154
  voltage equations, 149–153
transformation of a balanced set, 126
transformation between, 124–125
transformation equations, 111–115
transformation to, 111–113

variables observed, impact of, 133–136
voltage equations:
  balanced steady-state, 130–133
  for 3-phase capacitive circuits, 119–123
  for 3-phase inductive circuits, 117–119
  for stator windings of synchronous machines,
      248–250
  for 3-phase resistive elements, 115–117
Armature, defined, 69
Asynchronously rotating reference frames, 128,
    130–132
Average output voltage, 3-phase load commutated
    converter, 410–411, 416
Average-value model:
  current-regulated inverter drives, brushless dc
      motors, 595–597
  3-phase load commutated converter, 417–424

Balanced steady-state operation, synchronous
    machines, 210–214
B-H curve, 8–9
Bipolar junction transistors (BJTs), 481
Block diagrams:
  permanent-magnet dc machine, 95–98
  shunt-connected dc machine, 92–95
  symmetrical induction machines, 186–187
  synchronous machines, 248, 250, 253
Blocked-rotor test, symmetrical induction
    machine, 163–164
Bode characteristic, 584–585

Brereton, D. S., 110
Brushless dc machines:
  dynamic performance:
    sinusoidal phase voltages, 275–279
    six-step phase voltages, 279–280
  flux linkage equations, 261, 263
  free acceleration characteristics, 278
  inductances, 263–264
  motor drives:
    current command synthesis, 591–595
    current-regulated inverter drives, 586–590,
      595–600
    current-source inverter drives, voltage
      limitations, 590–591
    steady-state harmonics, 578–582
    voltage-source inverter drives, 558–578,
      582–586
  steady-state harmonics, analysis of, 279
  steady state operation analysis:
    common operating mode, 267–269
    overview, 266–267
    phase shifting applied voltages, 269–274
  torque-speed characteristics, 271–273, 275
  2-pole, 3-phase, 261–263
  voltage and torque equations:
    in machine variables, 261–264
    in rotor reference-frame variables,
      264–266

Capacitive elements, transformation to arbitrary
    frame reference, 119–123
Closed-loop frequency response, brushless dc
    motor drives, 585
Closed-loop voltage, 516–520
Coenergy:
  defined, 17
  multiexcited electromagnetic system, 26–27
  in singly excited electromagnetic systems,
    17–19
Commutation, elementary dc machines, 68–70
Compound-connected dc machine, 83–84
Conservation of energy, 12
Constant amplitude balanced set, 126
Constants and conversion factors, 604
Constant-slip frequency induction motor drives,
    532–540
Converters:
  ac-to-dc, 89
  analysis and operation of:
    single-phase, full, 395–406
    3-phase, full, 406–424
  equivalent circuit, for 3-phase load commutated,
    417

fully controlled 3-phase bridge:
  characteristics of, 481–487
  closed-loop voltage and current controls,
    516–520
  delta modulation, 512–513
  180 voltage source operation, 487–494
  hysteresis modulation, 510–512
  open-loop voltage and current control,
    513–516
  pulse-width modulation, 494–498
  sine-triangle modulation, 498–503, 515
  space-vector modulation, 506–510
  third-harmonic injection, 503–505
Corkscrew rule, 40
Coupling fields, energy in, 16–22
Critical clearing angle, 239, 246
Critical clearing time, 239–242, 246
Cross mutual reactance, 306
Current command synthesis, brushless dc motor
    drives, 591–595
Current-regulated inverter drives, brushless dc
    motors, 586–590, 595–600
Current-source inverter drives, voltage limitations,
    590–591

Damper windings, 52–53, 194–197, 203,
    284–286
Damping coefficient, 97, 264
Damping factor, 99
dc/dc converter:
  four-quadrant, 463–466, 469–473
  machine control:
    with current-controlled, speed control,
      468–475
    with voltage-controlled, speed control,
      466–468
  one-quadrant:
    average-value analysis, 451–455
    characteristics of, 443–444
    continuous-current operation, 444–449
    discontinuous-current operation, 449–451
    operating characteristics, 455–460
  two-quadrant, 460–463
Dc motor drives:
  one-quadrant dc/dc converter drive:
    average-value analysis, 451–455
    characteristics of, 443–444
    continuous-current operation, 444–449
    discontinuous-current operation, 449–451
    operating characteristics, 455–460
  solid-state converters:
    dc/dc, 429–431
    overview, 427–438

single-phase ac/dc, 428
3-phase ac/dc, 428–429
two-quadrant dc/dc converter drive, 460–463
dc test, symmetrical induction machine, 163
Delta modulation, 3-phase bridge converters,
    512–513
Dielectric losses, 12
Direct current (dc) machines:
  ac/dc converter drives, steady-state and dynamic
      characteristics:
    single-phase full-converter, 432–438
    3-phase full-converter, 438–442
  block diagram:
    permanent-magnet, 95–97
    shunt-connected, 92–95
  compound-connected, 84–88
  defined, 67
  dynamic characteristics, solution by Laplace
      transformation, 98–104
  elementary, 68–76
  equivalent circuit for:
    compound-connected, 85
    separately excited, 78
    series-connected, 83
    shunt-connected, 79
  four-quadrant dc/dc converter drive, 463–466,
      469–473
  interpoles, 74–75
  low-power motors, 88
  machine control with current-controlled dc/dc
      converter:
    characteristics of, 468–469
    four-quadrant dc/dc converter, 469–473
    speed control, 473–475
  machine control with voltage-controlled dc/dc
      converter, speed control:
    with feedforward voltage control,
        467–468
    overview, 466–468
  motor drives, *see* Dc motor drives
  motors, permanent-magnet and shunt, dynamic
      performance during:
    starting, 88
    sudden load torque changes, 89–92
  permanent-magnet, 74–76, 79, 88–92,
      95–97
  pole pairs, 75
  separately excited, 78–79
  series-connected, 82–84
  shunt-connected, 79–82, 88–92
  voltage and torque equations, 76–78
  windings:
    lap, 74

  uniformly distributed armature, 74
  wave, 74
Direct rotor-oriented field induction motor drives,
    544–546
Doherty, R. E., 229–230
Double-cage rotor machine, 144
Duty cycle:
  brushless dc motor drives, 578, 585
  modulation in 3-phase bridge converters,
      499–500, 503, 514–515
Dynamic electromechanical systems, performance
    of, 28–35

Eddy current, 12, 26
Eigenvalues:
  applications, 323–324
  induction machines, 324–327
  synchronous machines, 327–329
Electric field theory, 162
Electromagnetic forces, 25–28
Electromechanical energy conversion:
  air-gap MMF, 35–47
  coupling fields, energy in, 16–22
  electromagnetic forces, 25–28
  electromechanical systems, steady-state and
      dynamic performance of, 28–35
  electrostatic forces, 13, 25–28
  energy relationships, 11–16, 25
  graphical interpretation of, 22–24
  machine windings, 35–47
  steady-state electromechanical systems, 28–35
  voltage equations, 47–58
  winding inductances and voltage equations:
    induction machine, 53–58
    synchronous machine, 47–53
Electrostatic forces, 13, 25–28
Elementary electromechanical system:
  energy relationships, 11–16, 25
  graphical interpretation, 22–24
  steady-state and dynamic performance, 28–35
  voltage equation, 14
Energy, in electromechanical systems:
  electrostatic forces, 25–28
  graphical interpretation, 22–24
  relationships, 11–16, 25
Equal area criteria:
  input torque change, 243–244
  3-phase fault, 244–246
Equivalent circuits:
  dc machines:
    compound-connected, 85
    separately excited, 78
    series-connected, 83

Equivalent circuits: (*Continued*)
  shunt-connected, 79
  3-phase bridge converters, 483–484
  3-phase load commutated converter, 417
Excitation voltage, 212

Field rheostat, 78
Field-oriented induction motor drives:
  characteristics of, 540–544
  direct rotor-oriented, 544–546
  indirect rotor, 550–554
  robust direct, 546–550
Flat-compounded dc machine, 85
Flux:
  electromechanical energy conversion, 18–19
  magnetically coupled circuits, 2–3
Flux linkage:
  arbitrary reference frames, 149–150, 152, 185–186
  synchronous machines, 195–197, 203–207, 246
Forward-field impedance, 380
Four-quadrant dc/dc converter drive, 463–466, 469–473
Fourier analysis, 44, 502
Free acceleration characteristics:
  induction machines, reduced-order equation predictions, 344–347
  for 3-phase symmetrical induction machine, 165–178
Full-pitch winding, 37

Gauss's law, 41
Generalized rotating real transformation, 110
Graphical representations, energy conservation, 22–24

High-voltage ac-dc systems, 111
Hunting mode eigenvalue, 328
Hydro turbine generator, 232–236, 238–240, 334
Hysteresis:
  characteristics of, 7, 12, 16, 470
  in current-regulated inverter drives, brushless dc motors, 591–592
  modulation, 3-phase bridge converters, 510–512

Impedance:
  forward-field, 380
  input, 10, 164, 166
  no-load, 164
  operational, 284–288, 291
  stator, 187

Indirect rotor field induction motor drives, 550–554
Induction machines:
  inductance matrix, 118
  linearized equations:
    applications, 312, 315–316
    eigenvalues, 324–327
    transfer function formulation, 333–334
  motor drives, *see* Induction motor drives
  reduced-order equations:
    applications, generally, 340–342
    large-excursion behavior predicted by, 343–350
    simulation model, 356–357
  2-pole, 3-phase, wye-connected symmetrical, 53–55, 143
  winding inductances, 53–58
Induction machines, single-phase:
  capacitor-start, 387–388
  capacitor-start, capacitor-run, 389–393
  split-phase, 386–387
  stator winding, 384–385
Induction machines, 3-phase, symmetrical:
  characteristics of, 141–142
  computer simulation of:
    arbitrary reference frames, 184–187
    dynamic performance during:
      sudden load torque changes, 174, 179–181
      3-phase fault at machine terminals, 181–184
    flux linkage equations for:
      arbitrary reference frames, 149–150, 152, 185–186
    free acceleration characteristics:
      overview, 165–172
      viewed from various reference frames, 172–178
    per unit system, 155–157
    steady-state operation analysis, 157–165
    torque equations:
      arbitrary reference-frame variables, 153–154
      overview, 146–147
    transformation equations, for rotor circuits, 147–149
    voltage equations:
      arbitrary reference-frame variables, 149–153
      overview, 142–145
Induction machines, 2-phase symmetrical:
  balanced steady-state operations, 363–364
  linearized equations, 364
  torque equations:
    in arbitrary reference-frame variables, 362–363

in machine variables, 362
transfer function formulation, 364
unbalanced steady-state operation, 364–371
voltage equations:
in arbitrary reference-frame variables,
362–363
in machine variables, 362
Induction machines, 2-phase unsymmetrical:
balanced operation, 382–383
equivalent circuit:
in stationary reference-frame, 374, 376
for steady-state operation, 379
flux linkage equations, 372–373, 375
overview, 371–373
steady-state operation:
balanced operation, 382–383
overview, 377
torque equation, 382
voltage equation, 378–379
torque equations:
in stationary reference-frame variables,
373–377
in steady-state operation, 382
voltage equations:
in stationary reference-frame variables,
373–377
in steady-state operation in component form,
379–382
in steady-state operation in qs and ds
variables, 378–379
Induction motor drives:
constant-slip current control, 532–540
field-oriented control:
characteristics of, 540–544
direct rotor-oriented, 544–546
indirect rotor, 550–554
robust direct, 546–550
volts-per-hertz control, 525–532
Inductive elements, transformation to arbitrary
frame reference, 117–119
Injection, third-harmonic in 3-phase bridge
converters, 503–505
Input torque changes, synchronous machines,
219–225, 350
Insulated-gate bipolar junction transistors
(IGBTs), 481
Interpoles, 74–75
Inverse Laplace transform, 101, 297

Kron, G., 110

Laplace transformation, 67, 98–104, 295
Linear control theory, 330

Linear magnetic systems:
coupled circuits, 3–8
reference frames, 154–155
voltage equations, 142
torque equations, 146
Linearized equations:
eigenvalues:
applications, 323–324
induction machines, 324–327
reduced-order equations predictions,
354–355
synchronous machines, 327–329
for induction machines, 312, 315–316
reduced-order, 354
for synchronous machines, 313, 316–323
Taylor's expansion, 313–314
transfer function formulation, 330–335
2-phase symmetrical induction machines, 364
Line-to-neutral voltage, 3-phase bridge converters,
487, 489–490, 504, 512
Load torque changes:
induction machines, reduced-order equation
predictions, 347
permanent-magnet dc machine, 89–92
shunt-connected dc machine, 89–92
in 3-phase symmetrical induction machine,
174,179–181
Long-shunt field connection, compound dc
machine, 85–86
Losses, 12
Lossless coupling fields, 17

Magnetically coupled circuits:
characteristics of, 1–3
linear systems, 3–8
nonlinear systems, 8–11
Magnetization curve:
shunt-connected dc machines, 81–82
of transformers, 10–11
Magnetomotive force (MMF):
air-gap, in synchrous machines, 35–47,
215–217
in dc machines, 77–78
in 4-pole, 3-phase machine, 45–47
magnetically coupled circuits and, 5
in P-pole, 46
rotating air-gap, 1, 45–46, 54–55
sinusoidally distributed air-gap, 37, 42–47
synchronous machine torque produced by,
215–217, 253
in 3-phase symmetrical induction machine, 144
Metal-on-silicon controlled thyristors (MCTs),
481

Metal-on-silicon field-effect transistors (MOSFETs), 481
Modulation:
  current-regulated inverter drives, brushless dc motors:
    hysteresis, 591–592
    sine-triangle, 587–588
  fully-controlled 3-phase bridge converters:
    delta, 512–513
    hysteresis, 510–512
    pulse-width, 494–498
    sine-triangle, 498–503
Multiexcited electromagnetic systems, 26–27
Multipole dc machines, 74–75

Neglecting electric transients theory, 337–339
Newton's law of motion, 14
Nickle, C. A., 229–230
No-load test, symmetrical induction machine, 163–164
Nonlinear magnetic systems, coupled circuits, 8–11

Open-circuit tests, 7, 250
Open-loop voltage, 513–516, 518
Operational impedances, *see* Synchronous machines
Oscillation mode, 328
Overcompounded dc machine, 85
Overmodulation:
  brushless dc motor drives, 585
  3-phase bridge converters, 502, 504

Park, R. H., 109–110, 203–204, 283
  equation, 342
  transformation, 154, 194
  voltage equations in synchronous machines:
    characteristics of, 200–206, 210–217
    in operational form, 284
Permanent-magnet dc machine:
  block diagram, 95–97
  motors:
    commutation of, 75–76
    dynamic performance during starting, 88
    dynamic performance during sudden load torque changes, 89–92
    low-power, 74–75, 89
  voltage equation, 79
Per unit system:
  induction machines, 155–157
  synchronous machines, 209–210
Phase voltage, brushless dc machines:
  sinusoidal, 275–279

  six-step, 279–280
Phasor relationships, reference frames, 127–130
P-pole machine, 46, 56–58, 198
Pulse-width modulation, 3-phase bridge converters, 494–498

Reduced-order equations:
  induction machines:
    applications, generally, 340–342
    free acceleration characteristics predicted by, 344–347
    load torque changes predicted by, 347
    3-phase fault at machine terminals, 347–350
  linearized:
    characteristics of, 354
    eigenvalues predicted by, 354–355
  simulation models:
    induction machines, 356–357
    synchronous machines, 357
    types of, 355–356
  synchronous machines:
    applications, 342–343
    input torque changes predicted by, 350
    3-phase fault at machine terminals predicted by, 350–353
  theory of, 338–340
Reference coil, 5–6
Reference frame(s):
  commonly used, 123–124, 154–155
  theory development, 109–123
  3-phase bridge converters, 501–503, 507–508
Reluctance torque, 214
Resistive elements, transformation to arbitrary frame reference, 115–117
RL circuit, six-step bridge converters, 491–494
Robust direct field induction motor drives, 546–550
Rotor angle, *see* Synchronous mahcines
Rotor circuits, transformation equations, 147–149
Rotor phase windings, 55–56
Rotor reference frame, *see* Reference frame(s)
Rotor voltages, symmetrical induction machines, 155
Round rotor synchronous machine, 120–121

Salient-pole synchronous machine, 38, 46, 49, 119, 192, 194
Saturation:
  computer simulation, in synchronous machines, 250–254
  in series-connected dc machine, 84

simulation of:
    induction machines, 162–163, 184, 254
    synchronous machines, 248
    in transformers, 10–11
Second-order equation, 96
Second-order polynomials, 291
Self-excited dc generators, 81–83
Semicontrolled bridge converters:
    single-phase load commutated converter,
        395–406
    3-phase load commutated converter, 406–424
Separately excited dc machines, 78–79
Series capacitance, 14
Series-connected dc machine, 82–84
Settling out, 328–329
Short-circuit test, 7, 283
Short-shunt field connection, compound dc
        machine, 85–86
Shunt capacitance, 14
Shunt-connected dc machine:
    block diagram, 92–95
    characteristics of, 79–82
    motors, dynamic performance during:
        starting, 88
        sudden load torque changes, 89–92
Silicon-controlled rectifiers (SCRs), 80, 428,
        432–434
Sine-triangle modulation:
    current-regulated inverter drives, brushless dc
        motors, 587–588
    3-phase bridge converters, 499–503
Single-phase ac/dc converter, 428
Single-phase full-converter dc drive, 432–438
Single-phase load commutated converter:
    characteristics of, generally, 395–396
    modes of operation, 402–406
    operation with commutating inductance and
        phase delay, 400–404
    operation with commutating inductance and
        without phase delay, 397–399
    operation without commutating inductance and
        phase delay, 396
    operation without commutating inductance and
        with phase delay, 399–400
Sinusoidally distributed winding, 37, 42–47, 192
Six-step bridge converters, 491–493
Six-step modulated brushless dc motor drives,
        578–582
Solid-state converters:
    dc/dc, 429–431
    overview, 427–438
    single-phase ac/dc, 428
    3-phase ac/dc, 428–429

Space-vector modulation, 3-phase bridge
        converters, 506–510
Speed currents, 120, 124
Speed voltages, 120, 124
Stanley, H. C., 110, 154
State equations:
    permanent-magnet dc machine, 96–97
    shunt-connected dc machine, 94–95
Stationary circuits, 1–2
Stator eigenvalues, 328
Stator electric transients neglected, *see* Reduced-
        order equations
Stator voltages, symmetrical induction machines,
        155, 168
Steady-state electromechanical systems,
        performance of, 28–35, 46–47
Steady-state harmonics, brushless dc motors,
        578–582
Steam turbine generator, 232, 235, 237, 240, 244,
        329, 335
Straight-line energy relationships, 18–20, 82
Subtransient reactances, *see* Synchronous
        machines
Swing mode eigenvalue, 328–329
Symmetrical stator windings, 37
Synchronous condensers, 215
Synchronously rotating reference frames, 128,
        132–133
Synchronously rotation reference frame, *see*
        Reference frames
Synchronous machine(s):
    angle between rotors, 207–208
    balanced conditions, 255
    computer simulation:
        arbitrary reference frame, stator voltage
            equations, 248–250
        balanced conditions, 155
        reduced-order machine equations, 357
        rotor reference frame variables, 246–248
        saturation, 250–254
        unbalanced stator conditions, 255
    equal area criteria:
        input torque change, 243–244
        3-phase fault, 244–246
    equivalent circuit, in rotor reference-frame
        variables, 201–202
    excitation voltage, 212
    flux linkages:
        in machine variables with damper windings,
            194–197
        reduced-order machine equations, 342, 357
        in rotor reference-frame variables, 203–207,
            246–248

Synchronous machine(s): *(Continued)*
  inductance matrix, 118
  inductances of 3-phase stator windings:
    with damper windings, 194–197, 203
    derivation of 2-pole, 47–53
    P-pole, 46
  inertia constant (*H*), 209
  for input torque changes, 219–225
  linearized equations:
    applications, 313, 316–323
    eigenvalues, 327–329
    small displacement dynamics, 319–320
    transfer function formulation, 333–335
  operational impedances:
    defined, 284
    from frequency-response characteristics, 301–307
    from short-circuit characteristics, 294–301
  per unit system for, 209–210
  reactances of:
    subtransient, 289–290, 296
    overview, 203–205
    transient, 230, 288–290
  reduced-order machine equations:
    applications, generally, 342–343
    input torque changes predicted by, 350
    simulation model, 357
    3-phase fault at machine terminals predicted by, 350–353
  reference frame, 514
  rotor angle, 207–208, 212–213, 215
  round rotor, 120–121
  salient-pole, 38, 46, 49, 119, 192, 194
  saturation, computer simulation of, 246–248
  short-circuit characteristics, 294–301
  steady-state analysis, 210–219
  for 3-phase fault at terminals, 225–229
  time constants:
    derived, 291–294
    standard, 290–291
  torque angle characteristics, comparison with actual for:
    critical clearing time, 239–242
    first-swing transient stability limit, 232–239, 244
  torque equations:
    in arbitrary reference-frame variables, 207
    in machine variables, 197–198
    in rotor reference-frame variables, 206
    for steady-state operation, 213–214
  transient stability limit, 232–239
  transient torque, rotor angle *vs.*, 229–232
  unbalanced stator conditions, 255

  voltage equations:
    in arbitrary reference-frame variables, 198–200
    in machine variables, 192–197
    in rotor reference-frame variables, 200–206
    for steady-state operation, 211–212
  winding inductances, 47–53
  windings of 2-pole, 3-phase, 35–47

Taylor's expansion, 313–314
Thevenin equivalent inductance, 519
Thomas, C. H., 246, 250
3-phase ac/dc converter, 428–429
3-phase bridge converters, fully-controlled:
  characteristics of, 481–487
  closed-loop voltage and current controls, 516–520
  delta modulation, 512–513
  180° voltage source operation, 487–494
  hysteresis modulation, 510–512
  open-loop voltage and current control, 513–516
  pulse-width modulation, 494–498
  sine-triangle modulation, 498–503
  space-vector modulation, 506–510
  third-harmonic injection, 503–505
3-phase capacitive circuit, 119–123
3-phase fault:
  induction machine terminals, reduced-order equation predictions, 347–350
  synchronous machine terminals, reduced-order equations predictions, 350–353
3-phase full-converter dc drive, 438–442
3-phase inductive circuit, 117–119
3-phase load commutated converter:
  dynamic average-value model, 417–424
  modes of operation, 414–416
  operation with commutating inductance and phase delay, 411–414
  operation with commutating inductance and without phase delay, 408–411
  operation without commutating induction and phase delay, 406–408
  operation without commutating inductance and with phase delay, 411
3-phase resistive circuit, 115–117
3-phase RL circuit, 120–122
Thyristor:
  defined, 80
  in single-phase load commutated converter, 395–396, 403
Time constant, inertia:
  induction machines, 156–157
  in permanent-magnet dc machine, 97

Torque angle characteristics, *see* Synchronous machines
Torque calculator, robust direct field-oriented control, 547, 548
Torque constant, defined, 86
Torque control loop, robust direct field-oriented control, 547–548
Torque equations:
  brushless dc machines, 261–266
  for 3-phase symmetrical induction machine, 146–147, 153–154
  in 2-phase symmetrical induction machines:
    in arbitrary reference-frame variables, 362–363
    in machine variables, 362
  in 2-phase unsymmetrical induction machines:
    in stationary reference-frame variables, 373–377
    in steady-state operation, 382
Torque *vs.* speed characteristics, *see* Free acceleration characteristics
Torque-speed characteristics:
  brushless dc machines, 271–273, 275
  induction machines, 344, 390–391, 526
Transfer function formulation, linearized equations, 330–335
Transformation:
  equations, arbitrary reference frames, 111–115
  Park's, 154, 194
Transformers, *see* Magnetically coupled circuits
  magnetization curve, 10–11, 81
  open-circuit tests, 7
  short-circuit tests, 7
Transient reactances, *see* Synchronous machines
Transient stability limit, *see* Synchronous machines
Triezenberg, D. M., 253
Trigonometric relations, 603
Two-pole dc machines, 68–69
Two-quadrant dc/dc converter drive, 460–463

Unbalanced conditions, symmetrical induction machines, 187

Vector matrix, 95

Vector rotator, 125
Voltage equations:
  brushless dc machines, 261–266
  magnetically coupled circuits, 3, 6–7
  permanent-magnet dc machine, 79
  reference frames, 130–133
  in 3-phase load commutated converter, 408, 410
  for 3-phase symmetrical induction machine, 142–145, 149–153
  in 2-phase symmetrical induction machines:
    in arbitrary reference-frame variables, 362–363
    in machine variables, 362
  in 2-phase unsymmetrical induction machines:
    in stationary reference-frame variables, 373–377
    in steady-state operation, 378–382
  wind inductances, 47–58
Voltage-source inverter drives, brushless dc motor:
  average-value analysis, 568–571
  characteristics of, 558–560, 582–586
  equivalence of schemes to idealized source, 560–568
  speed control case study, 582–586
  steady-state performance, 571–574
  transient and dynamic performance of, 574–578
Volts-per-hertz induction motor drive, 525–532

Waveform(s):
  reference frame transformation, 113
  3-phase bridge converters:
    using hysteresis, 511–512
    line-to-neutral voltage, 487, 489–490, 504, 512
    sine-triangle modulation, 501–502, 504
    six-step inverter feed, 491–492
  3-phase symmetrical induction machine, 144
Winding(s):
  damping, 52–53, 194–197, 203, 284–286
  inductances and voltage equations:
    induction machine, 53–58
    synchronous machine, 47–53
  sinusoidally distributed, 37, 42–47, 54
  types of, overview, 35–47
  uniformly distributed in dc machines, 72–74